Sound Reproduction

Sound Reproduction

The Acoustics and Psychoacoustics
of Loudspeakers and Rooms

Floyd E. Toole

Focal Press
Taylor & Francis Group

NEW YORK AND LONDON

First published 2008

This edition published 2013
by Focal Press
70 Blanchard Road, Suite 402, Burlington, MA 01803

Simultaneously published in the UK
by Focal Press
2 Park Square, Milton Park, Abingdon, Oxon OX14 4RN

Focal Press is an imprint of the Taylor & Francis Group, an informa business

British Library Cataloguing in Publication Data
A catalogue record for this book is available from the British Library

Library of Congress Cataloging-in-Publication Data
A catalog record for this book is available from the Library of Congress

ISBN: 978-0-240-52009-4 (pbk)

Transferred to Digital Printing in 2014

Contents

Part Two: Designing Listening Experiences

Introduction

Reproduced sound is everywhere. In the author's life, a tiny portable audio device supplies podcasts and music while he walks his dog. Premium car audio systems provide superb surround sound on the road. A whole-house system lets us have the music of choice in most rooms and in the garden. A seven-channel entertainment room system provides music and audio accompaniment to television and movies on large direct-view and front-projection screens. It is all very high-tech, all very convenient, and all very enjoyable. It is also all very good.

It has not always been so. As a teenager in the 1950s, I scrounged war-surplus parts to build pre- and power-amplifiers. The first loudspeakers were salvaged from beaten-up jukeboxes used in a World War II Air Force mess. Audio was a participatory hobby in those days. If you didn't build things, there were always maintenance issues with the tube electronics and phonographs. The sound? It was mostly bad by today's standards, but at the time it was revolutionary. The family played records when entertaining. It was a novelty, a status symbol. It was the beginning of high-fidelity—*hi-fi*, a term so corrupted by abuse that it is now rarely used.

What began as a hobby became a motivating factor in the educational process and, as if it were predestined, a lifetime occupation. Now, after 43 years in research and the application of research in product development, I am voluntarily unemployed, and this book is my first retirement project. Some of it is inevitably autobiographical because I have been one of the many contributors to the scientific foundation of this industry.

Audio—sound reproduction—engages both the emotions and the intellect. Understanding the process is challenging because it embraces domains with enormous contrasts: human perceptions in their manifold dimensions and technology with its own system of devices, functions, and performance descriptors. The subjective side is notable for its complexity, flexibility, adaptability, and occasional capriciousness. The technical side is characterized by the near-absolute reproducibility of the devices, the stability of their performance over time, and the reliability of their measured parameters. The interface of these two cultures has met with mixed success over the years. Both sides seek excellence in the final subjective experience, but there are fundamental differences in philosophy, metrics, languages, and the economic and emotional attachments to the results.

In the midst of all this are the recording and broadcast industries that generate program material for our sound reproducing systems: all manner of music, verbal discourse, and the audio accompaniment to movies, television, and games. These programs are artistic creations, both in their informational content and in the timbral and spatial aspects of the sound. As consumers of these programs, we cannot know what was intended for the sound of any of these programs. We were not there when they were created. We may have been at performances by similar, or even the same, musicians, but they were likely to have been in different venues and possibly amplified. None of us ever placed our ears where the microphones were located to capture the sounds, nor would we want to; we were almost certainly at a distance, in an audience. A simple reproduction of the microphone signals cannot duplicate the experience. That is where the professional recording industry steps in.

Microphones are selected from the many that are available, and they are carefully positioned relative to the performers; this is the first level of signal processing. The signals are stored in multiple tracks that are then mixed together in proportions that create a combined sound that is pleasing to musicians and recording engineers listening through monitor loudspeakers in a small room: a control room in a dedicated facility or a home studio. This is the second level of signal processing. While doing this, various knobs are turned, electronically or physically, to add or subtract energy at certain frequencies and to add multiple delayed versions of sounds in the tracks, all designed to enhance the sound: the third level of signal processing. Finally, a mastering engineer transfers the completed "master" recording to the delivery medium, CD, DVD, and so on, and almost always will make further changes to the program: the fourth level of signal processing. In LP mastering, the changes are substantial: mono bass, dynamic compression, rolled-off highs near the center of the disc, and so forth to cope with the limitations of the medium. Digital media need no such dramatic manipulations, but mastering engineers may choose to tailor the sound based on what they hear through their own monitor loudspeakers in their own rooms, and possibly to adjust the dynamic range and bandwidth to be suitable for the intended audience. That is four clearly defined opportunities for signal processing of some kind; everyone in this chain of events is authorized to fine-tune the result. None of this is a problem. This is the creation of the art. It only becomes a problem when somebody, somewhere, presses a "play" button and what they hear is not the same as any of the preceding circumstances. Sadly, this is the norm.

The point of this story is that the program is a huge variable in the process of sound reproduction and that sound reproduction (in the control room, home studio, or mastering lab) is a factor in the creation of the program. This situation is well described as a conundrum; it certainly is not a linear, stable, and predictable process. So for the vast majority of recordings, the sound delivered to our ears is a new experience. It is nothing we have heard before, allowing us to establish even a faulty memory for reference purposes. Knowing that the playback devices are different from those used in the creative process, we cannot be certain what is responsible for what is heard; if we hear something we don't like, is it because it is in the program itself? Is it something in the playback system that has been revealed by the program? Or is it an unfortunate destructive interaction that is unique to these two factors and may not happen in other circumstances? In normal listening situations, we cannot know and are therefore left in a position of forming opinions on the basis of whether we like the combination of content (tune, musicianship, etc.) and the sound (timbre, directional and spatial impressions, etc.) and whether it moves us emotionally (how it "feels").

The origin of emotion in a listener is the art itself—the music or movie—and not the audio hardware. It is inconceivable that a consumer could feel an emotional attachment to a midrange loudspeaker driver, yet without good ones, listening experiences will be diminished. Since the true nature of the original sound cannot be known to listeners, one cannot say "it sounds as it should." But listeners routinely volunteer opinions on scales that are variations of like-dislike, which frequently have a component of emotion.

Descriptors like *pleasantness* and *preference* must therefore be considered as ranking in importance with *accuracy* and *fidelity*. This may seem like a dangerous path to take, risking the corruption of all that is revered in the purity of an original live performance. Fortunately, it turns out that when given the opportunity to judge without bias, human listeners are excellent detectors of artifacts and distortions; they are remarkably trustworthy guardians of what is good. Having only a vague concept of what might be correct, listeners recognize what is wrong. An absence of problems becomes a measure of excellence. By the end of this book, we will see that technical excellence turns out to be a high correlate of both perceived accuracy and emotional gratification, and most of us can recognize it when we hear it.

The following concepts form the basis for this book:

- Identifying the perceptual dimensions underlying listener responses

- Understanding the psychoacoustic rules governing them

- Developing record/reproduction methods and devices to maximize those dimensions that listeners respond to in a positive sense and to minimize those that are detractions

- Encouraging the music and film production and the consumer reproduction portions of the industry to share the same performance standards and system architectures so the art has a higher probability of being heard as intended

It is hoped that readers from many backgrounds will find the content accessible. Audiophiles and music lovers, film buffs, reviewers, and journalists may find that they better understand the psychoacoustic underpinnings of what they hear. Recording, mixing, and mastering engineers may find some thought-provoking perspectives on how their activities and practices fit into the grand scheme of sound reproduction. Loudspeaker designers will find no help what-soever in the design of magnets, voice coils, and enclosures, but they may find inspiration in how to better integrate collections of transducers into systems that provide more favorable interfaces with room boundaries and the listeners within them. Finally, acoustical consultants may find some explanations for old practices, suggestions for new ones, and justification for those all-important "billable hours."

Part One starts with a historical perspective on sound reproduction. It then addresses the scientific background, examining relevant acoustics and psycho-acoustics. The treatment is deliberately nonmathematical. Illustrations and graphs are used to explain the relationships between what is heard and what is measured. The intent is to convey ideas and an intuitive understanding of why things happen and result in certain kinds of perceptions.

Scientists often seek mathematical descriptions for relationships, including relationships between what is measured and what is heard. An equation does not add information; it attempts to describe information in a different form. In fact, almost always the equation is a simplification of the raw data that emerge from psychoacoustic examinations of a phenomenon. But such attempts are important in modeling more complex aspects of perception. Several simplified relationships may be combined into an explanation of something complex. The hope is that it can be done well enough that the input of technical data can yield an output that is a good prediction of a human perception. The long-term objective in the context of sound reproduction is to find technical metrics that usefully evaluate the physical world of electronic and transduction devices, operating in rooms.

The author tried to keep the discussions on topic and brief, but it is a multidimensional subject, and a certain amount of explanation is essential. There is also some redundancy because it is assumed that the book will not necessarily be read in a linear fashion. It is fully anticipated, for example, that many readers will start with Part Two, perhaps even the last chapter, seeking the answers without the explanations. That's fine, but please go back and learn the rest, because only by understanding the principles can you hope to deal with the infinite variations in the real world. In some cases, it may be found that there is no remedy because there is no problem. Human adapt-ability is an often ignored factor. We have a remarkable ability to "listen through" rooms to hear the essence of sound sources; that is, after all, the basis of live performances.

In recent years, the audio industry has seen the evolution of the custom installation business, driven by "home theater," but embracing all aspects of home automation, convenience, and entertainment. This business has energized an audio industry that was, after about 50 years with two-channel stereo, facing a lackluster future. Stereo has not gone away, and it will never disappear completely because of the huge collection of valuable legacy music that exists. Interestingly, it was film that provided inspiration for stereo, and it is film that has paved the way for multichannel sound to enter homes. For that we can be grateful.

The term *home theater* says it all: movies in the home. Movies are an important component of home entertainment, no doubt, but the facilities we call home theaters should not end there; that would be a sad compromise. Some consultants promote the notion that these spaces should emulate what is heard on a film-dubbing stage and, by inference, a cinema. In the important spectral, dynamic, directional, and spatial aspects, there is some merit in this proposition. These are basic dimensions of sound reproduction, and they apply to music and games as well as to movies. But the small physical size of the domestic installation allows us the freedom to use components with greater finesse and the flexibility to provide for different kinds of entertainment. So go ahead and call it a home theater, but design it for many uses, and, if it is a good one, be prepared for it to sound better than most (any?) large-venue presentations.

This is the first time in the history of audio that there has been a significantly capable service industry operating in the space between audio manufacturers and customers. These designers and installers are in a position to offer informed advice on the purchase of audio and video equipment and to complete the installation and calibration of it. With the complexity of contemporary A/V equipment and the need for centralized control, that is not a trivial advantage.

Part Two addresses this custom installation audience and anyone else who wants to create a state-of-the-art home theater or sound reproduction system. It is hoped that the presentation style and language serve their needs. This book is not a cookbook. Sadly, such simplifications—and several have been attempted—end up being oversimplifications. In an attempt not to scare off readers, the "keep it simple" principle has been applied to a topic that in reality isn't *that* simple. Only a very tolerant consuming public could be pleased with many of the home theaters constructed according to these compromised principles. Yet, there is nothing in the design of a home theater that is beyond the capabilities of a determined do-it-yourselfer.

With good design, absolutely superb sound reproduction can be achieved in rooms used for normal living and decorated as such. However, if isolating the dynamics of rock-and-roll or a blockbuster movie from the rest of the dwelling is an issue, then a dedicated home theater is needed, as well as, most likely, the services of a competent acoustical consultant to design effective sound isolation. A dedicated room will not have normal furnishings, and therefore interior acoustical treatment is necessary. This is an opportunity to design the "listening experience," optimizing it for the programs and playback formats favored by the customer. In this book we will review the accumulated wisdom of the industry. We will add to this the results of some scientific investigations that are relatively unknown in the audio business and, out of the combination, arrive at guidelines for the design of entertainment systems and rooms.

"Between believing a thing and thinking you know is only a small step and quickly taken." Mark Twain.

Audio is a mature industry, and as a result of countless repetitions, certain beliefs have come to have a status comparable with scientific facts. Many of these ideas have been altruistically well intended, others commercially motivated. Some of them are wrong. This is a wasteful situation because the casualty is so often the art itself.

Readers will undoubtedly find conflicts between some of the recommendations in this book and ideas published or promoted elsewhere. If so, try to understand why the recommendations came about by tracing the story through this book and to the references if necessary. Inevitably, parts of the story are incomplete. If there is a better or more correct way, now or in the future, the scientific literature provides opportunity for a free exchange of information: new evidence and new interpretations of evidence. This is the scientific method.

Science in the service of art is the only sustainable position. In this book, we "follow the science" to see where it leads us. In the end, it will be found that in some ways it points to where we already are, but in other respects, some course adjustments are necessary.

Acknowledgments and dedication

Thanks to the National Research Council of Canada and Harman International Industries, Inc. who, for 27 and 15 years respectively, provided funding and facilities for me and my colleagues to investigate the physical and psychoacoustic factors involved in sound reproduction. Harman engineering staff generously provided many of the loudspeaker measurements seen here. The numerous researchers around the world whose efforts have contributed substantially to the contents of this book have both my respect and gratitude.

I am grateful to Sean Olive, Todd Welti, and Marshall Buck who provided critical comment on portions of the manuscript, and to my extremely supportive wife Noreen who shared in the creative process on a daily basis.

A special thanks is due to John Eargle, who provided encouragement to undertake the project and whose continued involvement was prevented by his untimely death in May 2007. In John I lost a good friend and professional colleague, and audio lost a generous contributor with a rare combination of artistic, engineering and scientific skills. I dedicate this book to the memory of this fine human being.

PART ONE

Understanding the Principles

The purpose of this book is to understand how to design what I call "listening experiences," also known as home theaters, home cinemas, stereo systems, or just entertainment rooms. Video may or may not be a part of the experience, but audio is always there. But, what kind of audio? What sounds need to be delivered to listeners' ears in order to create the appropriate perceptions? What properties do loudspeakers need to possess in order to generate the right sounds, and what acoustical characteristics of rooms provide optimal propagation conditions for those sounds? What is it that we mean by "good sound," and is there agreement on what it is?

In this part of the book we review the history of sound reproduction because it will help explain how we got to where we are now. Some of the patterns set years ago continue to be daily influences in what we hear, for better and worse. We will examine the physical sound fields in rooms, and show that from concert halls to homes and cars there is a continuum of gradual change in how they are structured and behave. Part of this change is reflected in what we hear, and in what we need to measure to describe these sound fields in different situations. A constant factor throughout all of these situations is the perceptual process. We carry the same two ears and brain with us to classrooms and concerts, to cinemas, circuses, and jazz clubs. They are there while watching a televised football game with family and friends, and while listening to satellite radio in the car. If we understand the basic dimensions of perception, we should be in a good position to design sound reproduction systems that provide maximum gratification for listeners whatever the program and listening circumstances may be. Let us start with some definitions.

CHAPTER 1

Sound Reproduction

Sound reproduction. When we use these words, we assume that everyone knows what they mean. And it is very likely a safe assumption, but to be sure, let us begin with the perspective provided by the traditional source of meanings: dictionaries. The following definitions are based on those found in *Merriam-Webster's Collegiate Dictionary*, 11th ed.; *The Shorter Oxford English Dictionary*, 3rd ed.; and *The American Heritage Dictionary of the English Language*.

Reproduction
1. Imitating something closely, making a representation, an image or copy, of something
2. To translate a recording into sound
3. To cause something to exist again

The word has another meaning that has to do with the perpetuation of species. Although there is no known science, there is abundant anecdotal evidence suggesting that, for humans, moods cultivated by reproduced sound are often significant factors in this process. One would like to think that settings enhanced by the highest-quality reproduced sound might be the most conducive, even in these distracting situations. One could hardly imagine a better reason to pursue research in this topic than the continuation of humankind.

In definition 1, imitation requires that there is an original, a reference, and the task of the reproducing system is to create a close copy of the original. If we think of concert hall performances as the reference, convincing imitations might be challenging in small living spaces using conventional audio hardware. However, most of the music and movies we hear are created in recording control rooms and film-dubbing stages. They are abstract artistic impressions of any live performance or sound event. This form of imitation requires a similarity of loudspeakers and room acoustics used for professional playback and in home audio systems. Anyone involved in audio knows that we are far from this ideal situation, there being significant departures from any reasonable sound quality standard on both professional and consumer sides of the audio industry.

The word *closely* is part of this definition. How similar does the imitation need to be? How much beyond recognizing a melody and rhythm does it go? Because the true nature of the original is almost always unknown, it is clear that much of what is being endeavored in sound reproduction is affective—on an emotional level.

Definition 2 demands nothing in the way of sound quality, only creating sound from a recording. Cynics have suggested that this is what we all experience in much of contemporary audio. Definition 3 is even easier, requiring only that when a "play" button is pressed, definition 2 is satisfied.

The word *sound* has many meanings, several having nothing at all to do with music or movies, but if we isolate the meanings that do have a link, we find that *sound* embraces all of the following:

1. The act of making or emitting a sound

2. The physical event of sound waves propagating in a medium, air being but one of many media through which sound can pass

3. The perceptions arising from sound waves that cause the eardrums to vibrate, which can be (a) meaningless noise, (b) recorded material, or (c) the particular musical style characteristic of an individual, a group, or an area—for example, the "Nashville" sound.

The inclusion of sound as both a physical event and a perceptual event is notable. It answers the popular riddle "If a tree falls in a forest and nobody is there to hear it, does it make sound?" The answer is both yes and no. When the tree falls, it creates sound—the physical energy—propagating away in all directions. However, with no ears in the vicinity, there can be no perception of the physical event.

To be rewarding in our context, we need to focus on sound waves propagating in air within the audible amplitude and frequency range so they can result in perceptions. This is physics, pure and simple. The specific informational content of the sound waves is irrelevant to how they propagate. Viewed as a combination of both of these generous definitions, *sound reproduction* is a very imprecise statement of our objectives. It seems as though almost anything can pass as "reproduced" sound, and, actually, history bears this out.

But the system works. Look around. We live in a global marketplace that is replete with sound-reproducing equipment for public venues, cinemas, homes, cars, and portable use, at prices reflecting "down and dirty" basic usage up to status-symbol level. There are vast libraries of musical recordings and movies. Stylish periodicals in many languages feed the interest and curiosity of enthusiasts. All of it adds up to a multibillion-dollar industry. Multitudes enjoy these products. Not all reproduced sound is very refined, and it is clear that much of the time we are willing to suspend criticism of the sound itself to just enjoy the music, movie, or whatever program instigated the sound. All of us at one time or another have felt that chill running down our spine—that tingling sensation that tells us we are experiencing something special and emotionally moving. Was it "real"? Was it "reproduction"? Good sound or bad? Does it matter? The fact that these feelings happen confirms that the system works.

But—and this is the motivation for this book—if any sound is rewarding, better and more spatially complex sound may be more pleasurable. This is part of the ever-evolving entertainment industry. With the application of science and good engineering, it is reasonable to assume that we can enjoy better reproduced sound more often in more places. By understanding the perceptual dimensions and the technical parameters that give control over them, it may be possible to give the artists tools that allow them to move into new creative areas by expanding the artistic palette. If there can be some assurance that customers will hear a facsimile of the art that was created, the greater are the rewards for innovative musicians and recording engineers. Let's see where we are and how far we can go.

1.1 A PHILOSOPHICAL PERSPECTIVE

Before sound reproduction was possible, sound was a temporary phenomenon. When the last audible reflection of it flew past an ear, sound was lost. Now, recorded sounds can be made to last forever. There is a complex cultural interplay between live sounds and reproduced sounds. Sterne (2003) perceptively notes that "the possibility of sound reproduction reorients the practice of sound production; insofar as it is a possibility at all, reproduction precedes originality" (p. 221). Making recordings requires paraphernalia: microphones, stands, wires, electronic apparatus, recording devices, and people. It is not a spontaneous event. It may require that performers be physically arranged in unfamiliar ways in unfamiliar rooms to optimize the pickup of the sound. The contortions required of musicians to record in the days of Edison were enormous, but even today they are significant.

We have not yet reached the stage of invisible microphones that can be introduced into a live performance so subtly as to not alter the mood or break the spell of musicians who have found a good "groove" in a familiar habitat. Nevertheless, what can be done is impressive and hugely satisfying. Obviously, many musicians adapted to the context of performing in studio environments without an audience. Pianist and famous Bach interpreter Glenn Gould much preferred the control he could exercise in a recording studio to the pressures of performing live. He would rather be remembered for "perfect" massively edited recordings than, in his mind at least, imperfect, evanescent performances before audiences. He went so far as to predict that "the public concert as we know it today [will] no longer exist in a century hence, that its functions [will] have been entirely taken over by electronic media" (Gould, 1966).

Stravinsky offered the opinion, "How can we continue to prefer an inferior reality (the concert hall) to ideal stereophony?" (a 1962 remark, quoted in Dougharty, 1973). Milton Babbitt was similarly provocative:

I can't believe that people really prefer to go to the concert hall under intellectually trying, socially trying, physically trying conditions, unable to repeat something they have missed, when they can sit home under the most comfortable and stimulating circumstances and hear it as they want to hear it. I can't imagine what would happen to literature today if one were obliged to congregate in an unpleasant hall and read novels projected on a screen. (Gould, 1966)

The point here is that "reproduction does not really separate copies from originals but instead results in the creation of a distinctive form of originality: the possibility of reproduction transforms the practice of production" (Sterne, 2003, p. 220). Knowing that the production process will lead to a reproduction liberates a new level of artistic creativity. Capturing the total essence of a "live" event is no longer the only, or even the best, objective. Movies have taken this idea to very high levels of development. It is more than "high realism"; it includes aspects of extreme fantasy. If something can be done, someone will do it. A harpsichord, a feeble instrument, can be made to sound competitive with a 75-piece orchestra.

During a recording, microphones can sample only a tiny portion of the complex three-dimensional sound field surrounding musical instruments in a performance space. What is captured is an incomplete characterization of the source. During playback, a multichannel reproduction system can reproduce only a portion of the complex three-dimensional sound field that surrounds a listener at a live performance. What is reproduced will be different from what is heard at a live event.

Audiophile fans of "high culture" music have repeatedly expressed disappointment that what they hear in their living rooms is not like a live concert, implying that there is a crucial aspect of amplifier or loudspeaker performance that prevents it from happening. The truth is that no amount of refinement in audio devices can solve the problem; there is no missing ingredient or tweak that can, outside of the imagination, make these experiences seem real. The process is itself fundamentally flawed in its extreme simplicity. The miracle is that it works as well as it does. The "copy" is sufficiently similar to the "original" that our perceptual processes are gratified, up to a point, but the "copy" is not the same as the "original." Sterne (2003) explains that "at a very basic, functional level, sound-reproduction technologies need a great deal of human assistance if they are to work, that is, to 'reproduce' sound" (p. 246).

Sound reproduction is therefore significantly about working with the natural human ability to "fill in the blanks," providing the right clues to trigger the perception of a more complete illusion. It is absolutely *not* a mechanical "capture, store, and reproduce" process. In addition to the music itself, there is now, and probably always will be, a substantial human artistic, craftsmanship, component to the creation of musical product.

Sterne (2003) goes on to explain that "as many critics of film and photography have shown us, reality is as much about aesthetic creation as it is about any other effect when we are talking about media" (p. 241). And, in the context of sound recording, "far from being a reproduction of the actual event, the recording was a 're-creation'" (p. 242). The goal is not imitation but the creation of specific listener experiences. This certainly exists dramatically in the directional and spatial experiences in reproduced sounds.

For decades, society has been conditioned to derive pleasure from first single-channel sound (mono) and then two channels (stereo). Only recently has music been offered in multichannel formats that permit a somewhat realistic directional and spatial panorama. Impressed by the novelty that music and movies were available on demand, society appeared to lower its expectations and adapted to the inadequate formats. A great deal of enjoyment was had by all. So complete is this form of adaptation that significant new technical developments must go through a "break-in" period before there is acceptance. Having abandoned or forgotten the sound of a three-dimensional sound field, those who grew up with mono often argued that stereo was an unnecessary complication, adding little value. (I remember—I was there!) The same is now happening with respect to multichannel audio schemes. Part of the "break in" applies to the audio professionals, who must learn how to use the new formats with discretion and taste.

In terms of sound quality—fidelity—there have been claims of perfect sound since the very beginning of sound reproduction. In the earliest days, it seems that audiences were simply so amazed to be able to recognize pitch, tunes, and rhythm that they ignored huge insults to sound bandwidth, spectrum, dynamics, and signal-to-noise ratio. Now we do much better, of course, but then, as now, according to Sterne, "sound reproduction required a certain level of faith in the apparatus and a certain familiarity with what was to be reproduced" (p. 247). Expectations are a part of our perceptions—a fact well used by advertisers of audio appliances from the earliest times, and today. A 1908 advertisement for Victor Talking Machines asserted, "You think you can tell the difference between hearing grand-opera artists sing and hearing their beautiful voices on the Victor. But can you?" (Sterne, 2003, p. 217). The formula must work because, as this is being written 100 years later, boasts of sound quality still abound: "Everything you hear is true"; "Pro sound comes home" (www.jbl.com); "True sound" (www.bowers-wilkins.com); "Pure, natural, true-to-the-original performance™" (www.bostonacoustics.com). The suggestion that audio hardware is capable of a kind of acoustical transmigration is clearly attractive.

The introduction of digital audio sent ripples—actually, more like a tsunami—through the audio community. On the one hand were those who claimed that the perfect utterly transparent storage device had arrived. On the other hand, audio critics were hearing flaws. Although irritating swishes, ticks, pops, clicks, wow, and flutter were pleasantly absent, there were other problems. It turned out that early digital hardware did not always perform to claimed specifications. Arguments also went metaphysical, with assertions that in converting audio signals to numbers, a crucial link to the original sound was lost. The passage of time has brought immense improvements in all aspects of performance. Now, in a new calm, we discuss just how much digital bandwidth is needed to store "all" the music.

There is cause for some sadness, though, because some of the old analog tapes being converted for delivery through digital formats are not the original masters, which were lost or reused, but the LP master tapes used to drive the cutter heads for creating LP master discs. It seems that important decision makers thought the LP was the final development in the delivery of audio signals. These tapes have been skillfully manipulated—predistorted, in fact—to compensate for some of the significant limitations of LPs and therefore cannot sound their best when played through digital devices. The art has been compromised. This is an example of an old technology executing a strange form of revenge on the new. Perhaps there is something to the metaphysical argument after all.

1.2 RECORDINGS AND THE MUSIC BEING RECORDED

Finally, it is worth noting how sound reproduction has influenced music itself, especially jazz. Because they offer perfect repeatability, recordings became learning aids for musicians. According to Katz (2004), the widespread availability of recordings of major artists "gave budding musicians unprecedented access to jazz. Without this feature of recording technology, some jazz artists might never have pursued their careers" (p. 74). This is good, of course, but recordings of the pre-LP era had a severe restriction on playing time: a ten-inch, 78 rpm record had a maximum 3 m 15 s playing time. This forced a change in playing style, and long jazz improvisations were abridged for recordings. After much repetition, these improvisations became ritualized in performances of other artists, so they were no longer improvisations.

Even the vocal content changed because of recordings. Morton (2004) states, "Nearly all the early jazz captured on records was 'cleaned up' for white audiences. Live jazz was improvised and disorderly. Its lyrics and themes often had sexual overtones" (p. 62).

The early recordings that were formative in the development of jazz also suffered from recording difficulties. Spectral and dynamic limitations did not flatter instruments like pianos and drums, so substitute instruments were used. Live performances came to reflect some of these substitutions and sometimes even playing styles. For example "slap" bass playing was a means of minimizing bothersome low-frequency output but retaining some of the essential sound of the upright bass. Said Katz (2004), "The bass drum was a troublemaker even into the 1950s" (p. 81). (Wrapping it in a blanket was a common studio remedy.) Katz explained as follows:

Whether in France, the United States, or anywhere in the world, most listeners who knew jazz knew it through recordings; the jazz they heard, therefore, was something of a distortion, having been adapted in response to the nature of the medium. The peculiar strengths and limitations of the technology thus not only influenced jazz performance practice, it also shaped how listeners—some of whom were also performers and composers—understood jazz and expected it to sound. (p. 84)

Recordings also had significant effects on classical music. In a live performance, we wait intently while a musician pauses, lifts a bow, and leans forward to continue a work. In a recording, lacking the visual input, such a delay is "dead air." Recorded music has, accordingly, better continuity. When sound emerges from the violin, it will likely be played with more vibrato than was customary in earlier times. Katz (2004) makes the case that this is linked to influences of sound recording, and he includes a CD of some examples with his book.

Today, we make recordings from parts of other recordings. LP-era disc jockeys developed the technique of "scratching"—manually moving a disc under a phonograph cartridge, thereby repeating, reversing, and varying the speed of a musical excerpt. Now, in the original or in digital incarnations, it is used as a component in some kinds of recorded music. Digital "sampling"—excerpted recordings of anything that makes sound—is an enabling technology for some digital keyboard instruments, and, in editing, it is an ingredient in much popular music. Some compositions are clever compilations of nothing but sampled sounds. In musical terms, this is abstract or surreal art with no "live" equivalent. It is totally a studio creation.

And so it will continue in the unending interplay between musical creation, reproduction technology, and listener expectations and preferences. Katz states, "Recording is not a mysterious force that compels the actions of its users. Ultimately, they—that is, we—control recording's influence. Recording has been with us for more than a century; it will no doubt remain an important musical force, and users will continue to respond to its possibilities and limitations" (p. 191). The technologies discussed in the rest of this book are part of the future of our audio industry, our music, and our movies. We will see that there are several ways to improve the process, ensuring a superior and more reliable delivery of the art we know. We will also examine ways to introduce new ingredients into the art.

CHAPTER 2

Preserving the Art

A scientific understanding of the physical world has allowed us to do many remarkable things. Among them is the ability to enjoy the emotions and aesthetics of music whenever or wherever the mood strikes us. Music is art, pure and simple. Composers, performers, and the creators of musical instruments are artists and craftsmen. Through their skills, we are the grateful recipients of sounds that can create and change moods, that can animate us to dance and sing, and that form an important component of our memories. Music is a part of all of us and of our lives.

However, in spite of its many capabilities, science cannot describe music. There is nothing documentary beyond the crude notes on a sheet of music. Science has no dimensions to measure the evocative elements of a good tune. It cannot technically describe why a famous tenor's voice is so revered or why the sound of a Stradivarius violin is held as an example of how it should be done. Nor can science differentiate, by measurement, between the mellifluous qualities of trumpet intonations by Wynton Marsalis and those of a music student who simply hits the notes. Those are distinctions that must be made subjectively, by listening. A lot of scientific effort has gone into understanding musical instruments, and as a result, we are getting better at imitating the desirable aspects of superb instruments in less expensive ones. We are also getting better at electronically synthesizing the sounds of acoustical instruments. However, the determination of what is aesthetically pleasing remains firmly based in subjectivity.

This is the point at which it is essential to differentiate between the production of a musical event and the subsequent reproduction of that musical event. Subjectivity—pure opinion—is the only measure of whether music is appealing, and it will necessarily vary among individuals. Analysis involves issues of melody, harmony, lyrics, rhythm, tonal quality of instruments, musicianship, and so on. In a recording studio, the recording engineer becomes a major contributor to the art by adjusting the contribution of each musician to the overall production, adjusting the tonal balance and timbre of each of the contributors, and adding reflected and reverberated sounds or other processed versions of captured sounds to the mix. This too is judged subjectively, on the basis of whether it reflects the artists' intent and, of course, how it might appeal to consumers.

The evaluation of reproduced sound should be a matter of judging the extent to which any and all of these elements are accurately replicated or attractively reproduced. It is a matter of trying to describe the respects in which audio devices add to or subtract from the desired objective. A different vocabulary is needed. However, most music lovers and audiophiles lack this special capability in critical listening, and as a consequence, art is routinely mingled with technology. In subjective equipment reviews, technical audio devices are often imbued with musical capabilities. Some are described as being able to euphonically enhance recordings, and others to do the reverse. It is true that characteristics of technical performance must be reflected in the musical performance, but it happens in a highly unpredictable manner, and such a commentary is of no direct assistance in our efforts to improve sound reproduction.

In the audio industry, progress hinges on the ability to identify and quantify technical defects in recording and playback equipment while listening to an infinitely variable signal: music. Add to this the popular notion that we all "hear differently," that one person's meat might be another person's poison, and we have a situation where a universally satisfying solution might not be possible. Fortunately reality is not so complex, and although tastes in music are highly personal and infinitely variable, we discover that recognizing the most common deficiencies in reproduced sounds is a surprisingly universal skill. To a remarkable extent we seem to be able to separate the evaluation of a reproduction technology from that of the program. It is not necessary to enjoy the program to be able to recognize that it is, or is not, well reproduced.

How do listeners approach the problem of judging sound quality? Most likely the dimensions and criteria of subjective evaluation are traceable to experiences in live sound, even simple conversation. If we hear things in sound reproduction that could not occur in nature or that defy some kind of logic, we seem to be able to identify it. But, as in many other aspects of life, some of what we regard as "good" is governed by a cultivated taste. Factors contributing to the prevailing taste at different points in time are interesting to discuss, as they will ring bells of familiarity in the minds of many readers. Thompson (2002) states that "culture is much more than an interesting context in which to place technological accomplishments; it is inseparable from technology itself" (p. 9).

2.1 BACK TO THE BEGINNING: CAPTURING SOUND QUALITY

In terms of basic sound quality, claims of accurate reproduction began early. Edison, in 1901, claimed that the phonograph had no "tone" of its own. To prove it, he mounted a traveling show in which his phonograph was demonstrated in "tone tests" that consisted of presentations with a live performer. Morton (2000) reports, "Edison carefully chose singers, usually women, who could imitate the sound of their recordings and only allowed musicians to use the limited group of instruments that recorded best for demonstrations" (p. 23). Of a 1916 demonstration in Carnegie Hall before a capacity audience of "musically cultured and musically critical" listeners, the *New York Evening Mail* reported that "the ear could not tell when it was listening to the phonograph alone, and when to actual voice and reproduction together. Only the eye could discover the truth by noting when the singer's mouth was open or closed" (quoted in Harvith and Harvith, 1985, p. 12).

Singers had to be careful not to be louder than the machine, to learn to imitate the sound of the machine, and to sing without vibrato, which Edison (apparently a musically uncultured person) did not like. There were other consequences of these tests on recordings. The low sensitivity of the mechanical recording device made it necessary for the performers to crowd around the mouth of the horn and find instruments that could play especially loud. Because

FIGURE 2.1 *Singer Frieda Hempel stages a tone test at the Edison studios in New York City, 1918. Care was taken to ensure that the test was "blind," but it is amusing to see that some of the blindfolds also cover the ears. Courtesy of Edison National Historic Site, National Park Service, U.S. Department of the Interior.*

the promotional "tone tests" were of solo voices and instruments, any acoustical cues from the recording venue would reveal the recording as being different from the live performer in the demonstration room. Consequently, in addition to employing what was probably the first "close microphone" recording technique, Edison's studios were acoustically dead (Read and Welsh, 1959, p. 205).

Live versus reproduced comparison demonstrations were also conducted by RCA in 1947 [using a full symphony orchestra (Olson, 1957, p. 606)], Wharfedale in the 1950s (Briggs, 1958, p. 302), Acoustic Research in the 1960s, and probably others. All were successful in persuading audiences that near perfection in sound reproduction had arrived. Based on these reports, one could conclude that there had been no consequential progress in loudspeaker design in over 50 years. Have we come much farther a century later? Are today's loudspeakers significantly better sounding than those of decades past? The answer in technical terms is a resounding "Yes!" But would the person on the street or even a "musically cultured" listener be able to discern the improvement in such a demonstration? Are we now more wise, more aurally acute, and less likely to be taken in by a good demonstration?

When the term *high fidelity* was coined in the 1930s, it was more a wishful objective than a description of things accomplished. Many years would pass before anything resembling it could be achieved. Although recreating a live performance was an early goal, and it remains one of the several options today, the bulk of recordings quickly drifted into areas of more artistic interpretation. Morton (2000) states, "The essence of high fidelity, the notion of 'realism,' and the uncolored reproduction of music dominated almost every discussion of home audio equipment. However, commercial recordings themselves betrayed the growing divide between the ideals of high fidelity and the reality of what happened in the recording studio" (p. 39).

Whatever the musical content, high culture or low, there is still a reference sound that we must emulate in our listening spaces, and that sound is the sound of the final mix experienced in a recording control room. The sound of live performers is a kind of hidden reference, buried in all of our subconscious minds. It is relevant but not in the "linear" manner implied in many heated arguments over the years. Many believe that only by memorizing the sound of live performances can one judge the success of a reproduction. The principal issue is that few of us have ever heard the sound of voices and instruments with our ears at the locations of microphones. These are normally placed much closer to mouths and instruments than is desirable or even prudent for ears in live performances. The sounds captured by microphones are not the sounds we hear when in an audience. Microphones hanging above the string section of an orchestra inevitably pick up a spectrum that has a high-frequency bias; the sound is strident compared to what is heard in the audience due to the directional radiation behavior of violins (Meyer, 1972, 1978, 1993; see Figure 3.3). The total sound output of a piano cannot be captured by the close placement of any single microphone, which therefore means that such recordings cannot accurately represent what is heard by listeners in unamplified performances in natural acoustical spaces. Yet, they have been good enough approximations to give pleasure to generations of listeners.

The practical reality is that all recordings end up in a control room of some sort, where decisions are made about the blending of multiple microphone inputs, sweetening with judicious equalization, and enrichment with electronically delayed sound. This is the second layer of art in recordings, added by some combination of recording engineers and performing artists. There are many written discussions of how to "monitor" the progress of recordings—some in books and many more in magazine articles. Opinions cover an enormous range. Many mixers choose monitor loudspeakers that add a desired quality to the sound, instead of using a neutral monitor and achieving the same desired quality through signal processing. This attitude, which seems to be depressingly common, leads nowhere, because only those who are listening through the same loudspeakers will be able to hear that desirable sound quality. A good, recent perspective on the topic can be found in Owsinski (1999), where comments from the author and 20 other recording mixers are assembled. They all care about what their customers hear, but they differ enormously in how to estimate what that is. Using known "imperfect" loudspeakers, such as Yamaha NS-10Ms and Auratone 5Cs, is popular. Some also listen in their cars and through various renderings of inexpensive consumer products. The problem is that there are countless ways to be wrong (bad sound).

Choosing a single or even a small number of "bad" loudspeakers cannot guarantee anything. Nobody in this massive industry seems to have undertaken a statistical study of what might be an "average" loudspeaker. The author's experience suggests that the performance target for almost all consumer loudspeakers is a more-or-less flat axial frequency response. Failure to achieve the target performance takes all possible forms: lack of bass, excessive bass, lack of mids, excessive mids, lack of highs, excessive highs, prominent resonances at arbitrary frequencies, and so on. The only common feature that distinguishes lesser products seems to be a lack of low bass. In short, an excellent approach to choosing a monitor loudspeaker would be to choose a state-of-the-art "neutral" device, adjust it to perform in its specific acoustical environment, and then electronically introduce varying degrees of high-pass filtering to simulate reproduction through anything from a clock radio to a minisystem. It is very perplexing that no truly reliable technical standards for control room sound exist, making the reference a moving target.

2.2 BACK TO THE BEGINNING: DIRECTION AND SPACE

Sounds exist in acoustical contexts. In live performances we perceive sources at different locations, and at different distances, in rooms that can give us strong impressions of envelopment. A complete reproduction should convey the essence of these impressions. A moment's thought reveals that because our binaural perceptual mechanism is sensitive to sounds arriving from all angles, reproducing a persuasive illusion of realistic direction and space must entail multiple channels delivering sounds to the listener from many directions. The key questions are how many and where?

This aspect of sound perception has been greatly influenced by both recording technology and also by culture. Blesser and Salter (2007) discuss this in terms of "aural architecture," defined as those properties of a space that can be experienced by listening. This begins with natural acoustical environments, but nowadays we can extend this definition to include those real and synthesized spatial sounds incorporated in recordings and those that are reproduced through loudspeakers in our listening rooms. In this sense, all of us involved with the audio industry are, to some extent, aural architects.

From the beginning, we have come to associate certain kinds of sounds with specific architectural structures; for example, a highly reverberant spacious illusion is anticipated when we see that we are in a large stone cathedral or a multistory glass and granite foyer of an office building. Rarely are our expectations not met as we make our way through the physical acoustical world. However, in recordings, we now have the technology to deliver to a listener's ears some of the spatial sounds of a cathedral while seated in a car or living room. But are the illusions equally persuasive? Are they more persuasive if there is an image of the space on a large screen?

Auditory spatial illusions are no longer attached to visual correlates; they exist in the abstract, conceivably a different one for every instrument in a multi-miked studio composition. Traditionalists complain about such manipulations, but most listeners consider them just another form of sensory stimulation. Blesser and Salter (2007) said, "Novelty now competes with refinement" (p. 126).

All of this stands in stark contrast to the spatially deprived decades that audio has endured. It began with the first sound reproduction technology, monophonic sound, which stripped music of any semblance of soundstage, space, and envelopment. This was further aggravated by the need to place microphones close to sources; early microphones had limited dynamic range and high background noise. Adding further to the spatial deprivation was the use of relatively dead recording studios and film soundstages. Read and Welsh (1959) explained, "Reproduced in the home, where upholstered furniture, drapes and rugs quite often prevented such an acoustical development of ensemble through multiple reflections, the Edison orchestral recordings were often singularly unappealing" (p. 209). Recording engineers soon learned that multiple microphones could be used to simulate the effects of reflecting surfaces, so the natural acoustics of the recording studio were augmented, or even replaced, by the tools and techniques of the sound recording process.

With the passage of time, directional microphones gave further control of what natural acoustics were captured. With relatively "dead" source material, it became necessary to add reverberation, and the history of sound recording is significantly about how to use reverberation rooms and electronic or electromechanical simulation devices to add a sense of space. In the past, these effects were used sparingly and "the typical soundtrack of the early 1930s emphasized clarity and intelligibility, not spatial realism" (Thompson, 2002, p. 283).

A coincidental influence was the development of the acoustical materials industry. In the 1930s, dozens of companies were manufacturing versions of resistive absorbers—fibrous fluff and panels—to absorb reflected sound and to contribute to acoustical isolation for bothersome noises. Acoustical treatment became synonymous with adding absorption. Dead acoustics were the cultural norm—the "modern" sound—which aligned with recording simplicity, low cost, small studios and profitability (Blesser and Salter, 2007, p. 115). Thompson (2002) explains, "When reverberation was reconceived as noise, it lost its traditional meaning as the acoustic signature of a space, and the age-old connection between sound and space—a connection as old as architecture itself—was severed" (p. 172).

Read and Welsh (1959) recount the following statement, written in 1951 by popular audio commentator Edward Tatnall Canby in his "Saturday Review of Recordings":

"Liveness," the compound effect of multiple room reflections upon played music, is—if you wish—a distortion of "pure" music; but it happens to be a distortion essential to naturalness of sound. Without it, music is most graphically described as "dead." Liveness fertilizes musical performance, seasons and blends and rounds out the sound, assembles the raw materials of overtone and fundamental into that somewhat blurred and softened actuality that is normal, in its varying degrees, for all music. Disastrous experiments in "cleaning up" music by removing the all-essential blur long since proved to most recording engineering that musicians do like their music muddied up with itself, reflected. Today recording companies go to extraordinary lengths to acquire studios, churches and auditoriums (not to mention an assortment of artificial, after-the-recording liveness makers) in order to package that illusively perfect liveness. (p. 378)

This notion that reflections result in a corruption of "pure" music, and the apparent surprise in finding that musicians and ordinary listeners prefer "muddied up" versions, reappears in audio even today. We now have quite detailed explanations why this is so, but one can instinctively grasp the reality that, toward the rear of a concert hall, the direct sound (the "pure" music) is not the primary acoustic event. It may even be inaudible, masked by later acoustical events. Two ears and a brain comprise a powerful acoustical analysis tool, able to extract enormous resolution, detail, and pleasure from circumstances that, when subject to mere technical measurements, seem to be disastrous. Something that in technical terms appears to be impossibly scrambled is perceived as a splendid musical performance.

When Sabine introduced the concept of reverberation time into acoustical discussions of rooms at the turn of the last century, he provided both clarification and a problem. The clarification had to do with adding a technical measure and a corresponding insight to the temporal blurring of musical patterns that occur in large live spaces. The problem appeared when recordings made in spaces that were good for live performances were often perceived to have too much reverberation. A single microphone sampling such a sound field that was then reproduced through a single loudspeaker simply did not work; it was excessively reverberant. Our two ears, which together allow us to localize sounds in three dimensions to separate individual conversations at a cocktail party and to discriminate against a background of random reverberation, were not being supplied with the right kind of information. Multiple microphones that could convey information through multiple channels and deliver the appropriate sounds to our multiple (binaural) ears were necessary, but they were not available in the early years.

This disagreement between what is measured and what is heard has been the motivation for much scientific investigation of the acoustics of rooms, both large and small. In some ways, our problems with rooms, especially small rooms, began when we started to make measurements. Our eyes were offended by things seen in the measurements, but our ears and brain heard nothing wrong with the audible reality. As we will see, some of the resolution of the dilemma is in the ability of humans to adapt to, and make considerable sense of, a wide variety of acoustical circumstances. Separating sound sources from the spaces they are in is something humans do routinely.

Old habits die hard. The introduction of stereo in the 1950s gave us an improved left/right soundstage, but close microphone methods, multitracking, and pan-potting, did nothing for a sense of envelopment—of actually *being* there. The classical music repertoire generally set a higher standard, having the advantage of the reflectivity of a large performance space, but a pair of loudspeakers deployed at ±30° or less is not an optimum arrangement for generating strong perceptions of envelopment (as will be explained later, this needs additional sounds arriving from further to the sides). Perhaps that is why audiophiles have for decades experimented with different loudspeaker directivities (to excite more listening room reflections), with electronic add-ons and more loudspeakers (to generate delayed sounds arriving from the sides and rear), and with other trinkets that seem capable only of exciting the imagination. All have been intended to contribute more of "something that was missing" from the stereo reproduction experience. The solution to this is more channels.

2.3 A CIRCLE OF CONFUSION

When we listen to recorded music, we are listening to the cumulative influences of every artistic decision and every technical device in the audio chain. Many years ago, I created the cartoon in Figure 2.2 to illustrate the principle and suggest how we may break the never-ending cycle of subjectivity.

The presumption implicit in this illustration is that it is possible to create measurements that can describe or predict how listeners might react to sounds produced by the device being tested. There was a time when this presumption seemed improbable, but with research and the development of newer and better measurement tools, it has been possible to move the hands of the "doomsday clock" to the point where detonation is imminent. Figure 2.3 expresses the ideas of the "circle of confusion" in a slightly different form, one that more accurately reflects the impact on the audio industry.

2.4 BREAKING THE CIRCLE: PROFESSIONALS HOLD THE KEY

The audio industry has developed and prospered until now without any meaningful standards relating to the sound quality of loudspeakers used by professionals or in homes. The few standards that have been written for broadcast control and music listening rooms applied measurements and criteria that had no real

FIGURE 2.2 *The first version of the "circle of confusion," illustrating the key role of loudspeakers in determining how recordings sound and of recordings in determining our impressions of how loudspeakers sound. The central cartoon suggests how the "circle" can be broken, using the knowledge of psychoacoustics to advance the clock to the detonation time at which the "explosive" power of measurements will be released to break the circle.*

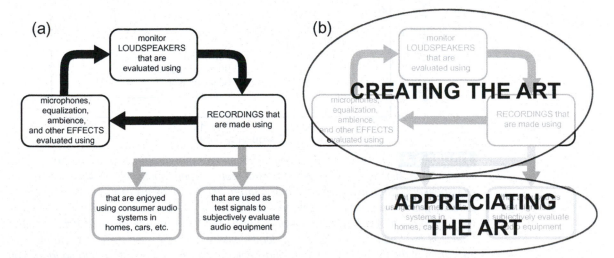

FIGURE 2.3 *(a) The "circle of confusion" modified to more accurately reflect its effect on the audio industry. The true role of loudspeakers is shown here. (b) Unless the loudspeakers involved in the creation of the recordings are similar in performance to those used in reproduction, the "art" is not preserved.*

chance of ensuring good, or even consistent, sound quality. Many years ago, the author participated in the creation of certain of those standards and can report that the inadequacies were not malicious, only the result of not having better information to work with. The film industry has long had standards relating to the performance of loudspeakers used in sound-mixing stages and cinemas. These too are deficient, but something is better than nothing.

A consequence of this lack of standardization and control is that recordings vary in sound quality, spectral balances, and imaging. Proof of this is seen each time a person reaches for a CD to demonstrate the audio system they want to show off. The choice is not random. Only certain recordings are on the "demo" list, and each will have favorite tracks. This is because the excitement comes not in the music—the tune, lyrics, or musical interpretation—but in the ability to deliver a "wow" factor by exercising the positive attributes of the system.

A recent survey of recording control rooms revealed a disturbing amount of variation in spectral balance among them (Mäkivirta and Anet, 2001a, b). Figure 2.4 shows that the differences were not subtle, especially at low frequencies.

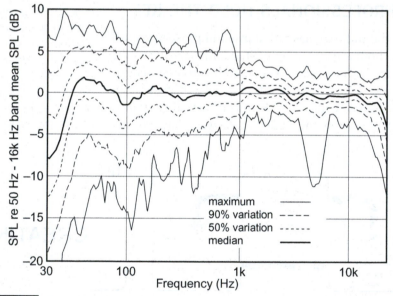

FIGURE 2.4 *The result of 250 frequency response measurements made at the engineer's location in many recording control rooms, using functionally similar Genelec loudspeakers. The curves show the maximum/minimum variations exhibited by different percentages of the situations. As we will see later, such "room curves" are incomplete data when it comes to revealing overall sound quality, but they are excellent indicators of sound quality at low frequencies. With similar loudspeakers, as is the case here, they are also indicators of the influences of room, loudspeaker mounting, equalization, and so forth. The real point of this display is to show the range of variations that occur in these critically important professional audio venues. The curves are 1/3-octave smoothed. From Mäkivirta and Anet (2001b).*

Recording engineers who work in these circumstances, presumably approving of them, are doing the art no favor. This is an excellent example of the circle of confusion in action because members of this group of audio professionals cannot even exchange their own recordings with a reasonable certainty of how they will sound in one another's control rooms.

Many audio professionals insist on using their own recordings when setting up new or alternative mixing or monitoring sites. All this does is to perpetuate whatever distortions were built into the original site. There is no opportunity for improvement, of moving to more "neutral" territory. It is disturbing to hear some such people argue that they attribute some of the success of their prior recordings to a monitoring situation that is clearly aberrant (one can only imagine that the composers and musicians feel insulted to hear that their contributions are subservient to a loudspeaker!). This is the kind of misguided argument that has led normally sensible people to promote the use of obviously "less than high-fidelity" loudspeakers for monitoring, on the basis that the majority of consumers will be listening through such loudspeakers.

It is true that the majority of consumers live with mediocre, even downright bad, reproduction systems. The problem is that it is possible to be "bad" in an infinite number of different ways, so any boom box or rotten little speaker that is chosen to represent "bad" is just one example of how to be bad, not a universal reference. The only aspect of their performance that is likely to be at all universal is the lack of low bass.

Looking at Figure 2.5, it appears that all of the designers tried to make a "flat" system, but each of them failed in a different way. Plotting an average of

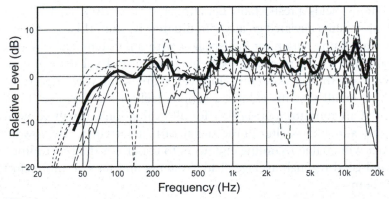

FIGURE 2.5 *On-axis frequency responses for six entry-level consumer "minisystems," incorporating CD/cassette/AM & FM tuners/amplifiers and loudspeakers. Prices ranged from $150 to $400 for these plastic-encased, highly styled units. These were among the most popular low-cost integrated systems in the marketplace in the year 2000. The heavy curve is the average response, which falls within a tolerance of ±3 dB from 50 Hz to 20 kHz.*

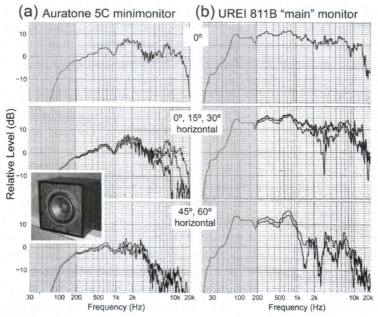

FIGURE 2.6 *(a) On- and off-axis frequency responses of an Auratone 5C, a five-inch (127 mm) full-range minimonitor. As can be seen in the photograph, it is in a small closed-box enclosure which contributes to seriously attenuated bass, and the single diaphragm shows evidence of resonances and strong directivity at high frequencies. (b) Comparable measurements for a UREI 811B large monitor loudspeaker of the same era, showing improved bass but similarly bad off-axis performance at middle and high frequencies.*

these systems yields a respectably flat curve (50 Hz to 20 kHz, ±3 dB), although individual systems deviated greatly, but differently, from this specification. This suggests that using a monitor loudspeaker with a flat frequency response might be a good way to please a large percentage of entry-level listeners, as well as those with superb audio systems. However, there is one very important proviso: all of these small systems exhibit a serious lack of low bass, so if one uses a state-of-the-art monitor loudspeaker for all evaluations, it will be necessary to incorporate a high-pass filter in the signal paths to attenuate the low bass. This simple act will enable recording engineers to condition their recordings so they will sound good through average "bad" systems and remove more clutter from control rooms.

Discussing this topic from the perspective of mastering engineering, Katz (2002) basically agrees:

Mastering engineers confirm that accurate monitoring is essential to making a recording that will translate to the real world. The fallacy of depending on an inaccurate "real-world-monitor" can only result in a recording that is bound to sound bad on a different "real-world monitor." Even the best master will sound different everywhere, but it will sound most correct on an accurate monitor system. Which leads us to this comment from a good client: "I listened to the master on half a dozen systems and took copious notes. All the notes cancelled out, so the master must be just right." (p. 82)

Figure 2.6 shows anechoic frequency-response measurements on two popular monitor loudspeakers of years past (some studios even today proudly advertise that either or both are still available). One is a small "near field" type, the Auratone, used to evaluate recordings as they might be heard out in the "real world." The UREI is a traditional large woofer, horn high-frequency configuration that typically would be built into a soffit or wall. The large loudspeaker has much better bass and can play at much higher sound levels, but in terms of the sound qualities of these two loudspeakers, there are more similarities than differences, and both have serious problems. Because of their distinctive imperfections, such loudspeakers are references only for themselves. The prime asset of the Auratone as a window into the world of bad loudspeakers is its lack of bass. As an indication of how far some of the guardians of our musical arts have strayed, it has been said that these minimonitors have "single-driver musicality." The mind is a marvelous instrument. We will see in Chapter 18 that the sound quality traditions of loudspeakers like these have been perpetuated in some contemporary products. Bad habits are hard to break.

Reflecting on all of this, it is easy to be pessimistic about the integrity of the circle of confusion as it applies to sound quality. Things are improving, however, as we will see in Chapter 18, where it is shown that, although the Auratone tonal personality can be seen in other newer products, there is another stream of superbly designed monitor loudspeakers that are remarkably neutral. In between, there is simply a boring collection of different versions of mediocrity. Evidence exists that audio professionals are as susceptible to a good marketing story as are consumers, and, without double-blind listening tests, their opinions are just as susceptible to bias.

With large, professionally designed studio complexes now being replaced or supplemented by home studios, the challenges have multiplied. We need to master how to reproduce good sound in relatively ordinary rooms of different configurations.

Returning to the concept of being aural architects, Blesser and Salter (2007) usefully summarize the situation:

Acoustic engineers determine the physical properties of the recording environment; design engineers develop the recording and reproduction equipment; recording engineers place the microphones; mixing engineers prepare the final musical product for distribution; interior decorators select furnishings for the listeners' acoustic space; and listeners position themselves and the loudspeakers within that space. Often acting independently, these individuals are members of an informal and unrecognized committee of aural architects who do not communicate with one another. With their divided responsibility for the outcome, they often create the spatial equivalent of a camel: a horse designed by a committee. (p. 131)

So listeners are merely the last in a long line of aural architects but with no influence on, or connection with, what has happened before.

No matter how meticulously the playback equipment has been chosen and set up, and no matter how much money has been lavished on exotic acoustical treatments, what we hear in our homes and cars is, in spatial terms, a matter of chance. Blesser and Salter conclude that "spatial accuracy is not a significant criterion for much of our musical experience" (p. 148). They go on to explain, "The application of aural architecture to cinema is a good example of aesthetically pleasing spatial rules that never presume a space as a real environment. Artistic space never represented itself as being a real space; it is only the *experience* of space that is real; and achieving artistic impact often requires spatial contradictions" (pp. 160, 161).

2.5 MEASURING THE ABILITY TO REPRODUCE THE ART

The contradiction implicit in the title of this section will reverberate through this book. How can we measure something that subjectively we react to as art. Measurements are supposed to be precise, reproducible, and meaningful. Perceptions are inherently subjective, evanescent, and subject to various nonauditory influences within and surrounding the human organism. However, perceiving flaws in sound reproducing systems appears to be an activity that we can substantially separate from our critique of the art itself. We can detect flaws in the reproduction of music of which we have no prior knowledge and in which we find no pleasure.

The audio industry uses—indeed, *needs*—measurements to define benchmarks of what is acceptable or not. Blesser and Salter (2007) contribute a simple, but not totally reassuring, perspective on the value of measurements. It begins with the recognition of a hierarchy in hearing:

- At the lowest level is *sensation*, an indication that the organism reacts to a sound—a detection threshold. This is probably quite well related to physical measurements of the sound.

- The next level is *perception*, which incorporates cognitive processes embracing cultural and personal experiences. Here we recognize what it is that we heard, and perhaps initiate a process of adaptation. This means that some features in measurements may be neutralized by adaptation, and no longer be relevant.

- At the highest level of response to sound, we attribute *meaning* to the recognition, and this can range from irrelevant to highly relevant, from undesirable to good. Depending on the informational content of the sound, we may choose to pay attention or to ignore it. In the latter case, it matters not what measurements tell us (p. 13).

As Blesser and Salter say, "Detectable attributes may not contribute to perceptible attributes, and perceptible attributes may not be emotionally or artistically meaningful. . . . Furthermore, affect can be at once meaningful and undesirable" (p. 14). What we, as individuals, consider to be meaningful and desirable is largely learned, although some of us show a more or less native ability to hear certain spatial and other attributes of sound. At this level of cognition, measurements are of dubious value.

Einstein's well-known quote is relevant: "Not everything that can be counted counts, and not everything that counts can be counted." The audio business hopes to convey much more than raw sensation; it aspires to perceptions and meanings as well. So, how do we quantify acoustical parameters in ways that correlate with the full panorama of subjective responses of individuals having wide-ranging personal and cultural characteristics? The hope is that we may be able to "connect" with some of the key underlying perceptual dimensions. The fact that several chapters follow this one signals that there has been some success at doing so.

CHAPTER 3

Sound in Rooms—Matters of Perspective

3.1 LIVE MUSICAL PERFORMANCES

The science of room acoustics developed in large performance spaces: concert halls. The sound sources, voices and musical instruments, were considered as a group to be approximately omnidirectional. Sounds from these sources radiate in all directions, being reflected by a few carefully positioned early reflecting surfaces, and then later by many somewhat randomly positioned reflecting, diffracting and scattering surfaces and objects, to create a uniformly mixed reflected sound field that makes its way to all parts of the audience.

All acoustical measurements done in these spaces begin with a sound source—a starter's pistol, explosive device, or a special loudspeaker—all of which are intended to radiate sound equally in all directions. So the sound source is a "neutral" factor, and this allows for some generalization in interpreting the meaning of whatever acoustical measurements are made. It is an examination of the room itself and how it modifies and manipulates the sound of a standard source and, by inference, a collection of musicians. Casual thought reveals that this is a great simplification because different musical instruments have significantly different directivities. The violin ensemble probably has, as a group, quite wide dispersion, with much of the high-frequency energy going upward. The brasses have a strong forward bias and are strongly directional at high frequencies, whereas the French horns have a backward directional bias responsible for their peculiarly spacious sound (Meyer, 1978, 1993).

So the basic assumptions underlying measurements in concert halls are a simplification of reality, meaning that they do not always convey a complete or correct meaning. In addition, low bass is as important to satisfying classical performances as it is to rock and roll, and none of that is revealed in the standard measurements. Perceptive concertgoers may notice that the number

In live performances:

- The sound sources are multidirectional, radiating sound in all directions, most of it away from individual listeners in the audience.

- Perceptions of timbre, space, and envelopment created by reflections within the room are essential parts of the performance.

- Musical performances are routinely adjusted—size, physical arrangement, and composition of the orchestra—to cater to the characteristics of individual halls.

- Classical composers sometimes wrote different versions of the same work for performance in specific halls, changing pace and instrumentation to compensate for the size and acoustical properties of the halls.

- Individual halls are more flattering to certain kinds of music than to others. A truly general-purpose hall is improbable without electronically-assisted reflections and reverberation.

FIGURE 3.1 *Walt Disney Concert Hall Auditorium, Music Center of Los Angeles County. Photo by Federico Zignani.*

In sound reproduction:

- Most loudspeakers have significant directivity and are aimed at listeners.
- Ideally, perceptions of timbre, direction, distance, space, and envelopment should be conveyed by multichannel audio systems delivering specific kinds of sounds to loudspeakers in specific locations.
- Ideally, what listeners hear should be independent of the room around them. In practice it is the required degree of independence that is under investigation.
- Conceivably with the right kind of recording and multichannel reproduction it would be possible to recreate the illusion of hearing musical performances or film sound tracks within any real or imagined acoustical spaces.

FIGURE 3.2 *The author's entertainment/family room ca. 2003. Note Pinot, the acoustically absorbent bichon frise, cleverly arranged with a large toy to help absorb the floor reflection during critical listening. His willingness to "stay" was found to be dependent on program selection.*

of bass viol players varies with musical performance and with the hall in which it is performed. Positioning of the players relative to adjacent boundaries is also a factor, although large auditoriums, because of their size, are missing the bass "boom" and seat-to-seat variation problems that plague us in small rooms.

Driven by habit and tradition, many of the same measurements have been carried over into small rooms used for recreational listening to music and movies. But, as we will see, acoustical events in a listening room are very different from those in a concert hall. The sound sources, the loudspeakers, usually have significant directivity, meaning they do not excite the room in a "neutral" manner. The rooms also have significant absorption, much of it concentrated in areas of carpet, drapes, and furniture.

Measurements made with such loudspeakers are measurements of the loudspeaker and room in combination, which is a different thing. And it is the sounds radiated by the multichannel sound system (two, five, or more directional loudspeakers all aimed at the listeners but from different directions) that deliver the impressions of direction and space, not the listening room itself, which is yet another different thing. Obviously the listening room modifies what is heard through the audio system, so the challenge is to be certain that it does not intrude excessively into the listening experience, or if it does, in a benevolent manner.

In contrast, a concert hall is designed to be a substantial positive contributor to a live performance. It is part of the performance to such an extent that classical composers often wrote in styles appropriate for specific types of halls, and a few had different versions of the same piece of music for performance in different halls. In an interesting parallel to the "circle of confusion" discussion in the previous chapter, it is fascinating to learn that some composers failed to adequately consider the effects of the performance environment when creating their works. As they toiled in their small studios, they did not make adequate allowances for what would happen when the music was performed by real orchestras in real halls. As a result, tempo markings were sometimes impossibly fast paced.

Conductors must try to take all of this into account as they prepare for performances in a specific hall, adjusting the orchestral technique to compensate for too much or too little reverberation, and so on. They change the number and physical arrangement of musicians and alter the playing style, tempo, and instrumentation to accommodate rooms of different size and acoustical character. Forsyth (1985), Beranek (2004), and Long (2006) discuss numerous interesting examples of the interplay between music and architecture over the centuries. Some of the comments are especially interesting in this age of sound reproduction, in which audiophiles seem to strive for increasing levels of detail and clarity. For an important period in the history of classical music, the opposite was the objective. Beranek explained this as follows:

From Haydn onward, each generation of composers increased the size and tone color of the orchestra and experimented with the expressive possibilities of controlled definition. The music no longer required the listeners to separate out each sound they heard to the same extent as did Baroque and Classical music: In some compositions a single melody might be supported by complex orchestral harmonies; sometimes a number of melodies are interwoven, their details only partly discernable in the general impression of the sound, perhaps rhythmic or dramatic, often expressive or emotional (p. 12).

Large halls are a challenge to much of the older musical repertoire. Forsyth (1985) explained it this way:

Consider, say, the mismatch of a 21-piece orchestra playing Haydn symphonies on baroque instruments in a concert hall holding 3000 listeners. The visual sense of involvement with the music is reduced, together with the emotional impact: the orchestra sounds "quieter" than in the small concert halls for which Haydn wrote his music. And not only is the lower sound level significant in itself: when an orchestra plays at *forte* level in a compatible-sized room, strong sound reflections can be heard from the side walls and to some extent the ceiling as the music "fills the hall." This criterion of "spatial impression" (*raumlichkeit*) has been identified as significantly important to the enjoyment of a live concert, and this is reduced when the orchestra is unable to achieve a full-bodied *forte*, as the early sound reflections seem to be confined to the stage area instead of coming from all directions. (p. 15)

It is common to increase the number of woodwinds, bass players, and so forth to meet the needs of the music in a particular hall. Nowadays, it is not unknown to electronically assist both the reflective acoustics of the hall and the acoustical output of portions of the orchestra—all done very discreetly, of course.

Thinking ahead, it is obvious that this perceptual phenomenon—"spatial impression"—has a parallel in sound reproduction. Turning up the volume greatly enriches the listening experience as more lower-level reflected sounds become audible because they are above the hearing threshold. The bass is fuller, and the overall sound more spacious, more enveloping. A symphony played at background listening levels can still be tuneful, but it is spatially and dynamically unrewarding. In cars, we experience a related effect when vehicles are in motion. Road, aerodynamic, and mechanical noises mask those same smaller sounds, shrinking the listening experience until, in the extreme, it is dominated by direct sounds. The enveloping sounds are swamped by those generated by the vehicle itself, and the useful dynamic range of the music is reduced. What we hear in a parking lot, or in stop-and-go traffic, will be very different from what we hear at highway cruising speeds.

The need for more sound power is seen in the evolution of musical instruments, which were progressively modified to produce more output. Most of us probably think of the Stradivari violins as instruments of great tonal refinement and subtlety. In the right hands, they are, but the original instruments were also characterized by being louder than the competition, a feature that was further enhanced by a longer neck (retrofitted to older instruments) to accommodate strings at a higher tension (Forsyth, 1985, p. 22).

Room acoustics have had an interesting interaction with religious services. The highly reverberant cathedrals of old were hostile to speech but powerfully enabling of certain kinds of music. Spoken passages were ritualized because of poor intelligibility; congregations memorized large portions of the services. Music was slowly paced and harmonically layered to take advantage of long reverberation times. When some newer religions demanded intelligible free-form services and sermons, churches changed, and modern structures are designed for high speech intelligibility and acoustics (often with sound reinforcement systems) that are well adapted to new, faster, rhythmic musical paradigms.

Looking back over the history of concert halls, acoustical specialists have come to identify only a small number of halls as being of special merit. Among these, the ones singled out for the highest praise are all rectangular (shoebox) halls: Musikvereinssall in Vienna, the Concertgebouw in Amsterdam, and Boston Symphony Hall. So why not just continue to build such halls? Part of the answer has to do with commerce: small halls don't pay the bills. Part of the answer has to do with architects: originality, signature artistic design, is a powerful driving force. So each new concert hall provides visual novelty, but acoustically it is a very expensive experiment. Cremer and Müller (1982) contribute the perspective that "variability in these facilities is attractive not only for the eye but also for the ear. This variability is also justified by the reasonable assumption that the acoustical optima, if they exist at all, are at least rather broad so that it becomes most important simply to avoid exceeding certain limits. Even the undoubted existence of different tastes supports the principle of variation in design."

Many modern halls have become wider, expanding into a fan shape to accommodate more people while retaining good sight lines. Famous acoustician Leo Beranek (1962) notes that "listening to music there is rather like listening to a very fine FM-stereophonic reproducing system in a carpeted living room." Is this reality imitating reproduction? Michael Forsyth (1985) comments on the tendency, especially in North America, to build "hi-fi concert halls," providing some of the impression of "front-row" close-to-microphone recordings favored in that region.

So is it a problem if audiences find it appealing that live classical concert performances sound like stereo recordings in some of these modern halls? It is, because their tastes have been cultivated by stereo recordings that are incapable of generating the full-scale envelopment of a great symphony hall. Well-traveled concertgoers know this, but, sadly, the populace at large seemed, at least for a while, to adapt to be satisfied with less.

Fortunately, nothing is forever, and in a recent analysis of trends in concert hall design, Kwon and Siebein (2007) note that "over the past two-decade modern period (Modern II: 1981–2000), seating capacity, width and length have all trended toward reduction even as the room volume has remained relatively constant. This resulted in an increasing room volume per seat ratio." They link these trends to advances in room acoustical technologies, and the result is a more spacious, enveloping, listening experience—less like stereo.

History shows that music, religion, and acoustical architecture have had complex interactive effects on each other. Reproduced sound also has had significant effects on our attitudes to the live experience, for both better and worse.

3.2 SOUND REPRODUCTION

What exactly are we trying to reproduce or imitate? There is no single reference or target to aim at. Some say that the live acoustical experience must be the ultimate standard, but typical recording techniques do not capture the performance as it is heard in the audience. It is usually a contrived blend of sounds picked up near to and above the orchestra and other sounds collected back in the hall.

At issue here is whether the record/reproduction system has the ability to capture and reproduce the *principal perceptual variables* that contribute to the live listening experiences. Sound field reconstruction is not the objective. That would be a great challenge in an anechoic laboratory space and impossible in homes and cars. Prevailing record/play systems are incapable of achieving anything so complex. They are not sound field encode/decode systems but only multichannel delivery systems with absolutely no rules or standards governing their use. Recording/mixing engineers may have some basic habitual practices in common, having found that they produce pleasant results, but they were the result of trial and error experiences over several decades. Little rigorous scientific analysis has been done on the choice and positioning of microphones or on design objectives for loudspeakers—their locations and the rooms within which they are used. The record/play systems we have enjoyed over the years, and still enjoy, have evolved without a complete underlying scientific basis or rationale. This does not mean they don't work, but it explains why they don't always work well or deliver predictable listening experiences and why it might be possible to make them work better.

If we cannot reconstruct a specific sound field, what is it that we need to reconstruct? The essential *perceptions* of complex sound fields is the answer:

- Direction—the ability to localize sound sources, beginning with a front soundstage to emulate the common paradigm of live concerts and movies. Specific sounds from the sides and rear are good for special effects in movies and can be used to create the illusion of being within a group of musicians, something not greatly appreciated by traditional-minded music lovers but definitely an attractive option for others. More channels permit more discrete localizations, but as we have learned from stereo, for listeners in the right locations, it is possible to create "phantom" sources between at least some of the real loudspeakers. Existing systems have been limited to horizontal localization—azimuth. Elevation remains a tantalizing option.

- Distance—a component of a recording delivered to only a single loudspeaker is perceived at the distance of that loudspeaker. If simulated or real reflections of that sound are added, it is possible to create the illusion of greater distance. It is exciting to perceive sounds originating outside the boundaries of a room or car. Under special conditions, it is also possible to create the impression of great intimacy, of proximity; it is also a worthy attention-getting device. However, it is a complicated perception, involving learning and adaptation in real circumstances, which makes creating illusions that are reliably perceived especially difficult.

- Spaciousness or spatial impression—perceptions associated with listening in a space, especially a large space. It has two principal perceptual components, ASW and LEV:
  —Apparent source width (ASW), a measure of perceived broadening of a sound image whose location is defined by direct sound. In live performances, it is the auditory illusion of a sound source that is wider than the visible sources; this is considered to be a strongly positive attribute of a concert hall. Perhaps because they lack other pleasures of live performances, many audiophiles have come to think that pinpoint localizations are a measure of excellence, so there are opposite points of view. It is a perspective also cultivated by the bulk of popular recordings, many of which are directionally uncomplicated: left, center, and right.
  —Listener envelopment (LEV) is a sense of being in a large space, of being surrounded by a diffuse array of sounds not associated with any localizable sound images. This is regarded as perhaps the more important component of spaciousness, differentiating good concert halls from poor ones. Envelopment was absent from monophonic reproduction and only modestly represented in stereo reproduction, so music lovers have experienced decades of spatial deprivation. Through multichannel audio systems, moviegoers have occasionally been exposed to better things for many years, and now, finally, the capability can be extended to the music repertoire.

Spaciousness is level dependent because the illusion requires that low-level reflected sounds be audible. The more of these sounds that are heard, the greater is the spatial impression. In live performances, profound spaciousness is a *forte* phenomenon. In reproduction, it is more apparent at high sound levels and in the absence of loud background masking noises (as in cars).

Listeners to the stereo and multichannel loudspeaker systems we are all familiar with are responding to a sound field that is *very* different from what they would experience at a live musical performance. Yet, from these very different physical sounds, we seem to be able to perceive what many of us judge to be very satisfying representations of our memories of concerts or other live sound experiences.

This could be the result of self-deception, the "willing suspension of disbelief" that is so fundamental to the enjoyment of unreal characters, scenery, and situations in poetry, literature, and movies. Add to this sometimes improbable plots and indifferent acting, and one has to believe that if movie audiences swallow all of this without complaint, their tolerance of the details of sound tracks must be considerable. Acoustical deception is possible, but it is deception

THE "WILLING SUSPENSION OF DISBELIEF"

Samuel Taylor Coleridge (1772–1834), English poet, critic, and philosopher, wrote "Biographia Literaria" (1817), in which he describes how, in his "Lyrical Ballads," his "endeavours should be directed to persons and characters supernatural, or at least romantic, yet so as to transfer from our inward nature a human interest and a semblance of truth sufficient to procure for these shadows of imagination that willing suspension of disbelief for the moment, which constitutes poetic faith." This quote, or at least the "willing suspension of disbelief" portion, has become a mantra of the entertainment industry. It means that it is possible to convince masses of people to accept premises about where they are (e.g., imaginary worlds) and what might be possible (e.g., assorted monsters and superheros) and to suspend their innate senses of reality and logic sufficient to experience fear, excitement, and pleasure as if it were actually happening—all in the interests of entertainment.

aided by some perceptual illusions that actually work quite well, providing persuasive reminders of acoustical circumstances that could be real. Our task is to identify the key acoustical cues and to create circumstances that allow them to be most persuasively presented to listeners in homes and cars.

3.3 RECORDING: MUSICAL INSTRUMENTS IN ROOMS

Musical instruments radiate sound in all directions with many frequency-dependent directional patterns—for example, high frequencies from violins radiate vertically away from the top plate of the instrument, and those from a trumpet project outward from the bell of the instrument. Other frequencies may radiate almost omnidirectionally, or with a pattern of a dipole radiator (figure eight). In a concert hall, one of the functions of the stage enclosure and its immediate surroundings is to collect these sounds and reflect some of them back toward the musicians so they can hear one another. The other function is to communicate a blended mixture of orchestral sounds into the audience through the reflected sound field.

Once past the first few rows in a concert hall, most of what one hears is reflected sound. This creates a problem in recordings made with microphones located at choice audience seating locations; the overall sound is dominated by reverberation. Two ears and a brain, operating in the live venue, are working with much more information than they are when listening through a small number of channels and loudspeakers at home. In the live situation, an energetic reflected sound field is an expected and important part of the acoustical context; in home audio, the same proportion of reverberation in a recording can be obtrusive.

One of the common criticisms of recorded music is that it just doesn't quite sound like the real thing. Of course, there are several opportunities for perceived

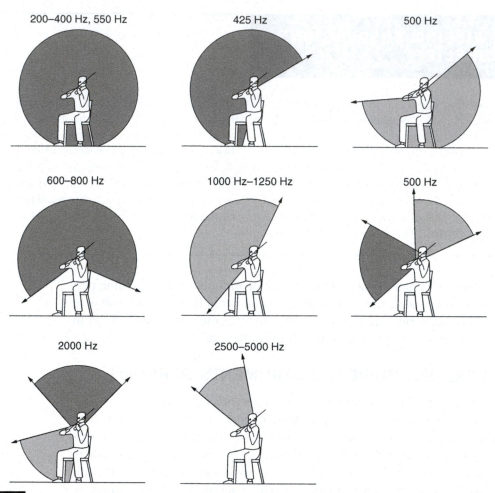

FIGURE 3.3 *Illustrations representing the sound radiated in different directions at different frequencies by a violin. It is clear that no single microphone location can capture a totally balanced spectrum and that a reflective room—like a concert hall or recording studio—plays an important role in allowing all of the sounds radiated in all directions to blend. The common practice of placing microphones above the violin section has been blamed for making them sound strident. Note the 2500–5000 Hz beam of sound. From Meyer, 1993, Figure 11.*

differences, such as timbre, localization, image size, spaciousness, and envelopment. In terms of sound quality—timbre—if the real thing is defined by what is heard in the audience at a concert, one need look no further than the locations of the recording microphones used to capture the orchestral sound. They are inevitably much closer to, and frequently above, the performers. What the microphones "hear" is only a small portion of the blended sounds that are delivered to the audience. Figure 3.3 shows the frequency-dependent directivity of a violin. Examining this and other musical instruments, it is evident that no single axis is an adequate representation of the timbral identity as heard by the audience in a reflective performance venue (Benade, 1985; Meyer, 1978, 1993). One of the great skills of good recording engineers is the choice, placement, and equalization of microphones so that what we hear in recordings is representative of what we hear in real life. Note the choice of the words "representative of," because obviously it cannot be identical. There is no possibility of anything like "waveform fidelity"; there simply is no single waveform that totally exemplifies the sound of real musical instruments. This fact presages discussions later in the book about the near inaudibility of phase shift compared to the dominant role of spectrum (frequency response, in loudspeakers); to have waveform fidelity, both are required.

Recording studios are deliberately reflective, and the best of them have quite diffusive sound fields to spatially integrate the differing off-axis sounds of musical instruments into a pleasing whole. Still, microphone placement is a factor, and some amount of equalization in a microphone channel is a common thing. The portion of the sound field sampled by the microphone often exhibits spectral biases that are not in the overall integrated sound of the instrument. Just think of a grand piano, a massive and massively complicated radiator of sound, and try to imagine how to capture a fully blended representation of its sound with a single microphone. Benade (1986) argued that recordings should provide listeners with a "room average," accomplished with what he called "reasoned miking." This, he said, would be preferred by concertgoers but not by audiophiles, who have been conditioned with close-miked, highly processed, reflection-free sound.

In a performance space, a rich reflected sound field can deliver more of the timbral and spatial signature of voices and instruments to our two ears and brain, which then make marvelously beautiful sense of it all. Early reflections allow us to accumulate information about onset, spectrum, pitch, and location. A complicated, three-dimensional sound field is part of natural listening. When we attempt to achieve the same effect with two or five loudspeakers in a small room, some acoustical factors are omitted. It can sound good, but it cannot sound the same.

3.4 HEARING: HUMAN LISTENERS IN ROOMS

At the receiving end, we have ears with frequency responses that are different for sounds arriving from different directions—the reverse of what we just discussed. Not only that, but each of us has differently shaped ears. It doesn't matter that we differ from each other because whatever we are born with is our lifetime reference for all sounds, whether live or reproduced. The directional characteristics of our ears are described by "head-related transfer functions" (HRTFs; Blauert, 1996). These are complex (amplitude and phase versus frequency, or impulse response) descriptions of how sounds arriving from different angles are modified on the way to the eardrum. Because of the detailed structure of our external ears and the placement of the ears on each side of an acoustically reflective head, HRTFs are unique characterizations of each of the incident angles of incoming sounds, helping us to localize where sounds come from.

In terms of timbre, it seems as though this could be a problem, making sounds coming from different directions take on distinctive sound qualities. Theile (1986) and Warren (1999) have investigated the effect, and Warren summarizes it as follows:

These position-dependent transformations do not interfere with the identification of a sound, but they do contribute to spatial localization. Thus, some spectral changes cannot be perceived as such. They are interpreted as changes occurring in an external physical correlate (azimuth and elevation of the source), while the nature of the sound (its quality or timbre) appears to be unchanged (p. 21).

Theile (1986) incorporates this in his "association model" of hearing, in which timbre is, in effect, associated with direction. We *expect* sounds from different directions to have somewhat different timbres because that is one of the clues helping to figuring out what direction it is coming from. Once the direction is identified, it seems that we apply a kind of correction, so we are not distracted by the timbre shift. Another factor that must be considered is localization, which is dominated by the direct, first arrival, sound. However, in normal rooms, reflected sounds arriving from other angles (and therefore modified by different HRTFs) will contribute to the perceived timbre. So it is entirely reasonable to think that the distinctiveness of HRTFs helps to identify the direction but that the "spatial average" of many HRTFs processing reflected versions of the sound helps to maintain timbral neutrality.

In a home theater, if broadband pink noise (not the common band-limited multichannel calibration signal) is switched to each of five identical loudspeakers, a listener facing forward will hear obvious changes in timbre. In jumping from the center to the left front or right front loudspeaker, a high-frequency boost may be heard. From there to either of the side surround loudspeakers, there are more timbral changes. This is as it should be. To confirm that all five loudspeakers in a system are similar in timbre, it is necessary to face each of the loudspeakers as they play the same test signal. Then, and only then, can identical loudspeakers have identical perceived timbres. This is the timbre matching important to sound reproduction, not what is heard while facing forward. If a five-channel system reproduced a quintet of identical musical instruments, one would want each one reproduced identically, even though we know that, depending on incident angle, the sound from each one will be distinctively modified on the way to our eardrums. After all, that is what would happen in a live situation.

One can reasonably speculate that a stereo phantom center image might present a problem because the perceived direction is 0° and the sound sources are at, say, ±30°, meaning that we may be applying the wrong directional "association" (0°) for sounds that are physically arriving from ±30°. This could be a contributing factor to the perceived difference between a real center loudspeaker and its phantom image version. The real one is correct, and Figure 9.7 illustrates a substantial physical reason for there being a difference.

To summarize, the perception of timbre in rooms has partly to do with (1) the identification of direction, (2) any compensation for perceived timbre that follows from this, and (3) the averaging of reflected versions of the direct sound arriving from many different directions. The effects of any individual HRTF are therefore completely intact for the first arrived sound, the direct sound that defines direction, and then it is progressively diluted by reflections from many directions, progressively building a short-term average that reveals the "generic" spectral identity of the sound source.

These reflected "repetitions" of the direct sound have a second benefit: They increase our sensitivity to the subtle medium- and low-Q resonances that give sounds their distinctive timbres (Toole and Olive, 1988). Music and voices are timbrally enriched by room reflections (e.g., singing in the shower). See Chapter 9 for more information.

3.5 REFLECTIONS: CONVEYERS, INTEGRATORS, AND DIFFERENTIATORS OF SOUND

In rooms for live performances, therefore, reflections are highly beneficial in integrating all of the sounds radiated from musical instruments and in conveying the essential timbres of those sounds to ears that themselves have strong directional properties. Reflections also contribute to perceived loudness, a good thing for voices and acoustical instruments like violins that have limited power. And, finally, reflected sounds convey to us important information about the size, reflectivity, and geometry of the performance space. In perceptual terms, though, these qualities appear as variations in impressions of reverberation, direction, distance, auditory image size, spaciousness, and envelopment, all of which, and more, combine to describe the essential qualities of concert halls and differentiate the good from the not so good.

As we get into the details, we will see that reflections from certain directions, at certain amplitudes and delays, are more or less advantageous than others and that collections of reflections may be perceived differently from isolated reflections. We humans like reflections, but there are limits (too much of a good thing is a bad thing) and ways to optimize desirable illusions.

3.6 AN ACOUSTICAL AND PSYCHOACOUSTICAL SENSE OF SCALE

Historically, our interest in room acoustics began with performance spaces. These days we think of the architectural monuments called concert halls, but the beginnings of public performances were much more modest—usually single

FIGURE 3.4

A graphic portrayal of the range of listening spaces relevant to our work and entertainment.

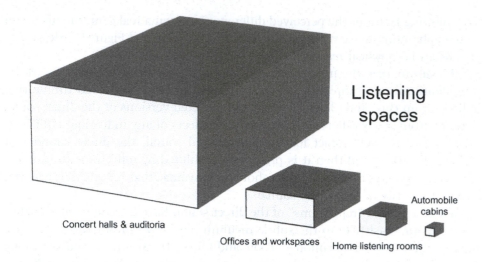

FIGURE 3.4

A graphic portrayal of the range of listening spaces relevant to our work and entertainment.

rooms attached to drinking establishments and eateries (Forsyth, 1985). By 1885, when Wallace Clement Sabine began his acoustical investigations, large, high-ceilinged, halls were the norm.

Figure 3.4 illustrates the range of shapes and sizes of spaces we live, work, and listen in. To the extent that the volume of the space is a factor, it can be seen that the range is enormous: from roughly 741 000 ft³ (21 000 m³) for a typical contemporary hall, through 3600 ft³ (100 m³) for a typical home theater, to 120 ft³ (3.5 m³) for a four-door sedan. Instinctively, it seems unlikely that the same psychoacoustic rules apply to auditory perceptions in all of them.

In reviewing the scientific research that has been done to understand sound fields within these spaces, the majority, by far, has been done in large auditoriums. The purpose of these spaces has traditionally been to ensure the delivery of unamplified musical sound with adequate loudness, sound quality, and musical integrity to a large audience. Today, amplified performances have become part of the entertainment mix.

Ironically, some of the best halls are among the oldest. So why don't they just copy them? It's because, in addition to being spaces for acoustical performances, concert halls are also opportunities for architects to exercise their visual imaginations and for budget-conscious administrators to argue for more seating capacity. Because each new hall involves an element of chance, it is important to understand the science so that the risks are minimized; this is the main motivator for continued research.

Smaller than concert halls and larger than home listening rooms are factories, machine shops, offices, and other work spaces. Acoustical investigations in these spaces have been driven by the need to understand the sound propagation of bothersome noise from HVAC systems and manufacturing machinery and by considerations of speech privacy in offices. These investigations have

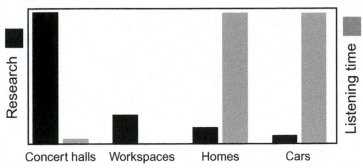

FIGURE 3.5 *An opinion about the relative amounts of scientific research dedicated to understanding sound propagation and psychoacoustics in each of the spaces shown in Figure 3.4 (dark bars). Also shown is an opinion about the relative time typical people spend in "foreground" listening in each of those spaces (light bars). Even allowing for large errors in these estimates, there is ample reason to believe that more work is needed where we listen most.*

tended to focus on the physical acoustics because the finer points of sound quality are subordinate to the more important issues of establishing a safe and comfortable work environment. Classrooms are also in this size category, and there the primary concern is speech intelligibility, a very special and important topic of investigation.

Domestic listening rooms, recording control rooms, and the passenger compartments of cars are still smaller, and they incorporate a new factor: the sound reproducing system. In these cases, the room plays a role, but it is a subordinate role. The recorded art is the star, and everything else—electronics, loudspeakers, and listening room—are merely a means of delivery. Listeners trust that what they hear is a reasonable facsimile of the recorded art, music, movies, games, television, or whatever. The purpose of this book is to assist in reaching that objective.

For all the pleasures of attending live, unamplified performances, such concerts are a small part of our collective entertainments. It is more than the "highbrow" music, which has limited appeal; it is cost, time, travel, and the fact that Tuesday night after a hard day at the office, one simply may not be in the mood for Stravinsky. Reproduced sound has been a great liberating factor in our lives, making background music nearly ubiquitous, sometimes annoyingly so, and our personal wishes for foreground music remarkably well gratified in our homes and cars and while we jog, work out, or walk the dog. It seems that almost anywhere it is possible to press a "play" button and hear the music of our choice.

Figure 3.5 illustrates the present situation. A great deal of research underlies our understanding of concert halls, but that is not where most of us spend most of our listening time. This book is an attempt to adjust the balance, just a little, toward the circumstances where we mostly live and listen.

CHAPTER 4

Sound Fields in Rooms

Physical measures of the sound fields in rooms are important because they can help us to understand the perceptual dimensions of speech, movies, and musical performances that we enjoy in those rooms. As in all psychoacoustic endeavors, not all physical measures are equally useful, however correct they may be in strictly physical terms. Also, a correlation does not imply causality, so although some measures may correlate with perceptions, they may not be the root cause of the effect. All of these measures form the foundation upon which the science of architectural acoustics has developed, and it is important to examine them even though some will turn out to be only of passing interest.

Nobody would pretend that small rooms for sound reproduction are intimately related to concert halls and that the same criteria for excellence apply. However, one of our goals is to reproduce the auditory illusions of concert hall experiences in our homes and cars. It is therefore necessary to understand the basic metrics of excellence for concert venues so that with multichannel record/reproduce systems, we may be able to optimize the experience in small listening spaces.

4.1 LARGE PERFORMANCE SPACES: CONCERT HALLS

Explanations of sound fields in concert halls begin with notions of ray (geometrical) acoustics, showing direct sound and discrete reflections from large surfaces. The rules are simple: The angle of incidence equals the angle of reflection. Greek and Roman open-air theaters relied solely on a few reflections to support the direct sound. In enclosed performance spaces, a new phenomenon appears: reverberation, which is caused by sounds being repeatedly reflected from all surfaces and objects in the room. If the sound source produces a sustained sound, a steady-state reverberant sound field builds up to a level where the sound

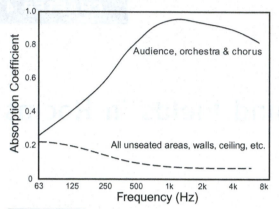

FIGURE 4.1 *The absorption coefficients for areas of a concert hall that are occupied by the audience and musicians and for all other areas. From Beranek, 1969, Figure 3.*

energy is absorbed at the same rate at which it is being created. When the sound source ceases, the reverberant sound field decays. The time it takes to decay by 60 dB is called the reverberation time (RT).

In the abundance of reflections that we collectively call reverberation, there are so many individual events that it has been common to think of them as a statistical entity distributed randomly in time and space. As a result, classic concert hall acoustical theory often begins with the simplifying assumption that the sound field throughout a large relatively reverberant space is diffuse. In technical terms that means it is homogeneous (the same everywhere in the space) and isotropic (with sound energy arriving at every point equally from all directions). That theoretical ideal is never achieved because of sound absorption at the boundaries, by the audience, and in the air, but it is an acceptable starting point.

Absorption in these large performing spaces is minimized to conserve the precious acoustical energy from musical instruments and voices. An active reflected sound field ensures the distribution of that energy to all seats in the house. The challenge is to preserve the sound energy in the reflections without obscuring the temporal details in the structure of music and speech. This is why reverberation time remains the paramount acoustical measure in performance spaces.

It is important to note that in the calculations of reverberation time, it is assumed that the acoustical activity—reflection, absorption, and scattering—occurs on the room boundaries and that the volume of the room is empty. In a concert hall, the height is such that the audience is essentially treated as a "layer" of material with a certain average absorption coefficient placed on the floor or distributed throughout the hall. Figure 4.1 shows representative absorption coefficients for areas occupied by audience, orchestra, and chorus, and for all other areas (Beranek, 1969). Obviously, audiences absorb a great deal of middle- and high-frequency energy. Consequently, to achieve the reverberation times necessary for music (typically 1.5–3 s), halls must have "other areas," such as walls and ceilings, that are much greater than the audience area. This requirement leads to high ceilings and, for large audiences, large volumes.

Figure 4.2 shows a familiar portrayal of idealized behavior in one of these halls. In this depiction, an omnidirectional sound source is located well away from the room boundaries, such as the center front of the stage. As a function of distance from the source, the level of the direct sound follows the inverse-square-law rate of decay (–6 dB per double distance, dB/dd) until it encounters the underlying steady-state reverberant sound field that is assumed to extend

CALCULATING REVERBERATION TIME

In large, highly reflective rooms, the reverberation time is often well predicted by the original Sabine formula:

$$RT = .049\,V/A,$$

where V is the total volume in ft^3 and A is the total absorption in the room in sabins. The total absorption, A, is calculated by adding up all of the piecemeal areas (carpet, drapes, walls, etc.) of the boundaries multiplied by their individual absorption coefficients:

$$A = (S_1a_1 + S_2a_2 + S_3a_3 \ldots),$$

where S is the area in square feet and a is the absorption coefficient for the material covering that area. Absorption coefficient is a measure of the percentage of sound that is absorbed when sound reflects from the material. The product of S and a is a number with the unit sabins. The absorption of some items, such as people or chairs, is sometimes quoted directly in sabins.

The metric equivalent of the Sabine formula is

$$RT = 0.161\,V/A,$$

where the volume is in m^3 and areas are in m^2 and A is in metric sabins.

As rooms get more absorptive and smaller and as the materials on the room boundaries begin to differ more from one another (e.g., wall-to-wall carpet on the floor), this equation becomes progressively less reliable. Over the past 100 years, several increasingly more complex equations have been developed to accommodate asymmetry in rooms and the fact that the sound field is not diffuse; Fitzroy (1959) and Arrau-Puchades (1988) contributed some of them. However, all of them, to be practical, make assumptions. Dalenbäck (2000) says, "These two formulas give a better estimate than the classical formulas [Sabine and Eyring] in *some cases*, but here a central question is: *How can one be sure they are better in a particular case?* So far, no equation with universal applicability has been shown" [his emphasis]. Fortunately, as we will see, in small rooms for sound reproduction, high precision is not required. If it were, it is likely that a computerized room model would be needed. In the meantime, the simple Sabine formula provides estimates that are adequate for our purposes in small listening rooms.

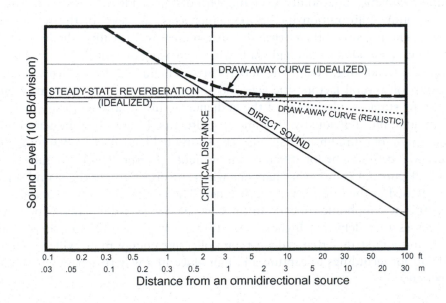

FIGURE 4.2

The structure of the steady-state sound field in an idealized large performance space. The dotted draw-away curve indicates the declining sound level that, in some form or other, is seen in real halls. Based on Schultz, 1983, Figure 10.

FIGURE 4.3

The large arrows indicate trends resulting from increasing amounts of absorption in the idealized room: the steady-state reverberation level decreases and the critical distance increases. Again, the dotted curve shows the additional effect that in real rooms the sound level falls with distance.

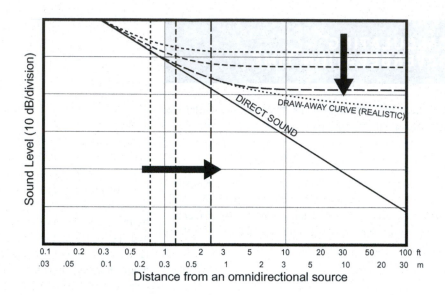

uniformly throughout the space (Beranek, 1986; Schultz, 1983). The distance from the source at which the direct sound equals the level of the reverberation is known as the critical distance (also known as reverberation distance or reverberation radius). The dashed-curve sum of these is what would be measured by a sound level meter as it is moved away from the source—a draw-away curve. In the ideal hall, the curve is horizontal at large distances, but in real halls it falls with distance, as shown by the dotted curve.

Because of sound absorption by the audience, the air (increasingly significant above about 1500 Hz), and the room boundaries, the level of the reverberant sound field varies in absolute level and with distance. Figure 4.3 shows that, as the amount of absorption in the room increases, the level of the reverberant sound field drops, and the critical distance increases. In reality, the level of the reverberant sound field gradually falls with increasing distance, as energy is dissipated, a trend suggested by the dotted curve. The rate of decline with distance depends on several factors: the size and shape of the hall, orientation of large reflecting surfaces, placement of the audience within it, and so forth.

In addition, individual voices and instruments do not obey the simplifying assumption of omnidirectionality, so the sound field at different listening positions will be different for different instruments (Meyer, 1993; Otondo et al., 2002). Figure 4.4 shows what happens when the directivity of the sound source is increased in the direction of the listener: The critical distance increases. We hear this at symphonic concerts in the contrast between the penetrating clarity of brasses that deliver a higher proportion of direct sound compared to the open and airy strings that radiate their collective energy more widely. This is difficult for recordings to capture and reproduce in a way that parallels the live experience.

Directional control is critical in designing sound reinforcement systems, the purpose of which is to deliver sound to the audience without exciting excessive reflections and reverberation within the room itself. The challenge is to put more of the audience in a predominantly direct sound field, precisely the opposite of a live concert hall experience.

In loudspeakers for sound reproduction, it will later be shown that constant, or at least only gradually changing, directivity over most of the frequency range is a desirable trait. Figure 4.4 shows one of the reasons. It is necessary to maintain a relatively constant direct-to-reflected sound proportionality as a function of frequency.

Combining the effects of increased absorption, increased source directivity, and a realistic attenuation with distance of the steady-state reverberant sound field, Figure 4.5 shows a hypothetical draw-away curve for a small room, assuming that no other factors are involved. It shows that at typical listening distances,

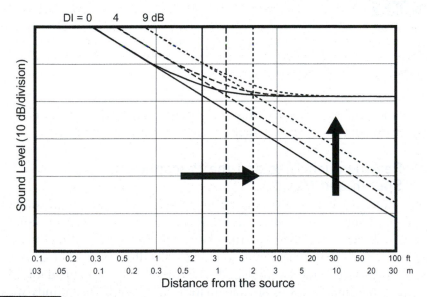

FIGURE 4.4 *The large arrows indicate trends resulting from increasing sound source directivity in the direction of the listener: the direct sound curve is elevated and the critical distance increases. The level of the steady-state reverberation is unchanged because the sound power output is constant in this example. DI (directivity index) is a common measure of directivity. A DI of 0 dB describes an omnidirectional source. In consumer and professional "cone and dome" loudspeakers, woofers typically exhibit DIs of 0 dB at frequencies below about 100 Hz, midrange drivers tend to be slightly directional at around 4 dB, and tweeters can reach 9 dB or more at the highest frequencies. Whether the sound source is a loudspeaker or a musical instrument, those with different DIs will project their sounds differently into rooms, and listeners will experience different proportions of direct and reflected sounds—different perceptions. Inspired by Beranek, 1986, Figure 10.23.*

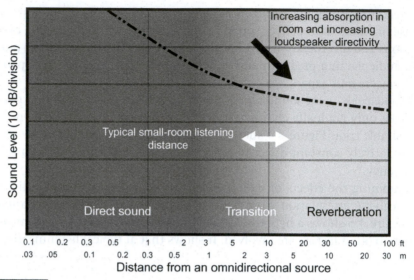

FIGURE 4.5 *An artistic interpretation of the transition from direct-sound dominance at short distances to reverberant-sound dominance at large distances. The transition region in between, where both are influential, appears to be that within which we find ourselves in small listening rooms because they are acoustically relatively absorbent and employ loudspeakers that are relatively directional.*

listeners may find themselves within the untidy transition region. This is interesting, but it isn't the final statement; we need to investigate further.

4.1.1 Reverberation Time and the Perception of Speech and Music

When a source of sound stops radiating energy, whatever reverberant sound field that exists begins to decay. Because reverberation has the effect of prolonging all acoustical events, reverberation time is an obvious influence on how we hear speech and music. Rapid changes in sound are difficult to hear when each sound is stretched in time by the reverberant decay. Although RT is defined for a 60 dB decay, it is the first 20 dB or so that matters most with the sounds we care about. What happens in the later portion of the decay has other effects, both desirable and undesirable. Consequently, there are several measures that attend to what happens in the early and late stages of the decaying sound field (Beranek, 2004; Long, 2006).

Speech intelligibility benefits from some control of reverberation time, so rooms for speech communication, such as classrooms, tend to have short RTs, around 0.5 s. However, there is more to the story. In terms of speech intelligibility in large spaces, it has long been recognized that the early reflections that are a component of the early portion of reverberation are important aids to speech intelligibility. Later reflections contribute nothing useful. In rooms where speech intelligibility is important, therefore, it is more important to pay attention to the reflection pattern in the room than to reverberation itself. Increased early-reflection energy has the same effect on speech intelligibility scores as an equal increase in the direct sound energy (Bradley et al., 2003; Lochner and Burger, 1958; Soulodre et al., 1989; see Chapter 10).

Opera houses have typical reverberation times of about 1–1.5 s, which is considered an acceptable compromise for understanding the spoken dialogue and singing, while providing some assistance to the instrumental accompaniment. However, this is considered a bit too "dry" for much of the classical music repertoire, so most dedicated concert halls are designed to have RTs in the 1.5–3 s range. The very long reverberation times in cathedrals, 5 s or more, allow a choir to effectively sing with itself, generating layers of harmonizing but severely limiting the musical repertoire suitable for the space. This is, of course, a great simplification, because in addition to these mostly temporal considerations, there are those related to perceptions of direction and space.

4.1.2 The Seat-Dip Phenomenon

Most of us think of live performances in good concert halls as "reference" experiences—not only greatly enjoyable but an opportunity to recalibrate our perceptual scales. That is because there are no technical devices to get in the way—no microphones, recordings, and loudspeakers. But that does not guarantee unimpaired sound transmission. There are acoustical phenomena, one of which has come to be known as the "seat-dip effect," wherein low-frequency sounds passing over an audience at low incident angles generate a substantial dip in the frequency response as measured at the head location. The exact nature of the dip is related to the geometry and acoustical treatment of the cavity formed by the rows of seat bottoms, backs, and the floor. Schultz and Watters (1964) and Bradley (1991), among others, have measured the effect as a function of source elevation, source and seat location, horizontal angle, and so on; the effect is not trivial, as shown in Figure 4.6. Not only is the dip 10–15 dB deep, but it is wide; showing significant attenuation over more than two octaves at the lowest incident angle.

The effect exists primarily in the direct and early-arriving sounds. Later reflections that arrive from other angles appear to alleviate the effect somewhat, but it remains an audible effect in the early-sound field in typical auditoriums. There is a suggestion that the preference for elevated RT at low frequencies is, at least in part, to compensate for the seat-dip losses (Beranek, 1962).

Davies et al. (1996) have measured detection thresholds for the dip of −3.8 dB for 0–80 ms early energy. It appears not to be influenced by reverberation, but it does seem to be less severe in halls with strong overhead reflections or steeply raked seating (Bradley, 1991). Holman (2007) found pronounced dips in the room curves of cinemas even with the elevated incident angle produced

FIGURE 4.6

The seat-dip effect for four angles of incidence of direct sound. From Bradley, 1991, Figure 1.

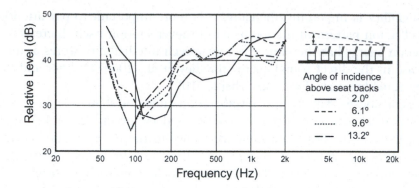

by raked stadium seating. Here the deliberate absence of reverberation and strong reflections would be an aggravating factor.

Naturally, in addition to a panoramic view, there is no seat-dip effect in the front mezzanine and balcony rows. Are these the "audiophile" seats? Just five rows back, however, the dip is very evident (Schultz and Watters, 1964).

4.1.3 The Effects of Early and Late Reflections

It is beyond the scope of this book to delve into the complex elements of acoustics and psychoacoustics of large concert venues. Beranek (2004) and Long (2006) provide excellent historical and contemporary perspectives. Still, it is important to understand what the important variables are because, ultimately, it is our goal to reproduce them for a few listeners in small rooms. In general, the main factors have to do with where reflections come from and when they arrive, early or late.

Several of the measures related to pleasurable perceptions in concert halls are related to the angles of incidence of reflected sounds relative to the direct sounds from the stage. In particular, those reflections arriving from the sides have been found to be especially useful contributors to what was originally called "spatial impression." Now it is recognized that there are two components to spatial impression (Bradley and Soulodre, 1995):

- **Apparent source width (ASW)**—a measure of perceived source broadening and defined as the width of a sound image fused temporally and spatially with the direct sound image (Bradley et al., 2000). Before 1990, this is what the literature generally referred to as spatial impression (Barron, 2000). It is associated with the level of early (<80 ms) lateral reflections, as measured by either a lateral energy fraction (LF) of the sound field or by interaural cross-correlations (IACC). It is also influenced by overall sound level.

- **Listener envelopment (LEV)**—a sense of being surrounded by a diffuse array of sound images that are not associated with particular source locations. It is associated with reflections arriving after about 80 ms.

According to Bradley et al. (2000), "It has been shown that all reverberant or late-arriving sound can influence LEV, but that late-arriving sound from the side of the listener is more important for creating a strong sense of LEV."

Figure 4.7 illustrates the concepts.

In typical halls, both ASW and LEV coexist in proportions related to the specific acoustics. In examinations of 16 halls, LEV was estimated to be the stronger influence. Bradley et al. (2000) report, "The highest values of both quantities are found in more reverberant smaller halls that would tend to have both strong early and late lateral sound energy." As we adjust our focus to concentrate on sound reproduction in small rooms, it will be seen that various interpretations of ASW, image broadening, will be contributed by the room itself, but the sound levels and delays required for LEV, envelopment, must be delivered through multichannel audio systems.

As noted in Section 3.1, Forsyth (1985) pointed out the importance of matching the size of the orchestra and the hall:

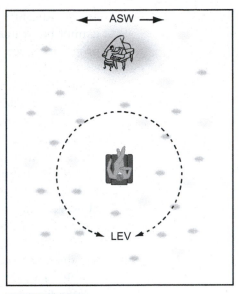

FIGURE 4.7 *An artistic impression illustrating apparent source width (ASW) and listener envelopment (LEV). Inspired by Morimoto, 1997, Figure 1.*

When an orchestra plays at forte level in a compatible-sized room, strong sound reflections can be heard from the side walls and to some extent the ceiling as the music "fills the hall." (p. 15)

This is the highly desirable illusion that is missing in many performances, both live and reproduced. When I had the good fortune to attend rehearsals and performances of a symphony orchestra in Vienna's Musikverein, one of the most celebrated halls in the world, I was frankly not ready for the intensity of the spatial impression—the hall was indeed "full" and the envelopment was profound. It was greatly pleasurable, but for a person habituated to more modern, larger, halls and after many years of exposure to two-channel reproduced sound, the first impression was one of surprise. It occurred to me that I and my audiophile acquaintances would probably consider such an illusion to be artificially overdone if we were to hear it through a multichannel audio system.

4.2 OFFICES AND INDUSTRIAL SPACES

Occupying the middle ground between large, high-ceilinged performance spaces and domestic rooms are those with large floor areas and lower ceilings: offices, factories, and the like. Most such spaces have significant amounts of absorption, much of it on the ceiling or floor, or both. They also have large sound absorbing and scattering objects distributed throughout the floor area, desks, people, office cubicles, machines, production lines, and so on. If the objects in these spaces are significantly large relative to the height and volume of the rooms, they cannot be treated as a "layer" of sound absorbing material on the floor. Sounds propagating across such spaces behave distinctively. At short distances from a source, the objects are obstacles to propagation, reflecting some portion of the sound back toward the source and causing the sound level to be higher than it otherwise would be. The objects themselves contribute to absorptive losses, as well as reflect and scatter sound into other absorbing surfaces. The result is an increase in overall sound attenuation with distance compared to what might be expected in concert halls.

Different dimensional ratios, differing deployment of absorbing materials, and scattering objects all result in different sound propagation characteristics. However, there are some strong common features. Close to the sound source, sound backscattered from objects in the space can cause the sound level to exceed that of the direct sound, especially at high frequencies. Over much of the distance, the draw-away curve falls at a rate of approximately −3 dB per double distance (at least for combined middle and high frequencies). Hodgson (1998) discusses several models for predicting the actual rate, which is frequency dependent. At longer distances, this trend may continue, or, depending on the room geometry, the distribution of absorbing material, and the presence of significantly large scattering objects, the rate of decay can accelerate (Hodgson, 1983). Figure 4.8 shows two simplified theoretical predictions for the tendencies

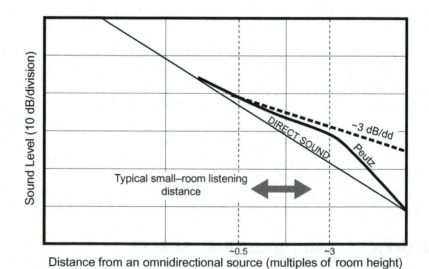

FIGURE 4.8 *Two predictive models showing anticipated shapes of draw-away curves in offices and industrial spaces. One predicts a progressive decline with distance at a rate of approximately −3 dB/double-distance (varying with frequency and the nature of the room). The other model, by Peutz, shows a similar trend but predicts a rapid decline beyond a distance of about three times the room height. The horizontal scale applies to the Peutz prediction. Adapted from Schultz, 1983, and Hodgson, 1998.*

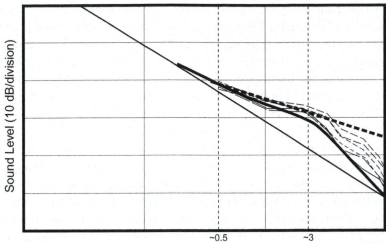

FIGURE 4.9 *Measured draw-away curves (thin lines) by Hodgson (1983) in one of his several investigations. These are from an industrial space 45 m by 42.5 m with an average height of 4 m and are shown for octave bands in the range of 125 Hz to 4 kHz. All fall between or close to the predictive dark lines from Figure 4.8.*

of draw-away curves: the popular −3 dB/dd and a more elaborate prediction by Peutz (1968) as compiled and reported by Schultz (1983) in a very insightful document. Continuing to speculate what may happen in small listening rooms, the range of typical listening distances is shown.

Real draw-away curves measured by Hodgson (1983) in several industrial spaces exemplify both trends, with a fair amount of scatter caused by differing behavior at different frequencies, Figure 4.9.

Late reflections are rapidly attenuated with distance from the source. Over almost the entire draw-away distance, including the range of listening distances typical of small rooms, listeners are in what can best be described as a prolonged transitional sound field, neither direct nor reverberant. This means that critical distance is not an appropriate concept in these spaces.

4.3 DOMESTIC LISTENING ROOMS AND CONTROL ROOMS

When the floor area shrinks from office/factory to domestic dimensions, it seems probable that this behavior will continue because key features of the commercial spaces are present. Large portions of one or more surfaces have significant absorption in the form of carpet, drapery, and, perhaps, acoustical ceilings. There are also sound-absorbing and scattering objects, such as sofas, chairs, tables, cabinets, and vertically stepped arrangements of bulky leather chairs in custom home theaters, all of which are large relative to the ceiling height in typical homes.

All of this continues the theme of reflected sounds filling the space with a sound field that has high diffusivity. We conceptualize what is happening using the notion of ray acoustics, geometric acoustics, separable direct and reflected sounds, and so forth. At middle and high frequencies, where wavelengths are short compared to the room dimensions, this is appropriate. However, as rooms shrink, the dimensions become significant when compared to wavelengths at low frequencies. At 20 Hz, the wavelength is 56.5 ft (36.7 m); at 50 Hz, it is 22.6 ft (6.9 m); at 100 Hz, it is 11.3 ft (3.45 m); and so on.

At these low frequencies, the behavior of small rooms is dominated by resonances (a.k.a. modes, eigenfrequencies, etc.) and the associated standing waves. This is best described in terms of sound waves, not rays. As frequency increases, there is a transition from the region in which wave motion and room resonances dominate to the region within which ray/geometric acoustics and reflections are better able to describe acoustical events. Consequently, the following discussion of the sound field in listening rooms is broken into three categories: events above the transition region, events within the transition region, and events below the transition region. First, though, it is necessary to identify at what frequency this transition takes place.

4.3.1 One Room, Two Sound Fields—The Transition Frequency

Ultimately, we are interested in knowing how loudspeakers interface with small rooms, so let us begin by putting a loudspeaker into a typical room and measuring what happens. Figure 4.10a shows the floor plan of the prototype IEC 268–13 (1985) listening room, indicating several possible loudspeaker locations and six listener locations. Because this room was used for subjective evaluations of loudspeakers, the mission was to find locations for both listeners and loudspeakers that would allow for reasonably equitable comparisons to be made between different products (Toole, 1986).

Figure 4.10b shows frequency responses measured at each of four listener locations for a loudspeaker in position C. This is a measurement with high (1/20-octave) resolution, so it shows a great deal of complexity (or "grass," as engineers call it), especially above about 200 Hz. This is acoustical interference between and among the numerous reflections arriving from many different directions at many different times. It is normal, but it is not instructive, especially because, as we will see later, we don't hear these details. Looking carefully, one can discern a central tendency among the curves, suggesting an underlying character that is more similar than is given by the first impression. Consequently, it is normal to spectrally average, or smooth, such curves. It is important to choose the smoothing function carefully because if it is too broad, one loses even the underlying trends and certainly the ability to examine what is happening in the simpler undulations at low frequencies. Although 1/3-octave smoothing is common, it is too broad for some purposes; 1/4-octave smoothing was used here. Notions that the smoothing must be associated with critical

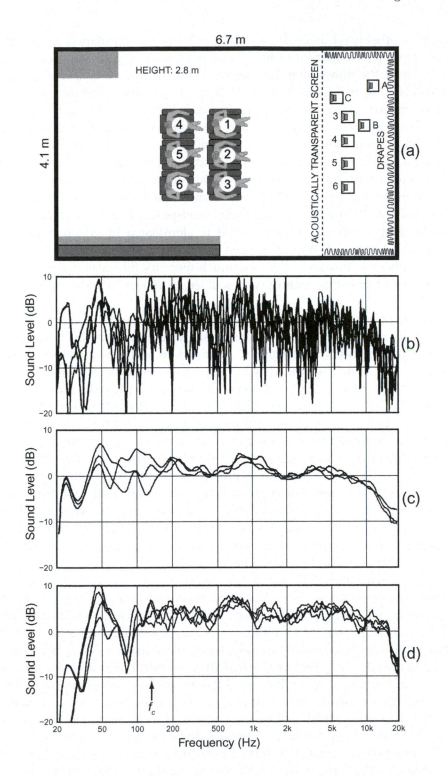

FIGURE 4.10

(a) Measurements were made in the prototype IEC 268–13 (1985) room. (b) Loudspeaker "X," located at position C, is measured at listener positions 1, 3, 4, and 6. 1/20-octave resolution. (c) Loudspeaker "X" at each of the locations A, B, and C averaged over all six listener locations, 1/4-octave smoothed. (d) Loudspeaker "Y," a different loudspeaker, averaged over loudspeaker locations 3 through 6, measured at each of the listener locations 1, 3, 4, and 6. 1/4-octave smoothed. The Schroeder crossover frequency (f_c) for this room is shown. All data from Toole, 1986.

bands or other psychoacoustic measures of loudness summation are irrelevant here, as we are looking for technical explanations of what is happening.

Figure 4.10c shows frequency responses for the same loudspeaker placed in locations A, B, and C. Each curve is the 1/4-octave-smoothed average of measurements made at all six listener locations. Below about 300 Hz, the frequency response is dominated by loudspeaker position, whereas at higher frequencies, the measurements follow a similar pattern. The fact that the loudspeaker positions each differ in distance to the side and end wall causes considerable variation up to at least 200 Hz. However, the very long wavelength of the first-order length mode, the bump at 26 Hz, is immune to these positional variations. Obviously, A, B, and C were not useful loudspeaker locations, as the balance and timbral quality of the bass would be dominated by which position the loudspeaker occupied. As can be seen, it ranged from inadequate to well balanced to boomy. Positions 3–5 were much more useful because they maintain a constant relationship with the end wall/length modes (not shown; see Toole, 1986).

However, where the listener sits also matters. Figure 4.10d confirms that reciprocity applies in these situations by showing measurements averaged over loudspeaker locations 3 through 6 for each of four well-separated listening locations, 1, 3, 4, and 6. Again, what happens at low frequencies is determined by location—this time it is the listener location—and, again, the curves follow a similar pattern at higher frequencies. The pattern is different from that in (c) because a different loudspeaker was used. The amount of bass around 50 Hz is dominated by front row versus back row listener locations interacting with the second-order length mode in the room; front row seats are close to the pressure peak, and the back row seats are approaching the null. Still, some side-to-side asymmetry can be seen. The consistent dip is the result of the loudspeakers all being at a constant distance from the end wall, close to 1/4 wavelength at 80 Hz, affecting the acoustic output because of destructive interference (an adjacent boundary effect; see Chapter 12), and failing to excite the third-order mode at $3 \times 26 = 78$ Hz (a standing-wave effect; see Chaper 13).

The acoustical explanation is the dominance of relatively isolated room modes and standing waves at low frequencies and of a complex collection of overlapping modes and reflected sounds at high frequencies. As will be explained in more detail in Chapter 13, room modes are the result of reflections that reinforce each other in an orderly fashion, but here we make a distinction because, at higher frequencies (shorter wavelengths), geometric and acoustic irregularities in the boundaries of normal rooms (doors, windows, fireplaces, furnishings, etc.) disrupt the orderly reflections necessary for the creation and support of room modes. As a result, as frequency increases, it becomes progressively less useful to think about regular patterns of standing waves in rooms but rather to think in terms of irregular patterns of constructive and destructive acoustical interference caused by numerous reflections traveling in many directions.

In between the orderly low-frequency room resonances and the disorderly higher-frequency acoustical behavior is a transition zone, the middle of which, in large rooms such as concert halls and auditoria, would be defined as the Schroeder frequency or, as Schroeder himself calls it, the "cross-over frequency f_c" (Schroeder, 1954, 1996).

$$f_c = 2000\sqrt{\frac{T}{V}}$$

where T is the reverberation time in seconds, and V is the volume of the enclosure in cubic meters. The multiplier constant changed from the original 4000 to 2000 in the 1996 paper.

Calculation of the Schroeder frequency assumes meaningful reverberation times, a strongly diffuse sound field, and an unimpeded volume. As we know, in small rooms, especially those with large furnishings, these are mismatched concepts, so the calculated value may be in error, as noted by Baskind and Polack (2000). For the room used in the measurements in Figure 4.10, the Schroeder frequency is 129 Hz ($T = 0.32$ s, $V = 76.9$ m^3). This would seem to be on the low side because the large undulations in the curves have not diminished, especially in Figure 4.10c, although some of these variations are likely to be associated with adjacent-boundary effects. However, no matter how it is identified or what it is called, the transition region is real, and it is necessary to take different approaches to dealing with acoustical phenomena above and below it.

Figure 4.11 gives us more insight into this topic. Here are shown, using an expanded frequency scale, high-resolution frequency-response measurements from each of the five loudspeakers in a surround-sound system at the prime listening position. The room is geometrically symmetrical, but differences in the curves reveal that it is not acoustically symmetrical. A door in one end wall causes it to flex more than the other one at certain frequencies, and a concrete wall behind, but not touching, one of the side walls gives it more stiffness than the opposite wall. The result is asymmetry in the standing wave patterns (shown in Figure 13.7). Five identical loudspeakers are arranged in the ITU-R BS.775–2 (2006) recommended arrangement (see Figure 15.10a), and measurements were made at the listener's head location.

The standing waves cause huge variations at low frequencies, covering the full 40 dB range of the display. Above about 100 Hz, the variations are attenuated, and above about 200 Hz, they seem to settle down even more. Looking at the details, below about 200 Hz, in spite of some obvious variations related to the very different loudspeaker locations, one can see evidence of relatively independent resonant peaks at clearly identifiable frequencies. Above 200 Hz, the pattern becomes less orderly, and the peak-to-peak variations are smaller. Yet, an underlying trend is visible, including the step at 500 Hz, which is obviously a characteristic of the loudspeaker, a large woofer running without a low-pass

FIGURE 4.11

(a) A listening arrangement according to ITU-R BS.775–2. (b) 1/20-octave steady-state measurements for each of the five loudspeakers, measured at the listening position. The Schroeder crossover frequency (f$_c$) is shown.

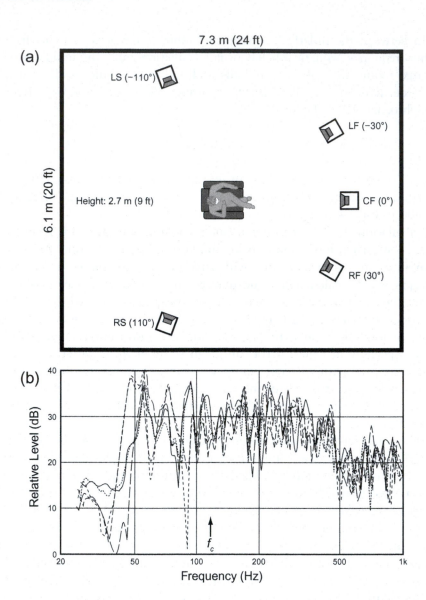

filter. The Schroeder (crossover) frequency (f$_c$) is 111 Hz ($T = 0.4$ s, $V = 128$ m^3), which again seems to be too low; a better estimate for this room would be 200 Hz, or even slightly higher.

Figure 4.12 is a stylized portrayal of this situation, indicating a region within which wave acoustics and room resonances are dominant factors, a region within which geometric/ray acoustics and reflections are dominant factors, and a transition region within which the two factors mingle in differing proportions at different frequencies. The position of the transition region on the frequency scale is dependent on room size, among other things, and this is shown. In very large auditoriums and concert halls, room resonances cease to be a problem.

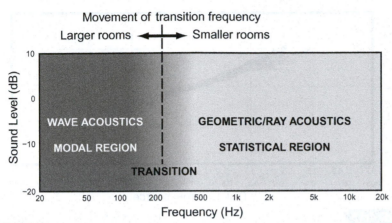

FIGURE 4.12 *An artistic interpretation of the transition between the low-frequency region dominated by room modes and the high frequency region dominated by reflected sounds. Also shown is the effect of room size on the position of the transition region in the frequency domain.*

Conversely, in very small spaces, like the interior of a car, the cabin resonances can be influential to much higher frequencies.

It may or may not be possible to find a simple calculation that will allow us to identify the "center" of the transition region, but it is a convenient concept, so further discussions of the phenomenon will often refer to the "transition frequency" as if it were a definable quantity. Right now, it is not. It is an empirical observation with a logical rationale, and finding it for a particular listening room requires acoustical measurements and visual inspection of the data.

4.3.2 Above the Transition Frequency

Picking up the story where it was left at the beginning of Section 4.3, let us continue to examine the behavior of steady-state sound fields as a function of distance from the source. Schultz (1983) measured draw-away curves in several living rooms. He used A-weighted measurements (see Figure 17.4) of broadband, omnidirectional or at least widely dispersing noise sources: an ILG fan (a calibrated noise source), a blender, a saw, and a drill. The sound field was found to decline at a rate of approximately −3 dB per double-distance. This was confirmed in draw-away measurements done by the author in two entertainment rooms using loudspeakers of various directional characteristics: omnidirectional, bipole, dipole, and forward firing.

The combined data from nine sound sources in six rooms are shown in Figure 4.13. The monotonic decline in sound level shown in all of the draw-away curves indicates a source-to-sink energy flow at increasing distance from the source, a confirmation of what was anticipated in Figures 4.8 and 4.9.

Variations in the curves at short distances are probably near-field effects caused by being so close to the sources, some of which (the electrostatic dipole

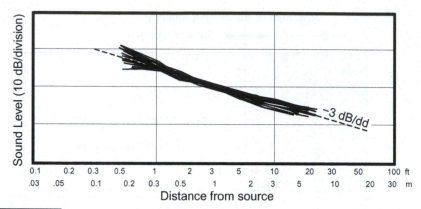

FIGURE 4.13 *Draw-away curves measured in four living rooms using four approximately omnidirectional sound sources (Schultz, 1983). Combined with these are measurements by the author using five loudspeakers with different directivities in two domestic listening rooms. All curves were adjusted to a similar middistance sound level to reveal the shape and slope tendencies.*

loudspeakers especially) were quite large (the meaning of "near field" is discussed in Section 18.1.1). At the other end of the curves, some of the measurements were made close to the back wall of the listening space where boundary effects may be expected. There may be rooms, unusually live or dead, or loudspeakers of sufficient directivity that could result in draw-away curves that fall outside this range, but that is precisely what would be expected in the real world, as was found and well noted by Hodgson (1983). In the cases shown here, the surprise is that the curves exhibit such similarity in spite of some real differences in source directivity and rooms.

The shapes and slopes of the draw-away curves suggest what may be going on in a room, but by themselves they are not proof of anything. Considering the distances at which we listen in our entertainment spaces and control rooms, it is clear that we are in the transitional region, where the direct and early-reflected sounds dominate and late-reflected sounds are subdued and progressively attenuated with distance. The sound field is not diffuse, and there is no critical distance, as classically defined. If we were to speculate at this early stage about loudspeaker performance in these rooms, it would seem that a combination of direct and early-reflected sounds would figure prominently in their potential sound quality and that sound power would not be the dominant factor.

4.3.3 Measuring the Lack of Diffusion in Small Rooms

Gover et al. (2004) provide hard evidence of what is going on in the sound fields in some small rooms. Using a novel spherical steerable-array microphone, the authors explored in three dimensions the decaying sound field in several small

THE LANGUAGE OF DIFFUSION

Diffusion is a property of a sound field. A perfectly diffuse sound field is *isotropic*: At any point within the sound field, sounds may be expected to arrive from all directions with equal probability. It is also *homogeneous*: It is the same everywhere in the space. Small listening and control rooms cannot have diffuse sound fields. In fact, true diffusion exists only as an academic ideal. Reverberation chambers used to measure the absorption of acoustical materials are designed to be diffuse and can come close, but as soon as a test sample of material is introduced into the space, it ceases to be. The result is measured absorption coefficients that exceed 1.0. Diffusion can be improved by using sound-*scattering* devices, irregular, curved and angled surfaces, and especially designed devices, often called *diffusers*. Perceptually, a diffuse sound field sounds spacious and enveloping. However, a diffuse sound field is not a requirement for the perception of *spaciousness* and *envelopment*. Much simpler sound fields also work, especially if multichannel sound reproduction is involved, because then it is possible to deliver sounds to the ears that are perceived to have those qualities—with or without a room.

rooms. None of them exhibited isotropic distributions at the measurement locations. Strong directional features were associated with early reflections. Small meeting rooms and a videoconferencing room with reverberation times of 0.36–0.4 s, in the range of typical listening rooms, had anisotropy indices and directional diffusion measures that fell roughly halfway between anechoic and reverberant conditions. Moreover, the values changed with time. Later sound showed increased anisotropy and even changed orientation in the room according to the surfaces that were more reflective (see Figure 4.14). This is interesting because, in physical terms, it means that in the initial interval after the source ceased output, there was a predominant front-back orientation to reflections in the room. However, in the decaying sound field, it can be seen that there is less overall sound absorption on the sides of the room, so, after a short interval, the reflection activity shifts 90° to a side-to-side orientation, and this pattern becomes even clearer with time.

None of this is necessarily bad. A highly diffuse sound field may be a worthy objective for performance spaces and recording studios, where the uniform blending of multiple sound sources and the reflected sounds from those multidirectional sources are desired. However, it is conceivable, indeed probable, that such a sound field may not be a requirement for sound reproduction through multiple, somewhat directional, loudspeakers surrounding and directed toward a listener. This becomes especially so when it is considered that, in popular applications like movie and television sound tracks and traditional music recordings, all of the loudspeakers are not allocated equivalent tasks. Front loudspeakers predominantly create real and phantom "soundstage" images, whereas side and rear loudspeakers provide occasional directional cues but are mainly utilized to create enveloping ambient and spatial illusions. This notion might need rethinking if "listener-in-the-middle-of-the-band" recordings become the norm.

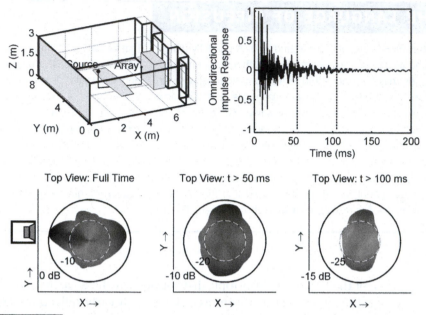

FIGURE 4.14 *Diffusivity measurements made in a videoconferencing room (7.23 m × 8.33 m × 3.01 m) with a midfrequency RT of 0.4 s. The omnidirectional source and the measurement microphone array were 2.03 m apart. The shapes across the bottom of the figure are the horizontal plane diffusivity patterns. The loudspeaker symbol shows the orientation of the direct sound. A perfectly diffuse sound field would show a circular pattern. The pattern on the left is for the entire time record, shown in the upper right. It shows prominent lobes for the direct sound, first-order lateral reflections, and a rear wall reflection. The middle and right patterns represent diffusivity of the later portions of the impulse response: The diffusivity rotates to a side-to-side orientation. From Gover et al., 2004.*

4.3.4 What Is a "Small" Room?

Diffuse-field theory may not apply perfectly to concert halls, but it applies even less well to other kinds of rooms. In the acoustical transition from a large performance space to a "small" room, it seems that the significant factors are a reduced ceiling height (relative to length and width), significant areas of absorption on one or more of the boundary surfaces, and proportionally large absorbing and scattering objects distributed throughout the floor area.

Different combinations of these characteristics result in basically similar acoustical behavior in large industrial spaces (Hodgson, 1983, 1998) and, with minor adjustment, in domestic listening spaces. Sound radiating from a source is either absorbed immediately on its first encounter with a surface or object or the objects redirect the sound into something else that absorbs it (see Figure 21.2). Thus, the late reflected sound field is greatly diminished with distance from the source. These are not Sabine spaces, and it is not appropriate to employ calculations and measurements that rely on assumptions of diffusivity. Schultz (1983) states, "The amount of sound-absorbing material in the room cannot be accurately determined by measurement, either with the decay-rate (reverbera-tion time) method or the steady-state (reference sound source) method. . . . One cannot trust the predictions of the Diffuse Field Theory for a non-Sabine room."

In the small listening rooms of interest to us, another distinguishing factor exists: the dominating presence at low frequencies of room modes. They are a major problem when attempting to communicate low-frequency musical sounds with important information in both the time and frequency domains.

4.3.5 Conventional Acoustical Measures in Small Listening Rooms

A measurement of reverberation time in a domestic-sized room yields a number. When the number is large, the room sounds live, and when the number is small, the room sounds dead. The implication is that there should be an optimum number. In spite of this, many thoughtful people believe that RT is unimportant or irrelevant (D'Antonio and Eger, 1986; Geddes, 2002; Jones, 2003; Kuttruff, 1998). The numbers measured are small compared to those in performance spaces, and so the question arises if the late-reflected sound field in a listening room is capable of altering what is heard in the reproduction of music. Yet, RT is routinely included as one of the measures of small listening and control rooms for international standards, even to the point of specifying allowable variations with frequency.

Reverberation time is a property of the room alone, and a correct measure-ment of it should employ an omnidirectional sound source capable of "illumi-nating" all of the room boundaries. The reason for this is that it is assumed that the boundaries consist of areas of reflection and absorption and that the central volume of the room is empty. The several formulae by which we estimate RT confirm this, and the values of absorption coefficient for the materials are "random incidence" values, meaning that there is an assumption of some con-siderable diffusivity in the sound field. Some practitioners incorrectly use con-ventional sound-reproduction loudspeakers as sources. The directivity of these is such that the resulting reflection patterns and decays are not properties of the room but of the room and loudspeaker combination—a very different situ-ation. Also, as we will see in Chapter 20, absorption at specific angles is quite different from random-incidence absorption. Figure 4.15 illustrates the funda-mental difference between a proper RT measurement and what it is that we listen to.

The result of a correct RT measurement is a number or a set of numbers for different frequency bands that describes the decay rate over a range of sound levels, maybe 20 or 30 dB (usually limited by background noise), and

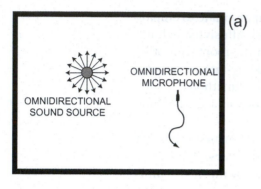

(a)

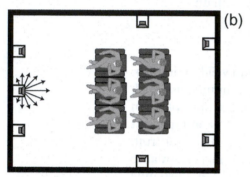

(b)

FIGURE 4.15 *(a) How RT should be measured, using an omnidirectional source aiming its sound at all of the room surfaces, and an omnidirectional microphone. Typically, several different setups would be used and the results averaged. (b) How we listen. A moderately directional loudspeaker directs most of its sound toward the audience and some of its sound toward some of the room surfaces and furnishings.*

then extended by multiplication to give a number for a 60 dB decay. It is common to look at the midfrequency reverberation time and the variations with frequency. The former is a measure of the suitability of a performance space for different styles of music. The variations with frequency are important because it is undesirable to change the spectral balance of voices and musical instruments by excessive absorption in narrow frequency bands. This is critical in large performance spaces because almost all of the listeners are in a sound field dominated by reverberation.

In a small listening room, we are in a transitional sound field that consists of the direct sound, several strong early reflections, and a much-diminished late-reflected sound field. What we hear is dominated by the directional characteristics of the loudspeakers and the acoustic behavior of the room boundaries at the locations of the strong early reflections. RT reveals nothing of this. As a measure, it is not incorrect, but it is just not useful as an indicator of how reproduced music or films will sound. Nevertheless, excessive reflected sound is undesirable, and an RT measurement can tell us that we are in the ballpark, but for that matter, so can our ears or an "acoustically aware" visual inspection.

This transitional sound field appears to extend over the entire range of listening distances we commonly employ in small rooms. It is therefore necessary to conclude that the large-room concept of critical distance is also irrelevant in small rooms. This said, there is still a perceptible transition that occurs as a function of distance, beyond which the front soundstage—real and phantom images—appear to change. Because critical distance is not the appropriate measure, a new one is needed. A reasonable hypothesis is that it is related to the ratio of direct to early-reflected sound and the extent to which laterally reflected sounds, especially, contribute to a perception of ASW, image broadening, frontal spaciousness, and so on.

All of the other acoustical measures employed in evaluating performance spaces: early/late-decay rates, energy ratios, lateral fractions, and others having to do with impressions of articulation, direction, image size, apparent source width, and spaciousness, could be applied to sounds reproduced over a multichannel reproduction system. However, in doing so, one is also evaluating the recording and the manner in which it captured or was processed to simulate those attributes. That is a worthy topic for investigation, and it could conceivably lead to improvements in recording technique and multiple-loudspeaker configurations. But, again, so far as the performance of the listening space itself is concerned, these are more traditional acoustical measures that find themselves in the wrong place.

The numbers produced by traditional acoustical predictions and by measuring instruments, while not totally irrelevant, are simply not direct answers to the important questions in small rooms used for sound reproduction. What, then, are the important questions? The accumulating evidence suggests that they have to do with reflections but not in a bulk, statistical, sense. We need to understand the influences of early reflected sounds. This means that the knowledge base must include the directivity and off-axis frequency responses of loudspeakers and the directional reflective, diffusive, and absorptive characteristics of materials at the points of first reflections. Only with this information can we predict the sounds that might arrive at listening locations in rooms, and only with careful experimentation can we understand the perceptual effects that they cause. This is very different from traditional acoustics.

CHAPTER 5

The Many Effects of Reflections

The next several chapters examine how sound propagates from a source to a listener in a reflective space and how those sounds are perceived when they arrive at the ears. We will find that technical measurements of the propagation path show enormous "flaws" that, over the years, made people believe that reflected sound is an "error" in need of immediate and expensive elimination. As discussed earlier, the development of porous absorbers in the 1930s led to a popular belief that acoustical room treatment begins with a large stack of fiberglass. In the author's opinion this approach has some value and should be applied to the interior of many popular restaurants within which conversation is all but impossible, especially for those with deteriorating hearing. However, for normal living and listening spaces, time has shown that a certain amount of reflected sound is not only welcome but expected. In performance spaces, reflections are essential. Arthur Benade (1984) had a wonderful clarity of insight into sound in rooms, some of which is embodied in Figure 5.1, which is a clear statement that no linear relationship exists between what we measure in a room and what may be perceived in that room.

As the story develops, it will be evident that measurements are indeed relevant, but some measurements are much more useful than others. It will also become clear that humans are wonderfully adaptable: We can usually compensate for things we can measure and for things we can hear while we are moving or while they are changing but fade away once stability is established. It is almost as if when we walk into a room, we hear all the reflections, and this gives us a great deal of information about the acoustical nature of the space. Then, when we sit down, within a very short time the perceptual effects of the reflections are attenuated, some more than others, and we settle in to listen to the sound sources, whatever they may be.

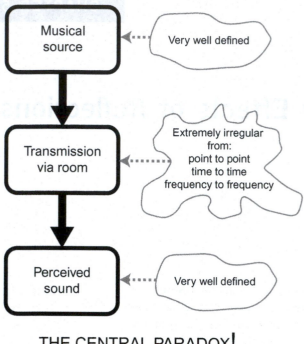

THE CENTRAL PARADOX!

FIGURE 5.1 *"The central paradox" of sound in rooms.*

Other perceptual effects are enhanced by room acoustics, and nowhere is this more apparent than in a good concert hall. The same sound source—a voice or musical instrument, for example—will take on some of the character of different rooms, and yet they can still be recognized as being the "same" sound sources. Within some range of "normal" rooms, we seem to have a built-in ability to "listen through" a room to attend to even minute details of the sound source. It is an interesting tale.

Reflections have many effects on the perception of sound in rooms; some are interactive with each other, and others are relatively independent. Before we get into an examination of experimental evidence of acoustical events in small rooms, it is advantageous to have a perspective on the factors involved with our perceptions of direction and space. To that end, let us very briefly summarize the dominant effects of sounds above the bass-frequency range:

- **Localization** has two principal dimensions:
  - □ **Direction:** Identifying the direction from which sound appears to be coming. Because of the ear locations, we are much better at localizing in azimuth than we are in elevation, and we are more precise close to the median plane (a vertical forward/back plane running through the head) than we are to the sides. In rooms, the **precedence effect** allows us to localize in the presence of numerous reflections. It is a cognitive effect, occurring at a high level in the brain, meaning that it can be different at different times and places, and for different sounds. It changes as we gain experience listening within a space.
  - □ **Distance:** Reflections help us to determine distance. Distance perception also has a cognitive component, meaning that we can learn to recognize aspects of the sound field. *Question:* If we learn to recognize the distance of the real loudspeakers, does this inhibit perceptions of artificial distance in recordings?

- **Spaciousness or spatial impression** can be separated into two components:
  - □ **Image size and position:** Strong reflections have the ability to shift the apparent position of a source in the direction of the

reflection and/or to make the source appear larger. In live classical performances, this is called ASW (apparent source width), and audiences like it. In sound reproduction, there is evidence that the tendency continues.

☐ **Envelopment and the sense of space:** Also called listener envelopment (LEV), this is the impression of being in a specific acoustical space. It is perhaps the single most important perceived element distinguishing truly good concert halls. In music recordings and movies, it is arguably the greatest improvement contributed by multichannel audio.

■ **Timbre changes** have two basic components; one can be negative, and the other is mostly positive. Simply detecting a "difference" is not a sufficient criterion.

☐ **Comb filtering, repetition pitch:** Colorations can be created when a sound is added to a delayed version of itself. When the result is measured, we see a repeating pattern of peaks and dips in the frequency response, which is why it is called "comb" filtering. In some circumstances a pitch can be perceived that is associated with a frequency defined by the inverse of the delay. The effect is audibly obvious if it occurs in an electronic signal path or if there is a single, strong vertical (median plane) reflection in an otherwise echo-free environment. However, for reflections that arrive from large horizontal angles and in normally reflective spaces where there are multiple reflections, the effect ceases to be a problem.

☐ **The audibility of resonances:** Resonances are the "building blocks" of voices and musical instrument sounds. Reflective spaces enhance our ability to hear resonances, making these sounds timbrally richer and more interesting. In loudspeakers, resonances are huge problems because they monotonously color all reproduced sounds. Listening in reflective rooms is more likely to reveal inferior loudspeakers.

■ **Speech intelligibility** is important because if the perception of speech is poor, our ability to be informed and entertained has been seriously compromised. Fortunately, the human hearing system is not just remarkably tolerant of reflections typical of small rooms. In fact, almost without exception, we make use of them to assist us in understanding speech.

Our understanding of these perceptual factors is not yet complete, but there is a lot of information in the accumulated literature of architectural acoustics. Complicating the situation is the fact that several of these effects can coexist, interacting with each other, and that the relationships can be different, at least

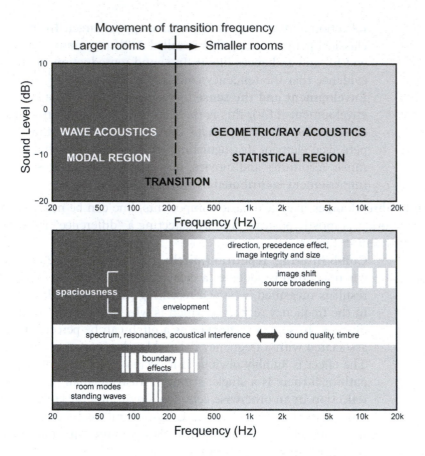

FIGURE 5.2

The approximate frequency ranges over which various audible effects of reflected sound may be heard. These will be discussed in the introductions to the relevant chapters that follow.

in some degree, for different kinds of sounds. A lot of the pioneering work was done using speech at the test signal and, although it is fundamentally important, it is not the only sound we listen to. Similarly, many experiments examined the effects of a single reflection auditioned in an otherwise reflection-free environment. It will be found that some conclusions need to be modified for normally reflective spaces. When looking at the results of data gathered in "scientific" circumstances, it is essential to think carefully before drawing conclusions about what may or may not be important in real-world situations.

We know that in real rooms there are multiple reflections. However, to understand the influence of many, it is useful to begin by understanding the influence of a few, or even one. It also makes experiments practical and controllable. As will be seen, there is a logical progression of effects from a single to multiple reflections, giving us, in the end, a better insight into the perceptual mechanisms at play.

All of the effects being discussed have portions of the frequency range over which they are most noticeable. Figure 5.2 includes a repetition of Figure 4.12, which illustrates that, in terms of physical acoustics, the frequency range is divided into two regions connected by a broad transition zone. Under it is an attempt to show the frequency ranges over which various audible effects of reflections are most likely to be heard. As we will see, these are very approximate divisions, subject to variations with different program material, reproduced in different environments, and so on. They will be shown at the beginning of each relevant chapter and will be discussed at that point.

CHAPTER 6

Reflections, Images, and the Precedence Effect

6.1 AUDIBLE EFFECTS OF A SINGLE REFLECTION

Investigations of these effects go back many decades, and observations of our ability to localize a source of sound in an acoustically hostile—that is, reflective—environment were first recorded more than a century ago.

In audio in the past, the terms *Haas effect* and *law of the first wavefront* were used to identify this effect, but current scientific work has settled on the other original term, *precedence effect*. Whatever it is called, it describes the well-known phenomenon wherein the first arrived sound, normally the direct sound from a source, dominates our impression of where sound is coming from. Within a time interval often called the "fusion zone," we are not aware of reflected sounds that arrive from other directions as separate spatial events. All of the sound appears to come from the direction of the first arrival. Sounds that arrive later than the fusion interval may be perceived as spatially separated auditory images, coexisting with the direct sound, but the direct sound is still perceptually dominant. At very long delays, the secondary images are perceived as echoes, separated in time as well as direction. The literature is not consistent in language, with the word *echo* often being used to describe a delayed sound that is not perceived as being separate in either direction or time.

Haas was not the first person to observe the primacy of the first arrived sound so far as localization in rooms is concerned (Gardner, 1968, 1969, 1973 describes a rich history), but work done for his PhD thesis in 1949, translated from German to English in Haas (1972), has become one of the standard references in the audio field. Sadly, his conclusions are often misconstrued. Let us review his core experiment.

Figure 6.2 shows the essence of the experiment. On the hemi-anechoic space provided by the flat roof of a laboratory building, a listener was positioned facing

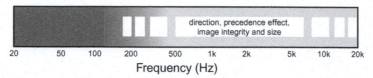

direction, precedence effect,
image integrity and size

20 50 100 200 500 1k 2k 5k 10k 20k

Frequency (Hz)

FIGURE 6.1 *The approximate frequency range over which reflections appear to influence perceptions of the direction of a sound source and the apparent size of that source. In some circumstances, reflections may be audible as separate "images" of the sound source.*

loudspeakers that had been placed 45° apart. The Haas (1972) translation describes the setup as "at an angle of 45° to the left and right side of the observer" (p. 150). This could be construed in two ways. Gardner (1968), however, in a translation of a different Haas document, reports "loudspeakers . . . at an angle of 45°, half to the right and half to the left of him. . . ." When Lochner and Burger (1958) repeated the Haas experiment, they used loud-speakers that were placed 90° apart. So there is ambiguity about the angular separation.

A recording of running speech was sent to both loudspeakers, and a delay could be introduced into the signal fed to one of them. In all situations except for Figure 6.2d, both signals were radiated with the same sound level.

Figure 6.2a shows summing localization. When there is no delay, the perceived result was a phantom (stereo) image floating midway between the loud-speakers. When delay was introduced, the center image moved toward the loudspeaker that radiated the earlier sound, reaching that location at delays of about 0.6–1.0 ms. This is called summing localization, and it is the basis for the phantom images that can be positioned between the left and right loudspeak-ers in stereo recordings, assuming a listener is in the "sweet spot" (Blauert, 1996).

Figure 6.2b shows the precedence effect. For delays in excess of 1 ms, it is found that the single image remains at the reference loudspeaker up to about 30 ms. This is the precedence effect—that is, when there are two (or more) sound sources and only one sound image is perceived. It needs to be noted that the 30 ms interval is only for speech and only for equal level direct and reflected sounds.

Figure 6.2c shows multiple images—the breakdown of precedence. With delays greater than 30 ms but certainly by 40 ms, the listener becomes aware of a second sound image at the location of the delayed loudspeaker. The prece-dence effect has broken down because there are two images, but the second image is a subordinate one; the dominant (louder) localization cue still comes from the loudspeaker that radiated the earlier sound.

Figure 6.2d shows the Haas equal-loudness experiment. In the first three illustrations, the first and delayed sounds had identical amplitudes. Obviously,

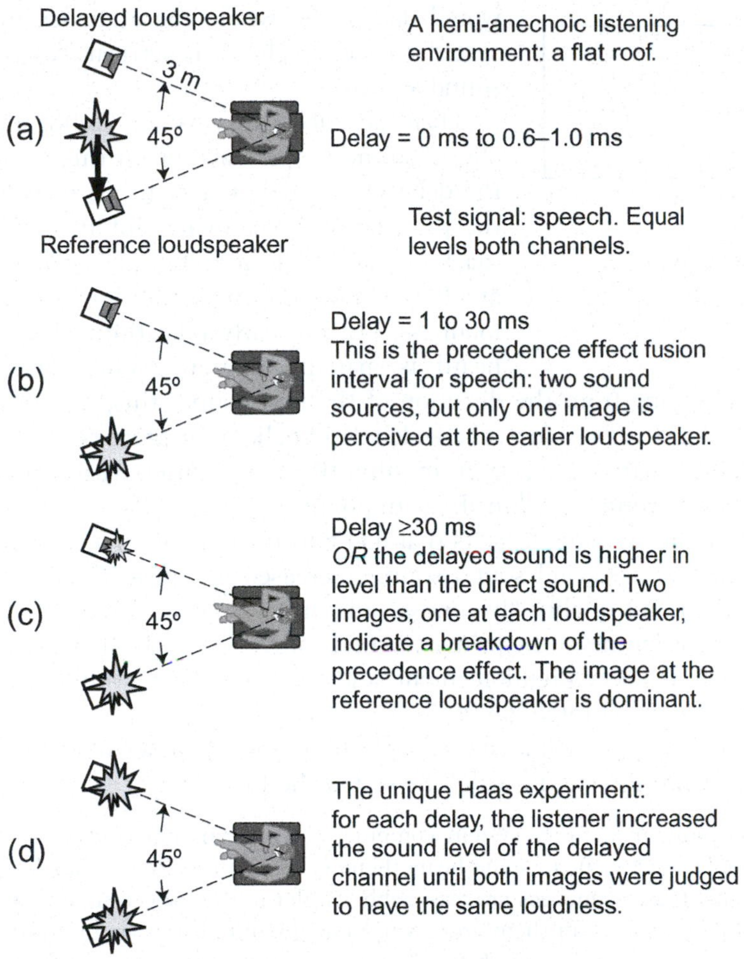

Delayed loudspeaker

(a)   3 m   45°

A hemi-anechoic listening environment: a flat roof.

Delay = 0 ms to 0.6–1.0 ms

Reference loudspeaker

Test signal: speech. Equal levels both channels.

(b)   45°

Delay = 1 to 30 ms
This is the precedence effect fusion interval for speech: two sound sources, but only one image is perceived at the earlier loudspeaker.

(c)   45°

Delay ≥30 ms
OR the delayed sound is higher in level than the direct sound. Two images, one at each loudspeaker, indicate a breakdown of the precedence effect. The image at the reference loudspeaker is dominant.

(d)   45°

The unique Haas experiment: for each delay, the listener increased the sound level of the delayed channel until both images were judged to have the same loudness.

FIGURE 6.2

A progression of localization effects observed in the experimental setup used by Haas, including stereo (summing) localization, the precedence effect, and the equal-loudness experiment. Because the experiments were done on a flat roof, to minimize the effect of the roof reflection, Haas placed the loudspeakers directly on the roof, aimed upward toward the listener's ears. He found, though, that there was no significant difference if the loudspeakers were elevated to ear level, and that is the configuration used for the experiments.

this is artificial because if the delayed sound were a reflection, it would be attenuated by having traveled a greater distance. But Haas moved even farther from passive acoustical realities and deliberately amplified the delayed sound, as would happen in a public address situation. His interest was to determine how much higher in sound level the delayed sound could be before it became the dominant localization cue—in other words, subjectively louder. To do this, he asked his listeners to adjust the sound level of the delayed loudspeaker until both of the perceived images appeared to be equally loud. This is the balance point, beyond which the delayed loudspeaker would be perceived as being domi-nant. The objective was to prevent an audience from seeing a person speaking in one direction and being distracted by a louder voice coming from a different direction. As shown in Figure 6.3, over a wide range of delays, the later loud-speaker can be as much as 10 dB higher in level before it is perceived to be equally loud and therefore a major distraction to the audience. Naturally, this

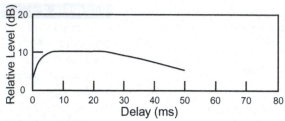

FIGURE 6.3 *The sound level of the delayed sound, relative to that of the first arrival, at which listeners judged the two sound images to be equal in loudness.*

would depend on where the audience member is seated relative to the symmetrical axis of the two sound sources.

Haas described this as an "echo suppression effect." Some people have taken this to mean that the delayed sound is masked, but it isn't. Within the precedence effect fusion interval, there is no masking—all of the reflected (delayed) sounds are audible, making their contributions to timbre and loudness, but the early reflections simply are not heard as spatially separate events. They are perceived as coming from the direction of the first sound; this, and only this, is the essence of the "fusion." The widely held belief that there is a "Haas fusion zone," approximately the first 20 ms after the direct sound, within which everything gets innocently combined, is simply untrue.

Haas observed audible effects that had nothing to do with localization. First, the addition of a second sound source increased loudness. There were some changes to sound quality "liveliness" and "body" (Haas, 1972, p. 150) and a "pleasant broadening of the primary sound source" (p. 159). Increased loudness was a benefit to speech reinforcement, and the other effects would be of concern only if they affected intelligibility.

Benade (1985) contributed a thoughtful summary under the title "Generalized Precedence Effect," in which he stated the following:

1. The human auditory system combines the information contained in a set of reduplicated sound sequences and hears them as though they were a single entity, provided (a) that these sequences are reasonably similar in their spectral and temporal patterns and (b) that most of them arrive within a time interval of about 40 ms following the arrival of the first member of the set.

2. The singly perceived composite entity represents the accumulated information about the acoustical features shared by the set of signals (tone color, articulation, etc.). It is heard as though all the later arrivals were piled upon the first one without any delay—that is, the perceived time of arrival of the entire set is the physical instant at which the earliest member arrived.

3. The loudness of the perceived sound is augmented above that of the first arrival by the accumulated contributions from the later arrivals. This is true even in the case when one or more of the later signals is stronger than the first one to arrive—that is, a strong later pulse does not start a new sequence of its own.

4. The apparent position of the source of the composite sound coincides with the position of the source of the first-arriving member of the set, regardless of the physical directions from which the later arrivals may be coming.

5. If there are any arrivals of sounds from the original acceptably similar set which come in after a delay of 100–200 ms, they will not be accepted for processing with their fellows. On the contrary, they will be taken as a source of confusion and will damage

the clarity and certainty of the previously established percept. These "middle-delay" signals that dog the footsteps of their betters may or may not be heard as separate events.

6. If for some reason a reasonably strong member of the original set should come in with a delay of something more than 250–300 ms, it will be distinctly heard as a separate echo. This late reflection will be so heard even if it is superposed on a welter of other (for example, reverberant) sounds.

It is important to notice that these very strongly worded categorical statements all emphasize that there is an *accumulation of information* from the various members of the sequence. It is quite incorrect to assume that the precedence effect is some sort of masking phenomenon which, by blocking out the later arrivals of the signal, prevents the auditory system from being confused. Quite to the contrary, those arrivals that come in within a reasonable time after the first one actively contribute to our knowledge of the source. Furthermore, members of the set that are delayed somewhat too long actually disrupt and confuse our perceptions even when they may not be consciously recognized. If the arrivals are later yet, they are heard as separate events (echoes) and are treated as a nuisance. In neither case are the late arrivals masked out.

6.1.1 Effects of a Single Reflection
This is the "begin at the beginning" experiment, in which the number of variables is minimized. The listening environment is anechoic, the signal is speech, and only a single lateral reflection is examined. It is not data that can be applied to real-world circumstances listening to music or movies, but it is scientific data that establishes a baseline for further research.

In Figure 6.5, the lowest curve describes the sound level at which listeners reported hearing any change attributable to the presence of the reflection. This is the "absolute threshold"; nothing is perceived for reflections at lower levels. Most listeners described what they heard as a sense of spaciousness (Olive and Toole, 1989). Although the experiment was conducted in an anechoic chamber, a single detectable reflection was sufficient to create the impression of a (rudimentary) three-dimensional space. Throughout, listeners reported all of the sound as originating at the location of the loudspeaker that reproduced the first sound,

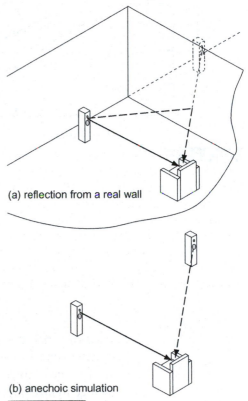

(a) reflection from a real wall

(b) anechoic simulation

FIGURE 6.4   *An explanation of how an anechoic simulation can imitate a reflection from a real—flat and perfectly reflective—wall. The anechoic setup uses a real loudspeaker to simulate the "mirror image" loudspeaker in the room situation. This is the experimental method that has been used in numerous experiments conducted over the decades. Electrical adjustments of delay, amplitude, and frequency response of the signal sent to the "reflection" loudspeaker allow the simulation of different geometries and reflective surface types.*

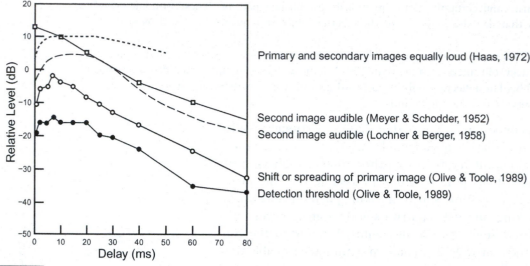

FIGURE 6.5 *An illustration of the several audible effects that occur when a single lateral reflection is added to a direct sound, in an anechoic simulation similar to that shown in Figure 6.4b. All of these curves were determined using speech as a signal. In the experiments, at each of several delays, the sound level of the reflected sound was adjusted to identify those levels at which each of the described perceptions became apparent. The bottom two curves are from Olive and Toole (1989), in which the direct sound was at 0° and the lateral reflection arrived from 65°. Meyer and Schodder (1952) had their reflection arrive from 90°, and listeners reported when the echo was not perceived at all. Lochner and Burger (1958) employed a direct sound arriving from −45° and a delayed sound from +45°, and their listeners reported when the second source was just audible. Adapted from Toole (1990), with additional information from Cremer and Müller, 1982, Figure 1.25.*

meaning that the precedence effect was working. As the sound level of the delayed sound was increased, the impression of spaciousness increased.

The next higher curve is the level at which listeners reported hearing a change in size or position of the main sound image, which the precedence effect causes to be localized at the position of the loudspeaker that reproduced the earlier sound. This was called the "image shift" threshold. In general, these changes were subtle and noticeable in these controlled A versus B comparisons, but it is doubtful that they would be detected in the context of a multiple-image music or movie soundstage. As the sound level of the delayed sound was further increased, the impression of spaciousness also increased.

With the two curves that portray the third perceptual category, a major transition is reached, because it is at this sound level that listeners report hearing a second sound source or image, simultaneously coexisting with the original one (we have not reached the long delays at which there is a sense of a temporally as well as a spatially separate echo). Data from Lochner and Burger (1958) and Meyer and Schodder (1952). This means that the precedence-effect directional "fusion" has broken down. Although the original source remains the perceptually louder, spatially dominant source, there is a problem because two spatial events are perceived when there should be only one.

The top curve is from the well-known work by Haas (1972) in which he asked his listeners to adjust the relative levels of the spatially separate images associated with the direct and reflected sounds until they appeared to be equally loud. This tells us that in a public address situation, it is possible to raise the level of delayed sound from a laterally positioned loudspeaker by as much as 10 dB above the direct sound before it is perceived as being as loud as the direct sound. It is important information in the context of professional audio, but it is irrelevant in the context of small-room acoustics.

All of the data points are thresholds—the sound levels at which listeners detected a change in their perceptions. As we will see later, some of the perceived changes are beneficial and, up to a point, listeners find that levels well above threshold provide greater pleasure. For example, the perception described at threshold as "image shift or spreading" may seem like a negative attribute, but when it is translated into what is heard in rooms, it becomes "image broadening" or apparent source width (ASW), which are widely-liked qualities. Even "second-image" thresholds can be exceeded with certain kinds of sounds, expanding the size of an orchestra beyond its visible extent in a concert hall or extending the stereo soundstage beyond the spread of the loudspeakers. In reproduced sound, the picture is more confused because some techniques in the recording process can achieve similar perceptions. Because all of these factors are influenced by how the recordings are made as well as how they are reproduced, these comments are observations, not judgments of relative merit. Some evidence suggests that even these small effects might be diminished by experience during listening within a given room (Shinn-Cunningham, 2003), another in the growing list of perceptual phenomena we can adapt to.

6.1.2 Another View of the Precedence Effect

If we extract from Figure 6.5 those things that are relevant to sound reproduction from a single loudspeaker, we end up with Figure 6.6. The Haas data have been removed because amplified delayed sounds do not exist in passive acoustics. The "second-image" data (Lochner and Burger, 1958; Meyer and Schodder, 1952) have been combined into a single average curve for simplicity. There is some justification for doing this, as one curve expresses a "just audible" criterion and the other a "just not audible" criterion.

The area under the "second-image" curve has been shaded. This is the real-world precedence-effect fusion zone for speech, within which any delayed sound will not be perceived as a spatially separate localizable event. This perspective is very different from most discussions of the precedence-effect fusion interval. Normally, only a single number is stated, and that number normally relates to direct and delayed sounds at the same sound levels. This is a correct description

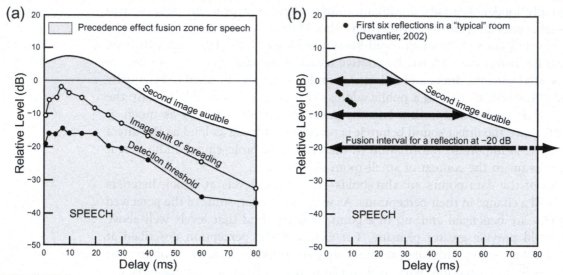

FIGURE 6.6 *(a) A simplification of Figure 6.5 in which only data that are relevant to sound in small rooms are preserved and a shaded area representing the precedence-effect fusion zone for speech is identified. This is the range of amplitudes and delays within which a reflected sound will not be identified as a separately localizable event. From Toole, 2006. (b) The precedence-effect fusion intervals for delayed sounds at three sound levels. The classic experiments much quoted from psychoacoustic literature generally used equal-level direct and delayed sounds. This is the highest large arrow at 0 dB, showing an interval of about 30 ms. In rooms, delayed sounds are attenuated by propagation loss, typically –6 dB/double distance, and sound absorption at the reflecting surfaces. As the delayed sound is reduced in level, the fusion zone increases rapidly. The set of black dots show the delays and amplitudes for the first six reflections in a typical listening room (Devantier, 2002), indicating that in such rooms, the precedence effect is solidly in control of the localization of speech sounds.*

of the results of a laboratory experiment but is simply wrong as guidance about what may happen in the real world.

The fusion interval for speech is often quoted as being around 30 ms. This is true for anechoic listening to a single reflection that has the same sound level as the direct sound, as can be seen in Figure 6.6b. This is how the classic psychoacoustic experiments were conducted, but these circumstances are far from the acoustical realities in normal rooms. For reflections at realistically lower levels, the fusion interval is much longer. So far, in small rooms, the precedence effect is undoubtedly the dominant factor in the localization of speech.

6.1.3 Reflections from Different Directions

Figure 6.7 shows more data from Olive and Toole (1989), in which it is seen that the thresholds for the side wall and the ceiling reflections are almost identical. This is counterintuitive because one would expect a lateral reflection to be

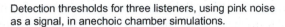

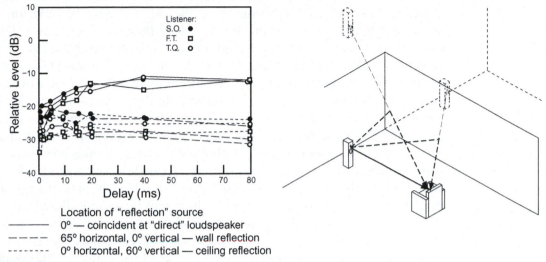

Detection thresholds for three listeners, using pink noise as a signal, in anechoic chamber simulations.

Listener:
S.O. ●
F.T. □
T.Q. ○

Location of "reflection" source
—————— 0° — coincident at "direct" loudspeaker
— — — — 65° horizontal, 0° vertical — wall reflection
- - - - - - - 0° horizontal, 60° vertical — ceiling reflection

FIGURE 6.7 *The detection thresholds for delayed sounds simulating a wall reflection, a ceiling reflection, and one arriving from the same direction as the direct sound. The test signal was pink noise. Adapted from Olive and Toole, 1989.*

much more strongly identified by the binaural discrimination mechanism because of the large signal differences at the two ears. For sounds that differ only in elevation, we have only the spectral cues provided by the external ears and the torso (HRTFs). Although the threshold levels might be surprising, intuition is rewarded in that the dominant audible effect of the lateral reflection was spaciousness (the result of interaural differences) and that of the vertical reflection was timbre change (the result of spectral differences). The broadband pink noise used in these tests would be very good at revealing colorations, especially those associated with HRTF differences at high frequencies. On the other hand, continuous noise lacks the strong temporal patterns of some other sounds, like speech.

This makes the findings of Rakerd et al. (2000) especially interesting. These authors examined what happened with sources arranged in a horizontal plane and vertically on the front-back (median sagittal) plane. Using speech as a test sound, they found no significant differences in masked thresholds and echo thresholds sources in the horizontal and vertical planes. In explanation, they agreed with other referenced researchers that there may be an "echo suppression mechanism mediated by higher auditory centers where binaural and spectral cues to location are combined." This is another example of humans being very well adapted to listening in reflective environments.

Another surprise in Figure 6.7 is that delayed sounds that come from the same loudspeaker are more difficult to hear; the threshold here is consistently higher than for sounds that arrive from the side or above, slightly for short delays, and much higher (10+ dB) at long delays. Burgtorf (1961) agrees, finding thresholds for coincident delayed sounds to be 5–10 dB higher than those separated by 40–80°. Seraphim (1961) used a delayed source that was positioned just above the direct-sound source (~5° elevation difference) and found that, with speech, the threshold was elevated by about 5 dB compared to one at a 30° horizontal separation. The relative insensitivity to coincident sounds appears to be real, and the explanation seems to be that it is the result of spectral similarities between the direct sound and the delayed sound. These sounds take on progressively greater timbral differences as they are elevated (or, one assumes, lowered) relative to the direct sound. For those readers who have been wondering about the phenomenon of "comb filtering," which will be specifically addressed in Chapter 9, it is worthy of note that this evidence tells us that the situation of maximum comb filtering, when the direct and delayed sounds emerge from the same loudspeaker, is the one for which we are least sensitive. (Encouraging news!)

All this said, it still seems remarkable that a vertically displaced reflection, with no apparent binaural (between the ears) differences, can be detected as well as a reflection that arrives from the side, generating large binaural differences. Not only are the auditory effects at threshold different—timbre versus spaciousness—the perceptual mechanisms required for their detection are also different.

6.2 A REFLECTION IN THE PRESENCE OF OTHER REFLECTIONS

Working with a single reflection allows for intensely analytical investigations, but, inevitably, the tests must include others to be realistic. A long-standing belief in the area of control room design is that early reflections from monitor loudspeakers must be attenuated to allow those in the recordings to be audible. Consequently, embodied in several standards, and published designs, are schemes to attenuate or eliminate the first reflections from a loudspeaker using deflecting reflectors, absorbers, or scattering surfaces (diffusers).

Olive and Toole (1989) appear to have been the first to test the validity of this idea. Figure 6.8 shows the results of experiments that examined the audibility of a single lateral reflection simulated in an anechoic chamber with 3 ft (1 m) wedges. For the second experiment, the same physical arrangement was replicated in a typical small room in which the first wall, floor, and ceiling reflections had been attenuated using 2-in. (5 cm) panels of rigid fiberglass board. A third experiment was conducted in the same room with most of the absorption removed (midfrequency reverberation time = 0.4 s). The idea was to show the

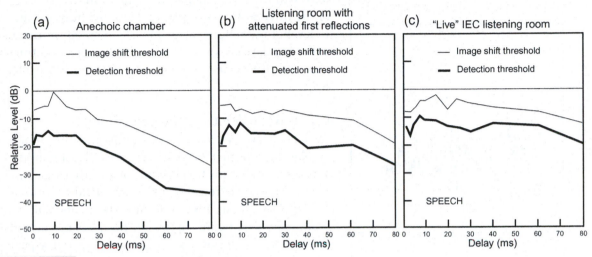

FIGURE 6.8 *Detection and image-shift thresholds as a function of delay for a single reflection auditioned in three very different acoustical circumstances: (a) Anechoic. (b) A normal room in which the first-order reflections were attenuated with 2-in. (50 mm) fiberglass board. (c) The same room in a relatively reverberant configuration (midfrequency reverberation time = 0.4 s). From Olive and Toole, 1989.*

THE IEC ROOM

The listening room used in these experiments was the prototype room underlying IEC 268–13–1985. It was constructed at the National Research Council, in Ottawa, within an existing laboratory space (which explains the dimensions). There was little real science to guide the choice of dimensions and acoustical treatment, so the resulting room became one of the variables in ongoing experiments. Of course, at that time stereo was the standard reproduction format. The room was 6.7 m × 4.1 m × 2.8 m (22 ft × 13.5 ft × 9.2 ft) with a midfrequency reverberation time of 0.34 s. More description and measurements are shown in the appendix of Toole (1982). The original concept of the standard was to specify a room that could be duplicated so test results from different laboratories could be compared. In subsequent editions of the standard, the requirements were relaxed so more rooms could qualify, which is a different and significantly less useful objective but much more popular among users who want to claim IEC compliance.

effects, on the perception of a single reflection, of increasing levels of natural reflected sound within a real room.

The large changes in the level of reflected sound had only a modest (1–5 dB) effect on the absolute threshold or the image-shift threshold of an additional lateral reflection occurring within about 30 ms of the direct sound. At longer delays, the threshold shifts were up to about 20 dB, a clear response to elevated late-reflected sounds in the increasingly live rooms. This is a good point to remember, as we will see it again: the threshold curves become more horizontal when the sound—in this case, speech—becomes prolonged by reflected energy (repetitions).

ETCs at the listening location compared with the detection thresholds for a single reflection in that same space. Signal: speech.

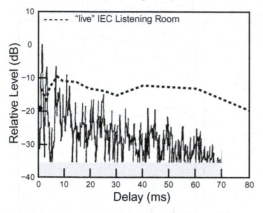

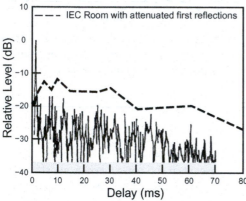

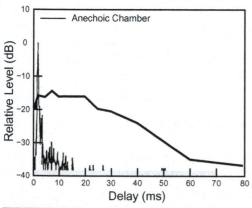

FIGURE 6.9 *A comparison of the absolute thresholds shown in Figure 6.8, with measured energy-time curves (ETCs) for the three spaces within which the tests were done. All data from Olive and Toole, 1989.*

Figure 6.9 shows a direct comparison of the thresholds with the ETC (energy-time curve) measured in each of the three test environments. Here the huge variations in level of the reflections can be clearly seen, in contrast with the relatively small changes in the detection thresholds within the first 30 ms or so. Section 6.6 explains why.

6.2.1 Real Versus Simulated Rooms

In a large anechoic-chamber simulation of a room of similar size, Bech (1998) investigated the audibility of single reflections in the presence of 16 other reflections, plus a simulated "reverberant" sound field beginning at 22 ms. One of his results is directly comparable with these data. The figure caption in Bech's paper describes the response criterion as "a change in spatial aspects," which seems to match the image shift/image spreading criterion used by Olive and Toole. Figure 6.10 shows the image-shift thresholds in the "live" configuration of the IEC room for two subjects (the FT data are from Figure 6.9; the SO data were previously unpublished) and thresholds determined in the simulated room, an average of the three listeners from Bech (1998). The similarity of the results is remarkable considering the very different physical circumstances of the tests. It suggests that listeners were responding to the same audible effect and that the real and simulated rooms had similar acoustical properties.

Bech separately examined the influence of several individual reflections on timbral and spatial aspects of perception. In all of the results, it was evident that signal was a major factor: Broadband pink noise was more revealing than male speech. In terms of timbre changes, only the noise signal was able to show any audible effects and then only for the floor reflection; speech revealed no audible effects on timbre.

Looking at the absorption coefficients used in modeling the floor reflection (Bech, 1996, Table II) reveals that the simulated floor was significantly more reflective than would be the case if it had been covered by a conventional clipped pile carpet on a felt underlay. Further investigations revealed that the detection was based mainly on sounds in the 500 Hz–2 kHz range, meaning that ordinary room furnishings are likely to be highly effective at reducing first reflections below threshold, even for the more demanding signal: broadband pink noise (see Section 21.3).

In terms of spatial aspects, Bech (1998) concluded that those sounds above ~2 kHz contributed to audibility and that "only the first-order floor reflection will contribute to the spatial aspects." The effect was not large, and, as before, speech was less revealing than broadband noise. Again, this is a case where a good carpet and underlay would appear to be sufficient to eliminate any audible effect. See Figure 21.3 for data on the acoustical performance of floor coverings.

In conclusion, it seems that the basic audible effects of early reflections in recordings are well preserved in the reflective sound fields of ordinary rooms. There is no requirement to absorb first reflections to allow recorded reflections to be heard.

6.2.2 The "Family" of Thresholds

Figure 6.11 shows a complete set of thresholds, like those shown in Figure 6.5, determined in an anechoic chamber but here determined in the "live" IEC listening room. The curves are slightly irregular because the data were based on a small number of repetitions. As expected, the curves all have a more horizontal appearance than for speech auditioned in an anechoic environment. It is significant that all the curves have the same basic shape from detection at the bottom to the Haas-inspired equal-loudness curve at the top.

6.3 A COMPARISON OF REAL AND PHANTOM IMAGES

A phantom image is a perceptual illusion resulting from summing localization when the same sound is radiated by two loudspeakers. It is natural to think that these directional illusions may be more fragile than those created by a single loudspeaker at the same location. The evidence shown here applies to the simple case of a single lateral reflection, simulated in a normally reflective room with a loudspeaker positioned along a side wall, as shown in Figure 6.12. When detection threshold and image-shift

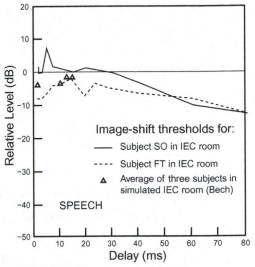

FIGURE 6.10   *Image-shift thresholds as a function of delay for two listeners in the "live" IEC room (FT data from Figure 6.8) and averaged results for three listeners in a simulation of an IEC room using multiple loudspeakers set up in a large anechoic chamber (Bech, 1998).*

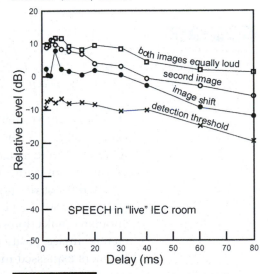

FIGURE 6.11   *The full set of thresholds, as shown in Figure 6.5, but here obtained while listening in a normally reflective room rather than an anechoic chamber. One listener (SO). Unpublished data acquired during the experiments of Olive and Toole, 1989.*

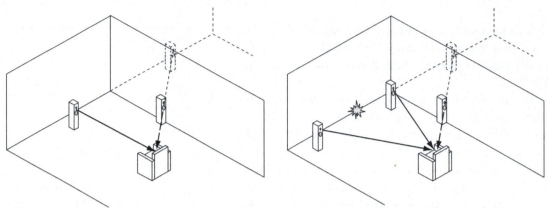

In a normally-reflective room, a single added "reflection" interacts with:

(a) a center image from a loudspeaker

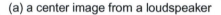

(b) a "phantom" center image

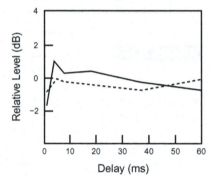

The *differences* between reflection thresholds as observed in a real and a phantom (stereo) center image.

NOTE: the vertical scale is greatly expanded.

- - - - - image–shift threshold

———— absolute detection threshold

FIGURE 6.12 *An examination of how a real and a phantom center image respond to a single lateral reflection simulated by a loudspeaker located at the right side wall. The room was the "live" version of the IEC listening room used in other experiments. Note that the vertical scale has been greatly expanded to emphasize the lack of any consequential effect. The signal was speech.*

threshold determinations were done first with real and then with phantom center images, in the presence of an asymmetrical single lateral reflection, the differences were insignificantly small. It appears that concerns about the fragility of a phantom center image are misplaced.

Examining the phantom image in transition from front to surround loudspeakers (±30° to ±110°), Corey and Woszczyk (2002) concluded that adding simulated reflections of each of the individual loudspeakers did not significantly change image position or blur, but it did slightly reduce the confidence that listeners expressed in the judgment.

6.4 EXPERIMENTAL RESULTS WITH MUSIC AND OTHER SOUNDS

A good introduction to investigations that used music is Figure 6.13, the widely reproduced illustration from Barron (1971), in which he combines several sub-

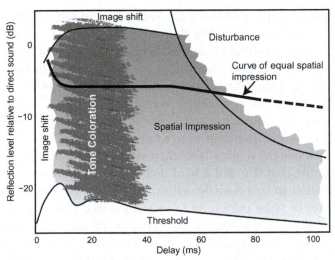

FIGURE 6.13 *Subjective effects of a single reflection arriving from 40° to the side, adjusted for different delays and sound levels. An important unseen effect is an increase in loudness, which occurs when the reflected sound is within what is colloquially called the "integration interval": about 30 ms for speech and 50 ms or more for music, all depending on the temporal structure of the sound. In this figure, the lowest curve is the hearing threshold. Above this, at short delays, listeners reported various forms of image shift in the direction of the reflection. At all delays larger than about 10 ms, listeners reported "spatial impression" wherein "the source appeared to broaden, the music beginning to gain body and fullness. One had the impression of being in a three-dimensional space" (Barron, 1971, p. 483). Spatial impression increased with increasing reflection level, a fact illustrated in the figure by the increased shading density. The "curve of equal spatial impression" shows that at short delays, levels must be higher to produce the same perceived effect. At high levels and long delays, disturbing echoes were heard (upper right quadrant). At intermediate delays and at all levels, some degree of tone coloration was heard (darkened brushstrokes). The areas identified as exhibiting "image shift" refer to impressions that the principal image has been shifted toward the reflection image. At short delays, this would begin with summing localization—the stereo-image phenomenon in which the image moves to the leading loudspeaker. At longer delays, the image would likely be perceived to be larger and less spatially clear. Finally, at longer delays and higher sound levels, a second image at the location of the reflection would be expected to add to the spatial illusion. From this presentation it is not clear where these divisions occur. From Barron, 1971, Figure 5, redrawn.*

jective effects for a single lateral reflection simulated in an anechoic chamber using a "direct sound" loudspeaker at 0° (forward) and a "reflected sound" loudspeaker at 40° to the left, both at 3 m distance. For different electronically introduced delays, listeners adjusted the sound level of the "reflection," reporting what they heard while listening to an excerpt from an anechoic-chamber recording of Mozart's *Jupiter* symphony. They heard several identifiable effects, as shown in the figure and described in the caption. There is more to this matter, but this important paper provides a good summary of research up to that point and some new data contributed by Barron.

There is a lot of information in this diagram, but most of it is familiar from the discussions of perceptions in experiments using speech. In the Barron paper, much emphasis is placed on spatial impression because of the direct parallel with concert hall experiences. These days, discussions of spatial impression would be separated into listener envelopment (LEV) and apparent source width (ASW). The discussions here appear to relate primarily to ASW, but the quote in the caption includes the remark "the impression of being in a three-dimensional space," indicating that it is not a hard division. In any event, Barron considers spatial impression to be a desirable effect, as opposed to "tone coloration."

On the topic of "tone coloration," it was suggested that a contributing factor may be comb filtering, the interference between the direct and reflected sound, but Barron further noted that this is mostly a "monaural effect" because "the effect becomes less noticeable as the direct sound and reflection sources are separated laterally." The "tone coloration . . . will frequently be masked in a complex reflection sequence," meaning that in rooms with multiple reflecting surfaces, tone coloration is not a concern. More recent evidence supports this opinion.

We will discuss the matter of timbre changes later, and we will see that tone coloration can be either positive or negative, depending on how one asks the question in an experiment. Again, we will go back to the quote in the caption that with the addition of a reflection, "the music [begins] to gain body and fullness," which can readily be interpreted to be tonal coloration but of a possibly desirable form.

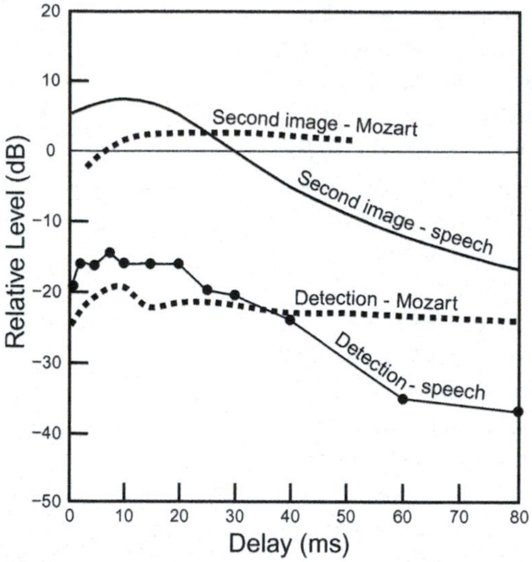

6.4.1 Threshold Curve Shapes for Different Sounds

It is useful to go back now and compare the shapes of the threshold contours determined by Barron for music with those shown earlier for speech, both in anechoic listening conditions. Figure 6.14 shows such a comparison, and it is seen that curves obtained using the anechoically recorded Mozart excerpt are much flatter than those for speech.

These data suggest two things. First, it appears that the slower paced, longer notes in the music cause the threshold curves to be flatter than they are for the more compact syllables in speech. This "prolongation" appears to be similar in perceptual effect to that occurring due to reflections in the listening environment (Figure 6.8).

FIGURE 6.14 *Data from Figure 6.6a showing thresholds obtained using speech and data from Figure 6.13 showing thresholds obtained using Mozart. The upper curve for music was described as that at which the "apparent source moved from direct sound loudspeaker toward reflection loudspeaker." This could be interpreted as being equivalent to the Olive and Toole "image shift" threshold, but the pattern of the data in the comparison suggests that it is more likely equivalent to the "second image" criterion.*

Second, it appears that the slope of the absolute threshold curve is similar to that of the second-image curve, something that was foreshadowed in Figure 6.11.

Figure 6.15 shows detection thresholds for sounds chosen to exemplify different degrees of "continuity," starting with continuous pink noise and moving through Mozart, speech, castanets with reverberation, and "anechoic" clicks (brief electronically generated pulses sent to the loudspeakers). The result is that increasing "continuity" produces the kind of progressive flattening seen in Figures 6.8 and 6.9. The perceptual effect is similar if the "continuity" or "prolongation" is due to variations in the structure of the signal itself or due to reflective repetitions added in the listening environment. In any event, pink noise generated an almost horizontal flat line, Mozart was only slightly different over the 80 ms delay range examined, speech produced a moderate tilt, castanets (clicks) with some recorded reverberation were even more tilted, and isolated clicks generated a very compact, steeply tilted threshold curve.

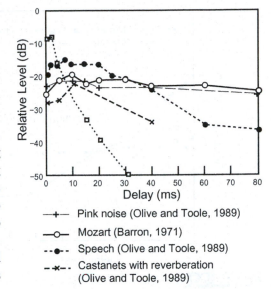

—+— Pink noise (Olive and Toole, 1989)

—o— Mozart (Barron, 1971)

--●-- Speech (Olive and Toole, 1989)

--×-- Castanets with reverberation (Olive and Toole, 1989)

···□··· Clicks (Olive and Toole, 1989)

FIGURE 6.15 *Detection thresholds for a single lateral reflection, determined in an anechoic chamber for several sounds exhibiting different degrees of "continuity" or temporal extension.*

Assuming that the patterns seen in previous data for speech and Mozart apply to other sounds as well, Figure 6.16 shows a compilation and extension of data portraying detection thresholds and second-image thresholds for Mozart, speech, and clicks. To achieve this, the second-image curve for clicks had to be "created" by elevating the click threshold curve by an amount similar to the separation of the speech and music curves. Absolute proof of this must await more experiments, but it is interesting to go out on a (strong) limb and speculate.

Looking at the 0 dB relative level line—where the direct and reflected sounds are identical in level—it can be seen that the precedence-effect interval for clicks appears to be just under 10 ms. According to Litovsky et al. (1999), this is consistent with other determinations (<10 ms), and the approximately 30 ms for speech is also in the right range (<50 ms). They offer no fusion interval data for Mozart, but it is reasonable to speculate based on the Barron data that it might be substantially longer than 50 ms. The short fusion interval for clicks suggests that sounds like close-miked percussion instruments might, in an acoustically dead room, elicit second images.

6.5 SINGLE VERSUS MULTIPLE REFLECTIONS

So far, we have looked at some audible effects of single reflections when they appear in anechoic isolation and when they appear in the presence of room

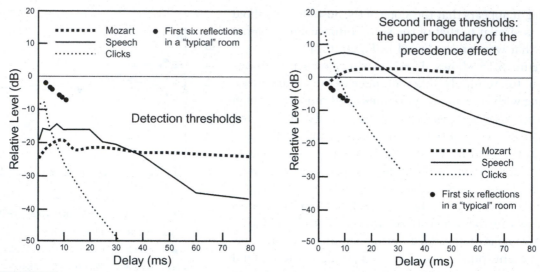

FIGURE 6.16 *Using data from Figures 6.16 and 6.17, this is a comparative estimate of the detection thresholds and the second image thresholds (i.e., the boundary of the precedence effect) for clicks, speech, and Mozart. The "typical room reflections" suggest that in the absence of any other reflections, the clicks are approaching the point of being detected as a second image. However, normal room reflections would be expected to prevent this from happening because the threshold curve would be flattened (see Figures 6.8 and 6.9).*

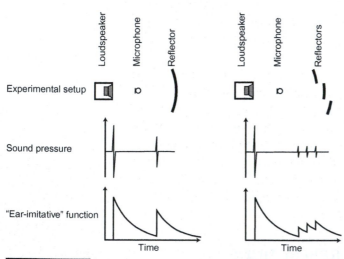

FIGURE 6.17 *A comparison of a single large reflection with a sequence of three lower-level reflections. From Cremer and Müller, 1982, Figure 1.16.*

reflections. Now we will look at some evidence of how a sequence of reflections is perceived.

Cremer and Müller (1982) provide a limited but interesting perspective. Figure 6.17 shows a microphone picking up the direct sound from a loudspeaker and either a single large or three smaller reflections in rapid sequence. The middle layer of images displays sound pressure, showing the direct sound followed by the reflections. The bottom layer of images portrays what Cremer and Müller call an "ear-imitative" function, which is a simple attempt to show that the ear has a short memory that fades with time—a relaxation time. The point of this illustration is that events occurring within short intervals of each other can accumulate "effect," whatever that may be. The sequence of three smaller reflections can be seen to cause the "ear-imitative" function to progressively grow, although not to the same level as that for the single reflection.

However, when the authors conducted subjective tests in an anechoic chamber, they found that the sequence of three low-level reflections and the large single reflection were "almost equally loud." The message here is that if we believed the impulse response measurements, we might have concluded that by breaking up the large reflecting surface, we had reduced the audible effects. This is one of the persistent problems of psychoacoustics. Human perception is usually nonlinear, and technical measurements are remarkably linear.

Angus (1997, 1999) compared large, single lateral reflections from a side wall with diffuse—multiple small—reflections from the same surface covered with scattering elements. There were no subjective tests, but mathematical simulations showed some counterintuitive results—namely that although the amplitudes of individual reflections were attenuated (as seen in an ETC), the variations in frequency responses measured at the listening position were not necessarily reduced. If the Cremer and Müller perceptual-summation effect is incorporated, the multiple smaller reflections seen in the ETC may end up being perceived as louder than anticipated. It is suggested, however, that a diffuse reflecting surface may make listening position less critical.

So there are both subjective and objective perspectives indicating that breaking up reflective surfaces may not yield results that align with our intuitions. It is another of those topics worthy of more investigation.

6.6 MEASURING REFLECTIONS

It seems obvious to look at reflections in the time domain, in a "reflectogram" or impulse response, a simple oscilloscope-like display of events as a function of time or, the currently popular alternative, the ETC (energy-time curve). In such displays, the strength of the reflection would be represented by the height of the spike. However, the height of a spike is affected by the frequency content of the reflection, and time-domain displays are "blind" to spectrum. The measurement has no information about the frequency content of the sound it represents. Only if the spectra of the sounds represented by two spikes are identical can they legitimately be compared.

Let us take an example. In a very common room acoustic situation, suppose a time-domain measurement reveals a reflection that it is believed needs attenuation. Following a common procedure, a large panel of fiberglass is placed at the reflection point. It is respectably thick—2 in. (50 mm)—so it attenuates sounds above about 500 Hz. A new measurement is made, and—behold!—the spike has gone down. Success, right? Maybe not.

In a controlled situation, Olive and Toole (1989) performed a test intended to show how different measurements portrayed reflections that, subjectively, were adjusted to be at the threshold of detection. So from the listener's perspec-

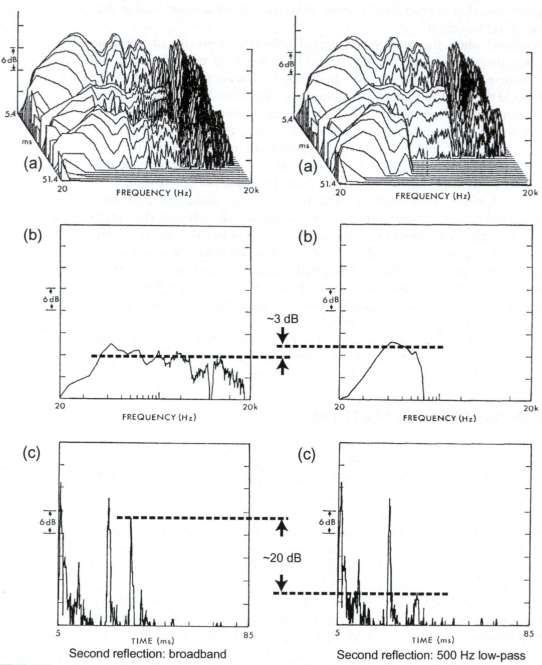

The left column of data shows results when the second of a series of reflections was adjusted to the threshold of detection when it was broadband; the right column shows comparable data when the reflection was low-pass filtered at 500 Hz. (a) Shows the waterfall diagram, (b) the spectrum of the second reflection taken from the waterfall, and (c) the ETC measured with a Techron 12 in its default condition (Hamming windowing). The signal was speech. The horizontal dotted lines are "eyeball" estimates of reflection levels. From Olive and Toole, 1989, Figures 18 and 19.

tive, the two reflections that are about to be discussed are the same: just at the point of audibility or inaudibility.

The results are shown in Figure 6.18. At the top, the (a) graphs are waterfall diagrams displaying events in three dimensions. At the rear is the direct sound, the next event in time is an intermediate reflection, and at the front is the second reflection, the one that we are interested in. It can be seen that the second reflection is broadband in the left-hand diagram and that frequencies above 500 Hz have been eliminated in the right-hand version. When that particular "layer" of the waterfall is isolated, as in the (b) displays, the differences in frequency content are obvious. The amplitudes are rather similar, although the low-pass filtered version is a little higher, which seems to make sense considering that slightly over 5 octaves of the audible spectrum have been removed from the signal. Recall that these signals have been adjusted to produce the same subjective effect—a threshold detection—and it would be logical for a reduced-bandwidth signal to be higher in level.

In contrast, the (c) displays, showing the ETC measurements, were telling us that there might be a difference of about 20 dB in the opposite direction; the narrow-band sound is shown to be lower in level. Obviously, this particular form of the measurement is not a good correlate with the audible effect in this test.

The message is that we need to know the spectrum level of reflections to be able to gauge their relative audible effects. This can be done using time-domain representations, like ETC or impulse responses, but it must be done using a method that equates the spectra in all of the spikes in the display, such as bandpass filtering. Examining the "slices" of a waterfall would also be to the point, as would performing FFTs on individual reflections isolated by time windowing of an impulse response. Such processes need to be done with care because of the trade-off between time and frequency resolution, as explained in Section 13.5. It is quite possible to generate meaningless data.

All of this is especially relevant in room acoustics because acoustical materials, absorbers, and diffusers routinely modify the spectra of reflected sounds. Whenever the direct and reflected sounds have different spectra, simple broadband ETCs or impulse responses are not trustworthy indicators of audible effects.

CHAPTER 7

Impressions of Space

In the results from Barron (1971) that were shown in Figure 6.13, this description was given of what a listener perceived when a single reflection was added to a direct sound: "The source appeared to broaden, the music beginning to gain body and fullness. One had the impression of being in a three-dimensional space." One can imagine that more reflections from different directions would intensify the effects. Barron bundled the perceptions under one name—spatial impression—but it obviously embraced multiple audible effects related to the following:

- Image broadening—"the source appeared to broaden"

- Timbral enrichment—"body and fullness"

- Spaciousness and envelopment—"the impression of being in a three-dimensional space." The experiments were done in an anechoic chamber, and the single reflection gave the listener a sense of being in a reflective room.

If a microphone had been placed at the listening location and a frequency-response measurement had been made, it would have shown the familiar pattern of a comb filter, something that popular audio culture has conditioned us to consider a problem. This is a superb example of Benade's paradox (see Figure 5.1). What appears to be a simple technical flaw turns out to be perceptually complex and beneficial.

An impression of space is a quality that most listeners, of whatever background, could recognize, although they may not be able to describe what they heard with the detail of Barron's sophisticated listener. Decades of thought and scientific investigation have still not created an unambiguous definition because multiple, somewhat separate but interactive, perceptions come into play, some of which correlate with the same physical measures of the sound field. Of course, the wild card in this scenario is the program material that, in the venue of greatest interest—concert halls—is music in all of its compositional and instrumental sound variations. This means that the same acoustical space, with the same measured parameters, can exhibit different perceptions for musical sounds of different frequencies and different temporal structures and pacing.

Impressions of space are the paramount audible factors that distinguish good spaces for live performances. They contribute much of the interest and identity to all the large, reverberant spaces we encounter whether they are office building foyers, gymnasiums, museums, cathedrals, or caves. A natural compulsion upon entering such a space is to clap hands or shout just to hear the enveloping barrage of reflected sounds and the gentle decay of the reverberation. All spaces, large and small, and the associated perceptions within them constitute a valuable personal library of acoustical mementos. Movies were the first to utilize multiple channels and loudspeakers to deliver these effects to audiences, adding audible support and excitement to the visual images. It is not necessary to replicate the sound field of a real space in a listening room; it is sufficient only to provide key cues to elicit a recollection or an emotion.

With good two-channel stereo recordings, one can get impressions of these types. With multichannel audio, such illusions can be delivered in any amount—including an excess amount. Understanding what causes these spatial illusions allows us to better tailor recorded sounds, to design more effective multichannel audio record/playback systems, and to extract the maximum effect from existing systems.

Most of the past research has focused on understanding what happens in concert halls, in the hope that the perceived acoustical performance of these spaces might be more predictable from model studies and measurements. All that can be said at the moment is that patterns are emerging that seem to make sense, but there is work yet to be done.

There is general agreement that there are two separable components to the perception of what is broadly called *spaciousness:* apparent source width (ASW) and listener envelopment (LEV), as described in Section 4.1.3 and Figure 4.7. ASW has to do with the perceived horizontal spatial spread of the orchestra, which can be much greater than the physical spread of the instruments. ASW is therefore a phenomenon associated with the sound source and the extent to which it is perceived to be broadened. This can also happen in sound reproduction in small rooms, where, in some circumstances, a certain amount of loudspeaker-image broadening is a good thing. This is especially true for movies with prolonged passages of on-screen dialogue and action accompanied by a monophonic center-channel sound track. It is also true for music recordings in which multiple instruments, a section of an orchestra or band, are delivered to a single loudspeaker. Pinpoint, sharply localized, spatial illusions are incongruous in these situations, and a directionally "softer" image can be an improvement. It needs to be stated, as it will be again, that such instances have nothing to do with imperfect loudspeakers or rooms. These are attributable to the recording techniques, the simplicity of which often leaves much to be desired.

Griesinger (1989, 1997, 1998, 1999) has spent more time than most people analyzing spatial effects in concert halls and in their domestic counterpart: multichannel audio systems. His insights are thought provoking; the 1997 paper is a good summary. We will not get into the minutiae of concert hall acoustics and the perceptions they elicit, as interesting as they are. They are not irrelevant, because one hopes that it might be possible to capture the directional and spatial essences of such acoustic events and to subsequently reproduce them through multichannel audio systems in homes and cars. This topic is discussed in Griesinger (2001).

In the context of small rooms, Griesinger points out that because most of the reflected sound energy occurs in the first 50 ms, the most that this can achieve, perceptually, is what he calls ESI (early spatial impression). Such a spatial impression is predominantly frontal, "closely associated with the direct sound." Multiple reflections in this interval, and beyond, are not perceived as discrete events but rather as a "fully enveloping surround" accompanying a sharply defined source, although, as Griesinger (1997) explains, musical sound that has "slow note onsets—such as legato strings . . . produces considerable source broadening (ASW). However the spatial impression remains frontal."

Figure 7.1 shows the approximate frequency ranges over which perceptions of envelopment, image shift, and broadening may occur. In addition to these frequency divisions, there is also time. Image shift and broadening effects are influenced by reflections that arrive within approximately the first 80 ms, as are some "early spatial impressions" restricted to the frontal hemisphere, whereas true envelopment tends to be created by later arrivals—those 80 ms and beyond. Obviously, the longer delays are far beyond those that can be generated by strong individual reflections within small listening rooms. Envelopment therefore requires multiple loudspeakers delivering recorded sounds containing the appropriately delayed sounds from the appropriate directions. It is possible that reflections within the listening room may assist in impressions of envelopment by adding repetitions, but they must be initiated by recorded sounds having the large initial delays. Therefore, what happens with sound

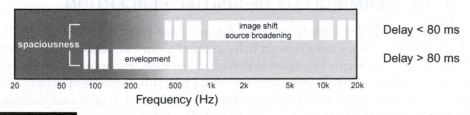

FIGURE 7.1 *The approximate frequency ranges and delay ranges over which reflected sounds contribute to the perceptions of different spatial effects.*

reproducing systems in small listening rooms must be carefully separated into two domains:

1. The interaction of a single loudspeaker and the listening room. Much of what we hear in movies, on TV, and in music is monophonic: an isolated microphone pickup delivered to a single loudspeaker or a more complex signal downmixed and panned to a single channel for delivery. It has been estimated that about 80% of a movie is dialogue (Allen, 2006), and experience tells us that virtually all of that emerges from the front center loudspeaker along with many other on-screen sounds.

2. The interaction of multiple loudspeakers and the listening room when those loudspeakers are reproducing a multichannel recording. Surround sound is more than being "surrounded by" sounds.
   a. The left-right, front-back sound effects—gunshots, ricochets, assorted thumps and bangs—used in movies to put us in the middle of an action sequence are easy. For these effects, loudspeakers must deliver a strong, direct sound for localization; from that point onward, the precedence effect takes over.
   b. More difficult and perhaps even more important is the sense of envelopment—of being there—that accompanies scenes that take place in large and small spaces, like caves, corridors, gymnasiums, and so forth. It also is part of the atmospheric musical accompaniment that often appears to come from everywhere and anywhere. It is directional ambiguity of a special kind. For this to happen, the right kinds of sounds must arrive at the ears from the right directions, in the right quantities.

It is obvious that item 1 is where one must begin with investigations of this complicated phenomenon. It is equally obvious that item 2 describes where we are going and what we need to know for thoroughly entertaining sound reproduction. It is to be expected that a listener's perceptions from the sound source(s) interacting with the room will be different in the two circumstances. In the end we need to understand both.

7.1 THE TERMINOLOGY OF SPATIAL PERCEPTION

The literature in this topic is unfortunately not consistent in the words used to describe perceptions within concert halls in other rooms. Gradually, researchers have come to use terms in a more standardized fashion, but some investigators, such as Griesinger (1997), have been inventing new subcategories of perceptions that seem to better describe the interactions of physical spaces and different kinds of musical sounds. As just mentioned, these details are better left out of the present discussions, but it is still necessary to establish the meanings of a few basic terms. The following is the set of meanings that the author will attempt to adhere to in the remainder of this book.

ASW = image broadening: The perception that a sound source is wider than the physical extent of the source. This can be a broadening of the sound from a single loudspeaker or the broadening of a sound stage created by a stereo pair, or by a multichannel trio of left, center, and right front loudspeakers. This is almost exclusively an effect within the frontal sound field, and it may include what Griesinger calls "early spatial impressions," in which a moderate sense of space becomes associated with the front loudspeakers.

Spaciousness = envelopment: The perception of being surrounded by a large and enveloping space. This is a directionally ambiguous spatial impression, although impressions of localizable sound sources may coexist within the illusion. An orchestra within a concert hall is probably the best real-world example, although simulations allow many other scenarios to be created for multichannel reproduction.

Diffusion: This is a property of the physical sound field; it is *not* a perception, nor is it a property of loudspeakers (that is dispersion). Perfect diffusion describes a situation in which sounds arrive at a point from all directions with equal probability. There is widespread confusion beginning with the belief that a diffuse sound field is necessary to perceive spaciousness or envelopment. This is not so. One will perceive spaciousness and envelopment within a diffuse sound field, but a diffuse sound field is not necessary to perceive spaciousness and envelopment. The only requirement is that the appropriate sounds are delivered to the listeners' ears. This can be satisfactorily achieved with a much simpler sound field—if it is of the right kind. This simplification is the essence of good–practical–multichannel reproduction.

7.2 LISTENERS AND THEIR "PREFERENCE" FOR REFLECTIONS

It is accepted that a reflective sound field is flattering to the sound of music. We like to listen in reflective spaces, not outdoors. The same is true for speech, and even in casual conversation, an excessively "dead" environment can be tiresome. The elaborate portrayal of detection and other perceptual thresholds, discussed in Chapter 6, lays an important foundation for understanding how we perceive reflections. It may seem logical to use these data, such as the curves in Figure 6.6a, as the basis for setting requirements for allowable reflection levels in listening rooms. However, hearing a change tells us nothing about whether the change is good, bad, or neutral. What happens when listeners are allowed to choose the level of a single reflection, based on what they perceive as a sense of pleasantness—a preference? Ando has provided some answers.

Figure 7.2 shows levels for a single delayed sound, from a horizontal angle of 36°, that listeners reported as enhancing the sound of speech. Since the early

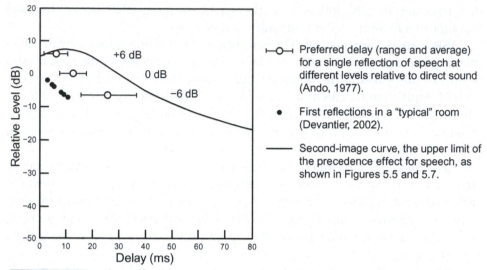

FIGURE 7.2 *With speech as the signal, the delays at which listeners expressed a preference for the perceived sound experience when a single reflection was added from a horizontal angle of 36°. Results are shown for three sound levels relative to the direct sound. Adapted from Ando, 1977, Figure 7.*

reflections in typical real rooms are so low in level, the result suggests that multichannel audio is needed to provide added, stronger and later, reflections for our listening pleasure (Ando, 1977; Ando, 1998). Note that the preferred levels just avoid the "second image" curve, indicating that the preferred reflections were all within the precedence effect fusion zone, thereby not generating distracting second images. The inevitable conclusion is that in listening to live speech, and for a single loudspeaker reproducing speech, individual room reflections are not problems. In fact, they are not loud enough. However, it is also evident that listeners do not wish to hear any secondary images; one person speaking is enough.

Figure 7.3 shows similar experimental results when the test sound was music. This figure needs to be examined in parts. First, three horizontal bars at +6 dB, 0 dB, and −6 dB show the delays at which listeners preferred single lateral reflections added to anechoic-recordings of classical music at three different sound levels (Ando, 1977, 1998). Second, intermingled with these data are thin dashed lines indicating the levels at which different percentages of listeners preferred the addition of the single lateral reflection, again with classical music (Ando 1985). It is clear that high-level long-delayed reflections are not as desirable as those with shorter delays and lower levels. However, all preferred levels are far above the natural reflections provided by small rooms, indicated by the six dots—they are simply not consequential factors in this matter.

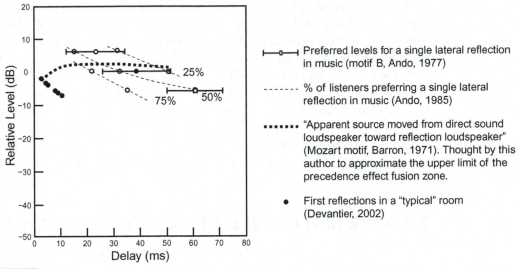

FIGURE 7.3 *A comparison of preferred levels for single lateral reflections when listening to music. Early reflections in a typical listening room are shown, as is the estimated upper limit of the precedence effect.*

Finally, the heavy dotted curve is what is suspected to be close to the upper limit of the precedence effect for Mozart, as provided by Barron (1971), although he did not ask that question specifically (see Figure 6.14). Obviously, listeners were willing to add reflections having sound levels and delays that would cause strong image shift and broadening, and perhaps second images, and still indicate a preference for listening to music in that state. In contrast, Figure 7.2 indicates that this was not the case when listening to speech. Why? Possibly a second image of a voice is unwelcome, because it is so extremely unnatural. However, when listening to a group of musicians, especially classical musicians as in these examples, it may not be considered unwelcome if the violin, cello, or bass sections get slightly expanded. The very different time-domain structures of speech and the chosen classical music examples are other factors. Different kinds of classical music or the many kinds of popular music might therefore yield different results.

The high levels required for reflections before they contribute substantively to positive listener reactions were not totally surprising. In Section 3.1, it was mentioned that spatial impression in concert halls is a *forte* phenomenon, being much attenuated in *piano* passages. Loud music "fills" the hall perceptually as well as physically, as more of the later, lower-amplitude reflections are audible. The importance of this cannot be overemphasized in evaluations of multichannel audio systems. Playback sound levels must be closely monitored if the comparisons are to have meaning, especially the relative levels of the front and surround channels.

In his pioneering investigations of reverberation about 100 years ago, Sabine experimented with different amounts of absorption in rooms of different sizes, asking listeners for opinions about how a live piano sounded. In rooms of typical living-room, home-theater sizes (74 to 210 m³) the most favored conditions were rooms that had reverberation times in the range 1 to 1.1 s (reported in Cremer and Müller, 1982, p. 528). These had to be sparsely furnished rooms to achieve such high RTs, meaning that reproducing a close-miked recording of a piano in a normally furnished domestic room (RT 0.3 to 0.5 s) would be less than fully rewarding. We need to have additional "reflected sounds" in the recordings and, ideally, reproduced from multiple loudspeakers in the optimum locations.

7.3 SOME REFLECTIONS ARE BETTER THAN OTHERS

Many reflections have a positive contribution to listener preferences, but some reflections are more desirable than others, depending on timing and direction. The deciding factor is which of them are most effective at generating a sense of spaciousness that, in turn, follows from low interaural cross-correlation (IACC). The greater the proportion of sound arriving from the sides, the greater the differences in sounds at the two ears, and the lower the IACC. Start to think in terms of "preference," "spaciousness," "low interaural cross-correlation (IACC)," and "lateral reflections" as positively correlated with each other. IACC is a statistical measure, and although it is a correlate of the desirable perception of spaciousness and envelopment, it needs to be remembered that correlation does not explain causality. Griesinger (1997), for example, is inclined to talk about fluctuations in interaural intensity differences (IID) and interaural time differences (ITD) as a better way of explaining certain spatial perceptions. These quantities are responsible for our ability to localize sounds in space, and when they are randomly varying, fluctuating, we are logically unable to do that. Instead, we perceive something directionally and spatially ambiguous. Randomly fluctuating IID and ITD at the two ears will result in a reduced IACC when this is measured. One metric is more closely allied to the binaural perceptual mechanism, and the other is a statistical measure of similarity, but both, it seems, are useful measures of the perception of interest.

Figure 7.4 is a compilation of data from Ando (1977) and Barron and Marshall (1981), in which a little showmanship has been employed to illustrate how certain factors appear to be related to each other. The persuasion employs no statistical calculations, only visual pattern recognition. These factors are involved:

- Preference (a subjective judgment)

- IACC (a technical measure). This is a measure of the similarity of sounds arriving at the two ears: 0 = different; 1 = identical.

■ Spaciousness or spatial impression (a subjective judgment)

■ Lateral-to-frontal energy ratio (a technical measure). This compares the proportion of sound arriving at a listener along the side-to-side axis to that which arrives along the front-back axis.

The caption explains Figure 7.4, which, from visual inspection alone, displays a strong relationship among the measured, calculated, and subjectively judged quantities just listed. For maximum "preference" from the Ando (1977) data, it seems that reflections from about 30° to 90° are most effective. When IACC is measured, a broad minimum is seen around 60°, corresponding to a maximum in the preference ratings. Preference, therefore, is associated with low interaural cross-correlation.

In Figure 7.4b, Barron and Marshall (1981) evaluated "spatial impression," concluding that it is strongly related to the proportion of sound arriving from the side, compared to that arriving from the front. This led to a "lateral fraction" method of examining sound fields. Since sound from the sides generates low IACC, the relation with the Ando results is established. Their results showed strong front-back symmetry; reflections from the front and rear were about equally effective at generating spatial impression (although not close to the medial plane). Those that arrived from the rear "were not perceived as coming from behind, but just produced a pleasant sense of envelopment" (p. 218). This symmetry is apparent in the subjective data and, of course, in the sine-relationship calculated prediction. There was no indication that 60° was in any way special. Why? Perhaps a judgment of "preference" is in some way different from one of "spatial impression." A second, more plausible explanation has to do with the spectrum of the test sounds, since the IACC versus angle relationship is somewhat frequency-dependent (Hidaka et al., 1997).

Barron and Marshall (1981) described spatial impression as "the sensation of . . . feeling inside the music" in contrast with "looking at it, as through a window" (p. 214). This is precisely the problem that beset two-channel stereo, leading Toole (1985) to include it as one of the factors to be used by listeners when interrogating spatial aspects of stereo reproduction. A continuum called listening "perspective" was created at one end of which was a "you are there" (at the performance) complete with enveloping ambient sound. This transitioned through "close, but still looking on," to an "outside looking in" (through an opening between the loudspeakers) impression in which there is no sense of being within the ambient sound, although you can hear it. There was also a "they are here" category, describing close-miked recordings without ambience, in which the listening space provided the acoustical setting (p. 31). Obviously, recording technique figured prominently in the results of these tests, but, in the end, there was a strong correlation between overall ratings of sound quality and overall ratings of spatial quality. "Preference," therefore, seems to incorporate elements of both sound and spatial quality, and it may be challenging for listeners to consciously separate the two.

The most important message from Figure 7.4 is that "preference" is associated with a strong "spatial impression." Technically, it seems to be possible to find correlation with a measure of the sounds arriving at the ears (a low IACC) and also a measure of the physical sound field in which the listener is immersed (a high proportion of lateral vs. frontal sound in the room).

So, summarizing these results, listeners showed a preference for sounds that had a strong sense of spatial extent. Such sounds were the result of reflections

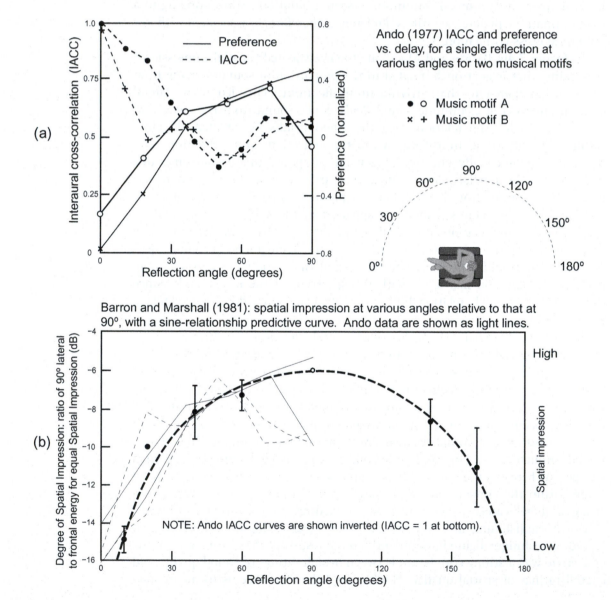

(a) Ando (1977) IACC and preference vs. delay, for a single reflection at various angles for two musical motifs

● ○ Music motif A
× + Music motif B

(b) Barron and Marshall (1981): spatial impression at various angles relative to that at 90°, with a sine-relationship predictive curve. Ando data are shown as light lines.

NOTE: Ando IACC curves are shown inverted (IACC = 1 at bottom).

that arrived from the sides that, when measured at the ears, resulted in a low IACC. Including the information from Figure 7.3, it can be added that complete listener gratification is likely to require reflections that are higher in level and later in time than those naturally occurring in small listening rooms. This is where multichannel sound reproduction systems enter the picture.

Bringing the discussion closer to the small room topic, where reflections within the room are most capable of influencing ASW/image broadening, there is the relevant data from Hidaka et al. (1997), who concluded that the apparent source width (ASW) was most strongly influenced by reflected sounds in the octave bands 500, 1000, and 2000 Hz (embracing frequencies from about 350 to 2800 Hz) occurring within the first 80 ms. They created the measure $IACC_{E3}$ to reflect this.

Figure 7.5 shows a form of this measure (1-$IACC_{E3}$) that in effect inverts the IACC vertical axis so it matches the "preference" and "spatial impression" subjective data. Also shown are the Ando data in the same form. In contrast to the Barron and Marshall data, the results are asymmetrical front and back, a fact explained on the basis of the lack of symmetry of the human head and the placement of the ears on it. Other than a vertical scaling difference (different sounds and measurement methods), the shapes of the curves reflect a similar trend. Because of this and the fact that the new data cover both front and back, the next step was taken.

In Figure 7.5b, the Hidaka et al. data from (a) were plotted in polar graphical form. This gives a more intuitive expression to what seems to be a reasonable measure of "potential ASW/image broadening" for reflected sounds arriving from

FIGURE 7.4 *(a) The data from Ando (1977) shows judgments of preference when a single reflection is presented from various horizontal angles for two musical motifs. In general, preference rises as the reflection angle increases away from the forward direction almost to the 90° limit of the experiment. Also shown on the same graph is interaural cross-correlation (IACC), a measure of the similarity of the sounds at the two ears. This is seen to almost perfectly mirror the shape of the preference curve, indicating that high preference is associated with low IACC. IACC exhibits a broad minimum around 60°. In (b), Barron and Marshall (1981) looked at judgments of spatial impression for pairs of lateral reflections arriving from various angles symmetrically left and right of the forward axis, shown by black dots with vertical error bars. They associated this perception with the proportion of lateral versus frontal energy arriving at the listener and calculated a predictive curve based on a sine relationship. This is shown by the dashed semicircle. As with the Ando results, there was a progressive rise in spatial impression with increasing angle up to 90°. The author has taken the liberty of superimposing the Ando curves on the Barron and Marshall data, inverting the IACC curve for clarity, and quite freely adjusting the vertical scaling for maximum effect. All of these data are on different vertical scales, one of them subjective and not all of them linear. The point of the comparison is simply to show that a comforting visual correlation exists among the data, even though they have different origins.*

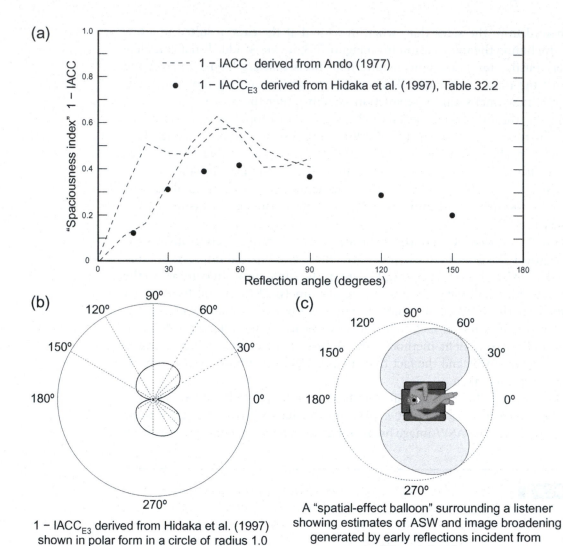

(a) The inverted IACC data from Ando in Figure 7.4a so that it relates to "preference" on the vertical scale. This is calculated as 1-IACC. This quantity is also the one shown for the Hidaka et al. (1997) measurements of $IACC_{E3}$ so it relates to the ASW (image-broadening) component of spaciousness. It is only important to observe that all curves reach a maximum around 60°; differences in vertical scaling are unimportant in this context. (b) The Hidaka et al. data reformatted as a polar plot and mirror-imaged to show left and right hemispheres. Shown in this manner, the relative importance of different reflection angles can be clearly seen. (c) The intention is made even more clear with the addition of the image of a listener. A circle of constant effect is shown, confirming that for a direct sound arriving from 0°, a reflection from about 60° is likely to generate the strongest impression of ASW/image-broadening, although those from many other angles make substantial contributions.

different directions. If the direct sound arrives from 0°, it is clear that reflections arriving from about 60° will have maximal effect, but there is a wide range of lateral angles over which reflected sounds will contribute to the illusion. In contrast, sounds that arrive from close to the forward and rear direction will have little effect.

All of this takes on greater meaning when, in Figure 7.5c, the shape is superimposed on the plan view image of a listener. Put this into a room plan, as we will, and it should be possible to see where reflections need to come from for maximal "preference," "spaciousness," "spatial impression," or, most precisely, "ASW/image broadening." Thinking ahead, those reflections can be the natural reflections of loudspeaker sounds in the listening room, reflections included in a multichannel recording and reproduced through one of several loudspeakers, or, as will inevitably be the case, a combination of the two.

It is not difficult to understand why this happens. First, because of where the ears are located on the sides of the head, differences in the sounds at them will be greatest for

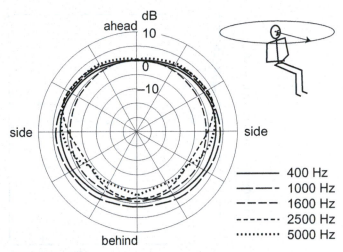

FIGURE 7.6 *Binaural equal loudness contours for sounds arriving from different horizontal angles (azimuth). Narrowband sounds from different directions were adjusted to have the same loudness as the same sound from the front. The graphic display shows relative loudness: the inverse of the amount by which the sound from other directions must be amplified or attenuated to have the same loudness as one from the front. This means that a positive level (longer radius, re 0 dB) indicates a perceived loudness increase (a lower sound level is required to achieve a loudness match to a frontal sound). Data at 1600 and 2500 Hz are from Robinson and Whittle (1960); those at 400, 1000, and 5000 Hz are from Sivonen and Ellermeier (2006).*

sounds that arrive from the sides and least for sounds that arrive on or close to the front-back axis. Second, the asymmetrical shapes of the external ears provides some acoustical gain for sounds, especially short-wavelength/higher-frequency sounds, that arrive from the forward hemisphere. This is inevitably linked to HRTFs, but it must also incorporate the manner in which the auditory processes combine the sounds at both ears. A measure of this is available in studies of directional loudness, in which listeners compare the loudness of sounds arriving from different directions. Robinson and Whittle (1960) and Sivonen and Ellermeier (2006) are good examples.

Figure 7.6 indicates that sounds arriving from the side-front, about 60° will appear to be louder than those from the front and louder than those arriving from the symmetrical directions in the rear hemisphere. One presumes that this relationship to the shape of the "spatial-effect balloon" in Figures 7.5b and 7.5c is not accidental.

Finally, we need to examine the effect of delay on IACC and on subjective impressions. Figure 7.7a shows that a lateral reflection arriving with a delay

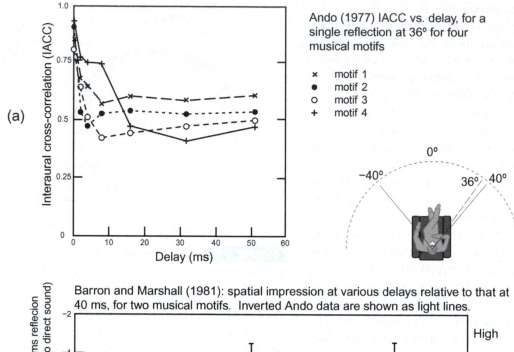

(a)

Ando (1977) IACC vs. delay, for a single reflection at 36° for four musical motifs

× motif 1
● motif 2
○ motif 3
+ motif 4

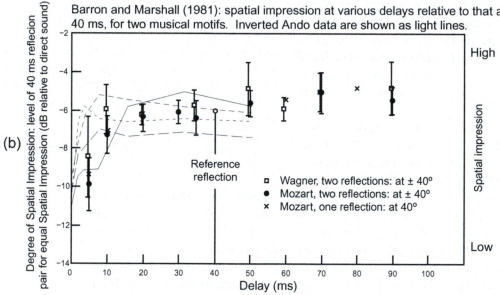

(b)

Barron and Marshall (1981): spatial impression at various delays relative to that at 40 ms, for two musical motifs. Inverted Ando data are shown as light lines.

□ Wagner, two reflections: at ± 40°
● Mozart, two reflections: at ± 40°
× Mozart, one reflection: at 40°

greater than about 2–3 ms is effective at reducing IACC and therefore will contribute to increased preference (Ando, 1977). With the exception of one of the musical motifs, reflections arriving later than about 4–10 ms have similar effects for each of the motifs. However, different motifs have different terminal values of IACC. It is not clear that this translates into different degrees of preference for those specific selections of music.

Figure 7.7b shows results from Barron and Marshall (1981) where listeners reported relative levels of spatial impression. This follows the trend in the Ando

FIGURE 7.7 *(a) Ando's (1977) measurement of interaural cross-correlation (IACC) as a function of reflection delay for four selections of music. A single reflection was presented from an angle of 36°. It shows that, in general, reflections arriving before 10–20 ms are more highly correlated with the direct sound than those arriving later. The nature of the sound (the motif) clearly has an effect. (b) Barron and Marshall's (1981) subjective determinations of spatial impression for one and two reflections, for two musical motifs. The results show that reflections occurring at delays less than about 10 ms produce less spatial impression than those at 40 ms (the reference). All others are roughly equal. This is the inverse of the "curve of equal spatial impression" shown in Figure 6.13, which indicated that to create the same spatial impression, reflections arriving in the first 10 ms or so would need to be higher in level than those at longer delays. That curve was developed from much less data than are shown in the present figure. The Ando data are superimposed as light lines, and the curves have been inverted for the comparison; elevated spatial impression is associated with reduced IACC. As in Figure 7.4, one can see a similar pattern in a comparison of the two sets of data, adding visual confirmation that IACC and spatial impression are objective and subjective correlates.*

data, in that listeners reported that reflected sounds gained rapidly in their ability to generate a sense of spatial impression up to about 10 ms, at which point the strength of the effect reached a long plateau. It is interesting that the effect of two symmetrical reflections was very similar to that of a single reflection, although, as with Ando, the musical motif made a difference, albeit only at short delays.

7.4 SUMMARIZING AND CHARTING THE WAY FORWARD

A persistent ambiguity exists in terminology relating to the perceptions of spatial effects. It is tempting to apply a single descriptor to it all, like "spaciousness" or "spatial impression," and just be done with it. That is what Barron did, but who knows what aspects of the complex spatial picture his individual listeners were attending to and what manner of scaling they applied to their judgments? It matters.

All that really can be said with certainty is that reflections that reduce the interaural cross-correlation increase something called "preference," and it is difficult to have negative thoughts about that. When pushed to describe what led to the preference, listeners described various kinds of spatial effects relating to the apparent lateral spread, a broadening, of the sound source or to the impression of being immersed in a large reflective space.

There were no complaints about "comb filtering," the universal justification for absorbing strong reflections, and nothing could be more starkly displayed than a single simulated reflection in an anechoic chamber. It is as though these listeners were in a topsy-turvy world, responding with praise to combinations of sounds that, in a stereo world, have long been thought of as flaws; this gives us all the more reason to press on and find out what is really happening.

The concepts of ASW—image broadening, early spatial impression, spaciousness, and envelopment—evolved within the concert hall context. The challenge is to transfer, or translate, these into the context of small rooms fitted with multichannel audio systems.

Recall the earlier classification of loudspeaker-room interactions:

1. The interaction of a single sound source, a loudspeaker, and the listening room.

2. The interaction of multiple loudspeakers and the listening room when those loudspeakers are reproducing a multichannel recording.

Category 1—the soundstage illusions apply most directly to the front left, center, and right loudspeakers, those responsible for delivering frontally localized dialogue and supporting sound effects for on-screen action in movies and for giving us the orchestra, the band, and the featured artist in music recordings. Considering only the acoustical interaction between these *front-located* loudspeakers and the adjacent room boundaries, it is reasonable to assume that full-scale envelopment is not possible with conventional stereo and multichannel mixes (the invocation of binaural signal processing is not permitted in this discussion). So in practical terms we are talking about perceptions of image size; ASW; image broadening; impressions of height, distance, and depth; and perhaps some early spatial impression in the frontal hemisphere. These are primarily associated with reflections occurring at delays less than about 80 ms and therefore include all of the early reflections that occur naturally in small rooms, as well as those in recordings.

Separately and together these factors contribute to what might constitute listener "preference." In the concert hall context, it seems that bigger is better, fuzzy beats sharp, but it may be found that in the context of small-room sound reproduction, listeners want something different, perhaps compensating for being deprived of a full-scale concert or cinema experience. These are things to be looking for.

Category 2—the surround illusions involve discretely localized sounds from the sides and rear: sound effects supporting a movie story or vocalists and instrumentalists comprising a "middle-of-the-band" style of music recording. More important, though, is the ability of the combination of front and surround loudspeakers to generate a sense of listener envelopment—LEV—that sense of being in a different, larger, space. This is arguably the single most distinguishing feature of surround-sound systems, not occasional bullet ricochets and helicopter flyovers, as entertaining as they are.

Illusions of localized sound effects and musicians involve strong direct sounds and the precedence effect. Small-room reflections have little to no effect on these phenomena. Illusions of envelopment require delays in excess of about 80 ms, which can only be delivered by multichannel recordings. Small-room reflections may usefully embellish the effect but cannot originate it. If so, those reflections, along with the direct sounds from the loudspeakers, should arrive from the sides, directions generally described by the "spaciousness balloon" described earlier. Other than involving long-delayed sounds, the prime difference from ASW/image-broadening effects is that lower frequencies are involved (see Figure 7.1).

CHAPTER 8

Imaging and Spatial Effects in Sound Reproduction

We now have the basic "tool kit" that will help us understand the factors at play in the directional and spatial perceptions of sound reproduction. We saw in Chapter 4 that sound fields in small rooms are not diffuse and that reverberation and critical distance are not useful metrics in any of the traditional senses. However, there is an active reflected sound field, although it is subservient to the precedence effect in terms of localization. However, localization is not perfect; there can be localization "blur," a region of uncertainty, the size of which depends strongly on direction (Blauert, 1996). In live performances, there is visual information to substantiate localization (the ventriloquism effect), and generations of audiences have voted in favor not of pinpoint localizations of musicians but of spatially embellished sound images, called apparent source width (ASW). Stating this again, we know where the sound is coming from, *and* we derive pleasure from having the auditory directional information corrupted! Think about that for a moment, and you may begin to anticipate some of the results of investigations of listener preferences in sound reproduction. As a clue, Dougharty (1973) reports that musicians feel that stereo places "too much emphasis on directional information, which is allowed to thrust itself forward and to demand too great a share of the listener's attention."

8.1 FIRST-ORDER REFLECTIONS

From the listener's viewpoint, the most energetic sound "ray" is the direct sound from the source. Next are the first-order reflections that, after passing by the listener, go on bouncing around the room to create second-, third- and higher-order reflections. All of these sounds eventually make their way back to the listening position as components of a disorderly sound field that, by delivering different sounds to each of the two ears, reduces the interaural cross-correlation

113

(IACC). Hence, there has always been an interest in measures of "diffusivity" as a correlate of both IACC and spaciousness.

As an example of the wisdom of the pioneers in room acoustics, Erwin Meyer (1954) quantified the diffusivity of the sound field in a rectangular enclosure. Using a scale model of a room fitted with an omnidirectional sound source, a directional microphone was rotated to evaluate the sounds arriving at a listening location from various directions. From this data he calculated a diffusivity index. 100% represented equal sound arriving from all directions. This is what he found:

1. Bare room with smooth walls: diffusivity = 69%

2. Adding scattering gratings to all walls: diffusivity = 75%

3. Bare room but floor totally absorbent: diffusivity = 46%

4. Same as preceding but with scattering gratings on all other walls: diffusivity = 64%

5. Bare room, using same absorbent as number 3, but divided into pieces to suppress the first reflections between the source and the microphone: diffusivity = 26%

Figure 8.1 illustrates situations 1, 3, and 5, leading to the conclusion that absorbing the first reflections has a powerful effect on the diffusivity, the IACC, and thereby the perceived spaciousness, of sound in a room.

It is important to note that the sound source used here was omnidirectional, not the horizontal omnidirectionality we accept in audio loudspeakers with that claim, but truly omnidirectional. If the sound source had the significant directivity of conventional forward-facing cone/dome or cone/horn loudspeakers, the diffusivity numbers would have been much lower.

Missing from this perspective are some important details of the diffuse sound field. As we saw in Chapter 4, the reflected sound field decays very quickly in small rooms. Figure 4.14 illustrates how it decays in amplitude and changes in directivity in a small room with a midfrequency reverberation time of 0.4 s. The first thing to note is that at no time is the sound field diffuse—that is, the pattern is not circular. In the first illustration (0 to 200 ms), the dominant features in the directional diffusivity pattern are the direct sound, the first-order side-wall and the rear-wall reflections. All of these are within about 7 dB of the direct sound. Once these early sounds pass, the next collection of reflections (t > 50 ms) in the example room are 12 to 17 dB down, with the stronger components being sounds reflecting back and forth between the side walls. After 100 ms, the lateral directional bias remains and levels have dropped to about −20 to −27 dB.

It would be very interesting to see this kind of measurement done in a manner that imitates the Meyer experiment so that the contribution of the

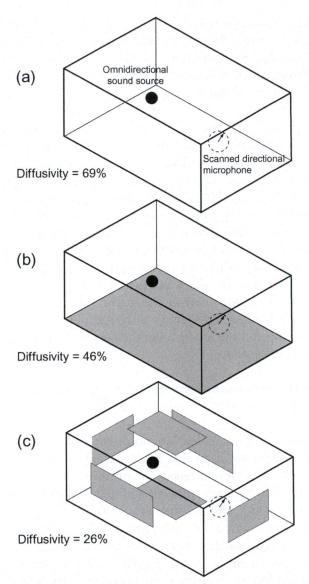

(a)

Omnidirectional
sound source

Diffusivity = 69%

Scanned directional
microphone

(b)

Diffusivity = 46%

(c)

Diffusivity = 26%

FIGURE 8.1 *Meyer (1954) scale model experiments in which the diffusivity of an omnidirectional sound source was measured using a rotating directional microphone with 10° as the half-energy value. The top figure shows an empty room. In the middle figure the floor has been covered by absorbing material. In the bottom figure the same absorbing material has been cut up and located so it would attenuate the first reflections between the source and the microphone. He claimed that the reverberation time was relatively unchanged in the two configurations of absorbing material, although the diffusivity changed dramatically. Unfortunately, Meyer published no drawings of the setups, so these are how this author imagined them to be from his verbal descriptions. The diffusivity number is a calculation from his measurements; 100% would describe a situation in which sounds arrived at the listening location equally from all directions—perfect diffusivity.*

first lateral reflections can be assessed directly. However, there is no doubt that the points of first reflection are the second-loudest "sources" of sound in the room and thereby contribute greatly to reflections occurring later in time. In a listening room, absorbing first reflections not only eliminates those specific components of sound but significantly alters all subsequent acoustical events.

Adding more to this perspective, Figure 8.2 shows a comparison of the IACC generated by a single lateral reflection in an otherwise anechoic space (Ando, 1977) and two measurements of IACC in a listening room, one with the side walls reflective and one with the side walls covered by absorbing material (Kishinaga et al., 1979). The Ando IACC values were calculated using music excerpts, whereas Kishinaga et al. used impulse responses; the values will not be exactly comparable because of this. When stereo listening tests were done in the two versions of the room, it was found that the condition with absorbing side walls was preferred for monitoring of the recording process and examining audio products, whereas reflective side walls (which reduced IACC) were preferred when listeners were simply "enjoying the music." As might be expected, reflective side walls resulted in a "broadening of the sound image." Adding absorption to the front wall, behind the loudspeakers, reportedly improved image localization and reduced coloration.

Memo for Listening room recommendations: add sound absorbing material to front wall.

Figure 8.3 offers a possible explanation, using the "spatial-effect balloon" from Figure 7.5. It shows in (b) that reflective side walls provide more lateral reflections and, thereby, several good opportunities for frontal spatial effects. Differences in the direct sounds from left and right channels would also generate these effects. A low IACC of 0.26 suggests a good listening situation for music and for any film or TV sound track with music or ambiance reproduced through the front left and right loudspeakers, as is very commonly done. If ambient sounds from the stereo music are delivered by surround channels, it is probable that they would dilute or possibly even overwhelm the lower-energy natural reflections in the room.

Figure 8.3c shows that attenuating the side-wall reflections reduces the prime cause of decorrelation. IACC rises to 0.44. The back wall (behind the listener) reflections have reduced decorrelation effect because they arrive close to the medial axis—almost mono. Not shown in (c) are the multitudes of second- and higher-order reflections, most of which would arrive at the listener from relatively unproductive incident angles, and, because of their long propagation paths, these reflections will be much reduced in sound level. Any absorbing material on the room walls would further diminish the levels.

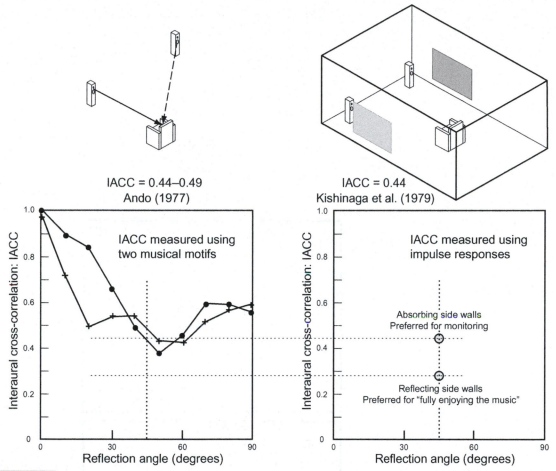

FIGURE 8.2   *A comparison of IACC measurements in two very different situations. First, Ando (1977) measured IACC from a single lateral reflection simulated in an anechoic chamber (this is taken from Figure 7.4, which also shows that the low values of IACC correspond to a maximum of "preference" when listening to music). Kishinaga et al. (1979) made measurements using a pair of stereo loudspeakers in a small listening room. IACC was measured with the side walls reflective and with absorbing material (apparently effective above about 500 Hz) located in the region of the first sidewall reflections. The dotted lines running between the boxes show that all the reflections remaining in the listening room after the side-wall reflections were absorbed yield an IACC that is comparable with that created by a single, strong lateral reflection from that same angle. The listening was done using both loudspeakers, and results indicated that for recreational purposes listeners preferred a more reflective sound field.*

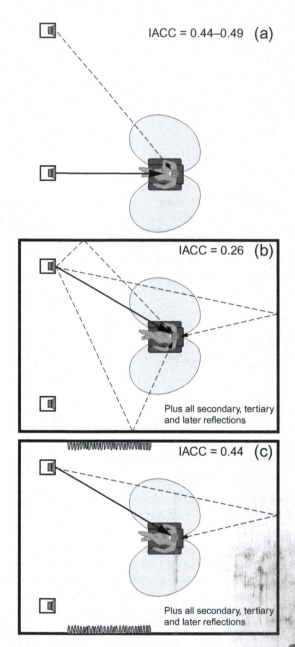

IACC = 0.44–0.49 (a)

IACC = 0.26 (b)

Plus all secondary, tertiary and later reflections

IACC = 0.44 (c)

Plus all secondary, tertiary and later reflections

Even though the data are not exactly comparable, it is tempting to speculate that the direct sounds from the stereo loudspeakers and the reflections remaining in a room after the first lateral reflections are removed appear to have about the same potential to generate ASW/image-broadening as a single, well-aimed, lateral reflection, as shown in Figure 8.3c. The huge reduction in diffusivity (seen in Figure 8.1 when the early reflections) adds credibility to this notion.

FIGURE 8.3   *A schematic description of what is happening in Figure 8.2 (not to scale). (a) An anechoic simulation of a direct sound from 0° accompanied by a side-wall reflection arriving from about 50°. The combination yields a moderate IACC (0.44 and 0.49 for each of two different musical motifs) and, it can be assumed, moderate spatial effects. (b) Measurements using one of a pair of stereo loudspeakers in a room with reflecting side walls yield a lower IACC (0.28), attributable to the multiplicity of lateral reflections arriving at the listening position. This would be expected to generate good impressions of ASW/ image broadening. (c) Placing absorption on the side walls attenuates the first-order lateral reflections, leaving the relatively ineffective rear-wall reflections (arriving from similar, mirror imaged, angles), yielding an IACC of 0.44. The IACCs in (b) and (c) were both computed from impulse responses, and they may be compared directly with each other, but there is some uncertainly about the comparison of either of them with (a).*

SENSITIVE LISTENERS?

Why do recording and mixing engineers prefer to listen with reduced lateral reflections (higher IACC)? Perhaps they need to hear things that recreational listeners don't. This is a popular explanation, and it sounds reasonable, but experiments reported in Section 6.2 indicate that we humans have a remarkable ability to hear what is in a recording in spite of room reflections—lots of them. But there is an alternative explanation, based on the observation that some listeners can become sensitized to these sounds and hear them in an exaggerated form. Ando et al. (2000) found that musicians judge reflections to be about *seven times greater* than ordinary listeners, meaning that they derive a satisfying amount of spaciousness from reflections at a much lower sound level than ordinary folk: "Musicians prefer weaker amplitudes than listeners do." It is logical to think that this might apply to recording professionals as well, perhaps even more so, because they create artificial reflections electronically and manipulate them at will while listening to the effects. There can be no better opportunity for training and/or adaptation. In fact, it is entirely reasonable to think that acousticians who spend much of their lives moving around in rooms while listening to revealing test signals can become sensitized to aspects of sound fields that ordinary listeners blithely ignore. This is a caution to all of us who work in the field of audio and acoustics. Our preferences may reflect accumulated biases and therefore may not be the same as those of our customers.

Memo for Listening room recommendations: for stereo listening, leave side walls reflective at first-reflection points. For multichannel listening it is optional. Audio professionals may have their own preferences—it's all right, they are just different.

8.1.1 Some Thoughts about Loudspeaker Arrangements

After listening to stereo for about 50 years, the author can reconstruct in memory the impressions of a great many recordings in which the front sound-stage consisted of hard-panned sounds to the left and right loudspeaker locations, combined with a phantom center image that was well defined in terms of direction but that had an attractive, space-filling body to it. There may have been other images panned to intermediate locations, and they too shared some of this quality. The problem with any of the phantom images was their sensitivity to listener location; there was only one sweet spot. Over the years, there have been several attempts to generate interest in a center channel, but it never caught on in the music domain.

Fortunately, movies were different. Now multichannel movie sound systems have migrated into homes and are, of course, available for music recordings with or without video accompaniment. It is not just a little frustrating to find that many music recordings substantially ignore the center channel. The featured artist may be presented as a stereo phantom image (the same signal sent to left and right loudspeakers) or combined in all three front loudspeakers. The center channel was intended as a more accurate sounding and much more spatially stable location for the center-located sound, but it was being avoided by recording engineers who learned their skills with two-channel stereo. Was this reactionary stubbornness, or is there more to it?

Figure 8.4 picks up on the current discussion of frontal spatial effects, ASW and image broadening, which listeners are partial to and examines how they apply to the three dominant components of a front soundstage: left- and right-located images and a center-located image. Figure 8.4a shows that, by itself, a sound panned to a left or right loudspeaker has a modest opportunity to generate ASW. Sounds from the two side walls have different trade-offs with respect to their ability to generate ASW. The negatives are that the adjacent side reflection has a small angular separation between it and the direct sound (16°) and that it happens early (delay = 2.7 ms). However, it has a high sound level (−2 dB relative to the direct sound). In contrast, the opposite-wall reflection is a generous 97° away (low IACC), and it has a good delay (12.3 ms), but the sound level is down a little (−6.3 dB) because of the propagation distance. The rear-wall reflection is of little value spatially; it is probably innocuous, but consideration may be given to absorbing or scattering it.

> **Memo for Listening room recommendations:** add sound absorbing material or diffusers to center portion of rear wall.

When we look at the situation leading to a phantom center image, the picture is much more complex (see Figure 8.4b). The left and right loudspeakers radiate identical sounds that arrive identically at the left and right ears. The sound quality will be degraded by the fact that there are two sounds combining at each ear (one from each loudspeaker, known as acoustical crosstalk; see Figure 9.7), but the fact that the left and right ear sounds are the same means that the listener cannot distinguish this symmetrical situation from that occurring when sounds arrive from a frontal sound source (the ears know only that the sounds are identical, left and right; it matters not how many sources created those sounds). Hence, we hear a "phantom" image. But it is a phantom image with a spatial effect associated with it because of the reflected sound field. As can be seen, several of these reflections arrive from directions that are very productive at creating impressions of ASW. A "direct" sound is shown arriving from the direction of the phantom image. This is because, from both the physical and the perceptual perspectives, this is the "effective" direction.

Generations of listeners have noted the obvious differences in directional and spatial impressions created by sounds panned to the real left and right loudspeakers and those panned to intermediate positions, including center. The difference is that the extreme left and right locations are created by monophonic signals, delivered to single loudspeakers, whereas the intermediate image locations result from "stereo" signals, delivered to both loudspeakers simultaneously, with amplitude biases and/or delays appropriate to define the direction. The common impression is that the left and right panned sounds appear to originate in the loudspeakers themselves, whereas the intermediate images appear to originate further back, in a more spacious setting, and sometimes elevated. Instead of a soundstage extending across a line between the loudspeakers, the center images tend to drift backward. Because the impression of distance is dependent on early reflections, this is a plausible perception.

Replacing the phantom center image with a center loudspeaker, as in Figure 8.4c, changes the situation. Now only two lateral reflections remain. The result is improved sound quality, with no acoustical crosstalk (which we will discuss in Figure 9.7). There is also directional stability (all listeners in the room hear the sound as arriving from the correct direction) and a degree of spaciousness that is appropriate for a real sound source in the listening room. Now, when compared to the phantom image situation, this could be thought of as a problem in that desirable ASW has been lost, and over the years, many have commented on the relative spatial "hardness" of the image presented by a real center loudspeaker.

Choisel and Wickelmaier (2007) found that listeners comparing a discrete center channel with a phantom center image generated by a stereo pair in a normal room consistently rated the phantom image higher in perceptual dimensions of width, elevation, spaciousness, envelopment, and

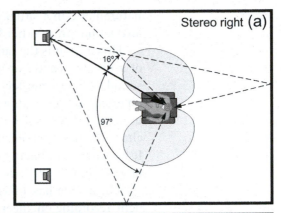

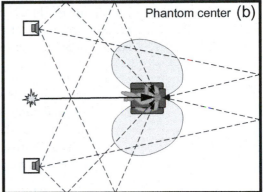

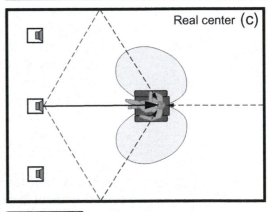

FIGURE 8.4 (a) Reflection diagrams for a single channel. (b) The situation of a phantom center image. (c) A center loudspeaker. The arbitrarily chosen room is 21.5 ft (6.5 m) × 16 ft (4.9 m).

naturalness. In a situation where the discrete center sound was unsupported by any sounds from other loudspeakers, this is consistent with expectations.

However, it is the task of the recording engineer to augment the spaciousness of a discrete center channel by using appropriately delayed and level-adjusted sounds sent to the left and right front channels and surround channels. If a phantom center is thought to have audible advantages, a real center channel, used in proper collaboration with processed signals delivered through other channels, has the potential to be better in every respect and much more flexible. It is a matter of having the necessary signal processing tools during the mixing process and the knowledge of how to use them.

The author has a vivid memory of an AES workshop in which several prominent recording engineers discussed the topic of multichannel microphone and processing methods and played illustrations of their work. Several of the presenters were adamant about not isolating the "talent" in the center channel, citing some unconvincing reasons for doing so, including the following:

- Customers could shut off all loudspeakers except the center channel and, in hearing the performer isolated, might realize that the "backing" disguised some serious limitations. In a recording studio, this isolation can be accomplished by pressing buttons on the console. At home, one must disconnect the front side and surround loudspeakers from the power amplifier. Who would bother to do this?

- Center channel loudspeakers are often not as good as the left and right speakers, and the featured artist will not sound good. Now, even in the least-expensive "home-theater-in-a-box" ensembles, all loudspeakers are identical. In more serious installations, the center channel is anything but an afterthought because in movies more than 80% of the sound is delivered by it, and in television the proportion is even higher. The center channel is the most important speaker in the entire installation! If there were widespread problems with center channel sound quality, it is the movie and television industries that would be raising the loudest objections.

In the demonstrations of material recorded by these engineers, programs recorded without benefit of a strong center channel were heard by the large audience as being biased to either the left or right front loudspeakers, depending on which side of the center aisle they were sitting, as in conventional stereo. Only those audience members who left their seats to crouch in the center aisle of the lecture room heard a phantom center image. However, there were two contributors, one pop and one classical, who had figured it out, presenting solid center localizations for the entire audience, placed in an acoustical setting that was compatible with the rest of the soundstage. A classical demonstration, with a solo voice against a spacious orchestral background, was especially convincing. It can be done, but it still rarely is.

The debate over a center channel started a very long time ago. In a classic paper, Steinberg and Snow (1934) explained, "The three-channel system proved definitely superior to the two-channel by eliminating the recession of the center-stage positions and in reducing the differences in localization for various observing positions." They add, "Although bridged systems [deriving the center signal from a combination of left and right] did not duplicate the performance of the physical third channel [a discrete center channel feed], it is believed that *with suitably developed technique* [author's italics], their use will improve two-channel reproduction in many cases." Their wise words, which predate commercialized stereo by about 20 years, appear to have been lost until much later, when a few people took another look at center channel configurations. For example, Eargle (1960) explored the effects of different amounts of center-bridged signal, and Torick (1983) proposed a three-channel system for television. None of the options caught on as an embellishment to stereo which has lumbered on in an elephantine manner for over 50 years.

Figure 8.5 shows the delays, sound levels, and incident angles of first-reflected sounds for the front channels in a five-channel system. Obviously, all of these dimensions are specific to this room 21.5 ft (6.5 m) × 16 ft (4.9 m). There is nothing special about these dimensions except that they are within the size range of domestic listening/home theater rooms. In Figure 8.5a, it can be seen that, as just mentioned, the reflection from the adjacent side wall, in addition to being from a direction similar to that of the direct sound, is delayed by only 2.7 ms. Neither the direction nor the delay is optimum for the creation of ASW/image broadening. The reflection from the opposite side wall is a better candidate, arriving from a large angular separation, and a good angle, and with enough delay to be effective. The reflection from the rear wall would seem to be of little value.

Figure 8.5b shows that the center channel has good possibilities for ASW, with arrivals from good angles and acceptable delays. If increased spaciousness is desired, one suspects that a deviation from the perfect lateral symmetry of this figure would help by causing

(a) Front right or front left

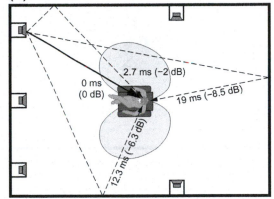

(b) Front center

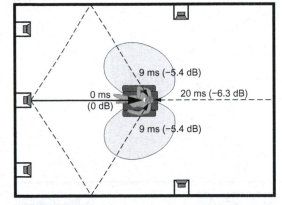

FIGURE 8.5 *Reflections generated by the front side and center loudspeakers of a typical five-channel arrangement. Delays and sound levels relative to the direct sound, assuming perfect reflections, are shown for a room 21.5 ft (6.5 m) × 16 ft (4.9 m).*

the lateral reflections to arrive at slightly different times. The reflection from the rear wall cannot add to spaciousness, and being a version of the direct sound differing only in time of arrival, it would seem like a very good candidate for elimination—absorb or scatter it.

> **Confirming memo for listening room recommendations:** add sound absorbing material or diffusers to center portion of rear wall.

(a) Discrete sound panned to a side surround channel

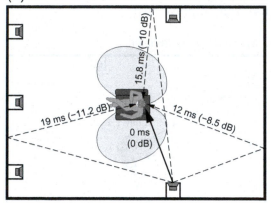

(b) Surround channels augmenting direct sound from any or all front channels. Delays are as in (a) plus the delays introduced in the recording process between the front channels and each surround.

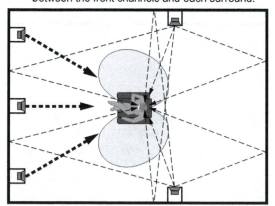

FIGURE 8.6 *Reflected sounds for two usages of surround loudspeakers: (a) as an independent sound source and (b) as a supplement to the front loudspeakers reproducing delayed sounds intended to yield impressions of distance and/or envelopment. Delays and sound levels relative to the direct sound, assuming perfect reflections, are shown for a room 21.5 ft (6.5 m) × 16 ft (4.9 m).*

Figure 8.6 shows what happens with the surround loudspeakers. These two conditions are considered:

1. **The loudspeaker is a discrete sound source:** Figure 8.6a illustrates the situation when the sound originates in the surround loudspeaker, as in a steered sound effect in a movie or a discrete-panned performer in a musical group. The direct sound is shown by the solid, heavy arrow emerging from the loudspeaker. The dashed curves describe early reflections related to this direct sound. The direct sound and the reflection from the opposite wall appear to be a highly productive source of decorrelation at the ears—that is, spaciousness—as together they constitute "lateral" sound. The fact that the reflection is 10 dB down in level reduces the likelihood of low IACC, but there should be strong localization, which is normally the purpose of these situations. The front and back wall reflections, by themselves, will contribute relatively little. In short, the contribution of room reflections would be normal for a hard-panned sound, as it would be for similar sounds directed to each of the front channels.

2. **The loudspeaker(s) reproduce delayed versions of frontal sounds:** Figure 8.6b illustrates what happens when the sound originates in the front soundstage and delayed versions of it are sent to the surround loudspeakers to generate a sense of distance or envelopment. In such instances, it is not uncommon for some or all of the front loudspeakers to contribute additional delayed sounds to enhance the sense of envelopment. The direct sound is shown to be arriving from one or

more of the front channels (thick dashed lines). In movies, the majority of the sound is delivered by the front-center channel. The dominant impression of envelopment will likely be delivered by the direct sound from the surround channels; they arrive at the listener from useful angles, but it is up to the recording engineer to optimize the amplitudes and delays relative to the front channels. Additional reflections contributed by the room will embellish the sense of space, making it more complex than that possible with a pair of surround loudspeakers. Here, the opposite wall reflection is likely to be the major contributor (good angle and delay). Reflections from the front and rear walls arrive from relatively unproductive directions and will contribute less to the effect.

8.1.2 Delayed Reflections and Reflections of Those Reflections

The condition described in Figure 8.6b is fundamentally different from all other situations discussed up to now because the delayed sounds originate from separate loudspeakers with amplitudes and delays defined by the recording with values representing spaces and distances that are much larger than the listening room. The delayed sounds themselves, and their subsequent first-order reflections within the room, can therefore be well above the levels of "typical," naturally occurring reflections shown in Figure 6.7. In multichannel recordings, the surround illusions that are among the principal objectives— image broadening (ASW) and listener envelopment (LEV)—are able to be created at will, without any help from the listening room. Additional surround channels simply add more capability and, very likely, greater independence from the room.

But the listening room is still there, and the reflections of those substantially delayed recorded reflections are separated from the recorded reflections by delays dictated by the listening room dimensions. It is reasonable to think that listening room reflections fall into that category of a desirable augmentation. In the example shown in Figure 8.6b, it is obviously the lateral, side-to-side reflections that should be encouraged. Looking at Figure 7.1, it is seen that it is the lower frequencies that contribute most to impressions of listener envelopment. Because most absorbing material used in home theaters and in cinemas is mostly effective at higher frequencies, being 2 in. (50 mm) or less in thickness (most effective above about 500 Hz), it means that the reflections containing frequencies useful to envelopment are substantially intact. What suffers is the spectral fidelity because the high frequencies have been disproportionately attenuated. As will be stated again later if one wishes to absorb sound, absorb all of it—at least down to the transition frequency—so as to preserve sound quality.

Let us now return to Chapter 4 and look again at Figure 4.14, which shows the changing pattern of directional diffusivity in a small room as a function of time. In the chosen example, the direct sound and early lateral reflections dominated the early diffusion pattern, but later reflections settled into a lateral pattern, indicating less absorption on the side walls. Hinted at in that chapter and in Toole (2006) was the notion of using directional diffusion to augment desirable effects in multichannel audio. Now, getting specific, it is reasonable to think that encouraging first-order side-to-side reflections from the side-located surround loudspeakers may be advantageous to the creation of envelopment when the number of channels and loudspeakers is limited. Obviously, one does not wish to create conditions for flutter echoes between the side walls. It will be left as a challenge to acousticians to utilize the numerous options—for example, angled or curved reflecting surfaces, scattering surfaces, and absorption—to deliver the desirable effects without aggravating such problems.

> **Memo for Listening room recommendations:** use reflecting or scattering surfaces on walls opposite side-surround loudspeakers to enhance envelopment. Be careful about flutter echoes between the side walls.

One relevant event that occurred in the evolution of home theater was the introduction of bidirectional surround loudspeakers. This occurred at a time when a single surround signal was fed to both side loudspeakers, and the notion was to add complexity to the reflected sound field by directing a greater proportion of the sound toward the front and back of the room. The implication of the ray diagrams in Figure 8.6 is that it would not be advisable to reduce the amplitude of the direct sound to the listener, either from the perspective of localizing sound effects or generating envelopment. Time has passed, and the addition of more discrete channels, now up to four surround channels in popular surround processors and receivers, has changed circumstances. Chapters 16 and 18 address some of the basic requirements for surround loudspeakers and consider the merits of several options.

> **Memo for Listening room recommendations:** think twice (or more) about using dipole surround loudspeakers. There seem to be better choices.

8.2 ASW/IMAGE BROADENING AND LOUDSPEAKER DIRECTIVITY

An obvious way to control reflected sounds reaching listeners is by using the directivity of loudspeakers to adjust the amount of sound reaching nearby reflecting surfaces. This can be done to either increase or decrease the reflections. In the monophonic era, some enthusiasts aimed the loudspeaker away from the audience, using sound reflected from room boundaries to create a directionally enriched reflected sound field.

When stereo arrived, it provided a second relatively discrete sound source, and, especially with the better microphone and mixing techniques, things greatly improved. However, many listeners still desired more directional and temporal diversity in the sounds arriving at the listening location, so various devices were contrived, some using additional loudspeakers, to create additional sounds, especially delayed sounds, from the two-channel mix. These could legitimately be considered to be the precursors to today's "upmixers" (algorithms that convert a two-channel input to a multichannel output). Not surprisingly, many loudspeaker designs came and (mostly) went. Among these were some openly contradictory and puzzling ideas.

On the one hand, there were some largish cone-horn designs that, at least above about 800 Hz, could exhibit significant directional control, preventing sound from energizing strong early reflections. Some persons considered narrow dispersion and the consequent avoidance of room reflections as a desirable objective—example, Kates (1960). On the other hand, some designs attempted to be omnidirectional, at least in the horizontal plane, such as Queen (1979), Moulton et al. (1986), and the mbl 101E loudspeaker (www.mbl-germany.de). Similar to these in directionality were so-called bipoles, bidirectional in-phase systems that from low through middle frequencies were almost horizontally omnidirectional. Between these extremes existed an infinite variety of forward-firing cone-dome and cone-small-horn systems with directivity that could vary significantly as a function of frequency, all exhibiting a forward directional bias. One ambitious design used a relatively directional arrangement aimed at the listeners, with a second arrangement aimed at the side-wall reflection point, delivering an electronically delayed lateral reflection (Kantor and de Koster, 1986). The limitations of stereo were evidently inspirational to creative minds.

Standing alone among these design variations is the dipole. In its classical form, it is a diaphragm that is allowed to radiate freely in both directions: a bidirectional out-of-phase loudspeaker. In that pure form, it has a directivity pattern that resembles a figure eight, with one lobe facing forward, the other backward, and the nulls looking sideways. Such loudspeakers had a fantasy factor, since many were electrostatically motivated; they had diaphragms of vanishingly low mass; they lacked a "resonating" box; and they had the potential of uniformly controlled directivity with nulls aimed at the side walls to eliminate lateral reflections popularly thought to be the cause of comb-filtering colorations (this will be discussed in Chapter 9). Since the 1940s, dipole loudspeakers have had a steady following of enthusiasts in the audiophile community. One of the advantages promoted for this loudspeaker configuration is that because of the directional pattern, there is less interaction with the listening room. Figure 8.7 shows an idealized radiation pattern for a panel—single diaphragm—dipole loudspeaker. This pattern would be expected to prevail at all frequencies up to the mid-kHz, when practical factors related to diaphragm size spoil the picture. It needs to be said here that this is *not* the

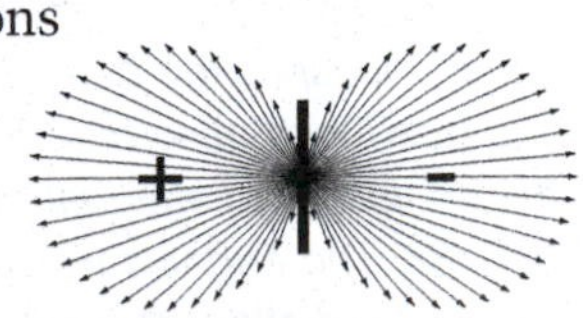

FIGURE 8.7 *The directional radiation pattern of an ideal dipole panel loudspeaker.*

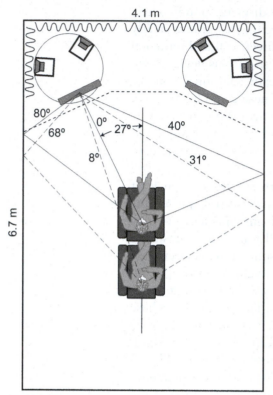

FIGURE 8.8 *The physical arrangement for the listening tests, showing the turntables used to rotate the three loudspeaker pairs into identical positions for listening. An acoustically transparent screen prevented listeners from seeing the loudspeakers. The geometry was such that the adjacent side-wall reflections occur at very large angles off axis (80° and 68° for front- and rear-row listeners, respectively), and those from the opposite wall are traceable to moderate angles (40° and 31°). Medium-weight drapes covered the walls behind the loudspeakers. This would have the greatest influence on the bidirectional dipole loudspeaker, but the product was already equipped with absorbing pads in the rear half of the enclosure to attenuate the output above about 500 Hz. The curtains would further attenuate the rear radiation. The side walls between the loudspeakers and the listeners were hard, flat broadband reflectors. The tests were done in stereo and mono, the latter using only the left loudspeaker.*

directional pattern exhibited by the bidirectional-out-of-phase wall-mounted surround loudspeakers commonly referred to as "dipoles" (which we will discuss later). This diagram shows that equal amounts of sound are radiated toward and away from the listener, and very little is radiated to the sides. Ideally, there is a complete null at 90°.

Thus, there appears to be a contradiction in listener preferences. Certain designs set out to minimize reflections from side walls and are apparently preferred by some listeners. Other configurations overtly generate reflections from all surfaces, and one must conclude that these are preferred by other listeners. Is there a real perceptual difference, one that matters, or is some other factor involved? Of course, there is a third possible perspective: that nobody had actually performed fair (i.e., controlled, blind) listening comparisons among the options, and as a result, the existence of different products in the marketplace has been the result of successful promotion or listener adaptation, not unbiased evaluation. An experiment was needed.

8.2.1 Testing the Effects of Loudspeaker Directivity on Imaging and Space

In 1984 the author conducted a series of experiments to explore the notion that loudspeaker directivity, and the variations in lateral room reflections that followed from this, were factors in listener opinions (Toole, 1985, 1986). The methodology was simple and the results instructive.

Figure 8.8 shows the room layout for stereo comparisons of two cone-dome forward-firing loudspeakers and a full-range electrostatic dipole loudspeaker. The program was muted while the turntables were rotated into position. Figure 8.9 is a photograph of the room, showing the acoustically transparent screens and the two chairs; the rear one was higher than the front one to reduce shadowing effects.

As can be seen in Figure 8.10, all three loudspeakers had comparably smooth and flat axial frequency responses. Off axis, AA shows the progressively increasing directivity of the woofer up to the crossover to the tweeter around 3 kHz. The tweeter then exhibits the

The front of the listening room setup, showing the acoustically transparent screens, fitted with a numerical scale to assist listeners in judging lateral dimensions of images and the front soundstage.

wide dispersion of a small diaphragm, until the eventual offaxis falls off the tweeter as it becomes more directional at short wavelengths. Loudspeaker E, the three-way design, exhibits this kind of undulating pattern twice, once when the woofer transitions to the midrange around 500 Hz and again when the midrange transitions to the tweeter around 3 kHz. Loudspeaker BB, the full-range dipole, is quite well behaved, showing a relatively uniform decrease in output with increasing angles off axis.

With what we know now about loudspeaker measurements and their correlation with subjective evaluations of sound quality/timbral accuracy (Chapters 18 and 19), a few things can be said. First, the axial frequency responses appear to be similarly good. Differences in sound quality are therefore likely to be determined by off-axis sound radiation and how it is perceived. Second, the desirability of relatively constant directivity is quite well satisfied by BB, but AA and E have what look like some significant problems in their off-axis frequency responses. Therefore, above the transition frequency, BB would seem to have the potential for more neutral timbre. Third, there is the matter of bass response, which depends on the manner in which the loudspeakers couple to the low-frequency room modes and where listeners sit. In this case, loudspeaker position was constant, which meant that AA and E, which had conventional omnidirectional woofers, would be similar to each other, but BB, a dipole, would couple in a different manner to the room modes (it is a velocity source, as opposed to the pressure source behavior of the others). Unfortunately, no in-room measurements were made at the time, so it is not known how much of a factor this might have been in the ratings. Apart from the uncertain behavior at low frequencies, and disregarding any influences of directivity, loudspeaker BB would appear to have a better potential for high sound quality.

Figure 7.1 shows estimates of the frequency ranges over which listeners perceive the elements of what is broadly called "spaciousness." Based on current understanding, the dominant perception that listeners would experience in this small-room situation would be image shift and broadening, ASW, and this is associated with reflected sounds containing frequencies in the range from about

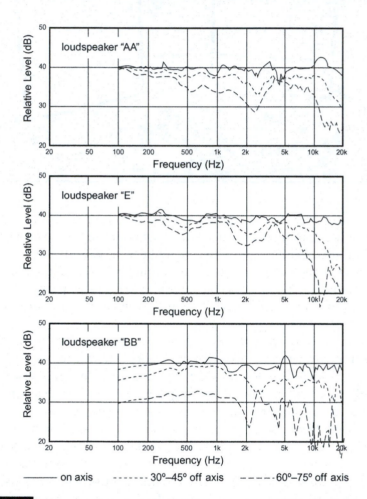

FIGURE 8.10 *These are the anechoic chamber measurements on the three loudspeakers (from Toole, 1986, Figure 24), with the on axis, 30° to 45° off axis and 60° to 75° off axis spatially averaged frequency responses shown.*

- *Loudspeaker AA was a two-way design, 8 in. (200 mm) woofer and 1 in. (25 mm) tweeter (Rega model 3).*
- *Loudspeaker E was a three-way design: 12 in. (300 mm) woofer, 5 in. (110 mm) midrange, and a 2 in. (50 mm) tweeter (KEF 105.2).*
- *Loudspeaker BB was a full-range electrostatic dipole, employing a diaphragm subdivided into areas driven in a manner to approximate a spherically expanding wavefront. The center circle, the "tweeter," was about 3 in. (76 mm) diameter (Quad ESL-63).*

500 Hz and higher occurring at delays less than about 80 ms. None of this was considered at the time of the experiments, 1984, but it is helpful to incorporate these notions into this discussion.

In terms of what might be expected, loudspeakers AA and E radiate higher sound levels at large off-axis angles (60° to 75°) than BB, although not uniformly at all frequencies. The poor off-axis frequency responses may well influence perceptions of timbral accuracy, but in terms of the potential for spaciousness from the adjacent side wall reflections, these loudspeakers seem to have the advantage. The 30° to 45° measurement window is descriptive of the opposite side-wall reflections, and here the differences among the three loudspeakers are smaller, although the wide dispersion of the tweeters around 5 kHz would be advantageous to AA and E. Let us see what listeners think when they are exposed to the choices using a questionnaire divided into two categories: sound quality (not shown) and spatial qualities (Figure 8.11).

In the results shown in Figure 8.12, the first surprise was the extent to which single loudspeakers elicited strong opinions about spatial quality. Those of us who had participated in many single-loudspeaker comparisons held the opinion that there were differences in the perception of the spatial extent and distance of the single sound source. To us, the most neutral-sounding loudspeakers tended not to draw attention to themselves. However, it was still a surprise that this was an impression shared by other listeners not accustomed to this form of critical analysis—in this case, a mixture of audiophiles and recording professionals.

In these results, spatial quality and sound quality ratings were obviously not independent—one tracks the other. Is it possible that listeners cannot separate them even though, consciously, most were confident that they could? If indeed they are separable factors, it is fair to consider which one is leading. In monophonic tests, listeners reported large differences in both sound quality and spatial quality, and, if anything, there were stronger differentiations in the spatial quality ratings. This was definitely not anticipated, but these listeners had little doubt that there were substantial differences in both rating categories. However, in stereo listening, most of the differences disappeared. The two highly rated loudspeakers (AA and E) kept their high ratings, almost identically in fact, but the loudspeaker (BB) that had a low rating in mono became competitive in stereo. In fact, looking at the stereophonic data, the scatter diagrams of judgments indicate a lot of indecision about the relative merits of these loudspeakers in both categories: sound quality and spatial quality.

With respect to BB, did stereo add something that was missing in mono? Did stereo mask problems that were audible in mono? Did stereo reveal a capability that could not be heard in mono? These are key questions. Earlier it was speculated, based on current understanding of loudspeaker measurements, that BB might have had an advantage in terms of sound quality—at the very least, not a disadvantage. That speculation was not borne out. It is difficult to conceive

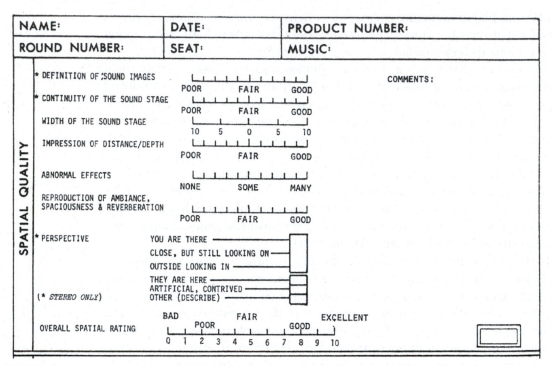

FIGURE 8.11 *The portion of the listener questionnaire that dealt with directional and spatial qualities. There was a similarly structured questionnaire pertaining to aspects of sound quality or timbral accuracy. It led to what was called a "fidelity rating." During the presentations, listeners were expected to respond to each of the "perceptual dimensions" and then, at the end, to arrive at a single number overall rating of spatial quality. Comments were encouraged. Loudness levels were carefully matched, and the music selection and presentation sequence were randomized. The experiment was done in stereo and also in monophonic form, using only the left loudspeaker. Half of the listeners started with the monophonic test and half with the stereo test. From Toole, 1985, Figure 2.*

of a mechanism that would cause the fundamental timbral character of the loudspeakers to change when they were used in stereo pairs. Therefore, the implication is that spatial factors were strongly influential, if not the deciding factors, in both tests. If the wider-dispersion loudspeakers benefited from superior spaciousness in the mono tests, adding stereo might enhance their performances more. But what about BB? It seems possible that the interchannel decorrelation generated by differences in stereo-miked and mixed sounds radiated from the left and right loudspeakers provided compensation for this loudspeaker in the stereo tests. This would be audible in the direct sounds alone, without support from lateral reflections.

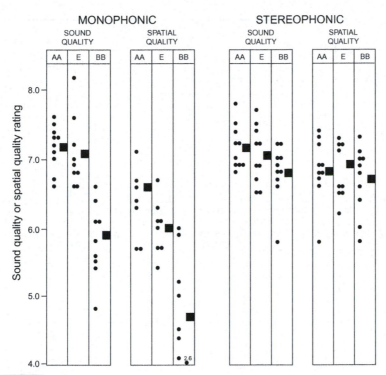

FIGURE 8.12 *Overall sound quality and spatial quality ratings for the three loudspeakers when auditioned in mono and in stereo. Each dot is the average of several ratings by a single listener. The ratings are averaged across the four music selections: choral, chamber, jazz, and popular. All were one generation from master-tape recordings done especially for this purpose, employing known microphones and documented processing (if any). On the vertical scale, 10 represented the best imaginable sound and 0 the worst. From Toole, 1985, Figure 20.*

Digging further into the raw data, Figure 8.13 shows histograms of judgments in the various categories of spatial quality listed in the questionnaire (see Figure 8.11). According to these data, in all but one category, loudspeaker BB received lower (only slightly but consistently lower) spatial ratings than the other two products. Only in the category "abnormal spatial effects" was the rating higher, and for this category, that is a problem (most often sounds were criticized for being inappropriately close to the listener and occasionally inside the head; see "In-Head Localization"). The generous scatter in all of the ratings indicates that listeners' opinions varied considerably, but there does seem to be a relative lack of scores in the highest categories for BB. In the overall description of listening perspective, "for the music (choral, chamber, and jazz) recorded with a natural perspective, the modal listener response was "you are there" for AA and

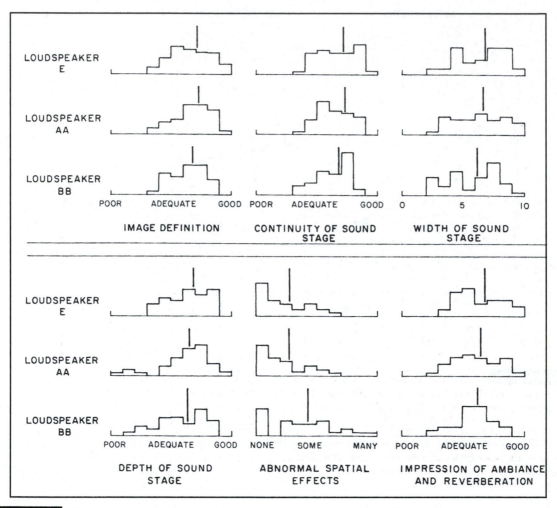

FIGURE 8.13 *Distributions of the analytical judgments in listener responses. The vertical line in each category indicates the mean response. From Toole, 1985, Figure 22.*

E and "close, but still looking on" for BB. According to the definitions of those phrases, "Loudspeakers AA and E gave listeners some impression of being enveloped by the ambient sound of the recording environment, with BB tending to separate them from the performance" (Toole, 1986, p. 342). This all sounds very much like the influence of lateral reflections, ASW, early spatial impression, and the associated variations in IACC. Because these, and the ratings of Figure 8.12, included ratings for all four musical selections, perhaps something more can be learned by examining the ratings for each of the musical selections (and the recording techniques).

IN-HEAD LOCALIZATION

In-head localization seems like the logical opposite of an enveloping, external, and spacious auditory illusion. Perceptions of sounds originating inside the head, which routinely occur in headphone listening, can also occur in loudspeaker listening when the direct sound is not supported by the right amount and kind of reflected sound. The author and his colleagues have experienced the phenomenon many times when listening to stereo recordings in an anechoic chamber, usually with acoustically "dry" sounds hard panned to center or, less often, to the sides. It prompted an investigation (Toole, 1970), the conclusion of which was that there is a continuum of localization experience from external at a distance through to totally within the head. It is often noted with higher frequencies, and it can happen in a normal room with loudspeakers that have high directivity or in any situation where a strong direct sound is heard without appropriate reflections. Moulton (1995) noted that "speakers with narrow high-frequency dispersion . . . tend to project the phantom at or in front of the lateral speaker plane." In an anechoic chamber, it can occur when listening to a single loudspeaker, especially on the frontal axis, in which case front-back reversals are also frequent occurrences. This phenomenon is so strong that it need not be a "blind" situation. Interestingly, a demonstration of four-loudspeaker Ambisonic recordings played in an anechoic chamber yielded an auditory impression that was almost totally within the head. This was a great disappointment to the gathered enthusiasts, all of whom anticipated an approximation of perfection. It suggested that, psychoacoustically, something fundamentally important was not being captured or communicated to the ears. An identical setup in a normally reflective room sounded far more realistic, even though the room reflections were a substantial corruption of the encoded sounds arriving at the ears.

Figure 8.14 shows the spatial quality ratings for each of the music selections. *They are all different.* The two classical pieces, recorded in concert hall circumstances, were not able to conclusively rank the three loudspeakers according to listener preference. At best, gentle trends can be seen, not substantial differences. If BB is deficient in its ability to generate a sense of spaciousness, it does not show up in these judgments, perhaps because the contributions of the room reflections are swamped by the spatial information incorporated into the stereo recordings. The jazz selection revealed a fairly general dislike for AA. No explanation for this was found.

The pop music selection put loudspeaker BB in a position of disfavor. In fact, the subjective ratings in this stereo test are remarkably similar to those seen in monophonic listening (Figure 8.12). Why is this? Of all the recordings, the pop recording was the only one to have significant amounts of hard-panned —that is, monophonic—sound emerging from the left and right loudspeakers. It is conceivable and logical that listeners reacted to the relative lack of spatial accompaniment for these sounds when they were auditioned through BB.

If the loudspeakers are not capable of generating IACC, stereo recordings can do it, as in the cases of the choral and chamber recordings. If the recording presents essentially monophonic sounds emerging from left or right loudspeakers, they are on their own, and if they lack sufficient means of generating lateral reflections, they end up being judged as overly simple, point sources. In a stereo mix, with some amount of interchannel decorrelation (generated by recorded

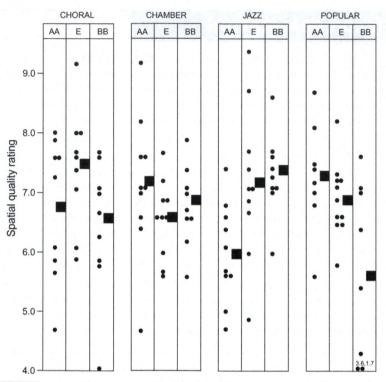

FIGURE 8.14 *Overall spatial ratings of loudspeakers in stereo for each of the musical selections. From Toole, 1985, Figure 23.*

reflections and delayed sounds) accompanying direct sounds, listeners would be hearing those sounds in a spatial context, but sounds hard panned to left and right would be heard from "naked" loudspeakers, dominated by direct sound. Obviously, recording and mixing methods greatly affect what is heard. Moulton et al. (1986) had similar experiences.

The principal conclusion is that recording technique is often the prime determinant of spatial impressions perceived in sound reproduction. The directivity of the loudspeakers is a factor, as is the reflectivity of the surfaces involved in the first lateral reflections, especially in recordings incorporating left or right hard-panned sounds. In other words, if simplistic "mono" hard pans had been ameliorated by the addition of recorded spatial cues, perhaps the criticism of BB might have been avoided. It remains a mystery why AA was disliked in the jazz recording; its directional behavior is by today's standards not very good, so there are reasons to suspect problems of one form or another.

Obviously, there is much yet to be investigated, including the tantalizing notion that wide-dispersion loudspeakers with what would appear to be compromised sound quality (AA or E) are given a higher sound quality and spatial quality ratings than a narrow-dispersion loudspeaker with potentially superior sound quality (BB). The provocative suggestion is that the two domains are interrelated and that the spatial component is greatly influential. Listeners appeared to prefer the sound from wide-dispersion loudspeakers with somewhat colored off-axis behavior to the sound from a narrow-dispersion loudspeaker with less colored off-axis behavior. In the years since then, it has been shown that improving the smoothness of the off-axis radiated sound pushes the subjective ratings even further up, so it is something not to be neglected.

Perhaps related to this is the acoustical crosstalk associated with the phantom center image (see Figure 9.7). This coloration cannot be ignored in a situation where the direct sound is strong (loudspeaker BB). Early reflections from different directions tend to fill in the interference dip, making the spectrum more pleasantly neutral (Figure 9.7e). This is a spectral reason to prefer wide-dispersion loudspeakers and to encourage reflections. Even with room reflections to moderate the interference dip, speech intelligibility is degraded (Shirley et al., 2007), and there are significant effects on instrumental and vocal timbre.

Returning for a moment to the classifications of directional and spatial illusions listed at the end of Chapter 7, what we have been discussing would be totally within Category 1: the soundstage illusions. These were stereo—two channel—recordings. What if they had been multichannel recordings that intrinsically can deliver a directionally and temporally enriched reflected sound field using only the direct sounds from the loudspeakers? The implication is that, in multichannel recordings where all channels are generously used for spatial enhancement, the nature of the loudspeaker off-axis behavior or listening-room acoustics may be perceptually even less important.

The exception is for hard-panned sounds. In movies, this means that the center channel certainly, and to a lesser extent the front left and right channel loudspeakers, must perform well in isolation. In "middle-of-the-band" music recordings, where individual performers are panned to individual channels including the surrounds, *all* loudspeakers must do well as "solo" performers. These results, extrapolated to multichannel audio, suggest that all of the loudspeakers in the array should have comparably good behavior and relatively wide dispersion. Wide dispersion is of no value if the reflections cannot be heard, so this means that lateral reflections should not be attenuated, a common practice in the domain of custom domestic installations and almost a rule in control room design.

And more thoughts to take away: A loudspeaker that sounds good in a monophonic comparison is likely to sound good in a stereo comparison, but the reverse is not necessarily true. Evaluate your loudspeakers in monophonic comparisons (to find out what you really have). Demonstrate your loudspeakers in stereo or, presumably, multichannel (to impress everybody). Choose the recordings carefully; they are a significant factor. Subsequent additional tests of this kind in the intervening years have not changed these conclusions.

8.2.2 The Audible Effects of Loudspeaker Dispersion Patterns—Other Opinions

The notion that monitoring the recording process is significantly different from recreational listening has already been introduced and that different criteria for lateral reflections apply (see Section 8.1 and Kishinaga et al. (1979)). There it was concluded that in the creation of recordings, engineers preferred to listen in rooms with attenuated lateral reflections. The year before, Kuhl and Plantz (1978) investigated "the directional properties of loudspeakers that would be most suitable for control-room monitoring. Using only professional sound engineers as listeners, they found that narrow-dispersion loudspeakers were required for good reproduction of voices in radio dramas; dance and popular music was also desirably 'aggressive' with 'highly directed' loudspeakers. The majority of these same listeners, however, preferred wide-dispersion loudspeakers for the reproduction of symphonic music at home. In the control room, though, only about half of them felt that they could produce recordings with such loudspeakers" (summary from Toole, 1986, p. 343).

Moulton et al. (1986) performed informal listening evaluations of forward-firing designs compared to a horizontally omnidirectional loudspeaker. It was concluded that with "the stereophonic omnidirectional playback system, the *musical* essence of the sound seems more palpable, more enduring, and more directly accessible than we have experienced with other loudspeaker systems." The content of the paper leading to this emotive conclusion includes technical discussions of aspects of direction and space that parallel some of those in this chapter. More recently, after more investigation, Moulton (1995) stated, "It appears that broad horizontal dispersion, with the engagement of a specularly responsive set of side walls, yields preferred sonic quality for the stereophonic playback of music, both in terms of spectral accuracy and also in terms of stereophonic illusion, image and entertainment quality." Most recently, Moulton (2003) discusses the loudspeaker as if it were a musical instrument, which is where the discussion becomes somewhat philosophical. However, as noted in the preceding discussion, when a voice or musical instrument is reproduced by a single loudspeaker, the comparison is not illogical. A human voice emerging from a single loudspeaker could be credible; a grand piano less so. In any event, it is evident from all of this that some serious-minded audiophiles and audio professionals are willing to concede that lateral reflections originating in wide-dispersion loudspeakers and delivered by reflective-room walls are highly pleasurable to listen to and possible to use as monitors for sound recording.

Augspurger (1990) describes a series of experiments with different loudspeakers and room acoustic treatments, varying the amount of reflected sound. He discusses the trade-offs in terms of image precision and spaciousness, and concludes, "After extensive listening to classical and pop recordings, I went back to the hard, untreated wall surfaces. To my ears the more spacious stereo image more than offset the negative side effects. Other listeners, including many recording engineers, would have preferred the flatter, more tightly focused sound picture."

Flindell et al. (1991) used anechoic chamber simulations, direct sound, and five reflections for each stereo loudspeaker to investigate listener preferences for different loudspeaker directivities achieved by simplistic filtering of the simulated reflections. The filters simulated no real loudspeakers. Ten of the listeners were experienced audio industry persons, and ten were naive. In general, the naive listeners preferred the widest possible high-frequency dispersion; the experienced listeners liked it, too, but also about equally liked a configuration that simulated a dominant direct sound above 500 Hz. Perhaps the listening circumstances allowed professionals to shift between listening modes—recreational and working (in which they would typically be in a dominant direct sound field). All other settings were rejected by both groups. The natural concern that wide dispersion and the attendant strong early reflections "would lead to degraded stereo imaging was not confirmed by the experienced listeners using rating scales and blind presentations of audio material."

Providing a contrasting point of view, Newell and Holland (1997) present a reasoned discussion of the requirements for control-room acoustical treatment (and, by inference, loudspeaker directivity). They favor the elimination of all lateral and vertical reflections—a near anechoic space, placing listeners in a direct-sound field. They conclude that "spaciousness and the resolution of fine detail are largely mutually exclusive. Spaciousness should . . . be an aspect of the final reproduction environment." There is no doubt that, listening to direct sound only, recording engineers may recognize the callously stark spatial presentation of hard-panned left and right stereo images and be motivated to remedy it, unless this turns out to be another preference associated with the professional side of the industry (see "Sensitive Listeners?").

Not to be ignored in any situation in which reflected sounds have been removed is the fact that the acoustical crosstalk that plagues stereo phantom images is present in its naked ugliness, without any compensation from reflected sounds (see Figure 9.7). One hopes that recording engineers in these situations do not attempt to remedy it with equalization. If they do, their compensation would be excessive for normally reflective rooms and totally wrong if ever the program is replayed through an upmix algorithm and the center image emerges from a center loudspeaker.

In summary, it is clear that the establishment of a subjective preference for the sound from a loudspeaker incorporates aspects of both sound quality and spatial quality, and there are situations when one may debate which is more important. The results discussed here all point in the same direction: that wide-dispersion loudspeakers, used in rooms that allow for early lateral reflections, are preferred by listeners especially, but not exclusively, for recreational listening. There appear to be no notable sacrifices in the "imaging" qualities of stereo reproduction. Indeed, there are several comments about excellent image stability and sensations of depth in the soundstage.

We are left, though, with a problem: how to explain why the often-mentioned comb filtering engendered by early reflections is not a problem. None of these listeners heard it, or at least they didn't comment on it except to say that they prefer sounds with reflections. It was not a factor in the anechoic experiments described in Chapter 6, where if a measurement had been made of the direct sound combining with a single reflection, it would have revealed a classic picture of a comb filter. If there is a subjective response to comb filtering, it is that it appears to have a beneficial effect. There is an explanation.

CHAPTER 9

The Effects of Reflections on Sound Quality/Timbre

In the perception of sound qualities, timbre is what is left after we have accounted for pitch and loudness. It is that quality of a sound that allows us to recognize different voices and musical instruments, and what allows us to distinguish the intonations of a superb musician from those of a learner. It is fundamental to the notion of "high fidelity," a much abused but still highly relevant concept.

When we talk of timbre change as a result of reflections, or anything else for that matter, the natural tendency is to think that any audible change is a negative thing—a degradation. However, as we will see, in some circumstances judgments can go either way—either for better or worse. Also, in some instances, a perceptible change may be expected and is a perfectly normal event. We experience this routinely when, for example, we are in conversation with a person as we walk through different acoustical spaces. There is no doubt that the sounds of both voices are modified by the changing patterns of reflections, but they remain the same voices—scoring 10 out of 10 on a scale of fidelity. Sounds arriving from different directions are modified differently by the head and ears, yet we don't make discriminatory judgments of sound quality as we rotate our heads while in conversation or at a concert. The physical sounds at the ears are changing dramatically, but it all is accommodated by two ears and a brain, functioning normally in normal acoustical situations—reflective spaces.

These are the two primary mechanisms for timbre change as a result of reflections:

1. *Acoustical interference*, constructive and destructive at different
   frequencies, when the direct and reflected sounds combine at the ears.
   Whether the acoustical interference is annoying, or even audible,
   depends on how many reflections there are and where they come from.

141

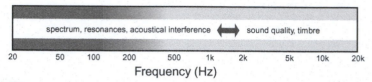

FIGURE 9.1   *There are influences on timbre at all frequencies.*

A special version of this takes place at low frequencies, where multiple reflections between and among the boundaries of rooms generate resonances within the volume of the room. This will be discussed separately in Chapter 13.

2. *Repetition*, the audible effect of the same sound being repeated many times at the ears of listeners. Reflections create new sound events, changing the temporal pattern of the original sound. This could be construed as an error, but in Chapter 8, we found that people like reflections—music in rooms is preferable to music outdoors. Repetition has another aspect, a more subtle one, in that it gives the auditory system more time to examine a sound, more individual "looks," making some aspects of complex sounds more audible, and, as will be shown in the following chapter, early reflections make speech more intelligible.

Figure 9.1 indicates that there are no frequencies where one or the other of these effects is not active. The audible effects appear not to be strongly frequency dependent except, perhaps, at the very lowest frequencies.

9.1 THE AUDIBILITY OF ACOUSTICAL INTERFERENCE—COMB FILTERING

The term *comb filtering* just in itself *sounds* ugly. And its physical appearance, a succession of deep notches, *looks* ugly. And ugly is bad, so comb filtering must be bad. But if this is the prosecution's argument, they lose! The defense can call witnesses, many who will have impressive academic credentials, and many, many more who are just ordinary listeners but can attest to the audible innocence of this phenomenon. Many of them will claim that, in some situations, comb filtering sounds *good*—and under oath, too!

The acoustical sum of a sound and a delayed version of the same single-frequency sound yields a result that depends on the period of the sound, the amount of the delay, and the amplitude of the delayed (reflected) sound. Figure 9.2 shows the two extremes that can occur only when the direct and delayed sounds have identical amplitudes. First, if the delay is a multiple of a whole period, the sounds add perfectly to produce double the amplitude. Second, where the delay is one-half period, the sounds cancel perfectly to yield a zero result. One can easily imagine that such dramatic level changes are audible. Consider, however, that the cancellation can occur only when the early (direct) and delayed (reflected) sounds coexist. These are steady-state examples. Transient events will interact differently and, if the delay is longer than the event itself, not at all.

The example uses single frequencies, pure tones, that could be one Fourier component of a complex sound, in which case there would be other frequency components that would be interacting with each other in very different ways. For a given delay, each of the components would combine in different states of constructive and destructive interference, changing the spectrum of the sound. All of this assumes a "steady-state" sound. Components of a complex sound that are transient in nature—and speech and music have many such components—would interact differently because they have no steady-state characteristics; the direct sound transient may be in decline before the reflected version arrives. In a complex sound, only a single frequency component of the many it incorporates will behave as shown here. This is *not* how a listener would perceive a normal sound with a complex spectrum. It is important to note that technical measurements will normally show the result as if it were a steady-state event.

Figure 9.3 shows what happens in the frequency domain. In (a), the solid black line describes the situation depicted in Figure 9.2, with equal-level direct and reflected sounds, with which perfect summation (+6 dB) and cancellation (−∞) occur. The delay is very small, only 1 ms, which yields alarmingly large fluctuations in the middle frequencies. This is the illustration found (with variations) in a number of popular texts describing the phenomenon of comb filtering, and it creates an impression that this is a major problem in audio. If this is a scare tactic, it works. This certainly looks ugly.

However, in listening rooms, the delayed sounds are reflections, and these will occur at much longer delays and be attenuated in amplitude by propagation loss (about −6 dB/double-distance) and absorption at the reflecting surfaces (which can be anything from near zero for a hard, flat surface to near infinity for a deep, fluffy resistive absorber). The dashed line in Figure 9.3a shows the result of a reflection reduced in amplitude by modest 6 dB. The peaks are a little lower, and the depth of the dips is greatly reduced. It looks much less alarming. This is important.

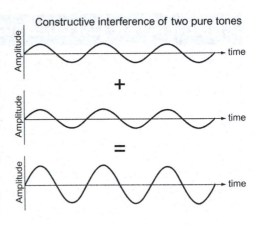

Constructive interference of two pure tones

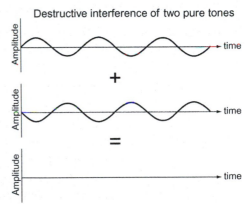

Destructive interference of two pure tones

FIGURE 9.2 *The idealized illustration of constructive (top) and destructive (bottom) acoustical interference. It is idealized because it employs only one frequency—a pure tone—and both the direct and delayed sounds are identical in amplitude. The direct and delayed components have equal amplitudes, resulting in perfect cancellation in the destructive interference illustration. For the same delay, the next cancellations will occur at three times the frequency of this tone, five times, and so on. When, as is most often the case, the delayed sound is lower in amplitude, the cancellation is incomplete, leaving a residue that is the difference between the two interfering components, and, correspondingly, the sum (constructive interference) will be reduced from the illustrated amplitude.*

COMB-FILTER CALCULATIONS

Obviously, a computer can do a pristine job of calculating the frequencies of peaks and dips in a comb filter, complete with high-resolution graphics. However, there are times when one needs a "back of the envelope" computational capability to reveal the key information. It is simple. Destructive interferences—the dips—occur when two sounds are out-of-phase—that is, one-half wavelength apart in time. If we know the delay, we know that the first destructive interference frequency will be that at which the period is twice the delay:

$$\text{Frequency} = 1/\text{period} = 1/[2 \times \text{delay (seconds)}]$$

The higher-order cancellations will occur at frequencies with an odd number of half-wavelengths in the delay interval, so the simple equation develops to

$$\text{Dip frequency}_N = N \text{ (odd integers)}/(2 \times \text{delay})$$
$$= 1,3,5,7,\text{etc.}/(2 \times \text{delay})$$

For the comb filter shown in Figure 9.3a, the delay is 1 ms (0.001 s), and the first dip occurs at $1/2 \times 0.001 = 500$ Hz. The second occurs at $3/0.002 = 1500$ Hz, and so on.

To find the frequencies at which the peaks occur, the key fact is that the delay is always a multiple of entire wavelengths. The equation is

$$\text{Peak frequency}_N = N \text{ (all integers)}/\text{delay (seconds)}$$

Again, for the example comb filter, the first peak occurs at $1/0.001 = 1000$ Hz; the second at $2/0.001 = 2000$ Hz; and so on.

As shown on a logarithmic frequency scale, we cannot see the equally spaced, comblike appearance that is obvious when using a linear frequency scale. But a log frequency scale is normal because it is more revealing of what we hear. The peaks and dips are all there; they are just progressively crammed closer together with increasing frequency. In this example, the first dip-peak-dip sequence between about 200 and 2000 Hz might be the most audible effect. A first dip at 500 Hz tells us that the delay is a half-period at that frequency: 1 ms. This corresponds to a path-length difference of about 1.13 ft (344 mm)—a circumstance that is improbable in most home listening situations but that could occur as a console reflection in a control room setup (although this is likely to be even shorter). Subsequent peaks occur at simple multiples of the frequency, meaning they are so close at high frequencies that it is unlikely they will be audible. If we were to listen through such a filter, the sound would be significantly colored, primarily because of the amplitude fluctuations at lower frequencies.

If the delay were longer, as would be more typical in real rooms, the whole pattern moves down the frequency scale. Figure 9.3b shows a realistic case for a first-order reflection in a listening room, a reflection arriving after a 10 ms delay (corresponding to a path length that is 11.3 ft (3.4 m) longer than the direct-sound path. This might easily happen for a center channel (see Figure 8.5). The corresponding sound level might be 6 dB lower than the direct sound, but let us begin by showing the equal-level "theoretical" case. Obviously, the large undulations have moved to the bass region—below the transition frequency in small listening rooms, where sound propagation is dominated by wave-acoustic phenomena. For this reason, it is necessary to diminish the importance of the curve at these low frequencies. In fact, we can ignore these effects because they will be merged with, and indeed swamped by, the standing-wave/room resonances active in this frequency region. (This will be discussed in Chapter 13.) At frequencies above about 200 Hz, it can be seen that the comb filtering undulations are so closely packed together that they cease to be the audible problem suggested by Figure 9.3a. For longer delays, the density of the undulations is even greater.

Completing the transition to reality, let us apply a realistic attenuation to the reflection. Figure 9.3c shows the result of a 6 dB level reduction for the 10 ms delayed reflection. Now, not only are the undulations more dense in the frequency domain, but the amplitude variations are much reduced. So in an examination of realistic sound fields, comb filtering would appear to be a factor but not the alarming circumstance portrayed by Figure 9.3a.

In terms of physical measurements of such a situation, it is highly improbable that the result would look like Figure 9.3c because this would require extremely high resolution in the frequency domain—for example, a slow single-tone frequency scan or a very large time window in an FFT-based measurement. What would typically be seen would be a single line that might follow the first few undulations at the low end of the frequency scale, gradually showing an inability to reveal the full depth of the dips, and from some frequency upward simply reverting to a straight horizontal line following the top of the black area.

Spectral smoothing produces even smoother-looking room curves. A measurement with 1/3-octave resolution would not show any of the detail in Figure 9.3c; it would deviate from a straight horizontal line only slightly at the lowest frequency in this display (200 Hz). This

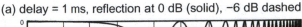

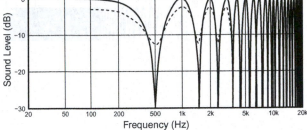

(a) delay = 1 ms, reflection at 0 dB (solid), −6 dB dashed

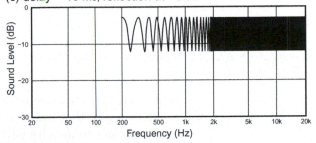

(b) delay = 10 ms, reflection at 0 dB

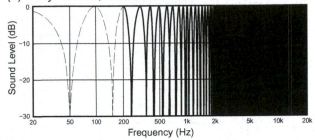

(c) delay = 10 ms, reflection at −6 dB

FIGURE 9.3 *Illustrations of comb filtering shown on a logarithmic frequency scale. (a) The solid line shows the interference pattern when the delay is 1 ms and the direct and delayed sounds have identical amplitudes. The dashed line shows what happens when the delayed sound is attenuated by 6 dB. (b) The delay is increased to 10 ms, moving the entire interference pattern down the frequency scale. Below 300 Hz, the curve is dashed to indicate that, in a small room, performance in this frequency range will be dominated by room resonances/standing waves. (c) The 10 ms delayed sound is attenuated by 6 dB, illustrating a situation realistic for a reflection in a normal room.*

is one reason why acousticians, for decades, have preferred smoothed versions of room curves: They look better and, typically, we are unable to hear the "grass"—the undulations at high frequencies. The explanation lies in the inability of the ear to separate spectral features that fall within a critical bandwidth

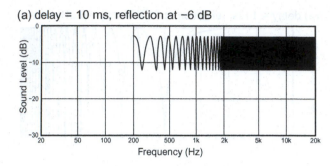

(a) delay = 10 ms, reflection at −6 dB

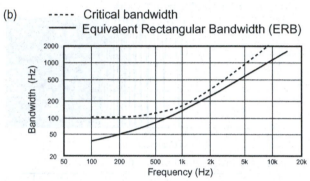

(b)
----- Critical bandwidth
—— Equivalent Rectangular Bandwidth (ERB)

FIGURE 9.4 *At the top is a repeat of Figure 9.3c. Below is the critical bandwidth and the equivalent rectangular bandwidth (ERB) taken from Moore, 2003, Figure 3.5.*

or its present-day variant, the equivalent rectangular bandwidth (ERB; Moore, 2003, Figure 3.5). Figure 9.4a repeats 9.3c and shows the critical bandwidths and ERBs at different frequencies. For the 10 ms delayed reflection illustrated in (a), spectral bumps and dips are separated by 100 Hz.

Obviously, when the ear is unable to perceptually separate such events, the details in a measurement are of little value. The implication of Figure 9.4 is that above 100 Hz, if one adheres to the traditional critical bands, or above about 500 Hz by the new ERB criterion, the ear cannot hear evidence of this comb filtering, and at lower frequencies the effects will be much less severe than the visual presentation suggests. As the delay increases, as would be the case for many small room reflections, the spacing between adjacent peaks and dips is reduced, more of them fall within the critical/ERB bandwidth, and the potential for audible effects is further lessened. In Section 19.2.2, the concept of critical/ERB bandwidth will be revisited, only from a different perspective: in terms of measurements that define the source of sound, the loudspeakers. There it will be argued that higher-resolution measurements are necessary. The reason has something to do with what is discussed in the following section: A problem with the sound source is present in all sound radiated into the room, in the direct sound that is heard and all reflections of it.

9.1.1 Very Audible Differences from Similar-Looking Combs

To add realism, just imagine that there is more than one reflection, some earlier and many later than the one we have been considering, each with a distinctive interference pattern when it arrives at the ears and, of course, slightly different at each ear. Clark (1983) conducted some listening evaluations and measurements that illustrate these effects very well.

Starting with a standard stereo arrangement in a normal listening room, frequency response measurements were made in three situations that yielded the same comb-filter interference pattern (see Figure 9.5). The first was a measurement at one ear location, with a microphone 4 inches (100 mm) off the stereo symmetrical axis. In a phantom center image situation, both loudspeakers radiate the same sound, so each ear location receives the direct sound from the nearer loudspeaker and then a slightly delayed version of it from the opposite

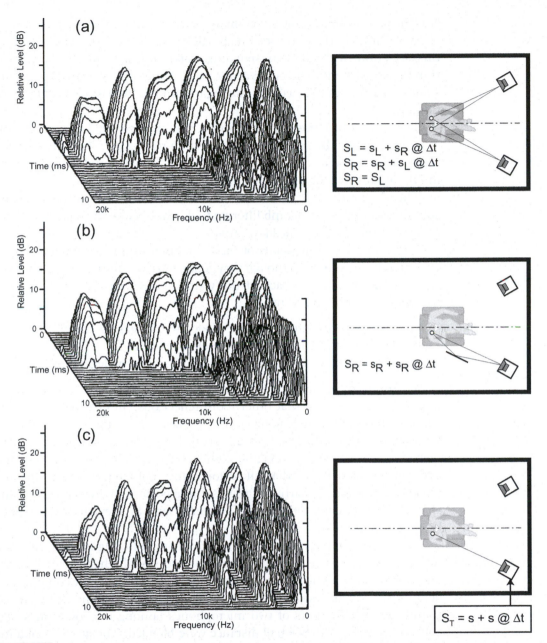

FIGURE 9.5 (a) A waterfall measurement for a microphone located at an ear position (no person was present) when a stereo pair of loudspeakers was radiating the same sound. The comb filter is the result of the direct sounds from both loudspeakers arriving at slightly different times. (b) The effect of a single lateral reflection having the same delay as (a). (c) The effect of adding the signal to itself, with a delay, before it is radiated into the room. Note the reversed frequency scale. Adapted from Clark, 1983.

loudspeaker. In this simple experiment, there was no head between the microphone locations—the effects of which will be illustrated later. Figure 9.5a shows this situation, with the corresponding waterfall (amplitude vs. frequency vs. time) measurement. The curve at the back (time = 0) is related to the steady-state measurement. The curves moving toward the front show events as a function of time.

The second situation involved sound from a single loudspeaker that arrived at an ear location directly, and after reflection from a (2 × 3 ft, 0.6 × 1 m) panel positioned to produce the same delay in the reflected sound path that occurred in (a). Figure 9.5b shows this.

The third situation involved electrically delaying the playback signal and adding it to itself so the comb filtering took place before the signal was radiated into the room. Figure 9.5c shows this.

The subjective impressions of the three circumstances shown in Figure 9.5 were greatly different even though they generated very similar frequency-response curves (the curve at the back of the waterfall diagrams). According to Clark, listeners found the following:

(a) The stereo phantom image: "moderate to pleasing effect"
(b) The reflector delay: "very small effect"
(c) The electronic delay in the signal path: "greatly degrading effect"

Returning to Figure 9.5 and observing what happens in the sound decay interval, it can be seen that other room reflections, all of which have different comb patterns, fill in the notches in the initial comb up to about 8 kHz (the loudspeakers that were used, UREI 813B, have diminished off-axis output at high frequencies, similar to Figure 2.6b). The same behavior can be seen in (a) and (b), both situations that allow for many room reflections from different directions and times to arrive at the listening position. Obviously, this, and the spatial effects of early reflections (Chapters 7 and 8) alleviate what might have been bad situations.

The worst situation for the audibility of comb filtering is when the summation occurs in the electrical signal path, as occasionally happens in live events, and in broadcasts, when a signal accidentally gets routed so it combines with a delayed version of itself. We hear these occasionally in news broadcasts; it sounds as if the outputs of two microphones (among the sometimes massive collections) separated by small distance were blended. The audible coloration is not subtle. The problem is that in the listening room, the direct sound and all reflected versions of it contain the same interference pattern. This explains the visibility of the notches throughout the duration of the decay in Figure 9.5c: There simply is very little energy in the room at the notch frequencies.

Large public-address/sound-reinforcement loudspeaker arrays, with many units spread out over a large area, all radiating the same sounds, obviously have the potential to be problematic. Dealing with these issues is fundamental to the

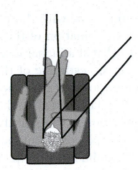

FIGURE 9.6 *An illustration of the individual paths from a front-center loudspeaker and a lateral wall reflection of that loudspeaker. The direct sound arrives identically at each ear, but the reflected sound travels farther to the opposite ear than it does to the near ear. This path-length difference translates into different arrival times, acoustical shadowing by the head translates into reduced acoustical interference at high frequencies, and, consequently, there are different interference patterns at each ear, all of which progressively diminish at high frequencies.*

design of such arrays, and elaborate measurements of the system components are used with mathematical models to predict the three-dimensional sound field radiated by arrays to anticipate and lessen these effects.

Another difficult situation is when there is only a single dominant reflection arriving from close to the same direction as the direct sound. In a control-room context, this could be a console reflection in an otherwise dead room.

9.1.2 Binaural Hearing, Adaptation, and Comb Filtering

Room reflections arrive from directions different from the direct sound, delivering different sounds at different delays to each of the ears. This means that the details of comb filtering will be different in each ear. Figure 9.6 shows that for a direct sound arriving symmetrically from the front, a single side-wall reflection will experience a significant path-length (i.e., arrival-time) difference between the two ears. The difference, about 0.4 ms in the example shown, is a significant fraction of a period at frequencies from the high hundreds of Hz upward. This means that the acoustical interference/comb filtering patterns will be different at the two ears, slightly at low frequencies and greatly at high frequencies. But there is more; because the head is a substantial acoustical obstacle between the ears, sounds arriving from angles away from the forward direction will be different in spectrum at the two ears. This factor was not considered in the data shown in Figure 9.5a, although it does not change the conclusions drawn from the measurements. The effect of acoustical shadowing by the head is that sounds that arrive at the more distant ear will have reduced amplitude at high frequencies, meaning that the comb filter acoustical interference will also be reduced. So what do we hear?

Interestingly, humans seem to cope with these situations very well because the spectrum we perceive is a combination of those existing at both ears. It is a "central spectrum" that is decided at a higher level of brain function. A microphone, at best, can give us a crude estimate of the sound that one ear might hear (crude because a microphone does not have the directivity pattern of human ears—described by the HRTFs). The fact that the perceived spectrum is the result of a central (brain) summation of the slightly different spectra at the two ears significantly attenuates the potential coloration from lateral reflections (Bilsen, 1977; Zurek, 1979). Krumbholtz et al. (2004) confirmed and extended Zurek's work, showing that the central spectral average of different sound events at the two ears can be mimicked by adding the stimuli from the two ears and presenting the sum identically to both ears.

If many reflections from many directions are present, the coloration may disappear altogether (Barron, 1971; Case, 2001; Moulton, 1995), a conclusion we can all verify through our experiences listening in those elaborate comb filters called concert halls.

Zurek observed another important effect: The spectral smoothing from multiple reflections occurs even when the delayed sounds are at levels 30–40 dB below the direct sound. This remarkable finding helps further explain why sound in rooms is so pleasant and adds weight to the reflected sounds shown in the measurements of Figure 9.5. Late, much-attenuated reflections appear to have their importance elevated by a kind of "automatic gain control" process and thereby are able to contribute significantly to the perceptual spectral smoothing. Blauert (1996) summarizes, "Clearly, then, the auditory system possesses the ability, in binaural hearing, to disregard certain linear distortions of the ear input signals in forming the timbre of the auditory event."

Superimposed on all of this is a cognitive learning effect, a form of "spectral compensation" wherein listeners appear to be able to adapt to these situations and hear "through and around" reflections to perceive the true nature of the sound source (Watkins, 1991, 1999, 2005; Watkins and Makin, 1996). Put differently, it seems humans have some ability to separate a spectrum that is changing (the program) from one that is stationary (the transmission channel/propagation path). This is a form of perceptual adaptation, which takes time to achieve full effect. It cannot happen if the situation is not stable, as when one is in motion.

It has been a habit of some acoustical practitioners to play pink noise and walk around the room listening for the telltale "swishes" of acoustical interference, but this is problematical for the following reasons:

- First, broadband noise is far more revealing of this kind of coloration than the music and speech we normally listen to.

- Second, a listener in motion cannot adapt. It is an unrealistic test. If one sits down, it is likely that the coloration will fade. If the listening begins from a seated position, which is the normal pattern of things, unless there is a truly unusual situation, the coloration will not be heard, especially in normal program material.

■ Third, this kind of activity is often performed closer to the loudspeakers than the listeners are seated. The demonstrative test may be performed in the "near field" of the source, which can be the loudspeaker combined with the reflection, when the listening positions may be in the far field, where the wavefront is more developed. Section 18.1.1 talks more about this.

So the conclusion is that in listening rooms, the potentially "ugly" effects of comb filtering are progressively alleviated by the following:

■ Natural acoustical events (rooms have many reflections, with differing delays and progressive amounts of attenuation)

■ The limited spectral resolution of the hearing system (familiarly known as critical bands, currently represented by equivalent rectangular bandwidth)

■ Binaural hearing (central spectrum smoothing of the different interference patterns at the two ears)

■ Spectral adaptation (an innate ability to separate out a constant acoustical coloration and to compensate for it)

The upshot is that, in any normal room, audible comb filtering is highly improbable. The less "live" the room, the more likely it will be that even a single reflection can be audible as coloration. This is a good point to look at again in Figure 9.3. Measurements don't lie, but some of them, like these, are not the most direct path to the truth that matters: what we *hear*. The reflections that cause comb filtering are the same reflections that result in the almost entirely pleasant, pleasurable, and preferable impressions of spaciousness discussed in the previous two chapters.

9.1.3 An Important One-Toothed Comb—A Fundamental Flaw in Stereo

On the matter of the sound quality of the center phantom image in stereo, I recommend a simple experiment: Arrange for monophonic pink noise to be delivered to both loudspeakers. When seated in the symmetrical sweet spot, this should create a well-defined center image midway between the loudspeakers. If it does not, something is seriously wrong. If it does, consider what you hear as you lean *very slightly* to the left and to the right of the symmetrical axis. The timbre of the noise changes and more obviously the closer you sit to the loud-speakers. In fact, it is possible to find the exact sweet spot by simply listening to when the sound is dullest. Moving even slightly left or right of the sweet spot causes the sound to get audibly brighter; there is more treble. It is much more exact to find the sweet spot by listening to the timbre change than by trying to judge when the center image is precisely localized in the center position. There is nothing faulty with the equipment or setup; this is simply stereo as it is—flawed.

In Figure 9.7a, a listener receives direct sounds, one per ear, from a center channel loudspeaker. Figure 9.7b is the situation for a phantom center image;

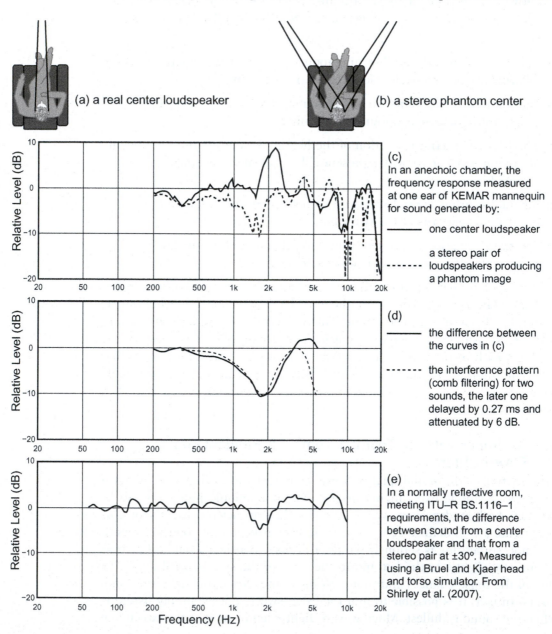

(a) a real center loudspeaker

(b) a stereo phantom center

(c) In an anechoic chamber, the frequency response measured at one ear of KEMAR mannequin for sound generated by:

—— one center loudspeaker

------ a stereo pair of loudspeakers producing a phantom image

(d) —— the difference between the curves in (c)

------ the interference pattern (comb filtering) for two sounds, the later one delayed by 0.27 ms and attenuated by 6 dB.

(e) In a normally reflective room, meeting ITU–R BS.1116–1 requirements, the difference between sound from a center loudspeaker and that from a stereo pair at ±30°. Measured using a Bruel and Kjaer head and torso simulator. From Shirley et al. (2007).

there are two sounds per ear, one of which is delayed because of the additional travel distance. Both of these are symmetrical situations, so the sound in both ears is essentially identical. Figure 9.7c shows the frequency responses measured at one ear when sounds were delivered by a center loudspeaker and then by a stereo pair of loudspeakers, set up in an anechoic chamber. Measurements were made using a KEMAR mannequin, an anthropometrically and acoustically correct head and torso with ear canals terminated by microphones and correct acoustical impedances. The curves are *very* different. The solid curve is the "correct" one. It shows what a real center sound source delivers to a listener. The dashed curve, for a phantom center, includes the effects of acoustical interference caused by the acoustical crosstalk from the two loudspeakers and, incidentally, the effects of the incident sounds arriving at the ears from the wrong directions: ±30° rather than straight ahead.

Figure 9.7d shows the difference between the curves, revealing the result of acoustical interference. This can be confirmed by a simple calculation. The time differential between the ears for a sound source at 30° away from the frontal axis is about 0.27 ms for an average head. A destructive acoustical interference will occur at the frequency at which this is one-half of a period: 1.85 kHz. It won't be a perfect cancellation because of a tiny propagation loss and a significant diffraction effect. The wavelength is just over 7 in. (178 mm), which, because it is similar in dimension to the head, will experience a substantial head-shadowing effect at the ear opposite to the sound source. There will be an interaural amplitude difference of the order of 6 dB in this frequency range. Taking a simplistic view, and following the pattern of Figure 9.3, the dashed curve in Figure 9.7d shows the first cancellation dip (the first "tooth") of a comb filter with these parameters. The fit is quite good. The cause of the "dullness" in the phantom center image is destructive acoustical interference. The rest of the comb is not seen because at higher frequencies the

---

FIGURE 9.7 *Anechoic frequency-response measurements (c) made at one ear of a KEMAR mannequin for sounds arriving from a real center loudspeaker, shown in (a), and from a stereo pair, shown in (b). The curves contain the axial frequency response of the loudspeaker used in the test, as well as the HRTFs for the relevant incident angles for that particular anthropometric mannequin. The important information, therefore, is in the difference between the curves that, around 2 kHz, is substantial. The smoothed difference is shown in (d). Nothing is shown above about 5 kHz because it is difficult to separate the effects of this specific acoustical interference from those of other acoustical effects. The dashed curve is the first interference dip (the first "tooth" of the comb filter), estimated in the manner of Figure 9.3, for an interaural delay of 0.27 ms (appropriate for a loudspeaker at 30° left or right of center) and for an attenuation of the delayed sound of 6 dB. (e) The same kind of measurement done in a normally reflective room, showing that early reflections within the room reduce the depth of the interference dip. Data from Shirley et al., 2007.*

whole situation is muddied by head-related transfer functions and rapidly increasing attenuation of the delayed sound at higher frequencies caused by head shadowing.

By any standards, this is a huge spectral distortion, a serious fault because it affects the featured "talent" in most recordings—the person whose picture is on the album cover. Under what circumstances are we likely to hear it as shown here? Obviously, only when the direct sound from the loudspeakers is the dominant sound arriving at the listener's ears. This is the situation in many recording control rooms and custom home theaters, where special care is taken to attenuate early reflections.

In normally reflective rooms, reflections that arrive from other directions at different times will help to fill in the spectral hole because there will be no acoustical interference associated with those sounds. Therefore, in normally reflective rooms, this will not be as serious as the curve in Figure 9.7d suggests—a fact confirmed by data from Shirley et al. (2007) in the data shown in Figure 9.7e.

It is a clearly audible effect. Augspurger (1990) describes how easy it is to hear the effect using 1/3-octave bands of pink noise, observing a "distinct null at 2 kHz" (p. 177). Pulkki (2001) confirmed that the comb filter was the dominant audible coloration in anechoic listening to amplitude-panned virtual images but that it was lessened by room reflections. Listeners in experiments by Choisel and Wickelmaier (2007) reported a reduction in brightness when a mono center loudspeaker was replaced by a stereo phantom image (their Figure 4). Shirley et al. (2007) measured the interference dip in a normally reflective room, Figure 9.7e, illustrating a substantial reduction in depth of the interference dip. Nevertheless, listeners in their experiments not only heard the dip but demonstrated that it had a significant negative effect on speech intelligibility.

Reflections and reverberation added to the mix will also help. However, the more acoustically "dead" the listening environment and the closer we sit to the loudspeakers, the more dominant will be the direct sound and the stronger will be the effect. This means that the common practice of eliminating reflections in control rooms and the common use of so-called "near-field" loudspeakers, sitting on the meter bridge of a console, both create situations where this problem is more likely to be audible. Taking a positive attitude to the effects of this on recordings, perhaps it will be the motivation to add delayed sounds to the vocal track, filling the spectral hole and in a very tangible manner "sweetening" the mix. On the other hand, if a recording engineer chooses to "correct" the sound by equalizing the 2 kHz dip, a spectral peak has been added to the recording that will be audible to anyone sitting away from the stereo sweet spot. Even worse, if the two-channel original is upmixed for multichannel playback, the center channel loudspeaker will not be flattered by the unnatural signal it is supplied with.

All of this should provide reasons to employ a real center channel in recordings, another point made by Augspurger (1990), who notes how very different, timbrally and spatially, a phantom image sounds in comparison to a discrete center sound source:

But, no matter what kind of loudspeakers are used in what kind of acoustical space, conventional two-channel stereo cannot produce a center image that sounds the same as that from a discrete center channel, even if it is stable and well defined. This leads to a certain amount of confusion in both the recording and playback processes. A dubbing theater that deals exclusively with motion picture sound does not have to worry about this problem. . . . A dialog track can be panned across the width of the screen without a noticeable change in tonal quality.

There is a parallel with an earlier discussion of the "seat-dip" effect in concert halls—another fundamental flaw (see Section 4.1.2). There too the fault was in the direct sound, and in many halls the audibility of the effect was diminished by reflections, often from the ceiling. It seems that reflections in rooms are coming to our rescue more often than they are creating problems.

9.2 EFFECTS OF REFLECTIONS ON TIMBRE— THE AUDIBILITY OF RESONANCES

Resonances are the "building blocks" of most of the sounds that interest, entertain, and inform us. Very high-Q resonances define pitches; they play the notes. Medium- and low-Q resonances add complexity, defining the character of voices or musical instruments. We learn to recognize patterns of resonances, including their relative amplitudes.

In sound reproduction, resonances are to be avoided. Added resonances alter the timbral character of voices and instrument in programs; they add coloration. The task of a sound reproduction system is to accurately portray the panorama of resonances and other sounds in the original sources, not to "editorialize" by adding its own.

In measurements of loudspeakers, it is common to find evidence of resonances, the normal clues being identifiable peaks in frequency response curves. In a single frequency response curve, a peak could be evidence of a resonance, or it could be the result of acoustical interference (as in the crossover region when two transducers are active). If a peak persists in a display of curves measured at different incident angles, on- and off-axis, it is highly probable that it is evidence of a resonance and not the result of acoustical interference that would be different in measurements made at different angles. All of this has been known for many years. What was missing was a perspective on how large a peak must be before it is evidence of a resonance that is audible as a coloration in audio programs.

Toole and Olive (1988) investigated and reported on the audibility of resonances using different program material, as well as expanding the investigation

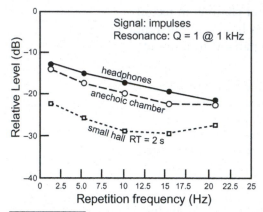

FIGURE 9.8 *Detection thresholds for a 1 kHz low-Q resonance using impulses, shown as a function of repetition rate and three different listening conditions: headphones, a single loudspeaker in an anechoic chamber, and in a small hall with a 2 s midfrequency reverberation time. From Toole and Olive, 1988, Figure 6.*

into an examination of how reflections in rooms influence that audibility. The core of this investigation will be discussed in Chapter 19. Here, we will look only at the influence of reflections on the detectability of resonances.

Figure 9.8 illustrates the crux of the matter: Repetition of a transient sound by reflections in the listening environment or by electronic regeneration of the signal makes low-Q resonances more audible; the threshold is lowered. The effect of electronic repetition is indicated by the downward slope of the curves. The exception is when there are many room reflections, the effect of electronic repetition reduces at high repetition rates, suggesting that the perceptual process has all the information it needs or can handle. It is somewhat surprising that there is a consistent difference between thresholds determined in headphone listening, where there can be no reflections, and those determined in an anechoic chamber, indicating that it is not perfectly anechoic and/or reflections from the listener's own body are sufficient to have an effect. The anechoic chamber had 3-ft (1 m) wedges and had been stripped of its floating mesh floor, so one must conclude that even a few reflections are audible contributors to this effect. That the target resonance was at 1 kHz (wavelength 13.6 in., 344 mm) implies that the reflecting surfaces must be of substantial cross section. In any event, the numerous repetitions in room reverberation have a dramatic effect, even with isolated 1 per second impulses, lowering the threshold by about 10 dB.

The first conclusion is that this helps to explain why live, unamplified music sounds better in a room than it does outdoors. In addition to the obvious spatial embellishments, it sounds tonally richer and more rewarding because we can hear more of the timbral nuances. In highly reverberant spaces, one can enjoy the lingering timbres of music in the decaying reverberant "tails" after the musicians have ceased to produce sound. These data suggest that timbre can be meaningfully enriched by fewer and much more subtle reflections.

In studio recordings, it is almost a ritual to add some amount of natural or electronically generated reflected or reverberated sound. Some call it "sweetening" the mix. Figure 9.9 shows the effect of added reverberation on the detection of the same low-Q resonance used in Figure 9.8. It shows that even at the lowest reverberation time setting available on the simulator used in the test, 0.3 s, the detection threshold fell by almost 10 dB. By 1 s, the threshold reached a plateau near −15 dB. It would be interesting to know just how few reflections are necessary to achieve useful improvements in the detectability of reflections, but implications from the data in Figures 9.8 and 9.9 are that it is not a lot. If any justification were necessary, this is proof that some amount of artificial reverberation is beneficial to the perception of resonances in voices and musical instruments, thereby clarifying their timbral signatures. It also points out that the effects of equalization will be more audible in reverberated sound, and it will be similarly effective whether it is done before or after the reverberation processing.

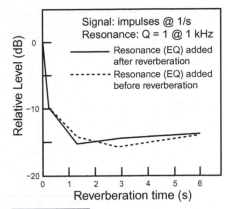

FIGURE 9.9 *The effects on detection thresholds of a low-Q resonance of adding electronically generated reverberation. From Toole and Olive, 1988.*

9.2.1 What Do We Hear—Spectral Bump or Temporal Ringing?

When we get to discuss the relative audibility of resonances having different Qs in Chapter 19, it may be surprising that, as they are revealed in frequency responses, we are more sensitive to the lower-Q phenomena—the ones that "ring" least. For years, we have seen oscilloscope presentations of resonant decays, and these days very attractive "waterfall" diagrams displaying the three axes of amplitude, frequency, and time are available. Both provide alarming information to our eyes, supporting notions that resonances (at least high-Q resonances) cannot be good for accurate sound reproduction. Subjectivists allude to the "smearing" of transient details, suggesting that they are able to hear these time-domain effects. But can they? Can anybody? It is interesting to test the notion.

Two experiments were devised to examine the extent to which time-domain information contributes to our perception of resonances. They were not very subtle, so the hope was that they should be persuasive. The first experiment began with the premise that transient sounds had the greatest potential to reveal audible ringing; the signal itself was very short—10 µs electrical pulses at 10/s—leaving the resonating "tail" starkly revealed to be heard. The detection tests were done in two very different environments: an anechoic chamber, where the temporal ringing should be clearly revealed, and a normally reflective listening room, where massive numbers of randomly occurring delayed versions of the transient would be superimposed, disrupting the tidy temporal pattern. If the resonance is audible because of its temporal misbehavior—the ringing—thresholds should be lower in the anechoic chamber.

Figure 9.10 shows that detection thresholds for the high-Q resonances, the ones that have most ringing, were only slightly changed by being auditioned in these two very different acoustical situations. On the other hand, the medium- and low-Q resonance thresholds were greatly changed, being audible at much lower levels in the reflective room. The 5 to 10 dB drop in the threshold is fairly persuasive.

The second experiment was conducted in an anechoic chamber, and this time it was the temporal structure of the signal that was changed. One test used the 10/s pulses, which would allow the resonant tails to ring down without

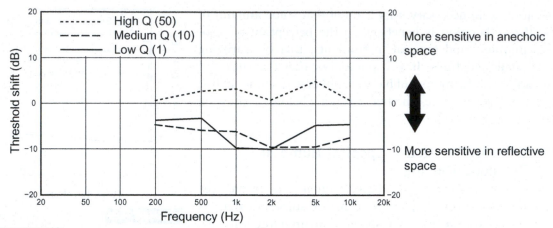

FIGURE 9.10 *Detection thresholds were measured for resonances added to 10 µs electrical pulses at 10/s, auditioned through a single loudspeaker in an anechoic chamber, and also in a normally reflective listening room. The results are shown here as the difference—the threshold shift—as the sound is moved from one listening environment to the other.*

interference. The other test used continuous pink noise, in which the impulses of random amplitude and timing would overlap and interfere, supposedly obscuring the ringing pattern.

Figure 9.11 shows no real change in detectability for the high-Q resonance using either signal, and a substantial 10 to 15 dB drop in the thresholds for medium- and low-Q resonances when pink noise was used.

Both of these tests make the argument that, at frequencies above 200 Hz at least, the detection process for resonances employs spectral information, not temporal cues. It seems that we are responding to the "bump" in the frequency response, an energy concentration, not ringing in the time domain. Repetitions, whether they are in the signal itself because of its temporal structure or added by the environment are obviously well used by the perceptual process in improved detection of medium- and low-Q resonances.

There is a possible explanation for the substantial contributions of signal repetitions to lowered thresholds. Viemeister and Wakefield (1991) investigated temporal integration or temporal summation, the phenomenon wherein the audibility threshold decreases with increasing signal duration. Conventional theories have been based on a leaky integrator concept, usually with a fairly long time constant—sometimes of the order of 300 ms. This work showed that the concept was fundamentally flawed, in that simple energy summation occurred only for pulse-pair separations of 5 ms or less. Beyond this separation the pulses were perceptually processed as if they were separate "looks." The thresholds for multiple pulses were lower than those for single pulses, but by an amount that

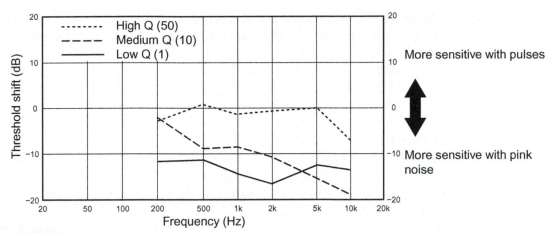

FIGURE 9.11 *Detection thresholds were measured for resonances added to 10 µs electrical pulses at 10/s and also to continuous pink noise, auditioned through a single loudspeaker in an anechoic chamber. The results are shown here as the difference, the threshold shift, as the sound is switched between pulses and pink noise.*

was relatively independent of delay (beyond a 3–5 ms interval). Wakefield (1994) developed a more elaborate model for multiple looks, and Buus (1999) provided more evidence in support of the multiple-look hypothesis.

9.2.2 Where Do We Find Timbral Identity?

The title of this section has to do with the portion of the sound from a musical instrument or voice that gives it a distinctive, recognizable character, in addition to the basic elements of pitch and loudness. Are some portions more consequential than others? The answer helps us to understand why reflections and repetition are so important.

The most distinctive timbral cues in the sounds of many musical instruments have been found to be in the onset transients, not in the harmonic structure or vibrato of sustained portions (Clark et al., 1963; Saldanha and Corso, 1964). According to Backus (1969), "The manner in which the various partials of the tone build up to their final amplitudes . . . is quite important in identifying the instrument; tones recorded without the initial transient are much harder to identify" (p. 102). The transient events being discussed are a mixture. Some are associated with the mechanical excitation of the resonant systems, hammers in pianos, fingers and plectrums in guitars, and others are the much more leisurely events associated with resonances building in milliseconds or tens of milliseconds after a string is plucked or struck, or puffs of air energize the resonant system of a horn with or without the added clatter of a vibrating reed. Strip away these transient events and the prolonged ringing decays of many stringed instruments, for example, may be confused with each other.

This being so, it is reasonable that repetitions of these transient onsets give the auditory system more opportunities to "look" at them and to extract more information.

To complete the story, one needs to examine what happens at frequencies below 200 Hz. It should be no surprise that at very low frequencies, the duration of ringing can be such that it becomes an audible extension of notes—bass "boom"—most easily detected in damped, impulsive sounds like kick drums, plucked bass, and so forth. It is reasonable to think, therefore, that in this frequency range, depending on the nature of the program, listeners at certain times may be sensitive to spectral characteristics and, at other times, to temporal characteristics. For now, it is sufficient to say that low-frequency resonances in rooms behave as minimum-phase phenomena, meaning that if there is a prominent "bump" in the frequency response, it is probable that this will be heard as excessive loudness at that frequency and that for transient sounds, there will be bass "boom" at that frequency. Using equalization to reduce the bump also attenuates the ringing so both problems are solved simultaneously.

Summarizing this chapter, on the topic of the role of reflections in the corruption or enhancement of timbre—sound quality—it is now evident that in normal listening rooms there is little risk of corruption (by comb filtering) and substantial evidence that resonances will be rendered more audible. If those resonances are in the program material, it is possible that the added tonal richness and timbral subtleties will be welcomed. If those resonances are in loudspeakers, it is possible that their enhanced audibility will not be welcomed.

A speculation: In Chapter 8, an experiment was described in which two wide-dispersion loudspeakers were compared to a loudspeaker with reduced lateral dispersion. The wide-dispersion loudspeakers were preferred, in spite of their both having irregular off-axis frequency responses. Even with this defect, the wide-dispersion loudspeakers were judged to be superior in terms of *both* sound quality and spatial quality. The stronger lateral reflections would generate a greater impression of ASW/image broadening, which is probably a positive attribute, and the same reflections, and those that follow them, will contribute to an enhanced sense of timbral richness, which is probably also beneficial. It is something to think about.

CHAPTER 10

Reflections and Speech Intelligibility

A sound reproduction system can have no greater fault than impaired speech intelligibility. Lyrics in songs lose their meaning, movie plots are confusing, and the evening news . . . well. In the audio community, it is popular to claim that reflected sounds within small listening rooms contribute to degraded dialog intelligibility. The concept has an instinctive "rightness," and it has probably been good for the acoustical materials industry. However, as with several perceptual phenomena, when they are rigorously examined, the results are not quite as expected. This is another such case.

10.1 DISTURBANCE OF SPEECH BY A SINGLE REFLECTION

When reflections intrude sufficiently to cause people to be "disturbed," something is seriously wrong. Muncey et al. (1953) and Bolt and Doak (1950) conducted studies on the disturbing effects of a single delayed sound on speech. This is an important issue in large venues as is evidenced by the attention it received at these early years. However, these investigations showed that natural reflections in small rooms are too low in amplitude and occur too soon to create problems of this kind (see Toole, 2006, Figure 8). We can move on to other issues.

10.2 THE EFFECT OF A SINGLE REFLECTION ON INTELLIGIBILITY

In the field of architectural acoustics, it has long been recognized that early reflections *improve* speech intelligibility. For this to happen, they must arrive within an "integration interval" within which there is an effective amplification

161

of the direct sound; it is *perceived* to be louder. For speech, reflections at the same level as the direct sound contribute usefully to the effective sound level, and thereby the intelligibility, up to about a 30 ms delay. For delayed sounds 5 dB lower than the direct sound, the integration interval is about 40 ms. Beyond about 95 ms, delayed speech components diminish intelligibility (Lochner and Burger, 1958). All of these experiments were done against a quiet background.

More recent investigations confirmed these findings and found that intelligibility progressively improves as the delay of a single reflection is reduced, although the subjective effect is less than would be predicted by a perfect energy summation of direct and reflected sounds (Soulodre et al., 1989).

Nakajima and Ando (1991) investigated the effect of a single reflection arriving from different directions on the intelligibility of speech. In this study, the reflection was at the same sound level as the direct sound, which makes this a "worst-case" test. The fact that it was done in a quiet anechoic chamber means that signal-to-noise ratio was not an issue. Within the time interval in which strong early reflections are likely to occur in listening and control rooms (about 15 ms), and adding the real-world fact that they will be attenuated by propagation loss and reflection attenuation, the evidence suggests that there would be no negative impact on speech intelligibility.

Summarizing the evidence from these studies, it seems clear that in small listening rooms, some individual reflections have a negligible effect on speech intelligibility, and others improve it, with the improvement increasing as the delay is reduced.

10.3 MULTIPLE REFLECTIONS, NOISE, AND SPEECH INTELLIGIBILITY

Following the pattern set by studies involving single reflections, Lochner and Burger (1958), Soulodre et al. (1989), and Bradley et al. (2003) found that multiple reflections also contribute to improved speech intelligibility. The most elaborate of these experiments used an array of eight loudspeakers in an anechoic chamber to simulate early reflections and a reverberant decay for several different rooms (Bradley et al., 2003). The smallest was similar in size to a very large home theater or a screening room (13 773 ft³, 390 m³). The result was that early reflections (<50 ms) had the same desirable effect on speech intelligibility as increasing the level of the direct sound. The authors go on to point out that late reflections (including reverberation) are undesirable, but controlling them should not be the first priority, which is to maximize the total energy in the direct and early reflected speech sounds. Remarkably, even attenuating the direct sound had little effect on intelligibility in a sound field with sufficient early reflections. Isolating reverberation time as a factor in school classrooms, it was found that optimum speech communication occurred for RT in the range

0.2 to 0.5 s (Sato and Bradley, 2008). This is conveniently the range of RT found in normally-furnished domestic rooms.

They also went further, looking at how multiple reflections were integrated. It was found that background noise disrupts the perfect integration of reflections, rendering them less effective aids to intelligibility than in the quiet (Soulodre et al., 1989). Signal-to-noise ratio is important, but the noise levels at which significant degradation occurs far exceed anything that would occur, much less be acceptable, in any home situation.

10.4 THE EFFECTS OF "OTHER" SOUNDS— SIGNAL-TO-NOISE RATIO

Background noise in a domestic room due to HVAC, outside traffic, and so on is not the main issue. To achieve high percentages in speech intelligibility, a signal-to-noise ratio of 5 dB is good, and 15–20 dB is nearly perfect. Noise, in this context, is everything other than the speech we want to hear. When several people talk at the same time, the noise is speech itself. In music, it is the sound of the band with which a vocalist is singing. In movies, it is everything else in a soundtrack occurring at the same time as the dialogue. For long passages in films and television programming, this is atmosphere-inducing music. When the action starts, all caution is abandoned, and things can get very noisy. In domestic listening situations, therefore, so far as speech intelligibility is concerned, ambient background noise is not a factor.

The biggest problems are intrinsic to the programs themselves. Obviously, the sound mixers pay attention to this, but they have two huge advantages over the rest of us. They have the script, and they get to hear each section of a film many times as they develop the audio design. They know the dialogue before they even hear it and understand it even if they don't pay attention.

The experience of a great many consumers of entertainment is that the ability to "rewind and play" and the options of having subtitles and closed captions exist for a reason. The reason has nothing to do with inferior loudspeakers or room acoustics. As we get older, we expect to have occasional difficulties, but one hears the same complaints from people who have no hearing problems.

Why is it that, within the same room and playback system, one can switch from highly intelligible "talking head" television programs to watch a movie, and dialogue is not always perfectly understood? Part of it is the dramatic effect of mumbles and whispers; movies have a significant dynamic range, whereas TV news and documentaries are highly compressed—always loud. Part of it is the inability to pick up clues from the lips, facial expressions, and so forth when the talker is not facing the camera. Part of it is the mixture of music and sound effects emerging from all of the loudspeakers in the room, creating atmosphere and supporting on-screen action. The latter degrades what technically is called "signal-to-noise" ratio, even though the sounds could hardly be described as noise in the normal sense. The result in the context of speech intelligibility is the same: more "noise," less intelligible speech. These extraneous, artistically, and aesthetically justified sounds create problems.

Shirley and Kendrick (2004) investigated the effects of different amounts of "extra" sound on listener impressions of dialogue clarity, overall sound quality, and enjoyment. Clarity is a quality that corresponds to intelligibility, although it is not a direct quantitative measure of that parameter. Some of the listeners had measurably normal hearing, whereas others had varying degrees of impairment.

The test conditions involved only the three front channels, LCR, replaying those components of 5.1-channel film clips. There were 20 excerpts, each 1–1.5 minutes long. The variable was the level of extraneous sounds delivered by the L and R loudspeakers, while the center loudspeaker delivered a constant level of dialogue. The first condition ran all three channels at reference level. The second condition attenuated the L and R channels by 3 dB, the third condition attenuated them by 6 dB, and the last condition ran the center channel alone.

The results shown in Figure 10.1 are interesting from several points of view; moving from left to right in the sequence of presentation styles results in a progressive increase in the dominance of the center channel. In terms of dialogue clarity, obviously *all* listeners thought this was a good idea, those with normal hearing and those with impaired hearing. This confirms that the structure of the soundtrack is a major factor in dialogue clarity and intelligibility (and remember, the surround channels are not involved in this test). The hearing-impaired listeners never got to the high levels of "clarity" reported by the normal-hearing group, but they could appreciate the improvement achieved by attenuating the L and R channels. So the lesson is, if you want to experience clear speech, listen in mono; turn off all the other channels.

In terms of "overall sound quality," there is some disagreement between the groups. Those with normal hearing preferred all the channels running at or close to reference levels; anything else was a degradation. Note that the difference between all three channels running at reference level, and the L&R loudspeakers attenuated by 3 dB, is trivial (and statistically insignificant). In total contrast, listeners with hearing disabilities, having difficulty understanding speech, clearly voted for progressively more monophonic sound. They obviously associated speech intelligibility with sound quality.

When we come to ratings of "enjoyment," the hearing impaired again placed great value on dialogue clarity because the pattern of ratings tracks the previous two rating categories. If the dialogue is not clear, then the movie is not enjoyable. It sounds entirely reasonable. Those listeners with normal hearing had trouble with this category because variations in the ratings were high, and differences between the averaged ratings shown here are not statistically reliable. To the extent that they may have meaning, it is interesting that these listeners entered *any* votes for turning the L&R channels down, but they did.

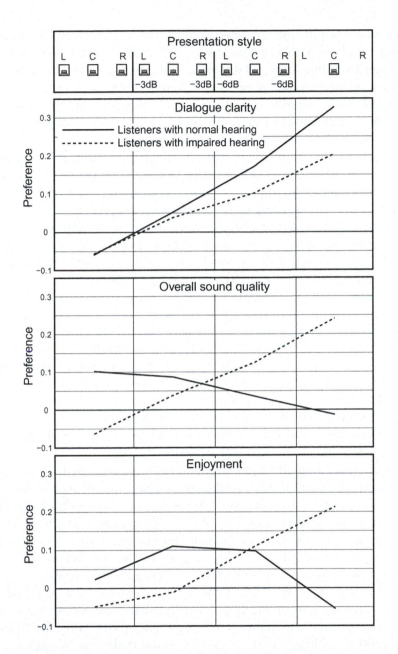

FIGURE 10.1

The subjective preferences for various LCR front-soundstage presentations judged in the categories of "dialogue clarity," "overall sound quality," and "enjoyment." From left to right, the presentation styles progressively emphasize the center channel, ending with it in isolation. Listener responses were divided into two groups: listeners with audiometrically normal hearing performance and listeners with hearing impairments. Compiled from data in Shirley and Kendrick, 2004.

So in summary, listeners with normal hearing find themselves conflicted. In terms of "dialogue clarity," things improved as the L&R channels were progressively attenuated and even turned off. In terms of "enjoyment," it seems that they debated about whether 3 or 6 dB attenuation of the L&R channels was an improvement, suggesting that they put substantial value in dialogue clarity, but there was not a clear result. However, in terms of "overall sound quality," 3 dB attenuation of the L&R channels was possibly acceptable, but more than that was rejected. Overall, the listeners with normal hearing seemed to be saying that they could live with a system in which the L&R channels were not running at reference levels but perhaps 3 dB lower. More than that was better for dialogue but was worse from other perspectives. Listeners with impaired hearing were utterly predictable. Anything less than mono was a degradation. This does not mean that they dislike the sound of a multichannel presentation but that they place a higher priority on the clarity of dialogue.

These data do not paint an attractive picture for multichannel audio for those among us with deteriorated hearing. The good news is that other tests done by these authors indicate an improvement in speech clarity when the person speaking is facing the camera; consciously or subconsciously, we all read lips. Perhaps changes in cinematic technique can compensate somewhat for the reduced speech clarity caused by the invocation of distracting sounds in other channels. Otherwise, it would seem that multichannel audio is a medium best suited for normal, normally younger, ears. It is interesting to note that in their population of 41 subjects, with ages ranging beyond 75, those exhibiting hearing impairment began showing up in the 30–44 age group, with progressive deterioration at more advanced ages. The message here is that this is a factor to consider in all home theater installations. The challenge is to decide what to do about it.

This is not news to the motion picture industry. Allen (2006) mentions examples of dialog intelligibility problems traceable to both the director of the film and the operators of the theaters; if film sound tracks are too loud, they get turned down. Dialogue levels drop along with those of the special effects. There have also been conflicts caused by differences in the listening circumstances—dubbing stages and screening theaters.

This is a fundamentally important topic for serious research, perhaps funded by the movie industry. If intelligibility is compromised by artistic effects, and intelligibility is linked to enjoyment, then perhaps we need to have a multichannel downmix option that is capable of tracking dialogue and effects levels and adjusting them to maximize intelligibility for audiences with less than normal hearing capability—an increasing percentage of the population based on age statistics and the abusive listening habits practiced by our youth.

A recent survey found that 14.9% of U.S. children 6–19 years old had at least 16 dB hearing loss in one or both ears (Niskar et al., 1998). Many possible causes were identified, in addition to noise-induced loss. Lonsbury-Martin and Martin (2007) emphasize that in addition to the traditional workplace risks of noise-induced hearing loss, other factors can be harmful. Other threats to healthy hearing have included loud leisure-time activities involving, for example, sporting events, live amplified music, and recreational shooting. However, it is only within the past few decades that the general availability of personal music-players has made the risk of hearing damage seem more menacing.

The point is that a significant fraction of present and future home theater users and owners can be expected to have hearing loss. A convenient way to adjust the center-to-other-channels level balance would be an excellent feature in households with mixed populations of younger and older listeners. In the meantime, there are always subtitles.

10.5 LISTENING DIFFICULTY—A NEW AND RELEVANT MEASURE

An important part of being entertained is being able to relax and become absorbed in music or the plot of a good movie. We know that intelligible film dialogue and vocalizations in music are crucial to those forms of entertainment, and we are reassured when tests indicate that circumstances yield near perfect intelligibility scores. And yet, we can all think of situations where understanding speech was anything but relaxing; it was actually hard work and required a level of focus and attention that detracted from the total experience.

Figure 10.2 shows a comparison of conventional word intelligibility scores and the new rating: listening difficulty (Sato et al., 2005). It can be seen that when the speech is at the same level as the noise (U50 = 0), word intelligibility is very high, nearing 100%. Under those conditions, though, listeners report great difficulty in understanding the words, reporting 90% listening difficulty. Elevating the speech (or reducing the noise) by 5 dB hardly changes the already excellent word intelligibility, but the listening difficulty drops dramatically—to

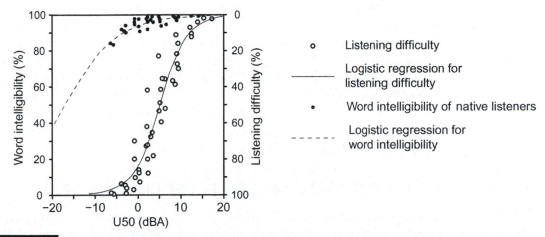

FIGURE 10.2 *A plot of conventional "word intelligibility" scores and of "listening difficulty"—a judgment of how much attention was required of the listener to recognize test words. The scale ranged from "not difficult: no effort required, completely relaxed listening condition" to "very difficult: considerable attention required." The horizontal scale, U50, in units of dBA, is the A-weighted useful-to-detrimental ratio within a 50 ms early time interval. It is a special evaluation of signal-to-noise ratio in which the useful sound (speech) is compared to detrimental sound (background noise), integrated over 50 ms, and A weighted so that low frequencies are discriminated against. At 0 dBA, the speech is at the same sound level as the noise. At 20 dBA, the speech is 20 dBA higher than the noise. From Sato et al., 2005, Figure 10.*

about 50%. For U50 of 10 dB and higher, listeners are understanding everything and are comfortably relaxed in doing so. Perhaps this was a factor in the results of Figure 10.1.

It is clear that "listening difficulty" ratings, are more sensitive indicators of problems than conventional intelligibility and word recognition scores and would seem to be more relevant to the assessment of entertainment content and reproduction systems. Intelligibility scores obviously are relevant, but they are most directly associated with the information content of a listening experience. "Listening difficulty" takes into account the entertainment value of an experience and the importance of allowing the participant to relax while understanding the message.

10.6 A REAL CENTER LOUDSPEAKER VERSUS A PHANTOM CENTER

It has long been noted that a phantom center has a sound accuracy problem, compared to a real center loudspeaker (discussed in Chapter 9). Although this leads to suspicions of reduced speech intelligibility—and, of course, altered perceptions of all sounds—it has never been put to a quantitative test until recently. Shirley et al. (2007) confirmed what seemed to be inevitable: the substantial comb-filter dip in the frequency response caused by the acoustical crosstalk results in a significant reduction in speech intelligibility, compared to a real center loudspeaker. They were able to relate the measured degradation in frequency response to a reduced ability to identify both vowels and consonants. As explained in the previous chapter and shown in Figure 9.7, the size of the interference dip at 2 kHz is reduced by room reflections, and so, presumably, would be the degradation in intelligibility. These experiments were conducted in an ITU-specified normally reflective room, so a more serious issue would be with listening environments in which early reflections have been eliminated by absorption, as in many professional recording control rooms. Preserving early reflections would appear to be a beneficial strategy in this respect.

10.7 A PORTABLE SPEECH-REPRODUCTION TEST

It is a convenient fact that the directivity of human talkers is not very different from those of conventional cone-and-dome loudspeakers (see Figure 10.3). The consequence of this is that if casual conversation is highly intelligible with one person in the location of the loudspeaker and another in the audience area, then it is probable that loudspeaker reproduction of close-miked vocals will be comparably intelligible. A large proportion of vocals in movies and television are close miked, containing little reflected energy. Because typical loudspeakers are more directional than human talkers, reflections will be at a *lower* level than for the real person. This is especially true for largish horn-loaded designs. The

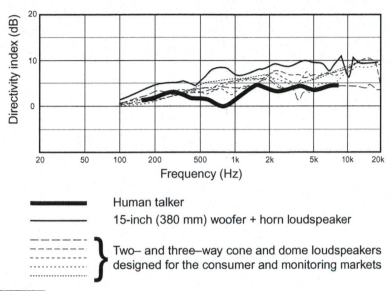

Human talker

15-inch (380 mm) woofer + horn loudspeaker

} Two– and three–way cone and dome loudspeakers designed for the consumer and monitoring markets

FIGURE 10.3 *The directivity index (DI) for a human talker, using data from Chu and Warnock (2002), reconfigured to be compatible with the Harman International format for loudspeaker measurements (see Figure 18.6). This is compared with directivity indices for several cone and dome loudspeakers of the kind found in homes and control rooms and for a large-format audiophile horn loudspeaker. A DI of 0 indicates an omnidirectional radiation pattern. A higher DI indicates a stronger forward bias in the sound-radiation pattern, generating lower-level reflections from room boundaries. These data indicate that at long wavelengths (low frequencies), all of the sources become omnidirectional and that the human talker does so again around 800 Hz.*

consequence is that to create comparable intelligibility, the loudspeaker reproduction may need to be at a higher sound level than a natural voice. If we add the further condition that there may not be the opportunity for lip reading, there is more justification for higher sound levels. Gilford (1959) noted that when asked to adjust a playback level to "correct loudness" for a colleague's voice, subjects "invariably set the level too high—sometimes by as much as 8–10 dB" (p. 258).

As discussed earlier in this chapter, none of this can guarantee highly intelligible dialogue in film and television programs. Other components of sound tracks emerging from all channels provide substantial interference. In addition, a high percentage of movie and television sound—dialogue, effects, and atmospheric sounds—emerges from the front-center loudspeaker. In this case, the human ability to binaurally discriminate against distracting sounds arriving from other directions cannot function. So, in any well-furnished domestic space or an equivalently treated dedicated home theater, it is highly improbable that the loudspeakers or the room acoustics contribute to problems in speech intelligibility in movies. If there are problems, ironically, the cause is most likely to be the sound track itself.

CHAPTER 11

Adaptation

"We humans adapt to the world around us in many, if not all, dimensions of perception— temperature, luminance level, ambient smells, colors and sounds, etc. When we take photographs under fluorescent or incandescent lighting or outdoors in the shade or direct sunshine, we immediately are aware of color balance shifts—greenish, orangish, bluish, etc. Yet, in daily life, we automatically adjust for these and see each other and the things around us as if under constant illumination. We adapt to low and high light levels without thinking. There are limits—very colored lighting gets our attention, we cannot look into the sun, or see in the dark, but over a range of typical circumstances we do remarkably well at maintaining a comfortable normalcy. Most adaptation occurs on a moment-by-moment basis, and is a matter of comfort—bringing our perception of the environment to a more acceptable condition. In the extreme, adaptation, habituation or acclimatization, whatever we call it, can be a matter of survival, and a factor in evolution."
—Toole, 2006

We have already seen a few examples of auditory adaptation. In the contexts of precedence effect (angular localization), distance perception, and spectral compensation (timbre), humans can track complex reflective patterns in rooms and adjust our processes to compensate for much they might otherwise disrupt in our perceptions of where sounds come from and of the true timbral signature of sound sources. In fact, out of the complexity of reflected sounds, we extract useful information about the listening space and apply it to sounds we will hear in the future. We are able, it seems, to separate acoustical aspects of a reproduced musical or theatrical performance from those of the room within which the reproduction takes place. This appears to be achieved at the cognitive level of perception—the result of data acquisition, processing, and decision making, involving notions of what is or is not plausible. All of it indicates a long-standing human familiarity with listening in reflective spaces and a natural predisposition to adjusting to the changing patterns of reflections we live in and with. The inevitable conclusion is that all aspects of room acoustics are not targets for "treatment." It would seem to be a case of identifying those aspects that we can, even *should*, leave alone and focusing our attention on those aspects that most directly interact with important aspects of sound reproduction: reducing unwanted interference on the one hand or enhancing desirable aspects of the spatial and timbral panoramas on the other.

11.1 ANGULAR LOCALIZATION— THE PRECEDENCE EFFECT

The topic was introduced in Chapter 6, but there is much more to the precedence effect. What Haas (1972) discussed in his 1949 thesis, as well as studies by his contemporaries and those that preceded him (well summarized in Gardner, 1968, 1969), was just the beginning. Recent research (e.g., Blauert, 1996; Blauert and Divenyi, 1988; Djelani and Blauert, 2001; Litovsky et al., 1999) suggests that the precedence effect is cognitive, meaning that it occurs at a high level in the brain and not at a peripheral auditory level. Its purpose appears to be to allow us to localize sound sources in reflective environments where the sound field is so complicated by multiple reflections that sounds at the ears cannot be continuously relied upon for accurate directional information. This leads to the concept of "plausibility" wherein we accumulate data we can trust—both auditory and visual—and persist in localizing sounds to those locations at times when the auditory cues at our ears are contradictory (Rakerd and Hartmann, 1985). Among localization phenomena encountered in audio/video entertainment systems are bimodal (e.g., hearing and seeing) interactions, including what we know as the ventriloquism effect, wherein sounds are perceived to come from directions other than their true directions.

At the onset of a sound accompanied by reflections in an unfamiliar setting, it appears that we hear everything. Then, after a brief build-up interval, the precedence effect causes our attention to focus on the first arrival, and we simply are not aware of the reflections as spatially separate events. Remember that this is not *masking*; in all other respects, the reflections are present, contributing to loudness, timbre, and so on. This suppression of the directional identities of later sounds seems to persist for at least 9 s, allowing the adaptation to be effective in situations where sound is not continuous (Djelani and Blauert, 2000, 2001). Figure 11.1 shows an exaggerated view of how we might localize a sound source in a small room in (a) the first impression and (b) after adaptation.

However, it would be wrong to think that this is a static situation. A change in the pattern of reflections, in number, direction, timing, or spectrum, can cause the initiation of a new build-up, without eliminating the old one. We seem to be able to remember several of these "scenes." All of this build-up and decay of the precedence effect must be considered in the design of experiments where spatial/localization effects are being investigated—namely, are the reported per-

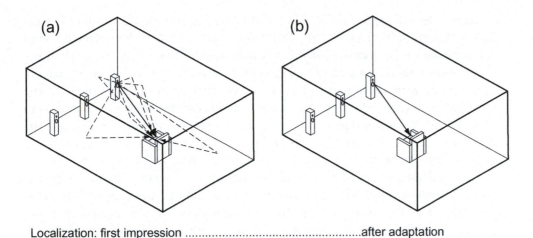

Localization: first impression ....after adaptation

FIGURE 11.1 *A simplistic illustration of how the precedence effect allows us to focus on the true direction of a sound source when listening in a reflective space. It takes time to develop, and it will fade from memory if the reflection pattern is changed or not reinforced from time to time.*

ceptions before or after precedence-effect build-up? In any situation where listener adjustments of acoustical parameters are permitted, adaptation may not occur at all. One has to think that this may be a factor in what recording personnel hear in control rooms—and it will be different from what is heard during playback at home.

This appears to be the essence of the observation by Perrot et al. (1989) and Saberi and Perrot (1990) that, with the practice that inevitably comes from prolonged exposure in experiments, listeners can ignore the precedence effect, detecting the delayed sounds almost as if they existed in isolation. In the context of the audio industry, it is conceivable that recording engineers, while focused on manipulating the ingredients of a mix, could find themselves hearing aspects of sounds that will be completely lost to those whose first exposure is to the final product. A component in this heightened sensitivity is likely to be the fact that a recording engineer can adjust the level of a sound component upward to the point where it is clearly audible, reduce it, and at any time turn it on or off. Under these circumstances, where the component can be aurally "tracked," it is highly probable that it can be heard at levels below those at which it is likely to be audible when listening normally to the completed mix. Thus, sounds that may be gratifying to the mixing or mastering engineer may be insufficient to reward a normal listener or, worse, simply not heard at all. Recall the observation in Chapter 8 that musicians can develop an elevated sensitivity to reflected sound, judging reflections to be about seven times greater than ordinary listeners.

Important for localization, and very interesting from the perspective of sound reproduction, is the observation that the precedence effect appears to be most effective when the spectra of the direct and reflected sounds are similar (Benade, 1985; Blauert and Divenyi, 1988; Litovsky et al., 1999). If the reflected sound has a "sufficiently" (not well defined at this stage) different spectrum from the direct sound, there is a greater likelihood that it will be separately localized and not merged with the direct sound so far as localization is concerned. In the extreme, if all early reflections had sufficient spectral differences, it would seem that the illustration in Figure 11.1a would not transition into (b). Because the precedence effect is more effective for broadband sounds than for narrow-band sounds, this seems to be a matter of importance (Braasch et al., 2003).

Spectral differences of this kind can originate in sound sources with frequency-dependent radiation patterns (many musical instruments) or frequency-selective reflecting surfaces. In live performances, this suggests that the precedence effect may not function perfectly, with a residue of at least some spatially distinct reflection "images" contributing to the illusion of image broadening or ASW (apparent source width). In the context of sound reproduction, where we might most wish for the loudspeakers not to draw attention to themselves, this appears to be an argument for constant-directivity loudspeakers (sending sounds with similar spectra in all directions) and frequency-independent (i.e., broadband) acoustical surfaces. If it is necessary to absorb, attenuate, scatter, or redirect reflections, the acoustical devices should be similarly effective over the entire spectrum above the transition frequency (say, 300 Hz), not part of it, so the sounds arrive spectrally intact at the listeners' ears. This is a non-trivial requirement, implying that resistive absorbers should be not less than 3 in. (76 mm) thick and that scattering devices/diffusers must be a significant fraction of a wavelength deep. See Chapter 21 for more detailed information on the performance of acoustical devices.

11.2 PERCEPTIONS OF DISTANCE

In thinking about sound reproduction and the perceptual dimensions that distinguish excellence, one of the foremost is distance. The idea that we can sit in our homes or cars and have a credible sense that recorded sounds are originating at points beyond the boundaries of our enclosures is highly desirable. It is a complicated process, relying a little (very little) on sound level, a lot on a running comparison between the direct sound and early reflections, and a lot on information about the environment and past experiences with familiar sounds. For example, a shouted voice is automatically assumed to be farther away than a conversational or whispered voice, irrespective of sound level (Philbeck and Mershon, 2002). In general, we tend to underestimate distance (Zahorik, 2002), which means that we humans have an innate bias against hearing the very illusion we want to create.

We pay attention to certain aspects of the sound field in rooms, accumulating contextual data within which to place sounds. Once learned, the knowledge transfers to different locations in the same room and, to some extent, to rooms having similar acoustical properties (Bronkhorst and Houtgast, 1999; Neilsen, 1993; Pellegrini, 2002; Schoolmaster et al., 2003, 2004; Shinn-Cunningham, 2001; Zahorik, 2002). Distance perception is another perceptual dimension with a cognitive component; it is not simply a mechanistic process.

All of this is clearly relevant to localizing real sources of sound in rooms; in audio these are the loudspeakers. However, successful localization of the loud-speakers may run counter to the objectives of music and film sound, which is to "transport" listeners to other, mostly larger, spaces. This requires impressions of distance that are not tied to loudspeaker locations, meaning that we need to know how listeners react to combinations of direct and reflected sounds associated with the loudspeakers in the listening room, together with combinations of direct and reflected sounds in the recordings being reproduced through those same loudspeakers. Inevitably, there will be a superposition of all of these sounds, meaning that each of the recorded direct and reflected sounds will have a set of accompanying reflections associated with the listening room. What are the rules governing perceptual dominance in this situation? A reasonable thought is that it is substantially determined by the manner in which the recording is made.

Listening through a typical multichannel audio system, the single-channel, hard-panned voice of a news reader is perceived as originating in the center-channel loudspeaker. This is localized as being a few feet from the listener, appropriately in the same direction and plane as some form of video display within the listening room. This would be true for any hard-panned voice or musical instrument in any channel. Yet, in the same situation, reflected sounds that have been artfully incorporated into a good multichannel recording can make all of the loudspeakers less obvious and cause the walls of the listening space to significantly recede.

If we understood the psychoacoustics of what is happening in these instances, we would be able to indicate the characteristic early reflections, whether generated within the room or incorporated into recordings, that are needed to establish dominance in our perception of distance. Hints that the perception of distance is more driven by monaural cues than binaural cues (Shinn-Cunningham, 2001) are encouraging, given the limited number of channels available in our audio systems. However, if there is even an element of "plausibility" in distance perception, it may be difficult not to be influenced by walls and loudspeakers that we can see. It is clear that more research is needed on this important topic.

11.3 SOUND QUALITY—TIMBRE

Section 9.1.1 introduced the notion of "spectral compensation" (Watkins, 1991, 1999, 2005; Watkins and Makin, 1996). It seems humans have some ability to separate a spectrum that is changing (the program) from one that is stationary (the transmission channel/propagation path) and to implement a form of correction, an adaptation that renders the perceptual distortion of the communication channel or the room less bothersome. The demonstrable fact that we have come to accept the gross insults to sound quality, the massive linear and nonlinear distortions present in cell phone communication, is evidence of some form of adaptation—or is it just toleration?

Many years ago, Gilford (1959) looked into the effects of listening-room acoustics on judgments of sound quality. Being with the British Broadcasting Corporation, he was primarily interested in voice. The tests consisted of recordings made in six studios, played in four different listening rooms to panels of engineers. Excerpts from the recordings were played, in paired-comparison fashion, so that each studio recording was compared to each other one. The results were statistically analyzed.

The results indicated that listening rooms with high reverberation, although not altering the order of preference for the recordings, increased the variability in the judgments. There was evidence of compensating errors: Listening rooms with excessive bass reverberation favored studios with less bass. A dead listening room favored a studio with long bass reverberation. Conclusion: There is an interaction between the bass characteristics of a recording due to the studio in which it was recorded and the bass characteristics of the listening environment. This appears not to be overcome by adaptation.

However, Gilford explains that "listening tests with recordings of speech from several studios played into the [different living] rooms showed that most of them allowed the characteristics of speech from the different studios to be distinguished, but some of the rooms introduced very severe colorations which entirely masked the other effects" (p. 254). Here, there appears to be some compensation for the individuality of rooms, but there is a limit to how much can be compensated for. Gilford goes on to say, "The fact that the listening room does not have a predominant effect on quality is very largely due to the binaural mechanism." James Moir, in discussion, added, "In my view, if a room requires extensive treatment for stereophonic listening, there is something wrong with the stereophonic equipment or the recording. The better the stereophonic reproduction system, the less trouble we have with room acoustics." Knowing what we do now, these were very insightful views. Both suggest some ability to "listen through" rooms to recognize intrinsic characteristics of the source(s).

Shinn-Cunningham (2003) takes a more academic view:

While the acoustic effects of reverberant energy are often pronounced, performance on most behavioral tasks is relatively robust to these effects. These perceptual results suggest that listeners may not simply be adept at ignoring the signal distortion caused by ordinary room acoustics, but may be adapted to deal with its presence. . . . Listeners are not only adept at making accurate judgments in ordinary reverberant environments, they are in fact adapted to the presence of reverberation and benefit from its presence in many ways.

Shinn-Cunningham used the word *reverberation* to describe all reflected sounds in a classroom that measured $16 \times 29.5 \times 11.5$ ft ($5 \times 9 \times 3.5$ m).

11.3.1 A Massive Test with Some Thought-Provoking Results

Olive et al. (1995) published results of an elaborate test in which three loudspeakers were subjectively evaluated in four different rooms. Figure 11.2 shows the rooms and the arrangements within them. The rest of this section is based on the account in Toole (2006).

In the first experiment, called the "live" test, listeners completed the evaluation of the three loudspeakers in one room before moving to the next one. The loudspeakers were all forward-firing cone/dome configurations, with similar directivities and similarly good performance, so it was not an obvious matter to make sound quality judgments. It was also a situation in which differences might have been masked by differences in room dimensions, loudspeaker location, or placement of acoustical materials.

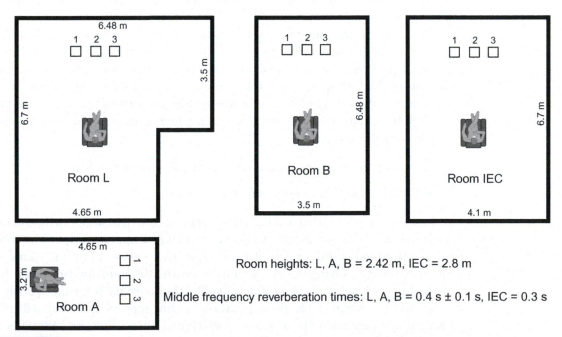

FIGURE 11.2 *Four listening rooms, showing arrangements in which listeners auditioned three loudspeakers. The loudspeakers were evaluated in each of the locations, using three different programs, by 20 listeners. Binaural recordings were made for subsequent headphone reproduction. Based on Olive et al., 1995, Figure 3.*

Binaural recordings were made of each loudspeaker in each location in each room, and the tests were repeated, only this time with listeners hearing all of the sounds through calibrated headphones. All tests were double blind. In each room, three loudspeakers were evaluated in three locations for each of three programs. The whole process was repeated, resulting in 54 ratings for each of the 20 listeners. These were the results from a statistical perspective:

- "Loudspeaker" was highly significant: $p = 0.05$.

- "Room" was not a significant factor.

- The results of live and binaural tests were essentially the same.

A possible interpretation is that the listeners became familiar with—adapted to—the room they were in and, this done, were able to accurately judge the relative merits of the loudspeakers. Since they were given the opportunity to become familiar with each of the four rooms, they were able to arrive at four very similar ratings of the relative qualities of the loudspeakers. Obviously, part of this adaptation, if that is the right description for what is happening, is an accommodation for the different loudspeaker locations. Different rooms and different positions within those rooms have not confused listeners to the point that they were not able to differentiate between and similarly rate the loudspeakers.

Then, using the same binaural recordings that so faithfully replicated the results of the live listening tests, another experiment was conducted. In this, the loudspeakers were compared with themselves and each other when located in each of the loudspeaker positions in each of the four rooms. Thus, in this experiment, the sound of the room was combined with the sound of the loudspeaker in randomized presentations that did not permit listeners to adapt. These were the results:

- "Room" became the highly significant variable: $p = 0.001$.

- "Loudspeaker" was not a significant factor.

It appears, therefore, that we can acclimatize to our listening environment to such an extent that we are able to listen through it to appreciate qualities intrinsic to the sound sources themselves. It is as if we can separate the sound of a spectrum that is changing (the sounds from the different loudspeakers) from that which is fixed (the colorations added by the room itself for the specific listener and loudspeaker locations within it). This appears to be related to the spectral compensation effect noted by Watkins (1991, 1999) and Watkins and Makin (1996).

It is not beneficial to overdramatize these results because, although the overall results were as stated, it does not mean that there were *no* interactions between individual loudspeakers and individual rooms. There were, and almost all of them seemed to be related to low-frequency performance. The encouraging part of this conclusion is that, as we will see in Chapter 13, there are ways to control what happens at low frequencies.

Still, there is more to consider because these tests involved single loudspeakers. The results are relevant because much of what we listen to can be considered monophonic (e.g., center channel sound in movies and TV and hard left-right panned sounds in stereo). In Section 8.2, we saw evidence that room interactions can have different effects on our perceptions of loudspeakers, depending on whether the test is done in mono or stereo and whether monophonic components are in the stereo program. The suggestion is that when listening in stereo to programs with strong decorrelated sounds in the two channels, the image-broadening effects of natural room reflections may be diluted or masked by spatial effects generated by the stronger direct sounds from the loudspeakers. If this is so, one might anticipate that tests performed using five channels would yield even smaller room effects. Nowadays we increasingly listen to multichannel audio, whether it is discretely recorded or upmixed from a two-channel source.

11.3.2 A Multichannel Test—And Something Is Learned

Olive (2007) and Olive and Martens (2007) describe experiments inspired by those in the previous section but conducted using a 3/2 multichannel audio system and five-channel discrete program material. The test was done in four rectangular rooms of different sizes and proportions, three of which had variations of normal acoustics and the fourth was deliberately made excessively reflective. The loudspeakers were, by normal standards, all excellent, having been especially constructed and equalized to have nearly identical, very flat, axial frequency responses, differing only in their off-axis (i.e., reflected sound) performance.

The test methodology was significantly different from that used in the previous tests, although the underlying process was the same: one set of loudspeaker evaluations in which all products were compared in the same room before moving to a different room, and a second set of evaluations in which the loudspeakers and rooms were combined in random sequence—all of which was possible using binaural room scanning (BRS) and headphone reproduction. The results showed that the effects of the rooms were stronger in the intermixed presentations, where adaptation would have been difficult, but there was a suspicion that the test presentations might have been long enough to allow some amount of adaptation in both tests. If so, room adaptation occurs in one or two minutes or less, which seems instinctively plausible. But there is the other factor: the program material was five-channel music selected because all the channels were generously used. It is also plausible that this resulted in the interactions of loudspeakers with naturally occurring room reflections being at least somewhat masked by the (higher-level) recorded reflections and decorrelated sounds. Obviously, a definitive test is needed.

Of interest was another finding: The experienced listeners were more discerning of loudspeaker differences (discriminating against the horizontal MTM—midrange-tweeter-midrange arrangement), whereas inexperienced listeners were more discerning of room effects (discriminating against the overly reflective room). The experienced listeners at issue gained their experience in evaluations of loudspeakers. It is a fascinating concept that one may be able to learn to become better at separating the sound of a source from the effects of the surroundings.

11.4 SUMMARY

It seems safe to take away from all this the message that listeners in comparative evaluations of loudspeakers in a listening room are able to "neutralize" audible effects of the room to a considerable extent. If residual effects of the room are predominantly at low frequencies, these differences, and also those in the reflected sound field, can be physically neutralized by employing a positional "shuffler" to bring active loudspeakers to the same location in the room (Olive et al., 1998).

There are everyday parallels to this. We carry on conversations in a vast range of acoustical environments—from cavelike to the near-anechoic—and although we are certainly aware of the changes in acoustical ambience, the intrinsic timbral signatures of our voices remain amazingly stable. The excellence of tone in a fine musical instrument is recognizable in many different, including unfamiliar, environments. Benade (1984) sums up the situation as follows:

The physicist says that the signal path in a music room is the cause of great confusion, whereas the musician and his audience find that without the room, only music of the most elementary sort is possible! Clearly we have a paradox to resolve as we look for the features of the musical sound that gives it sufficient robustness to survive its strenuous voyage to its listeners and as we seek the features of the transmission process itself that permit a cleverly designed auditory system to deduce the nature of the source that produced the original sound.

So we humans manage to compensate for many of the temporal and timbral variations contributed by rooms and hear "through" them to appreciate certain essential qualities of sound sources within these spaces. Because adaptation takes time, even a little, there is the caveat to acousticians not to pay too much attention to what they hear while moving around—either stop or sit down and listen!

With this in mind, the concept of "room correction'" becomes moot; how much and what really needs to change, and how much can the normal perceptual process accommodate to? What do we have the option of changing, and what should we simply leave alone?

In spite of the incomplete state of this area of work, there remains one compelling result: When given a chance to compare, listeners sat down in different rooms and reliably rated loudspeakers in terms of sound quality. Now we need to understand what it is about those loudspeakers that caused some to be preferred over others. If that is possible, it suggests that by building those properties into a loudspeaker, one may have ensured that it will sound good in a wide variety of rooms—a dream come true. The corollary to this is that we should be able to predict much about the sound quality of loudspeakers, at least above the transition frequency, by a thorough analysis of the sounds they radiate—that is, anechoic data.

CHAPTER 12

Adjacent-Boundary and Loudspeaker-Mounting Effects

Where a loudspeaker is placed in a room has a major effect on how it sounds, especially at low frequencies. Figure 4.10c is an excellent illustration of what happens when a loudspeaker is placed in three different, but entirely feasible, locations in a room. The measurable and audible differences are not subtle. What is seen in the figure is a combination of the effects of standing waves and adjacent-boundary effects. For this chapter, we focus on adjacent-boundary effects, which occur when the loudspeakers are less than a wavelength from one or more room boundaries. Then, depending on the distance from each boundary, a systematic acoustic interference causes fluctuations in the sound power radiated into the room.

The phenomenon was well known to acousticians (Waterhouse, 1958), but it was Allison and Berkovitz (1972) and Allison (1974) who introduced it to audio people in papers that described, in measurements, the significant dimensions of the problem. These will be discussed, but first it is necessary to begin at the beginning—by looking at what happens when loudspeakers are essentially "in" the boundaries: when the distance separating them is a small fraction of a wavelength.

12.1 SOLID ANGLES AND THE RADIATION OF SOUND

At 20 Hz, the wavelength is 56.5 ft (17.25 m). At this and similar frequencies, any practical separation between a loudspeaker and a room boundary is "small." In addition, any practical loudspeaker will be small relative to the wavelength, and therefore it will radiate sound in an essentially omnidirectional manner. For both of these reasons, the conditions are met for the classic set of relationships shown in Figure 12.1.

183

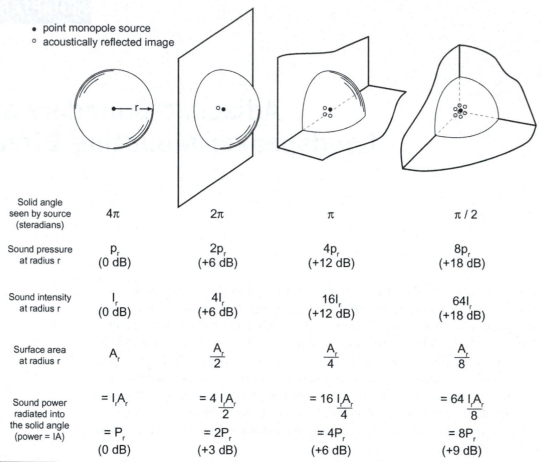

- point monopole source
- acoustically reflected image

Solid angle seen by source (steradians)	4π	2π	π	$\pi/2$
Sound pressure at radius r	p_r (0 dB)	$2p_r$ (+6 dB)	$4p_r$ (+12 dB)	$8p_r$ (+18 dB)
Sound intensity at radius r	I_r (0 dB)	$4I_r$ (+6 dB)	$16I_r$ (+12 dB)	$64I_r$ (+18 dB)
Surface area at radius r	A_r	$\dfrac{A_r}{2}$	$\dfrac{A_r}{4}$	$\dfrac{A_r}{8}$
Sound power radiated into the solid angle (power = IA)	$= I_r A_r$ $= P_r$ (0 dB)	$= 4\,I_r\dfrac{A_r}{2}$ $= 2P_r$ (+3 dB)	$= 16\,I_r\dfrac{A_r}{4}$ $= 4P_r$ (+6 dB)	$= 64\,I_r\dfrac{A_r}{8}$ $= 8P_r$ (+9 dB)

FIGURE 12.1 *Various measurable quantities at a fixed distance from a physically small omnidirectional sound source (a point monopole) in several basic locations closely adjacent to large, flat surfaces. This diagram was inspired by a simple one in Olson, 1957, Figure 2.1 and is based on a more elaborate version in Toole, 1988, Figure 1.11.*

The technical description of full spherical, omnidirectional radiation is that the sound source "sees" a solid angle of 4π steradians. It is a full space—a "free field" with no surfaces to reflect or redirect the radiated sound. Placing the sound source on or in a large plane surface reduces the solid angle into which the sound radiates by half—to 2π steradians—a half space. Energy that would have traveled into the rear hemisphere is reflected forward; there is a reflected acoustical "image" of the source. Additional surfaces, positioned at right angles, reduce the solid angle by half, to π steradians and then to $\pi/2$ steradians. The number of reflected images increases correspondingly. It can be seen that the sound pressure level, measured at a constant distance from the sound source in these otherwise reflection-free circumstances, goes up by 6 dB for each halving of the solid angle.

It is commonly heard in the audio industry that each factor-of-two reduction of the solid angle increases the sound level by 3 dB. This is not correct: sound pressure level increases in 6 dB increments, but sound *power* (the total sound energy radiated into and distributed over the solid angle) goes up in increments of 3 dB. Because we are interested in sound pressure levels measured at a point in space (a microphone) and heard at two points in space (our ears), it is important to remember that the relevant relationship is with sound pressure level and that is a 6 dB change per doubling or halving of the solid angle.

But there is more to be aware of. What is shown in Figure 12.1 happens only at wavelengths that are long compared to the source size and the separation distance. It also requires that there are no other reflecting surfaces in the vicinity.

All of this is predictable from modeling, but when an opportunity presented itself, it was nice to collect some "real" data. Figure 12.2 shows measurements made at the NRC in Ottawa, in the mid-1980s, in an anechoic chamber and outdoors in a parking lot adjacent to an isolated slab-sided building. The loudspeaker was a conventional 12-in. (305 mm) driver in a closed box, about 17 in. (432 mm) on each side. Noise problems prevented the acquisition of data below about 35 Hz, and a technical problem resulted in the corruption of the 2π steradians data, but the trend is very evident. First, at the lowest frequency, the curves are separated by very close to 6 dB, as Figure 12.1 predicts. The fact that there is a tiny discrepancy can easily be accounted for by the fact that the large powerful woofer is not exactly a vanishingly small "point monopole"; it is not located "in" the wall/floor, and the building was sheathed in corrugated metal siding, which is not a perfect reflector. In any event, the approximate 6 dB difference begins to diminish immediately, and just below 200 Hz all of the curves converge in a null. The reason is that the woofer was facing forward, with the rear of the enclosure close to the wall. This separation allowed the sound to travel backward from the driver to the wall and back again toward the driver, where it interacted with sound being radiated by the diaphragm. The round-trip distance, about 36 in. (914 mm), is one-half wavelength, at 188 Hz, which is where the destructive acoustical interference occurs. This is evidence of the next kind of boundary effect: a reduction in sound output from a loudspeaker related to the spatial relationship between the loudspeaker and adjacent boundaries.

It is now interesting to see how this translates into more normal rooms. Figure 12.3 shows the first step, a single measurement

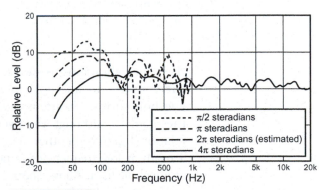

FIGURE 12.2 *The same loudspeaker measured when it radiates into different solid angles as approximated by an anechoic chamber (calibrated at low frequencies) for the 4π condition and by an outdoor parking lot adjacent to a rectangular building. All measurements were made at 2 m.*

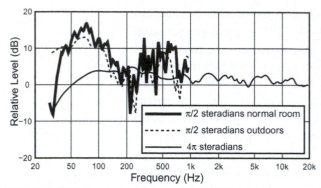

FIGURE 12.3 *The 4π and π/2 curves from Figure 12.1 compared to a measurement made at the same distance when positioned on the floor in the corner of a "hard" room, a laboratory space with concrete floor and masonry walls.*

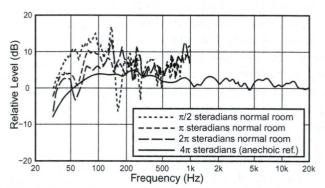

FIGURE 12.4 *The averages of four to six measurements made in the same loudspeaker/microphone relationship used in the previous two figures but moved to different locations within a normal listening room.*

made in a highly reflective, large rectangular room. Inevitably, the effects of room resonances and reflections are revealed in the cancellation dip at 32 Hz and in the general irregularity and elevation of the curve at higher frequencies. Overall, though, the effect translates quite well, including the interference dip around 188 Hz, which here is a bit wider.

Figure 12.4 shows evidence that normal listening rooms are different in several important ways. First, it can be seen that the solid angle gains are much reduced. Some of this is in the inevitable smoothing effects of spatial averaging, and some of it confirms that the vibrating room surfaces are both absorbing sound energy (membrane absorption) and allowing it to escape into other parts of the dwelling (transmission loss). The persistent dip around 60 Hz is evidence of the fact that the ear-level microphone was close to the first-order vertical standing wave dip (ceiling height = 2.8 m, first-order modal frequency = 62 Hz). The 188 Hz interference dip is gone; it is swamped by other resonant and reflective sounds in the room. There is a lot going on in this space—the kind we live in—so we need to get used to it.

Allison (1974, 1975) presents many examples of room curve shapes resulting from different arrangements of loudspeakers and corner boundaries. The curves are all significantly different from each other. An average of 22 of these, in eight rooms, is shown in Figure 12.5. To get an impression of how this curve relates to anechoic measurements, the figure also shows 2π and 4π anechoic curves. No attempt was made to find the correct vertical alignment of these curves, although the two anechoic curves are believed to be in a correct relationship to each other. The point was simply to show that what is heard in a room at low frequencies is not likely to be anticipated from an examination of anechoic data. Above about 200 Hz, there seems to be a passing resemblance to the 4π anechoic measurement, but at lower frequencies, the room seems to be in control—which is exactly what was predicted in Chapter 4.

Figure 12.6 takes advantage of computer modeling to illustrate the interaction between a loudspeaker and adjacent boundaries. In the model, the loudspeaker has a perfectly flat frequency response, so what is seen is due to the loudspeaker/boundary interaction. The dashed curve shows the predicted effect of the adjacent boundaries on in-room frequency responses. To assist in understanding what is happening, the figure shows a superimposition of many room curves and frequency responses calculated for many different locations within the listening area. Each one is different because of standing waves and reflected sounds. Also shown is the average of all of these measurements. The point of the figure is to show that the effects of room boundary interactions are in all room curve measurements, but in any individual one they may be obscured by other factors. However, by averaging several room curves, measured at different locations—a spatial average—the effects of the position-dependent variations are reduced, and evidence of the underlying adjacent-boundary effects is more clearly seen. The average of the room curves is obviously similar in shape to the predicted curve.

This tells us that eliminating the adjacent-boundary effect will not eliminate all problems, but it is definitely one of the problems. Because the adjacent-boundary effect is one that changes the acoustical radiation resistance experienced by the loudspeaker, the sound power radiated by the loudspeaker at different frequencies is altered. How can such a problem be addressed?

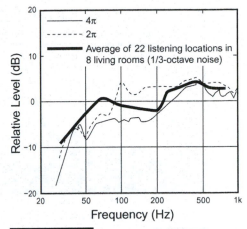

FIGURE 12.5   *A comparison of 2π and 4π anechoic measurements on a loudspeaker (an Acoustic Research AR-3A) compared with the average of 1/3-octave measurements made at 22 listening locations in eight living rooms. The anechoic measurements are correctly aligned with each other, but both have been arbitrarily adjusted vertically to show how the shapes compare with the averaged room data. Anechoic data from Allison and Berkovitz, 1972, Figures 4 and 9. The room data was also in this paper, but it was more conveniently presented in Allison, 1974, 1975.*

12.1.1 Correcting for Adjacent-Boundary Effects

The approach offered by Allison (1974, 1975) and Ballagh (1983) involves choosing the position of the loudspeaker with respect to the boundaries in a manner that minimizes the variations in frequency response at the listening locations. Equalization, changing the frequency response of the loudspeaker, is another one. The former is messy, possibly involving some trial-and-error in practical listening rooms. It may also result in visually unappealing, asymmetrical, or incorrect (in terms of stereo or multichannel imaging) locations for the loudspeakers. The attraction of equalization is that it allows the loudspeakers to be located according to other criteria, and then the performance is electronically optimized.

As shown in Figure 12.6, averaging several measurements within the listening area can reveal the underlying shape of the room-boundary effects and thereby provide a basis for correcting the frequency response to meet whatever target curve is decided on. However, there is another method, described by Pedersen (2003), in which a clever device measures the acoustic power output of the loudspeaker in situ and makes the appropriate correction to the frequency response.

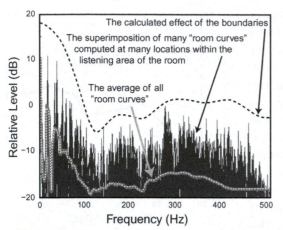

FIGURE 12.6 *A computed simulation of an omnidirectional loudspeaker situated in a "normal" relationship to the floor, ceiling, and wall boundaries, in a "normal" listening room. Simulation by Todd Welti, Harman International Industries, Inc.*

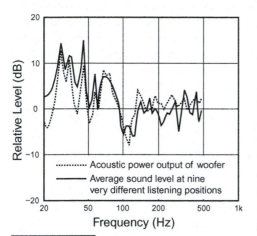

FIGURE 12.7 *A comparison of two methods of evaluating adjacent-boundary effects: a measure of the sound power radiated by the loudspeaker and the average of several room curves measured at different locations in the listening area. From Pedersen, 2003.*

Figure 12.7 shows a comparison of the two methods in the same room, one the result of making frequency response measurements at nine very different listening locations and the other the result of a measurement of the sound power radiated by the loudspeaker. They are remarkably similar and if any amount of spectral smoothing were incorporated, they would be even closer than shown. Obviously this means that a sound-power measurement can identify the adjacent-boundary problems. The same can be said for the spatial average of in-room measurements. The point is that both methods allow us to separate out the adjacent-boundary problems, but the solution is distributed uniformly over all listening locations, whether it is optimum for any one of them or not. Because the effects of room resonances are added to this problem, it is conceivable that for any single listener location the sound quality could get worse. If the equalization is performed at a single listening location, it will obviously be correct for that seat.

12.2 LOUDSPEAKER MOUNTING OPTIONS

Anechoic chambers are wonderfully useful devices, especially large chambers, and most especially those that have strong trampoline floors that allow measurements of loudspeakers mounted in a special "apparatus." The chamber used for the following highly instructive measurements was chosen because it could accommodate the apparatus, an 8-ft-square (2.44 m) section of "domestic frame wall" on and into which loudspeakers were mounted. This chamber was not perfectly anechoic at the lowest frequencies, and, although it had been calibrated for the measurement of loudspeakers in isolation, the presence of the massive wall structure introduced errors. Consequently, no claim is made for absolute accuracy at low frequencies in the measurements that follow, but they should be reliable in a comparative sense.

The tests were designed to show the effects on the acoustic performance of a small bookshelf loudspeaker, an Infinity Primus 160, when it is placed in progressively more complicated local environments. It starts with the loudspeaker in the free field, an anechoic chamber, as shown in Figure 12.8a. The acoustical measurements shown are the conventional on-axis frequency

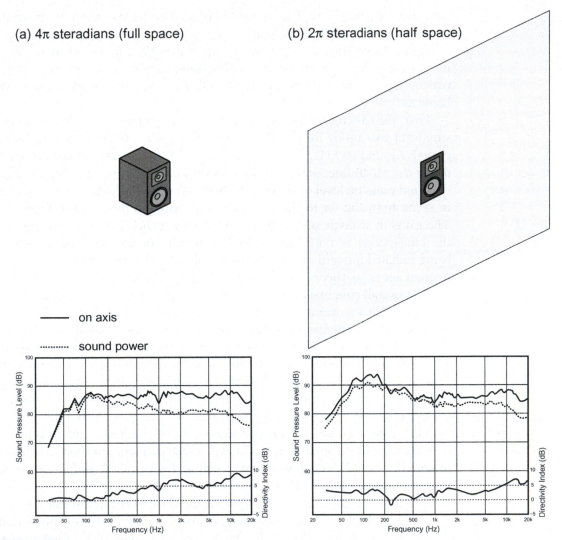

FIGURE 12.8 *The same small bookshelf loudspeaker measured in 4π, full anechoic conditions, and in 2π, half-space conditions. The enclosure was carefully flush mounted.*

response, measured at 2 m; the total sound power output; and the directivity index—the difference between the two (see Figure 18.6 for more detail on these measurements). Directivity index (DI) is 0 for an omnidirectional source, and here we see that this little 6.5-in. (165 mm) woofer approximates that quite well up to about 150 Hz. One can see the directionality progressively increase to about 7 dB at 2 kHz, above which the woofer is progressively turned off by the crossover network, and the tweeter takes over. Being small, the tweeter exhibits much better dispersion, about 4.5 dB around 4 kHz, and then it too becomes progressively more directional, reaching a DI around 9 dB at the highest frequencies. Just for perspective, this is extremely good performance for an inexpensive product, approximately $220/pair.

Figure 12.8b shows what happens when it is then mounted in a wall with its front face flush with the surface. This is the classic 2π, half-space, condition, which happens to be met by all in-wall and in-ceiling loudspeakers. What happens?

First, the bass increases, exactly as predicted by Figure 12.1. The sound pressure level goes up by roughly 6 dB (remember the measurements include some errors). Around 100 Hz, it can be seen that in (a) the on-axis curve is about 3 dB below the 80 dB line, and in (b) it is about 3 dB above the line, for an increase in sound pressure level of about 6 dB. The sound power increases by about 3 dB in going from the 4π to the 2π conditions, again as predicted in Figure 12.1. The gains in acoustic output drop at higher frequencies because, as seen in (a), the loudspeaker is no longer perfectly omnidirectional; more of the sound is being radiated forward and not being reflected by the boundary. The DI at low frequencies is, as theory would predict, 3 dB for a half space.

The overall conclusion is that mounting this excellent little loudspeaker in a wall has left its overall performance substantially intact, but the bass output has been greatly increased, making it sound fat, thick, and tubby. After all, it was not designed to be used in this manner; almost all "bookshelf" loudspeakers are designed to perform optimally in a free-standing mode—sitting on a stand some distance from the room boundaries. The solution in this case is to turn down the bass. Any competent equalizer can do it, or the old-fashioned bass control may just be optimum if the "hinge" frequency is around 500 Hz.

The proper solution, if a loudspeaker is to be used in this manner, is to design it from the outset so that it has a flat, axial frequency response when it is mounted in a wall. All in-wall and in-ceiling loudspeakers *should* be designed in this manner, but not all are.

Moving on, Figure 12.9 shows what happens when the loudspeaker is simply mounted on the surface of a wall. For comparison purposes, the half-space data from Figure 12.8 are repeated. It can be seen that there is a strong acoustical interference dip at about 220 Hz. We saw this kind of thing before, back in Figure 12.2, but at a lower frequency. This loudspeaker is smaller, and the round-trip distance from the woofer to the wall and back is shorter—about 31 in. (787 mm), which is one-half wavelength at about 220 Hz, the destructive-interference condition representing the first "tooth" in a comb filter. There is a hint of a second tooth, a partial cancellation, in the on-axis curve at 660 Hz, but one can presume that increasing source directivity at higher frequencies eliminates any higher-order cancellations. But why is there no corresponding "hole" in the sound power curve? It turns out that there is, but it is not as easy to see as the dramatic event in the on-axis curve.

Figure 12.10 offers a simplistic explanation. It begins with a comparison of the sound power measurements of the same loudspeaker mounted in a wall,

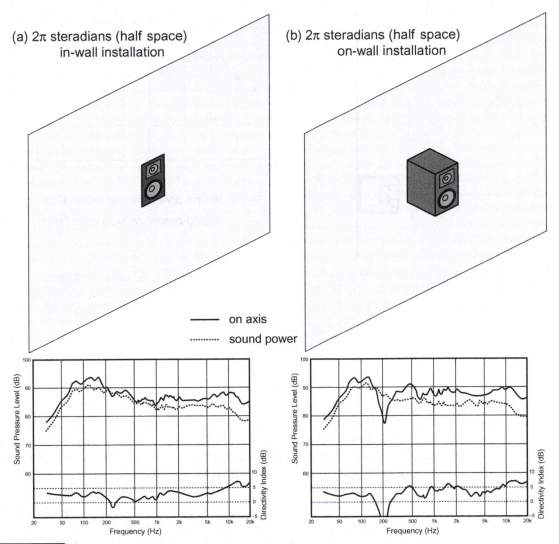

(a) 2π steradians (half space) in-wall installation

(b) 2π steradians (half space) on-wall installation

—— on axis

·········· sound power

FIGURE 12.9 *The same bookshelf loudspeaker is flush mounted in a wall (a), and attached to the surface of the wall (b).*

with the front panel flush with the surface and the loudspeaker mounted on a wall bracket. There is a difference, with the on-wall configuration showing a reduction in output over a range of frequencies, starting around 200 Hz, the frequency of the dip shown in the on-axis curve in Figure 12.9b, and extending to about 500 Hz.

The adjoining diagram shows the loudspeaker and its reflected image relative to the wall surface. The sound source, the "acoustic center" of the loudspeaker, is shown slightly forward of the diaphragm. This appears to be the case at very low frequencies, but it may or may not be applicable at the frequencies of interest here (Vanderkooy, 2006). Omnidirectional behavior can also be assumed at

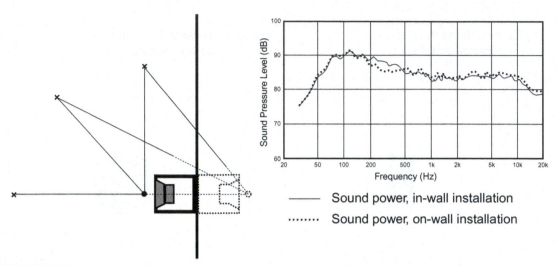

FIGURE 12.10 *A direct comparison of sound power measurements made with the same loudspeaker mounted in and on a wall. The diagram illustrates why the frequency at which the first destructive acoustical interference occurs rises as the measurement point moves away from the forward axis. The acoustical centers shown could be appropriate at low frequencies but may not be at the higher frequencies of interest here.*

low frequencies but not at frequencies above about 150 Hz (see Figure 12.8a). The point of the sketch is to show that the path length difference between the direct and reflected sounds reduces as the measuring point moves farther off axis, which causes the destructive interference frequency to increase. Sound power is an energy sum of measurements made over the entire frontal hemisphere and therefore includes the 220 Hz dip in the axial frequency response as well as many others, all occurring at higher frequencies as the measuring point moves farther off-axis. Increasing source directivity will diminish the effect at frequencies approaching 500 Hz. Consequently, adjacent-boundary effects are broad trends, not highly frequency specific, as seen in Figure 12.6.

The question of the moment is: Does it change how the loudspeaker sounds? The answer is yes, but the amount will depend on the relative contributions of direct and reflected sounds in the room. In a very dead room, the audible impression may be more influenced by the frequency response on the prime listening axis, such as the on-axis curve in Figure 12.9b. In a more reflective room, the smoother sound power curve may be more representative. The important fact is that both of these curves are compromised compared to the in-wall configuration, and some amount of equalization will be required to realize the performance potential of this mounting method.

We have all seen it: loudspeakers sitting in otherwise empty cavities in bookcases, entertainment furniture, and, embarrassingly, expensive custom installations. Figure 12.11 shows what can result: In this case, a perfectly good little

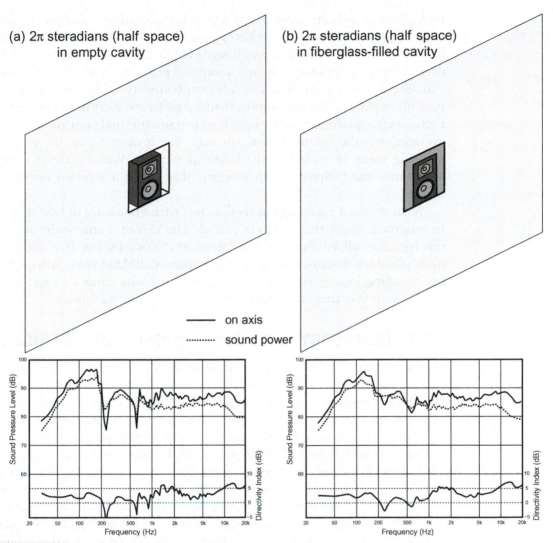

(a) 2π steradians (half space) in empty cavity

FIGURE 12.11 *A comparison of acoustical performance when the same small bookshelf loudspeaker is placed in a cavity—a bookshelf perhaps—as often happens in "entertainment system" furniture. In (a) the cavity is empty. In (b) the cavity has been filled with fiberglass.*

loudspeaker has been seriously compromised by the installation. There is evidence of high-Q cavity resonances and diffraction effects created by the edges of the cavity. The everyday remedy of filling the cavity with absorbing material helps, but the root problem is still in evidence. Figures 12.8b and 12.9a show the much-improved performance in a mounting where the cavity openings have simply been closed off with hard material.

Reflecting on what has just been discussed, it can be concluded that there are really only two locations in which a loudspeaker has the potential of performing at its best: free-standing, or flush-mounted in a wall (or ceiling). All other options involve compromises of some sort. The on-wall placement of this generic bookshelf loudspeaker is flawed, but loudspeakers can be specifically designed to perform extremely well as on-wall products. In practical terms, free-standing is also compromised by adjacent-boundary effects because it is not possible to place loudspeakers more than a wavelength away from room surfaces. Fortunately, equalization of the right kind (parametric) and properly done (spatial averages over the listening area) can help, but it cannot cure the problem of standing waves discussed in the following chapter. Without clever computer algorithms and foolproof setup routines, this is not a solution for the mass market.

What we need are loudspeakers designed with knowledge of how they are to be mounted, where they are to be placed. The idea of a universally applicable, one-type-does-all loudspeaker is a "steam-era" concept, but it is the basis of most of today's designs: bookshelf loudspeakers that don't work in bookshelves, free-standing loudspeakers that for practical reasons cannot "stand free." It would seem that there is an "opportunity" for something different.

12.3 "BOUNDARY-FRIENDLY" LOUDSPEAKER DESIGNS

Having expounded on the problems created by adjacent-boundary effects, it was no surprise that Allison (1974) proposed a loudspeaker design that minimized the effects. Figure 12.12 shows the configuration of the loudspeaker and how it was intended to be placed in a room. First, the woofers were located close to both the floor and front wall, eliminating issues with those boundaries and taking full advantage of solid angle gains at low frequencies. The effect of the side wall was minimized by placing the loudspeaker some distance away from it. Around 350 Hz the woofers crossed over to the midrange-tweeter array situated at the top of the enclosure, at ear level. The drivers radiating the lower middle frequencies were located close to the wall to minimize that boundary problem, and two sets of drivers at 90° to each other were intended to approximate a hemispherical radiation pattern. It was a very thoughtful design, but, unfortunately, we have no comprehensive measurement data on it.

Good ideas don't go away; they just morph, evolve, or get reinvented. In this case, the surround loudspeaker shown in Figure 12.13a is an example of a class of products developed to cater to home theaters. It was designed with on-wall mounting in mind, and considerable thought went into the physical layout and the crossover network to control the acoustical interactions of the drivers with each other and with the wall behind. It can be switched among three radiating patterns, but here we look

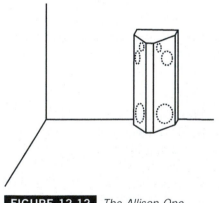

FIGURE 12.12 *The Allison One loudspeaker as described in Allison, 1974, Figure 16.*

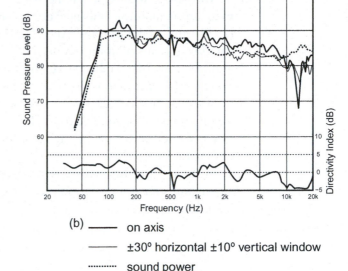

(a) A bidirectional in-phase
(aka "bipole") surround loudspeaker
designed for on-wall mounting

(b) ——— on axis

——— ±30° horizontal ±10° vertical window

············ sound power

FIGURE 12.13 *(a) An on-wall surround loudspeaker with switchable directivity, an Infinity Beta ES250. (b) Frequency response measurements on-axis and averaged over the listening window ±30° horizontal and ±10° vertical (nine curves) compared with the total sound power. The directivity index was computed using the listening window as the reference. For the measurements, the loudspeaker was used in the "bipole" mode: Both sets of drivers were radiating in phase with each other.*

at the most favorable one, called "bipole," meaning that both sets of drivers are radiating in phase with each other. This was the configuration it was optimized for, and the performance shown in (b) is excellent for a loudspeaker aimed at the mainstream market. Section 18.4.3 will discuss surround loudspeakers in more detail.

The measured curves in (b) are all very similar. As will be explained in Section 18.2.2, the spatially averaged listening window curve is used to compute the directivity index. Over most of the frequency range, this loudspeaker approximates a hemispherically omnidirectional radiator. As has been suggested before, and will be confirmed in later chapters, this is a desirable characteristic.

Obviously, to avoid adjacent-boundary problems, loudspeaker drivers must be less than a wavelength separation from large reflecting surfaces. In-wall, flush-mounting is excellent, but with good design, on-wall configurations work very well, and, as shown in this example, they allow for multidirectional or hemispherical radiation. Many surround loudspeakers are designed in this fashion—a welcome trend. Ironically, it is the front loudspeakers, arguably the most important ones, that routinely are designed with little or no regard for the acoustical settings into which they will be placed.

In control rooms, it has been common practice for decades to mount the main front monitor loudspeakers in some form of half-space mounting. Eargle (1973) and Makivirta and Anet (2001a, 2001b) are examples of advocates of this form of installation. Some rooms, because of the location of the viewing window into the studio, force the loudspeakers to be installed in a kind of "eyebrow," soffit-like overhanging structure placed against the ceiling and with the front some distance forward of the window and remainder of the wall. This might be an improvement on a free-standing loudspeaker, but it is not half-space mounting; consequently, nonoptimal boundary interactions may be anticipated.

A final thought: There is a somewhat reciprocal effect, imperfect, but significant, for the location of a listener's head with respect to adjacent boundaries.

CHAPTER 13

Making (Bass) Waves—Below the Transition Frequency

Bass is extremely important, and it matters in both quantity and quality. The phrase "deep, clean, and tight" comes to mind. We want to hear the lowest rumbles of the pipe organ and synthesizer, the body-shaking punch of a kick drum, the rhythmic throb of a good bass guitar riff, and the solid foundation of the acoustic bass in a jazz trio. We don't want individual notes to be omitted or accentuated. We don't want overhang on transient sounds. We want it *all*.

But we rarely get it. Real-world experience is that bass reproduction in small rooms is a game of chance because rooms are different from one another. Different listening positions in the same room can sound different. We get used to the rooms we live and work in, warts and all, and manage to find great pleasure. However, upon entering strange rooms, it is not uncommon to hear differences. They may or may not be problems, but they stand out. Even returning to one's own room after an extended absence may elicit feelings that something is not quite right, and a short "break-in" period may be required while we readapt. It is comforting that we are able to adapt to some extent, but it would be nicer if it were not necessary.

Later, when we get to the point of assessing the factors contributing to subjective judgments of sound quality, it will be shown that about 30% of the overall rating is attributed to factors associated with low-frequency performance (Olive, 2004a, 2004b). All of the listening tests in that study were done in the same room, which was equipped with apparatus to move the active loudspeakers to the same locations (Olive et al., 1998) and where listeners had ample time to adapt to the physical circumstances. All of this helped to neutralize—not eliminate—the room and loudspeaker location as factors in the evaluations; they were constants, not variables.

Our idealized objective is to achieve high subjective ratings for loudspeakers and to do it in different rooms. Up to a point, it happens now, as discussed in

Section 11.3.1, but even so, there was evidence of variations in judgments due to irregularities in bass performance. Again, humans seem to be able to adapt to a certain amount of bass misbehavior, but the more extreme the problem, the more difficult it is to completely ignore.

A strategy is needed that can ensure the delivery of similarly good bass to all listeners in all rooms. As discussed in Chapter 2, achieving such consistency is a necessary objective for the entire audio industry. Ideally, we want recording professionals and consumers to hear the same quantity and quality of bass. Let us see how far this idea can be taken.

13.1 THE BASICS OF RESONANCES

Resonances exist in many forms, in many devices and circumstances. In all of them, the existence of a resonance indicates that there will be at least one frequency where "activity" will be maximized compared to all other frequencies. The resonance frequencies can be calculated from the values of three parameters. In the physical world, there are many examples of resonances that are combinations of mass, compliance, and friction. School textbooks often use the example of a mass suspended on a spring, with frictional losses provided by "wings" attached to the mass that are dragged through the air. In electronics, resonances are created from combinations of inductors, capacitors, and resistors, or their functional equivalents: direct "lumped-element" analogs of the mechanical components. Acoustical resonances provide yet another set of analogs—example, the mass, compliance, and friction in Helmholtz resonances. Although it is convenient to think of these as lumped elements, it is easily understood that this is a simplification. An electronic capacitor that you can hold in your hand is a true lumped element, in that the total functional contribution it makes to a resonant system is contained in the physical device that has known and (relatively) constant properties. It can be described by a number. Likewise, a steel ball suspended from a spring—a classic example of a mechanical resonance—is another lumped-element component of a resonant system.

An acoustical Helmholtz resonance is commonly exemplified by a narrow-necked beverage bottle, where the mass of air in the neck bounces on the springiness of the air in the main volume to create a resonance that is typically excited by blowing across the mouth. Here, the mass of air in the neck is obviously not so well defined, and "end corrections" must be invented to take into account the "soft" transition to the outside world at the mouth and to the main volume (the compliance) of the container. At high sound levels, the flow of the air in and out of the neck becomes nonlaminar and turbulent, and the effective size of the mass reduces, changing the resonance frequency (a significant problem with bass reflex loudspeaker enclosures). Still, we persist with the notion that the air in the neck can be treated as a lumped-element mass; the air in the main body of the bottle can be treated as a lumped-element compliance—a spring; and some frictional losses are associated with the airflow in the neck—another lump. The lumped-element concept is useful only as long as the dimensions of the components in the Helmholtz resonance system are small compared to a wavelength. Everywhere within each element, it is assumed that conditions are identical. If there is discernable wave motion or turbulence within the element, the simplifying assumption breaks down.

There are other circumstances leading to resonances, which involve waves and propagation. Musicians may instantly think of instruments with strings— guitars and pianos, or pipes like organs, trumpets, and oboes. In musical instruments, the resonance frequencies can be changed by varying the length of a resonating pipe—organ, trombone, trumpet valves, and so on—or by changing the mass-per-unit-length of strings and their tensions—guitars, pianos, and so forth. Conditions are absolutely not the same at different points along a resonating string or tube; there are regions of higher and lower activity.

In rooms we also talk of resonances. They exist *because* of sound propagation within the space. This means, first of all, that conditions everywhere within the space *cannot* be identical. We talk of *room resonances, room modes, resonant modes, eigenmodes, eigenfrequencies*—these all describe the frequencies at which conditions in a room conspire to selectively accentuate sounds at those special frequencies. Evidence of a resonance can be found in three distinctive aspects of behavior:

- A narrow-band peak or dip in a frequency response measured at a point in the room

- Some amount of ringing in the time domain

- Changes in both of the above at different locations within the room

The peaks and dips in measurements of frequency response show up at frequencies below about 150 Hz (at higher frequencies they are harder to identify because they are so numerous and closely packed, although they exist). The audible consequences of the peaks in frequency response may be heard in sustained bass notes in keyboards, guitars, and so on. Some notes are too loud (one-note bass), and others not loud enough or, in sharp dips, even missing.

In the time domain, room resonances exhibit the same properties as any other resonances. They have Q, a quality factor reflecting the amount of acoustical damping or frictional loss in the total system. High-Q resonances have low loss; they exhibit a narrow "footprint" in the frequency domain, with narrow sharp peaks, and they exhibit prolonged "ringing" in the time domain. These are the "booms" we hear in kick drums, and, in a bad room, successive kicks can merge into a sustained drone at the resonance frequency. The more absorption there is in the room, in the boundaries themselves, in furnishings, or in acoustical devices, the lower the Q of the resonances. Low-Q resonances result from lossy systems; they exhibit wide "footprints" in the frequency domain and much shorter, well-damped ringing. Damping resonances is generally a good thing to do, but as we will see, damping low-frequency room resonances using passive acoustics is a nontrivial exercise. Fortunately, there are some electronic and electroacoustic alternatives that can assist in reducing the audibility of the ringing.

The mechanism responsible for the resonant behavior, a perfect constructive interference between sounds traveling between and among two or more room boundaries, creates what are known as *standing waves*. These exist at all resonance/modal frequencies and can be observed by examining the point-to-point variations in sound level between and among the room boundaries. Those associated with high-Q resonances will have higher, sharper peaks and narrower, more pointed dips. Low-Q resonances result in smaller, gentler undulations in sound levels within the room. All of these are responsible for the always audible, sometimes huge, seat-to-seat variations in bass quality we experience in small rooms.

At subwoofer frequencies the behavior of room resonances is essentially minimum phase (e.g., Craven and Gerzon, 1992; Genereux, 1992; Rubak and Johansen, 2000), especially for those with amplitude rising above the average spectrum level. This suggests that what we hear can substantially be predicted by steady-state frequency-response measurements if the measurements have adequate frequency resolution to reveal the true nature of the resonances. In minimum-phase systems, the magnitude versus frequency response (henceforth simply "frequency response") contains enough information to enable the phase response to be computed, and from those two data sets, the transient response can be computed.

It also means that both time- and frequency-domain correction is possible with appropriate equalization filters—but *only* at locations that have the *same* frequency responses. From this we can infer, and will later confirm, that low-frequency resonances can be attenuated by equalization at a single seat in any audio system. However, only in those audio systems that are capable of delivering very similar bass to multiple listeners can equalization provide comparable improvements to all listeners. As we will see, achieving more uniformity in bass at different seats substantially reduces the need for equalization.

As in any other aspect of audio, the critical issue is what we hear, or are able to hear. It is one thing to discuss the physical behavior of the sound field, and it is another to discuss what aspects of it matter to humans listening to movies and music. If it is possible to sufficiently manipulate the physical sound field, minimizing all problems, listeners will be pleased. If for some reason (e.g., economic or physical constraints) that is not achievable, it may be possible to find electronic solutions that leave residual imperfections in the physical sound field that are rendered more subjectively acceptable. Always, there is adaptation, and its effects tend always to be beneficial.

13.2 THE BASICS: ROOM MODES AND STANDING WAVES

All rooms, of all shapes and sizes, have resonant modes. Those in rectangular spaces are well understood and easy to predict. Modes in nonrectangular rooms are difficult to predict. The traditional method of predicting how a strangely shaped room might behave was to construct a scale model. These days, it can be done with acoustical modeling programs in computers. Both approaches involve assumptions—leaps of faith—about deviations from the shapes and acoustical properties of the surfaces. The predictions are helpful indicators, but real structures will differ in detail. In new construction, the driving force for using nonrectangular spaces often lies in the visual aesthetic—architects striving for the excitement of new forms. In practice, if there is a choice, acousticians tend to prefer working with rectangular spaces. However, when working within existing structures, there is often no choice. We must learn how to cope with all spaces.

Figure 13.1 explains the basics of how a standing wave is formed between two parallel surfaces. At the special frequencies for which the wall separation is a multiple of one-half wavelength, there will be a resonance and a standing wave. These are called axial modes, because they exist along each of the principal axes of a rectangular room: length, width, and height.

When there is a standing wave, it is obvious that the sound level at the frequency of the resonance will change as one moves around the room. In fact, a classic demonstration of the phenomenon is to set up a loudspeaker placed against an end wall radiating a pure tone at the first-order resonance frequency along that dimension of the room and have listeners walk from one end to the other. If all is well, the sound will be about equally loud at both ends of the room and will almost disappear at the halfway point. A variation of this is to have the listeners stand at a single point and scan the signal generator through a range of low frequencies; there will be huge fluctuations in loudness, high at some frequencies and low at others. It is a simple and persuasive demonstration of the problem confronting us.

Figure 13.2 shows for a rectangular room: (a) the orientations of the three axial modes, (b) one of the three tangential modes, and (c) one of many possible oblique modes. Because some energy is lost at every boundary interaction, modes that complete their "cycle" of the room with the fewest reflections are the most energetic. The axial modes are therefore the most energetic, followed by the tangential modes and the oblique modes. It is rare for an oblique mode to be identifiable as a problem in a room. Tangential modes can be found in rooms that have unusually hard, reflective boundaries or when multiple sources are appropriately located. Axial modes are omnipresent, and they are the usual culprits in bass problems in small rooms.

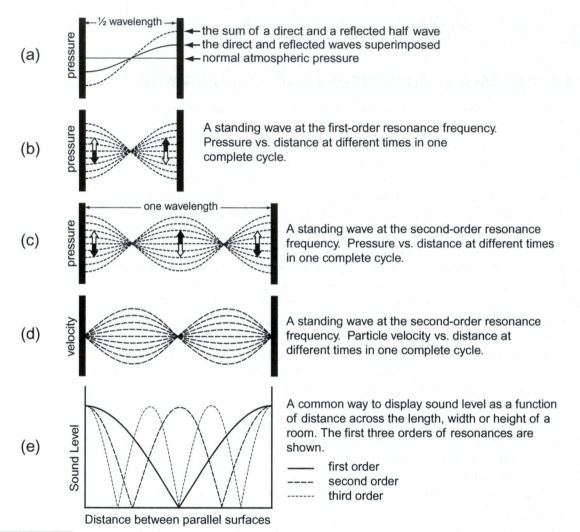

FIGURE 13.1 *(a) The mechanism by which a standing wave is formed, illustrated in a "stop-action" view. At the frequency for which the distance between the walls is exactly one-half wavelength, the direct sound wave traveling toward one wall is precisely replicated by the reflected sound traveling away from it. They combine, producing a higher-amplitude resultant, which is called a "standing wave" because it has a constant form. (b) What happens in a running situation. The "stop-action" resultant waveform shown in (a) cycles up and down with time, once per period of the signal, and it is shown at several points in the cycle. In this first-order standing wave along one axis of a room, there is one null, one location where it is possible to stand and hear almost nothing, in the center of the room. As one moves toward either wall, the sound gets louder. It is very important to note that, at any instant in time, at any one of the "stop-action" curves, when the sound pressure on one side of the null is increasing, the sound pressure on the other side is decreasing (shown by the white and black arrows). (c) The situation when the distance between the walls is one wavelength. This pattern could also exist between the walls in (a) but at double the frequency. (d) The distribution of particle velocity as a function of distance. (e) A common manner of representing the sound pressure distribution across a room that is graphically simpler, but it must be remembered that there is a polarity reversal at each null.*

"BACK OF THE ENVELOPE" ACOUSTICAL CALCULATIONS

To compute the frequencies at which axial standing waves occur, simply measure the distance between the walls (this is one-half wavelength of the lowest resonance frequency), multiply it by 2 (to get the wavelength), and divide that number into the speed of sound in whatever units the measurements were done (1131 ft/s, 345 m/s). The result is the frequency of the first-order mode along that dimension. All higher-order modes are the result of multiplying this frequency by 2, 3, 4, 5, and so on. *Example*: A room is 22 ft long. The first-order resonance occurs at 1131/44 = 25.7 Hz. Higher-order resonances exist at 51.4 Hz, 77.1 Hz, 102.8 Hz, and so on. Do this for the length, width, and height of the room, and all of the axial modes will have been calculated.

Figure 13.3 shows all of the modes for a small rectangular room. This is the kind of information yielded by any of several computer programs that run the following equation:

$$f_{n_x n_y n_z} = \frac{c}{2} \sqrt{\left(\frac{n_x}{l_x}\right)^2 + \left(\frac{n_y}{l_y}\right)^2 + \left(\frac{n_z}{l_z}\right)^2}$$

$f_{n_x n_y n_z}$ = the frequency of the mode defined by the integers applied to dimensions x, y, and z. Examples of mode identification:
$f_{1,0,0}$ is the first–order length mode (x dimension)
$f_{0,2,0}$ is the second–order width mode (y dimension)
$f_{0,0,4}$ is the fourth–order height mode (z dimension)
$f_{1,2,0}$ is a tangential mode (involving two dimensions)
$f_{1,3,2}$ is an oblique mode (involving all three dimensions)

n_x, n_y, n_z = integers from 0 to ∞ applied to each dimension: x, y, z.

l = dimension of the room in ft (m)
c = speed of sound: 1131 ft/s (345 m/s)

Not all of these modes will be problematic in practical circumstances, but they exist, and knowing the frequencies at which they occur can be helpful in analyzing specific installations.

13.2.1 Optimizing Room Shape and Dimensions

A recurring fantasy about rooms is that if one avoids parallel surfaces, room modes cannot exist. Sadly, it is incorrect. Among the few studies of this topic, Geddes (1982) provides some of the most useful insights. He found that "room shape has no significant effect on the spatial variations of the pressure

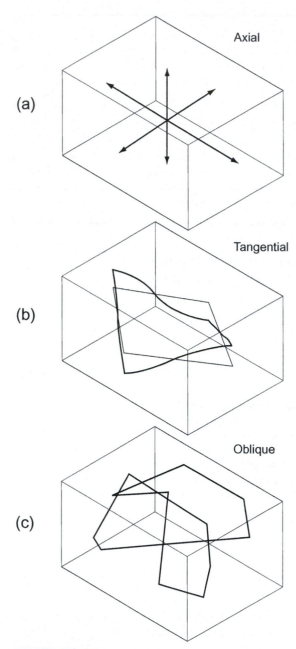

(a)

Axial

(b)

Tangential

(c)

Oblique

FIGURE 13.2 *The three classes of room modes: (a) axial: length, width, and height; (b) tangential, each of the three involving two pairs of parallel boundaries and ignoring the third pair. The other two orders of tangential modes would involve the ceiling and floor, combined with either the side walls or the end walls. (c) One of the many possibilities for oblique modes that involve all surfaces.*

response. . . . The spatial standard deviations of the p^2 response is very nearly uniform for all the data cases [the five room shapes evaluated in the computer model]." Source location was a factor in the behavior of the modes of course, as was the distribution of absorption. "Distribution of absorption was far more important in the more symmetrical shapes—[a nonrectangular] shape did help to distribute the damping evenly among the modes" (Geddes, 2005). It seemed that the most effective modification to a rectangular room was to angle a single wall.

The behavior of sound in nonrectangular rooms is difficult to predict, requiring either scale models or powerful computer programs. Figure 13.4 shows two-dimensional estimated pressure distributions for modes in rectangular and nonrectangular spaces. It is clear that both shapes exhibit regions of high sound level and nodal lines where sound levels are very low. The real difference is that in rectangular rooms, the patterns can be predicted using simple calculations.

Consequently, acousticians favor the simplicity of rectangular shapes. The question then becomes: Which rectangular shapes are most advantageous? A widespread assumption in the acoustics profession is that one should strive for a uniform distribution of room modes along the frequency axis, avoiding crowding, coincident frequencies, or large gaps. The specific dimensions of rooms determine the frequencies of resonances, but it is the ratios of length-to-width-to-height that determine the modal distribution in the frequency domain. Avoiding square rooms or rooms with dimensions that are simple multiples of each other has been customary because multiple resonances pile up at certain frequencies (though, as we shall see, this intuitive and logical restriction can be circumvented). Beginning many years ago, a lot of effort has been put into finding optimum dimensional ratios for reverberation chambers, where the sound power output of mechanical devices was measured and it was important to have a uniform distribution of resonance frequencies.

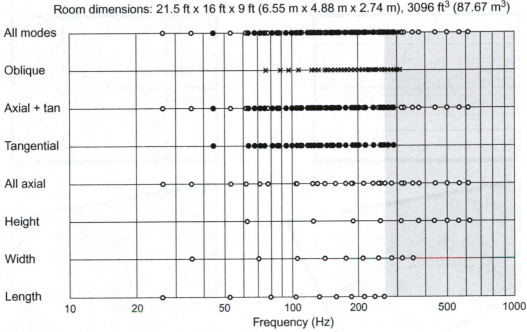

FIGURE 13.3 *All of the low-order modes for a small rectangular room. In the shaded area, all of the modes have not been calculated. This is based on the presentation of data by a Microsoft Excel program, which is available for download at www.harman.com. An Internet search for "room mode calculator" will reveal many more, along with some opinions about their value that may differ from those expressed here.*

These concepts migrated into the audio field, and certain room dimensional ratios have been promoted as having especially desirable characteristics for listening. In normal rooms, the benefits apply only to low frequencies. Bolt (1946), who is well known for his "blob"—a graphical outline identifying recommended room ratios—makes this clear in the accompanying, but rarely seen, "range of validity" graph (Figure 13.5). This shows that in an 85 m³ (3000 ft³) room, the optimum ratios are effective from about 40 to 120 Hz. This is similar to the room in Figure 13.3, which shows that this frequency range embraces six or seven axial resonances. This is consistent with the common experience that above the low-bass region the regularity of standing-wave patterns is upset by furniture, openings and protrusions in the wall surface, and so on, so that predictions of standing wave activity outside the bass region are unreliable. In fact, even within the low-bass region wall flexure can introduce phase shift in reflected sound sufficient to make the "acoustic" dimension at a modal frequency substantially different from the physical dimension.

Nevertheless, efforts to solve the riddle continued, with Sepmeyer (1965), Rettinger (1968), Louden (1971), and Bonello (1981) all making suggestions for superior dimensional ratios or superior metrics by which to evaluate the

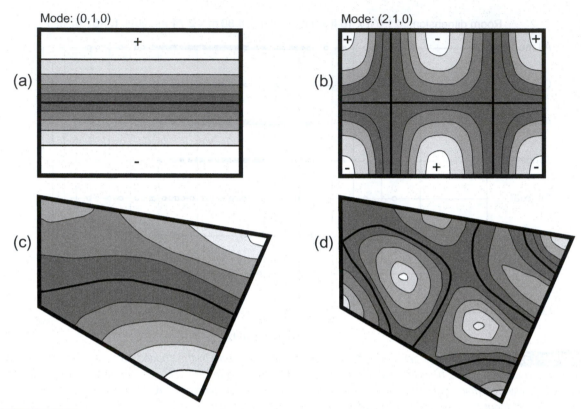

Mode: (0,1,0)   Mode: (2,1,0)

(a)   (b)

(c)   (d)

FIGURE 13.4 *Computed pressure distributions in rectangular and nonrectangular rooms, showing for each a simple mode and a more complex modal pattern. Pressure minima—nulls—are shown by heavy lines. Lighter shades indicate increasing sound pressure levels. Note that the instantaneous polarity of the pressure reverses at each nodal line (pressure minimum).*

distribution of modes in the frequency domain. Walker (1993) proposed some generous room ratio guidelines. All of them differ, at least subtly, in their guidance. Driven by the apparently undeniable logic of the arguments, information from these studies has been incorporated into international standards for listening rooms and continues to be cited by numerous acoustical consultants as an important starting point in listening room design.

However, more recent examinations have given less reason for optimism. Linkwitz (1998) thought that the process of optimizing room dimensional ratios was "highly questionable." Cox et al. (2004b) found good agreement between modeled and real room frequency responses of a stereo pair of loudspeakers below about 125 Hz, but they ended their investigation by concluding that "there does not appear to be one set of magical dimensions or positions that significantly surpass all others in performance." Fazenda et al. (2005) investigated subjective ratings and technical metrics, finding that "it follows that descriptions of room quality according to metrics relying on modal distribution or magnitude

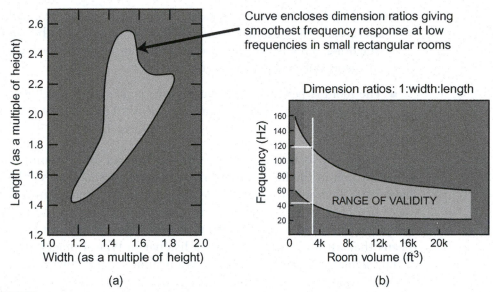

FIGURE 13.5 *(a) The Bolt "blob," a specification of room ratios, which are interpreted here as length and width, that yield the smoothest frequency responses at low frequencies in small rectangular rooms. (b) The frequency range over which the relationship has validity. For a 3000 ft³ (85 m³) room, the optimum ratios are effective from about 40 to 120 Hz, as shown by the white lines on the graph. Adapted from Bolt, 1946.*

pressure response are seriously undermined by their lack of generality, and the fact that they do not correlate with a subjective percept on any kind of continuous scale." These people seem to be saying that the acoustical performance of rooms cannot be generalized on the basis of their dimensional ratios and that reliably hearing superiority of a "good" one may not be possible.

There is a simple explanation. It is that there are problems with the basic assumptions underlying determinations of "optimum" room dimensions for domestic listening rooms or control rooms. The normal assumptions are as follows:

- All of the room modes are simultaneously excited, and by a similar amount. This requires that the sound source be located at the intersection of three room boundaries—for example, on the floor or at the ceiling in a corner. *Any* departure from this location will result in some modes being more strongly energized than others.

- The listener can hear all of the modes—equally. This requires that the head be located in another, preferably opposite three-boundary corner. Strictly, this could force either the head or the loudspeaker to be at ceiling level. *Any* departure from this listening position means that all of the modes will not be equally audible.

- All classes of modes—axial, tangential, and oblique—are equally energetic. In any evaluation of distribution uniformity, they have equal weighting. This is not the case; axial modes are typically the most energetic, and oblique modes the least.

- The room is perfectly rectangular, with perfectly flat, highly reflecting (rigid and massive) walls, floor, and ceiling. This concept does not describe most of the rooms in which we live and listen.

It is difficult to understand how this concept of an optimum room got so much traction in the field of listening room acoustics, and why it has endured. Figure 13.6 illustrates the principles. In (a) it is shown that, even with the greatest of determination, a listener is not likely to put ears in the ideal location, and practical loudspeakers do not radiate all of their sound into a corner. This means that with a loudspeaker and a listener in typical practical locations, all of the calculated modes will not be equally audible, and any of the measures of modal distribution will fail. In (b) there is another fatal flaw. We insist on listening to at least two loudspeakers, if not five or more. All of the calculations underlying the ideal dimensions come to naught. In (c) there is an idea that might work. Because we must deal with wave effects below the transition frequency, let us employ a separate sound system that is optimized for this purpose. In conclusion, it is not that the idea of optimum room ratios is wrong, just that as originally conceived, it is irrelevant in our business of sound reproduction.

With modifications, the idea can be made to work. However, doing so is not simple because one must take into account how many loudspeakers there are, where they are located, how many listeners there are, and where they are seated. This will be addressed soon, but first we must examine the phenomenon of standing waves and how real rooms may differ from the idealized spaces considered in mathematical models.

13.2.2 Standing Waves in Real Rooms

Some of us have, no doubt, tried to relate calculated resonance frequencies to those found in rooms. Usually, the correlation is reasonable, but occasionally there is a glaring error. There are many possible explanations, a common one being that domestic spaces often deviate from the "perfect rectangle" assumption. The conceptual rooms have perfectly reflecting boundaries, but real rooms are constructed of studs and drywall, bricks and mortar, lath and plaster. Real rooms have doors, windows, fireplaces, and cabinets and furnishings that cause errors in calculations and expectations.

The following is an example of a well-constructed, perfectly rectangular room in which the ears and measured acoustical evidence said that the first-order length mode was much lower in frequency than that determined using a tape measure and a calculator.

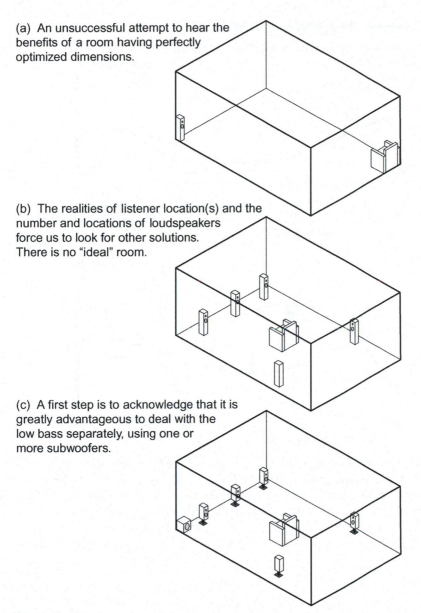

(a) An unsuccessful attempt to hear the benefits of a room having perfectly optimized dimensions.

(b) The realities of listener location(s) and the number and locations of loudspeakers force us to look for other solutions. There is no "ideal" room.

(c) A first step is to acknowledge that it is greatly advantageous to deal with the low bass separately, using one or more subwoofers.

FIGURE 13.6 *(a) Source and receiver locations necessary for all modes to be equally energized by the loudspeaker and equally audible to a listener—being practical and keeping them both at floor level. (b) How far a real situation is from the simplistic ideal of (a). (c) Separating the reproduction of low frequencies is a good way to address the room resonance problems in a manner that is independent of the number of reproduction channels.*

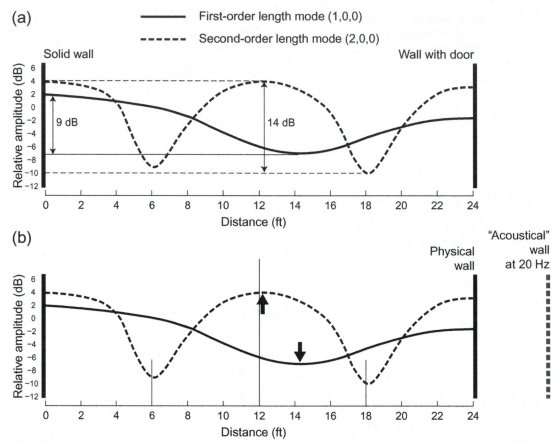

FIGURE 13.7 *Measurements of sound level as a function of distance along the length of a room, measured at frequencies representing the first- and second-order modes (1,0,0 and 2,0,0). (a) The difference between the maximum sound level and the minimum sound level for each of the modes. (b) The predicted locations for maxima and minima for the modes (fine vertical lines are shown at quarter- and half-wavelength distances from the boundaries); this illustrates that the measured pressure maxima and minima are displaced from their anticipated locations. The discrepancy is large for the first-order mode (solid line) and small for the second-order mode (dashed line).*

Figure 13.7 shows measurements of sound level versus distance along the length of a listening room, at frequencies representing the first- and second-order axial modes along the length of the room. The room was perfectly rectangular, constructed of 2 × 6-in. (50 × 150 mm) nominal-dimension studs with two layers of 5/8-in. (15.9 mm) gypsum board on the interior surface. Structurally, these walls are more massive and stiffer than those found in typical North American homes. The only opening was a heavy solid-core door located in the middle of one end wall, as indicated on the right in Figure 13.7a. Several important observations can be made about these diagrams:

- Figure 13.7a shows that the max/min, peak-to-trough, ratio of the first-order mode is only 9 dB. This indicates that the walls are absorbing substantial energy at this frequency—calculated to be 23.6 Hz for the 24 ft (7.32 m) long room. Energy lost in the boundaries causes the standing wave (constructive interference) peaks to be lower and the (destructive interference) minima to be higher. Further confirmation is in the shape of the sound pressure distribution, smoothly undulating, without a sharp null as shown in Figure 13.1e. The spatial Q of this resonance has been reduced compared to the second-order mode.

- The second-order mode exhibits a larger max/min spread of 14 dB, indicating that the walls are more reflective at this frequency. The curves also exhibit sharper, albeit not sharp, dips. This mode has a higher Q than the first-order mode.

- Figure 13.7b shows that the minimum for the first-order mode is not where it should be: at the halfway point down the length of the room. It is shifted by slightly more than 2 ft (0.6 m) toward the wall having a door. This behavior suggests that the room is "acoustically" longer than the physical dimension and that the extension is at the end where the door is located. A boundary that moves is a membrane absorber, absorbing a portion of the energy falling on it, and reflecting the remainder with a phase shift. In this case, at this particular frequency, the phase shift has the same effect as moving the wall by some distance beyond the physical location. If this is so, there should be a corresponding lowering of the frequency at which the resonance is observed. This is confirmed in Figure 13.8b, which shows that the first modal peak in frequency-response measurements is substantially lower than the predicted frequency shown in (a).

- The second-order mode is also shifted, as the nulls and the maximum are all displaced toward the wall with the door. However, the movement is small, indicating that at this higher frequency (47 Hz), the wall with the door is substantially less absorbing than it is at the lower first-order mode frequency. Figure 13.8b confirms that the shift in resonant frequency is not evident.

The simple explanation for all of this is that the structural integrity of one end wall of this room was changed by the presence of a door in the center of it. The wall became more flexible at very low frequencies. The performance at higher frequencies—above about 50 Hz—is very likely unchanged. It is possible that placing the door toward a corner might have been better, but, then again, it might just have been different; it all depends on structural technique. In fact, low-frequency absorption is a good thing, damping the mode, and the small frequency shift is normally of no consequence.

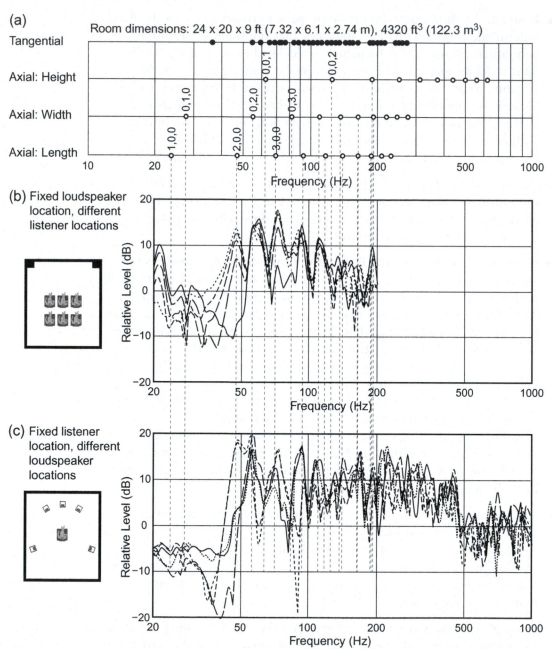

FIGURE 13.8 (a) Predicted frequencies of axial and tangential modes for a room. (b) Frequency response measurements made at five listening locations within the room for a fixed arrangement of two subwoofers. Some of the modes have been identified using the nomenclature described in Section 13.2. (c) Frequency-response measurements at a single listener location for five different loudspeaker locations (the loudspeaker was different from those used in (a), which is a variant of Figure 4.11b). Thin dashed lines have been drawn vertically at the calculated frequencies of some axial modes to help identify evidence of their existence in the frequency response measurements. Note that these measurements all have high resolution in the frequency domain, 1/20 octave, without which it would not be possible to see the important details that are discussed.

Not shown is a plot of the second-order width mode (0,2,0). In general form, it resembles the second-order length mode shown here, but it is not perfectly symmetrical; the overall sound level drifts substantially upward from one side of the room to the other. Why? There were no doors or windows, and the walls were identically constructed. The explanation that makes most sense is that one of the walls was placed close to, but not touching, a masonry exterior wall, and the opposite wall was an interior wall facing another larger room. It obviously was enough to disrupt acoustical symmetry, although the resonance frequency as shown in Figure 13.8 is correct.

There are many such examples that can be produced, all showing how physical rooms differ from the elegant simplicity of their modeled counterparts. In the end, only measurements can describe what is happening in a room. Used in conjunction with calculations or a model, it is possible to explain most of the strange happenings in real rooms.

13.2.3 Loudspeaker and Listener Positions, Different Rooms, and Manipulating Modes

Let us explore the problem of loudspeakers, listeners, and rooms at low frequencies by looking at some real-world data. Figure 13.8 shows measurements in the same room, but from two different perspectives. In Figure 13.8a, the predicted axial and tangential modes are shown, assuming no deviation from the typical mathematical simplifying assumptions, of a perfect rectangular shape and perfectly flat, near-perfectly, reflecting boundaries. Figure 13.8b shows a pair of subwoofers (connected in parallel, in phase) located in the left- and right-front corners, measurements being made at five different listening locations. Here are a few observations:

- The axial modes account for the dominant peaks in the frequency responses at low frequencies. Above about 150 Hz, the situation is so complex that individual modes do not stand out. It is possible that tangential modes have significant influence, but there is no unambiguous evidence in these data. In general, tangential modes are apparent only in rooms with significantly massive boundaries.

- The frequency shift of the first-order length mode (1,0,0), just discussed, can be seen; it is lower than predicted, as explained in the previous section.

- There are peaks at or close to the frequencies of most of the axial modes below about 150 Hz. Notable exceptions are the first- (0,1,0 @ 28 Hz) and third-order (0,3,0 @ 84 Hz) width modes, and the first-order height mode (0,0,1 @ 63 Hz). This is explained below.

- Very large seat-to-seat variations can be seen for some modes and very little for others. The variation at the frequency of mode 2,0,0 is about

20 dB, and at 3,0,0 it is close to 15 dB. Obviously a single "global" equalization in the signal path cannot work equally well for all listeners in the room.

■ None of this would be visible without high-resolution (1/20-octave in this case) measurements and the ability to display multiple curves superimposed on the same graph. Several computer-based measurement systems permit this kind of measurement and display. Traditional 1/3-octave measurements would be "blind" to the rich details in these data.

The location of the microphone at ear level, close to the midpoint between floor and ceiling, explains the lack of a peak at 63 Hz, the frequency of the first-order height mode (0,0,1).

The explanation for the missing odd-order width modes (0,1,0 and 0,3,0) is more complicated, but very important. They have been canceled by the use of two subwoofers, which radiate in-phase located in opposite-phase portions of the standing-wave patterns. For the first-order width mode, the phase opposition of lobes on opposite sides of the null is illustrated in Figures 13.1a and (b). A single subwoofer on one side of the room would excite this mode effectively, but two subwoofers on opposite sides, "destructively" drive it. In Figure 13.8b, where there should be a peak at 28 Hz (for the noncentral seats), there is, for all seats, a narrow cancellation—destructive interference—dip.

In contrast, the second (0,2,0 @ 56 Hz) and fourth (0,4,0 @ 112 Hz) and all even-order width modes are amplified in this two-subwoofer configuration. Figure 13.1c shows that the lobes in which the subwoofers are located are in phase, coinciding with the subwoofer output. This helps explain why these modes are so easily seen. Obviously, multiple subwoofers give us some control over which modes are attenuated and which are amplified by their coexistence. There will be more discussion and explanation of the multiple subwoofer effects in the following section.

Figure 13.8c shows a set of frequency-response measurements for five loud-speakers set up in an ITU surround-sound arrangement (ITU-R BS.775–2), with measurements made at a single location in the listening sweet spot (from Figure 4.11). The loudspeakers had limited low-frequency extension, so the two lowest-frequency modes are not excited. Otherwise, evidence can be seen for most of the axial modes, with the exception—again—of height mode 0,0,1. At low frequencies, large variations can be seen in the frequency responses associated with the five loudspeaker positions. For example, just below 50 Hz amplitude, swings in the 30–40 dB range can be seen. This is a strong reason to question the use of five full-range loudspeakers in such an arrangement. Bass management and subwoofers may have been popularized in low-cost audio systems, but they seem to have fundamental advantages for all sound reproduction systems. Everything in Figures 13.7 and 13.8 came from the same room, but what happens in different rooms?

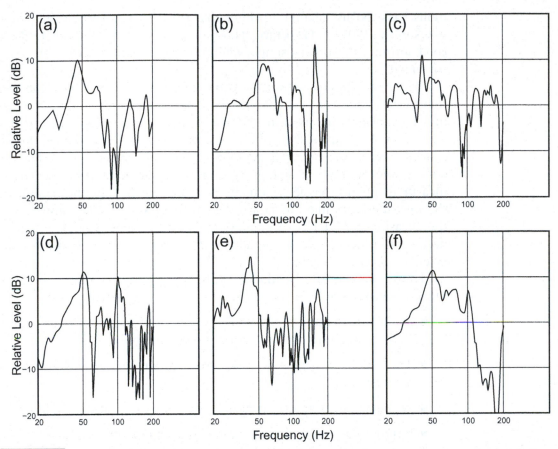

Measurements made in the sweet spot for the same loudspeaker set up as a stereo left channel in six different listening rooms.

Figure 13.9 shows measurements of the same loudspeaker, set up in a stereo-left location in six different rooms. No effort was made to find "difficult" rooms. In fact, these were simply the listening rooms in my home and in those of some colleagues at the time (ca. 1984). The purpose of the survey was to get a measure of the magnitude of the loudspeaker/listener/room interaction problem as it is experienced by consumers. It is huge! Binaural recordings were also made at the listening position in each room. Instantaneous comparisons of these recordings provided aurally subjective confirmation of the visually subjective differences seen in these curves. The differences were not subtle. And yet, we all went home and listened to music, and enjoyed it, in all of these rooms. And none of us would be considered naive. Is it true that we can adapt to such huge differences and derive comparable satisfaction? Chapter 11 confirms that considerable adaptation is possible, but there must be limits to our tolerance. In these examples, it is hard to imagine that listeners in room (e) did not hear an enormous bass "boom" around 40 Hz or that those in room (f) did not miss the notes in the 100–200 Hz octave. Those in room (c) should be grateful for their good fortune; the others simply fall into the general category of "typically corrupted."

In showing these data over the years, a number of people have commented that they had not seen such dramatic variations in bass. Upon further questioning the explanation was clear: They were measuring with reduced-frequency resolution. The curves shown here are steady-state measurements, done with a slowly swept pure tone and recorded on an analog pen recorder. These were done before the days of laptop-based FFT measurement systems. The same measurements are possible with the new equipment, but the parameters of the measurements must be properly configured, and it takes time to execute the measurement: long MLS durations (measurement windows) to get the necessary frequency resolution and possibly several averages to combat the inevitable background noise.

The summary of this section could be that, in understanding the communication of sound from loudspeaker(s) to listener(s) in small rooms:

- *Everything* matters: dimensions, placement of listeners, loudspeakers, wall construction, where you put a door, and on and on.

- There are no generalized "cookbook" solutions, no magic-bullet room dimensions.

- Without your own acoustical measurements, you are "flying blind."

- Without high-resolution measurements, you are myopic.

- With good acoustical measurements and some mathematical predictive capability, you are in a strong position to identify and explain major problems.

- There are indications that some combination of low-frequency acoustical treatment, multiple subwoofers, and equalization will be helpful.

- The idea of optimizing room dimensions has not been abandoned, but future investigations must take into account where the loudspeakers and listeners are located.

13.3 DELIVERING GOOD BASS IN SMALL ROOMS

Getting good bass in small rooms has traditionally been a hit-or-miss affair. Remedies for unacceptable situations typically included spending more money on a loudspeaker with a "better" woofer (without useful technical specifications, that was a lottery of another kind) and a bigger amplifier (for useless headroom or, equally useless, higher damping factor—see Section 18.6.3). Very occasionally, some form of passive acoustical treatment may have been employed, but most such devices were of little value at very low frequencies. Now there is a choice of effective products, but a few pretenders remain.

Equalization was always there as an option, but the lack of affordable and portable high-resolution measurements hobbled the efforts. Unsatisfactory results were widely attributed to phase shift or other nonspecific maladies supposedly introduced by the electronics. In the early days, the equalizers brought to bear on the task were often of the multifilter "graphic" type—typically octave- or 1/3-octave-band resolution, which matched the resolution of the real-time analyzers used to make the measurements. Room modes can have very high Qs, as can be seen in the narrow spikes in Figure 13.9. We realize now that much of the problem with equalization was that the industry had been performing surgery with a blunt instrument.

Figure 13.10 presents a self-explanatory guide to the author's view of where we are and where we have been in the process of delivering bass in rooms. The active manipulation of sound fields—using multiple, individually signal-processed subwoofers to control the energy delivered to individual room modes—is a recent development. For the first time it is possible to go a long way toward engineering a good bass-listening experience.

13.3.1 Reducing the Energy in Room Modes

In stereo it was common to think single-mindedly of a sweet spot, and to arrange for everything to be optimum for a single listener. At low frequencies, an equalizer can be used to reduce the audible excesses of objectionable room resonances, thus delivering respectable bass to a single listener. However, the existence of the standing waves between and among the room boundaries ensures that other seats experience different bass.

Delivering similarly good bass to several listeners simultaneously means that the room resonances must be physically manipulated in a manner that reduces the point-to-point variations in sound pressure. Conventional acoustics attacks the problem with absorption, damping the resonances by draining energy from the offending modes, resulting in lowered pressure maxima and elevated minima. Low-frequency absorption is always a good idea, but it can be difficult. Traditional low-frequency absorbers were bulky devices, some of which are hostile to even progressive concepts of interior décor. They still exist, but there are some new devices that are more elegant. The options fall into several categories, and the effectiveness of each depends on knowing where in the room to place the acoustical material or devices.

Resistive Acoustical Absorbers

These are devices that offer resistance to the vibrations of air molecules—the method by which sound propagates. They are porous materials, forcing air "particles" to dissipate energy in moving through the tortuous paths within fibrous tangles in fiberglass or heavy fabrics, or open-cell, reticulated, acoustic

(a)

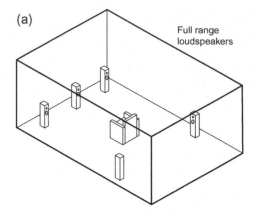

Full range loudspeakers

(b), (c), (d)

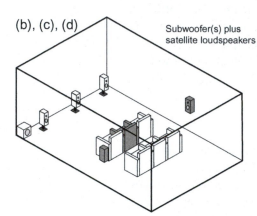

Subwoofer(s) plus satellite loudspeakers

(e), (f)

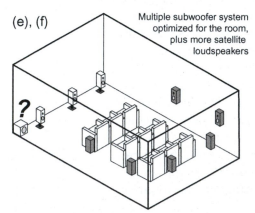

Multiple subwoofer system optimized for the room, plus more satellite loudspeakers

The evolution of bass in small rooms

(a) In the beginning, some number of full-range loudspeakers are located as required by the playback format (2, 5, 6, or 7 channels):
▶listener hears unpredictable and different bass quantity and quality from each channel.
▶bass shared among all channels is unpredictable and different from that in the individual channels.
▶other listeners in the room hear bass that is different from each other and from that at the sweet spot.
▶equalization of bass gives unpredictable results because bass treatment in recordings is not standardized: some is steered, some is shared among the channels.

(b) Bass management + one or more subwoofers located arbitrarily or experimentally:
▶each listener hears the same bass quality from each channel, and the same bass quality when it is shared among the channels.
▶the quantity and quality of the bass may differ from the program.
▶listeners in each seat in the room hear bass that is different from the others and from that at the sweet spot.

(c) As (b) above + equalization at the sweet spot (dark seat):
▶as above, except that the quantity and quality of the bass is substantially predictable for the listener in the sweet spot—and *only* for that listener.

(d) As (b) above + equalization employing measurements at several locations in the listening area:
▶because the physics of the room are unchanged, the seat-to-seat differences remain.
▶equalization in the signal path to the subwoofer involves a decision about which seats are most important so that the compromise solution can be configured to bring the greatest pleasure to the maximum number of listeners. No listener may receive the fully optimized bass delivered to the sweet spot in solution (c).

(e) Mode manipulation in simple rectangular rooms.
Bass management + multiple subwoofers + equalization:
▶ using multiple subwoofers in the appropriate locations, the sound field in the room is manipulated so as to reduce seat-to-seat variations within a designated area.
▶ *all listeners* within the specified area hear similar bass from each channel, and the same bass when it is shared among the channels. Equalization is required for good sound.
▶ equalization applies similarly to all listeners.

(f) Sound field management in rooms of arbitrary shape or rectangular rooms with acoustical asymmetry or restrictions.
Bass management + multiple subwoofers + electronic optimization + equalization:
▶ as (e) above, with smaller seat-to-seat variations; less global equalization is needed; systems typically have higher efficiency.

FIGURE 13.10 *A simplified explanation of how bass reproduction evolved to satisfy one, and then several listeners in small listening rooms.*

foam. Energy lost in this friction is converted to heat. Logically, these must be located where particle movement exists and preferably is at a maximum. Figure 13.1d shows that particle velocity is at a minimum (zero, in fact) at the reflecting surfaces and increases to a maximum at the quarter-wavelength point (for axial modes). So if the intent is to use fiberglass, acoustic foam, heavy draperies, and the like to absorb acoustic energy at bass frequencies, they must be located away from the room boundaries. How far? One-quarter wavelength at 100 Hz is 2.8 ft (0.86 m); at 50 Hz, it is 5.6 ft (1.7 m); and at 30 Hz it is 9.4 ft (2.9 m). It is clear that this is not a practical solution for homes, although some costly—and large—control room designs have employed it. A few creative purveyors of acoustical materials have marketed acoustic foam or fiberglass shapes that fit into corners, claiming that they are effective "bass traps." This is wishful thinking at its best, because it is sound pressure, not particle velocity, that is maximum at such locations. A significant advantage of these resistive absorbers is that they are inherently non-resonant, being similarly effective over a wide bandwidth. See Chapter 21 for more information.

Mechanically Resonant, Membrane or Diaphragmatic Absorbers

These devices have solid, somewhat flexible, surfaces that remove energy by moving in response to sound pressure, and dissipating the energy mechanically. To be effective, they should be located at the high pressure points within the sound field, a logical starting point being a corner. Figure 13.1c shows pressure maxima at the end boundaries, but this particular mode also has maxima at the midpoints along the side walls. Corners are end points for modes in two directions, and three boundary intersections are high pressure points for all modes. The pressure nulls, or minima, are locations to be avoided—precisely the opposite of desirable locations for resistive absorbers. A number of free-standing products have been designed for this application (the ASC Tube Trap, introduced in 1985, was one of the first; Noxon, 1985, 1986). It is worth remembering that suitably constructed room boundaries can themselves function effectively as low-frequency absorbers (Bradley, 1997; see also Figure 21.6). One of the common errors of listening room construction is to employ wall constructions that are excessively massive and stiff. It certainly helps in reducing the transmission of sound to adjacent spaces, but it does nothing to improve the sound inside the space itself. These devices are inherently resonant, being a mass (the membrane) spring (the air cavity behind the membrane) system, so the effective bandwidth (the Q) depends on the amount of mechanical and acoustical damping in the system. They can be tuned and positioned to address specific modes, in which case they can move nulls and modify the modal frequency (as in the case of the absorbing wall in Figure 13.7). Limp-mass diaphragms, such as that described by Voetmann and Klinkby (1993), can be effective over wide bandwidths. See Chapter 21 for more information.

Acoustically Resonant, Helmholtz Absorbers

In these devices, the mass consists of "lumps" of air in tubes or slots that interact with the compliance of a volume of air behind. These must be located in high-pressure regions to be effective, and, because they are frequency selective, they must be in the high-pressure regions for the modes in that frequency range. They are inherently resonant, and the effective bandwidth, as well as the absorption coefficient, are modified by the acoustical damping incorporated into the system. See Chapter 21 for more information.

Active Absorbers

These are loudspeakers, combined with microphones, amplifiers, and control circuits, configured to operate as absorbers of acoustic energy (Olson, 1957, pp. 417, 511). Darlington and Avis (1996) describe the results of model experiments confirming the effectiveness of the concept, but thus far, this is a technology yet to be seriously exploited. Placed in tricorner locations, such devices should be effective over a wide bandwidth.

13.3.2 Controlling the Energy Delivered from Loudspeakers to Room Modes

There are two ways to vary the amount of energy transferred from loudspeakers to modes:

- Locate the subwoofer at or near a pressure minimum in the offending standing wave. A subwoofer is a "pressure" source, and it will couple inefficiently when located at a pressure minimum (velocity maximum). Moving the loudspeaker closer to or farther away from the null location will vary the amount of energy coupled to the mode.

- Use two or more subwoofers to drive the standing waves constructively or destructively. This takes advantage of the fact that lobes of a standing wave on opposite sides of a null have opposing polarity; as the sound pressure is rising on one side, it is falling on the other. Two subwoofers connected in parallel, one on each side of a null, will destructively drive the mode, reducing its effects. Positioned two nulls apart, the same subwoofers will amplify the mode.

Figure 13.11 provides illustrations of these basic effects. They work, as is confirmed in the data in Figure 13.8b, but they are not applicable in a routine manner that leads to somewhat predictable benefits in consumers' homes. Some acoustical expertise can use these effects to great benefit, but we need specific methodologies to take advantage of them in a more general sense. These will be described in the following sections in a progression of sophistication, and success, in delivering predictably good bass to several listeners. These are the steps in the progression:

(a)

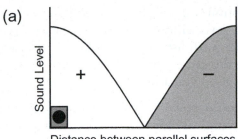

Distance between parallel surfaces

A single subwoofer excites the first-order width mode (shown) and all other modes (shown in Figure 13.1c). Lobes on opposite sides of the null have opposite polarity at any instant in time. In higher order modes the polarity alternates, changing each time a null is crossed.

(b)

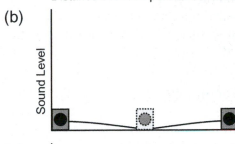

Two methods of attenuating room modes:
1. Two subwoofers, one on each side, radiating identical signals, destructively drive the mode reducing the amplitude. This is "mode cancellation."
2. Placing a single subwoofer at the null location minimizes the energy transfer from the (monopole) loudspeaker to the mode, attenuating it. These methods work for any order of mode.

(c)

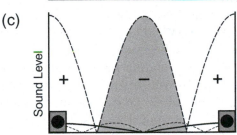

Looking at the second- and third-order modes, the third has also been attenuated because the lobes containing the subwoofers have opposite polarity. The second-order mode has been amplified because the subwoofers are in lobes having the same polarity.

———   first order
- - - -   second order
-------   third order

(d)

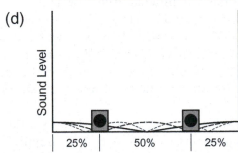

25% | 50% | 25%

Moving the subwoofers to the null locations for the second-order mode reduces the efficiency of energy transfer to that mode, attenuating it. The other modes remain attenuated because the subwoofers continue to be located in lobes having opposite polarity. These locations are the 25% points across the width of the room. Stereo setups in this form incorporate a mode attenuating scheme that minimizes the problem of the listener being in a position of left-right symmetry at low frequencies.

FIGURE 13.11 *Examples of the basic interactions between single and multiple subwoofers and axial standing waves.*

1. Recommendations of subwoofer arrangements that offer increased probability of similar bass in seats within a defined listening area in rectangular rooms of any shape.

2. As above, but we pay attention to the end result in terms of equalization needs and power-output requirements.

3. Selecting optimum room length-to-width ratios to maximize the benefits of specific multiple subwoofer installations.

4. Using measurements, an optimization algorithm, and signal processing to deliver predictably similar bass within rooms of arbitrary shape, rectangular or nonrectangular, with listeners and multiple subwoofers in arbitrary locations, or using the same process in an iterative manner, to maximize performance in a given situation.

All of these, especially number 4, are greatly different from traditional methods in that we are using increasing amounts of control over the extent and manner with which both the subwoofers and the listeners couple to room modes. The primary goal is to reduce seat-to-seat variations. However, it is found that in doing so, the modes are attenuated to such an extent that the need for global equalization is substantially reduced.

13.3.3 Step One: General Recommendations for Rectangular Rooms

The limited number of modes within the subwoofer frequency range, below about 80 Hz, suggests the possibility that, in small listening rooms, some generalized recommendations may be possible. Welti (2002a, 2002b), a member of the Harman International corporate research group, undertook a systematic examination. The first step toward a general solution was to restrict the application to simple rectangular spaces. Then, a 6-ft-square (2 m) seating area was defined in the center of the spaces, within which 16 measurement points were chosen, all at a fixed ear height. A computer model was used to anticipate the frequency response at each potential listening location for many different arrangements of subwoofers within the room. He examined results for a minimum of 1 to a maximum of 5000 subwoofers. For each configuration of subwoofers, Welti calculated a Mean Spatial Variance (MSV) metric, a figure of merit describing the frequency-dependent differences in sound level at the 16 locations within the seating area—the smaller the MSV, the more similar the sound at all locations.

It was not surprising that more subwoofers led to reduced seat-to-seat variations at low frequencies, but it was pleasantly surprising that the improvements were small for more than four subwoofers, assuming that they have been properly located. Figure 13.12 shows some of the preferred subwoofer arrangements resulting from Welti's investigation. All of the best arrangements have even numbers of subwoofers. There appears to be no advantage to using more than four. In fact, according to this metric of seat-to-seat variation, two midwall subwoofers work almost as well as four (front-to-back or side-to-side arrangements work equally well).

It can be seen that two of the four arrangements place subwoofers at the 25% locations, taking advantage of the mode control shown in Figure 13.11d. The arrangement on the far right is practical only in custom installations where they could be located in the ceiling or in the space under a seating riser. In spite

  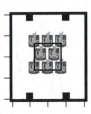 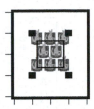

FIGURE 13.12 *The best arrangements for multiple subwoofers if the primary objective is to reduce seat-to-seat variations within the 6-ft-square (2 m) listening area in the center of the space. The only restriction on the room is that it is rectangular in shape. Variations calculated at 2-ft intervals, not at seats shown here. Derived from Welti, 2002a, 2002b.*

of instinct, there is no reason why low frequencies need to originate near the floor.

13.3.4 Step Two: Digging Deeper for Clarification

The data from this first investigation informs us that from the perspective of consistency, minimizing seat-to-seat variations, there are several options. Surprisingly, the two-subwoofer midwall arrangement is competitive with the others. In the interests of reducing costs and room clutter, this option is attractive.

Figure 13.13a shows that a single subwoofer in a corner energizes all of the horizontal-plane modes in the room. Moving the single subwoofer from the corner to a midwall position does precisely what was predicted in Figure 13.11b: The first-order width mode (0,1,0) at 28 Hz has been eliminated because the subwoofer is located at a pressure minimum. Both curves are very ragged, and the Mean Spatial Variance metric is very large for both of them. Because the first- and other odd-order modes have nulls running down the middle of the room—directly through the seating area—the seat-to-seat variations are high. These curves are really just references to indicate how bad things can be.

In Figure 13.13b, the addition of a second subwoofer at the center of the opposite wall cancels the first-order length mode (1,0,0) at 23.5 Hz (as shown in Figure 13.11b) and the accompanying odd-order mode (3,0,0) near 70 Hz. Eliminating these nulls in the listening area greatly improves the mean spatial

ROOM MODES AND EQUALIZATION

A fact of life: Global equalization in the signal path cannot change the seat-to-seat variations. Equalization, whether it is based on measurements made at a single seat or a combination of measurements made at several seats, changes the frequency response at all seats in exactly the same manner, but it cannot alter the seat-to-seat variations.

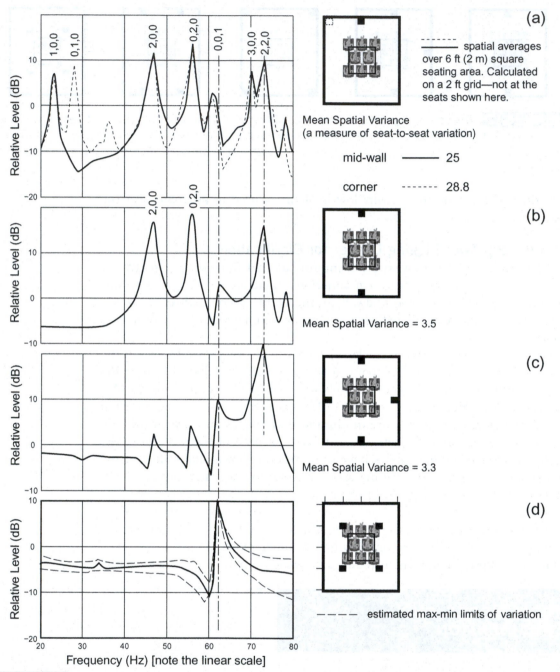

FIGURE 13.13 *"Average total" frequency response curves taken from Welti (2002a, 2002b). This is the combination of direct sound from the subwoofers and the modal energy within the room. They have been traced and vertically rescaled to conform to the standard used throughout this book.*

variance. But the two woofers amplify the second-order length mode (2,0,0) at 47 Hz, giving it more amplitude. The gain in the second-order width mode (0,2,0) at 56 Hz appears to be the result of having subs at both ends of the room, symmetrically driving this mode from the central high-pressure region. Because the pressure nulls of second-order modes are at the 25% locations, they miss the designated listening area, and this is a second reason for the Mean Spatial Variance (MSV) to fall. The amplitude gains in modes 2,0,0 and 0,2,0 help boost the bass efficiency metric that Welti calculates (the MOL). But the shape of the frequency response we are left with is not attractive—monster isolated peaks. Obviously, some acoustical damping would be greatly advantageous here to attenuate and broaden the resonance peaks. However, the MSV is attractively low, and equalization to reduce the resonance peaks in fact gives the system increased headroom at those frequencies, so this is a viable option.

In Figure 13.13c, it is seen that adding two more midwall subwoofers puts in-phase excitation in all three pressure maxima of the second-order modes on both length and width axes. No longer selectively energized, they are in fact substantially attenuated. But what is happening around 74 Hz? While we were attending to the lower-frequency axial modes, a tangential mode has been creeping up in amplitude. Now it is huge—24 dB above the average spectrum level! This is because we have placed energy sources at prime locations for this mode, and it is responding. Figure 13.14a shows what is happening. The

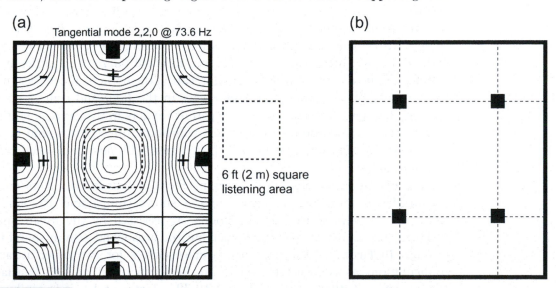

(a) Tangential mode 2,2,0 @ 73.6 Hz

6 ft (2 m) square listening area

(b)

FIGURE 13.14 *(a) In the style of Figure 13.4, the pressure contours for tangential mode 2,2,0 are shown for the test room. Subwoofers are shown to be located in high-pressure points that all share the same polarity, resulting in maximal stimulation of this mode. The listening area is shown to be fully within the central high-pressure region, resulting in this mode being strongly visible in the frequency response of Figure 13.13c and in the low seat-to-seat variation metric, MSV. (b) The subwoofers have been moved to locations that are 25% of the room dimensions from each wall.*

seat-to-seat variations are low because this mode includes all listeners within a broad high-pressure region, and this is now, by far, the strongest mode in the room. Seat-to-seat variations may be small, but we have been left with a serious "one-note" bass problem. The large peak generated by the tangential mode (2,2,0) must be attenuated by equalization, the consequence of which is that a lot of useful sound output has been removed. Compared to (a), three more subwoofers have been added, and the sound level, after equalization, is no higher, and is possibly lower. In moving from one corner subwoofer (a) to four midwall subwoofers (c), in a system in which the maximum sound output is displacement limited, it is clear that the maximum low-bass sound level is not higher. In fact Welti's MOL (mean output level) metric puts it about 6 dB down, (Welti and Devantier, 2006, Figure 5c).

Moving to Figure 13.13d, we see that placing the subwoofers at the 25% locations is a great improvement. We could have anticipated this result from Figure 13.11d, which is given a two-dimensional rendering in Figure 13.14b. Second-order length and width modes are gone, as is the bothersome tangential mode. The seat-to-seat variations have all but disappeared, but there is now a painfully obvious height mode: 0,0,1 at 63 Hz. As acoustical interactions with other modes have been removed, we are left with a height mode that is now much the same at all measurement locations within the seating area. Adjusting ear height and varying room height using a seating platform are possibilities (see Figure 16.4a). If a passive solution is sought, it is known where the absorption needs to go: ceiling or floor. Floor is impractical, but a dropped T-bar ceiling, with a generous fiber-damped cavity, is eminently practical in rooms with sufficiently high ceilings. Just be sure to avoid buzzes and rattles in the installation. If an active solution is sought, cancel the bothersome vertical mode by using mirror-image arrays of subwoofers in the floor and ceiling; custom theaters that have stepped seating may permit woofers in the elevated portions. In this case, the additional loudspeakers contribute directly to sound output. Low-profile woofers intended for wall installation will also fit between ceiling and floor joists.

But there is one more, obvious, arrangement of four loudspeakers not yet examined: the four corners. Figure 13.15a shows comparable data for this configuration. The seat-to-seat variations, as assessed by the Mean Spatial Variance metric, are very slightly higher, but the broadband gain in sound level is huge. In Figure 13.15b, the total average curves for two other attractive configurations are shown for comparison. Remembering that 3 dB is a factor of 2 in power, 6 dB a factor of 4, and 10 dB a factor of 10, it is easily seen that these differences are highly consequential to the demands made on a set of subwoofers. Indeed, multiple sets of subwoofers may be needed to reach equivalence in some of these comparisons. Observing that the two opposing midwall locations produced a curve that resembles the shape of the four-corner configuration, it is possible to achieve sound-level parity by using a pair of

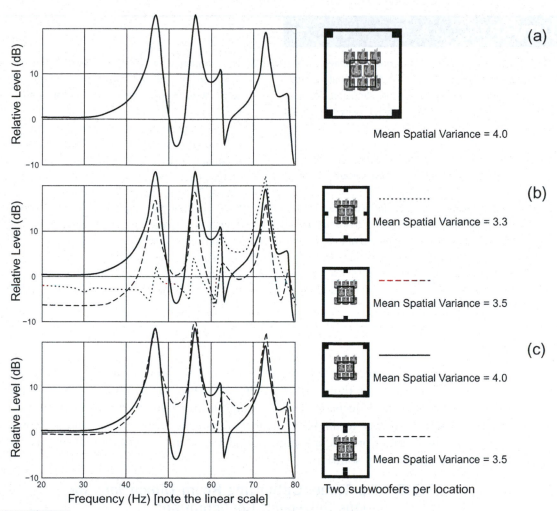

FIGURE 13.15 *(a) Results for a corner arrangement of four subwoofers in the format of Figure 13.13. Unpublished data provided by Todd Welti, Harman International. (b) This data with two other curves from Figure 13.13 for comparison. (c) What appear to be the two most attractive solutions from the perspective of low MSV and similarly high sound output: the two midwall locations with double-subwoofers at each location and the four corners. The difference between MSV values (3.5 and 4.0) is utterly trivial when compared to those for a single subwoofer (25 or 28.8) shown in Figure 13.13. NOTE: In principle, either end- or side-wall locations can be used. Obviously, four smaller subwoofers in the corners would be comparable to two appropriately large subwoofers in the midwall locations. It all depends on room size and sound level requirements.*

subwoofers at each location, thereby raising the sound level by 6 dB because their outputs add coherently. This option is shown in Figure 13.15c, and it is, by a narrow margin in both MSV and overall sound level, the prime choice for a four-subwoofer solution. Either side or end midwall locations can be used. If corner locations turn out to be more practical, use them. None of this has

SUMMARY

To summarize this section, in the quest of minimal seat-to-seat variations in bass, it was found that multiple subwoofers can be advantageous in simple rectangular rooms. However, there are several very important considerations:

1. Everything discussed so far assumes a rectangular room, with "reasonable" acoustical symmetry and so forth.
2. These results were calculated for a flat (not staged) square seating area in the center of the room. Loudspeaker configurations that reduce seat-to-seat variations do not attenuate all modes—only those that cause large variations within the designated seating area. If the seating area is moved, the results will be greatly different; see the next section.

3. The frequency response that results from the use of some of the advantageous arrangements of subwoofers may require substantial amounts of equalization.
4. Adding more subwoofers does not necessarily mean that the system can play louder. The loudspeakers are doing work just to attenuate the standing waves. Some arrangements are considerably more efficient than others. The four corner locations prove to be highly beneficial, but the simple two-location midwall configuration is very attractive if it is equipped to provide enough sound output.
5. Room proportions matter—*all of these conclusions are broad generalizations*, not solutions that may be optimum in any given rectangular room. There are no guarantees.

taken into account the size of the room, the power-output capabilities of the subwoofers, or their cost, so there are other factors to consider in arriving at an optimum decision. In this example, to achieve the same sound level, it was necessary to double the subwoofers at the midwall locations. However, if single units provide adequate sound levels in a given room, obviously that is a satisfactory solution.

13.3.5 Step Three: Optimizing Room Dimensions for Various Subwoofer Configurations

The concept of an "ideal" room does not work in any generalized sense, but the idea of a room that is better suited to certain combinations of loudspeaker and listener positions does. Welti and Devantier (2006) published contour plots showing a measure of seat-to-seat variation (MSV) for several common subwoofer arrangements, and for each arrangement, as a function of room length and width. Figure 13.16 shows results for three of the better-performing arrangements, for a listening area centered on the room, and for one moved back and centered at 1/3 room length.

With the listening area centered in the room (left column of data), the contours are closely symmetrical around the diagonal, which draws attention to the fact that with four subwoofers, the normally forbidden square rooms are not problematic, except perhaps to note that they must be *precisely* square to avoid beating or pitch-shifting modes (see Figure 13.22). It may be advisable to move significantly off-square but not for the traditional reason of summing modes. In

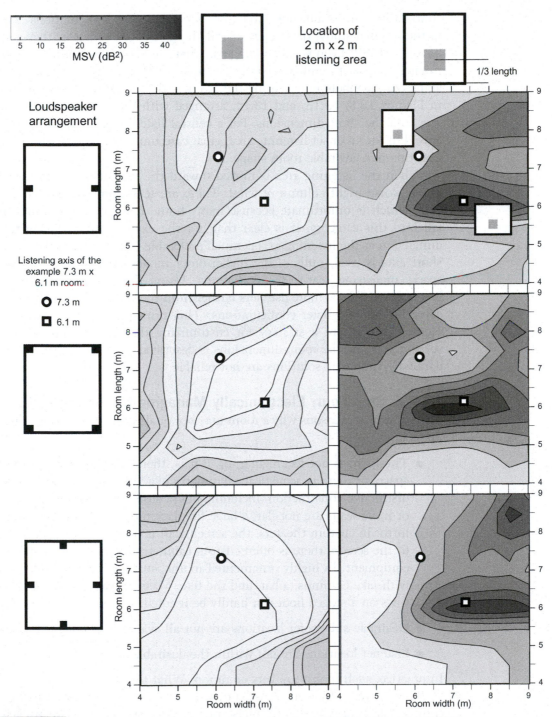

FIGURE 13.16 *A selection of the most attractive solutions from Welti and Devantier (2006).*

contrast to a single subwoofer and other two-subwoofer configurations examined in the original study, these plots are all "light" in color—that is, they generally have low MSV. Even so, there are areas that are better than others, which fortunately appear to embrace common ranges of room shapes. Shown on these graphs are symbols representing the 7.3 m × 6.1 m room used as an example in Figures 13.8, 13.13, and 13.14, arranged with the listening axis along the long and the short dimensions. For a central listening area, there is no difference, but for an offset listening area, great care must be taken. It also turns out that this is a favorable room shape.

When the listening area is moved toward the rear of the room, things get much worse (right column of data). There are few room shapes that perform well, which is unfortunate because many home theaters and listening rooms fall into this category. It is clear that, for the example room, using the long dimension as the listening axis is much preferable. The reason is that when the short axis is used, nulls of the higher-order modes intrude into the listening space, increasing seat-to-seat variations. Using the long dimension for the listening axis keeps more listeners farther from the walls. The conclusion is the same for all subwoofer configurations. This principle is clearly illustrated in Figure 22.4, where a strategy for customizing the relationship between the seating area and the room dimensions is described. It is easy to see from these data why simplistic solutions are not reliable.

13.3.6 Step Four: Electronically Managing the Sound Field

There are many reasons why a room may not qualify for the solutions discussed up to now:

- The room is not rectangular, or, if it is, there is an opening to another space, or there is acoustical asymmetry (e.g., one or more walls have distinctive construction, including brick, stone, expanses of glass). One or more walls are not flat: a large fireplace protrusion, alcoves, and so forth. In custom theaters, the screen wall is often a culprit; in addition to the screen, there is often custom cabinetry to house loudspeakers and equipment. In highly ornamented rooms, surfaces may be broken up with fake columns, a bar, and the like. Massive leather seats arranged in rows on a staged floor can hardly be ignored.

- Desirable subwoofer locations are not all available or practical to use.

- Listener locations are not within the desirable areas.

In practice, such rooms are very common. What then?

The answer is that one must start with real acoustical measurements—full complex data: amplitude and phase or impulse response—between each subwoofer location and each listener location. Then some form of signal processing is used to modify the signal sent to each subwoofer so that seat-to-seat variations are minimized. Welti and Devantier (2006) discuss some options, but we will discuss Sound Field Management (SFM) here.

In SFM, measurement data are plugged into an algorithm that combines the sound from all active subwoofers at each listening position (superposition). The algorithm allows for signal manipulation in the path to each subwoofer—gain, delay, and equalization—to which workable values have been chosen for the parameters and limits applied to the permitted ranges of the manipulation. Then, an optimization program is run, systematically varying the parameters of the signal processing and monitoring the frequency response at the listening positions, with the objective of minimizing the seat-to-seat variations. It is a brute-force trial-and-error system, but it is something that computers are very adept at, and with the power and speed of common laptops, solutions are typically reached in seconds or minutes. To summarize, Sound Field Management has the ability to modify signals fed to individual subwoofers with respect to (1) signal level/gain; (2) delay; and (3) one parametric filter, with values of center frequency, Q and attenuation (no gain).

The optimization process starts with the operator selecting some number of subwoofer locations and some number of listening locations. Limited only by patience and time, the operator can let the computer optimize different combinations to find the solution that best meets the needs and budget of the situation. Let us look at some examples.

Example 1. Figure 13.17 shows data from the room used in Figures 13.7, 13.8, 13.13, 13.14, 13.15, and 13.16. In these comparisons, listening positions were selected on the basis of how the room was used in "home theater" mode; these are not the listening locations shown in Figure 13.16. As discussed earlier, this is, in real-world practical terms, a well-constructed simple rectangular room. Other than some acoustical asymmetry associated with the two lowest-order modes, discussed in Section 13.2.2, this should behave in a predictable manner. To make the discussion interesting, let us assume that the subwoofer used is displacement limited at low frequencies; it will play sufficiently loud to be satisfying with a flat spectrum input, but no more.

Figure 13.17a shows that a single subwoofer in a corner yields a very high seat-to-seat variability. Bass falls off sufficiently, so that about a 10 dB boost around 20–30 Hz would be needed to bring the output to the overall average level around −5 dB. In practice, flat is not regarded as optimal, so more would be needed to provide gratification (although normally it is thought to be unnecessary to extend this requirement much below 30 Hz). This is a very large boost (10 dB is 10× power), and it would drive this displacement-limited unit into gross distortion, and possibly damage. In fact, it would compromise the maximum sound level from many more capable subwoofers.

Figure 13.17b shows an SFM-optimized installation of four of these subwoofers in the midwall locations. This is the real-room version of the modeled-room

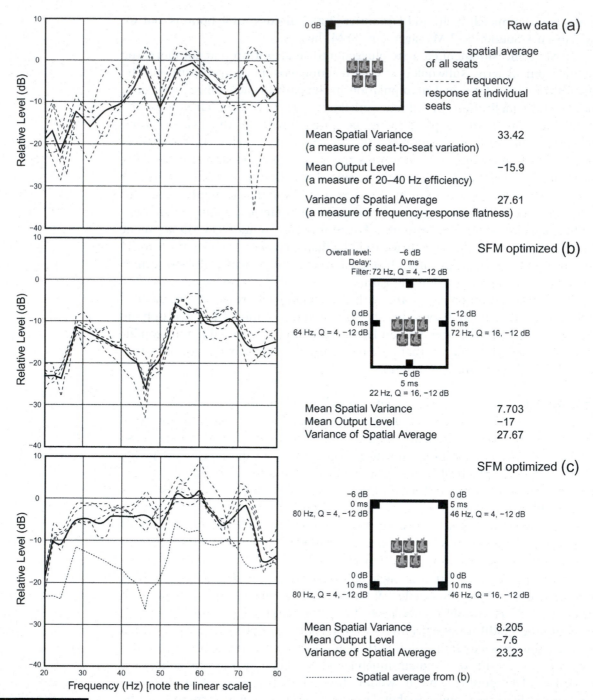

FIGURE 13.17 *Measurements made in the example room used in several earlier figures. Data from Welti and Devantier, 2006; (a) their Figure 20, (b) their Figure 24, and (c) their Figure 25.*

curve in Figure 13.13c—with two important differences that the seating arrangement has changed and SFM optimization has been performed. Interestingly, the algorithm found a need to *reduce* the output from three of the four subwoofers, two by 6 dB and one by 12 dB. This, plus the fact that three of the four filters are in the 64–72 Hz range suggests that the major problem addressed by the optimization routine was the monster tangential mode illustrated in Figure 13.14. The presence of this mode served the 16 positions sampled in the centralized listening area used in Figure 13.13 very well, but not the five offset listener positions in Figure 13.17. The filter at 22 Hz is curious, as this is the frequency of the first-order length mode, which should be cancelled by the front-back subwoofers. However, if we go back to Figure 13.7, it is seen that due to a flexural mode in the end wall of this particular room, the room is "acoustically" lengthened, meaning that the subwoofer is physically not optimally located to perform the task; the optimization routine delayed it by 5 ms, perhaps for that reason.

The curves measured at the individual seats end up being very nicely grouped, which is what was intended, but the overall spectral balance is far from ideal, and the overall sound level is low. One factor is that in killing the energetic tangential mode, a substantial portion of the total energy in the room was removed. Using the same "target" level of −5 dB, it is seen that there is need for boosts of the order of 10 dB from 20–50 Hz and 70–80 Hz, and 3–5 dB from 50 to 70 Hz. Because one subwoofer is already running at full 0 dB level, it cannot accept more input. The ones that have been attenuated can absorb some additional input but perhaps not the full amount required. So in addition to having just purchased three more subwoofers, the installer has to break the bad news to the customer that more are needed. This is not a good solution in any respect, except to minimize seat-to-seat variations.

Figure 13.17c shows an SFM-optimized installation of four corner subwoofers. This is the real-room version of the modeled-room curve shown in Figure 13.15(a), with a different seating arrangement and SFM optimization. In this case only one subwoofer was attenuated, by 6 dB. It is possible to see the filters at work addressing the 47 Hz (2,0,0) mode which is amplified by this loudspeaker arrangement. Beyond that speculation about the missions for the filters is made difficult by the large delays introduced into the signal paths. The very good news is that, overall, everything has improved. Spectral balance looks good. The acoustic gain due to the corner locations (see Chapter 12) contributes to significant sound level gains at very low frequencies—where subwoofers are most under stress. To achieve the −5 dB target only subtle amounts of global equalization are needed and those are mainly attenuation. The slight drop in output near 20 Hz is probably acceptable. Seat-to-seat variations are small. For this room, with these seating locations, this SFM optimized subwoofer arrangement is a very good solution.

It can be seen that, even in a seemingly straightforward rectangular room, substantial signal processing can be required to deliver the best possible sound to *all* of the listeners, especially when they choose to sit outside of an acoustically convenient symmetrical central area. Minimizing seat-to-seat variations is a key factor, but if it means not being able to achieve a satisfying sound level, it may well be considered an unacceptable trade-off. Finding the optimal solution for a particular room may involve some trial and error. However, several real-world installations have confirmed the good behavior and efficiency of four subwoofers in or near corners and SFM optimized. The following is another example.

Example 2. Figure 13.18 shows the results obtained in the author's entertainment/family room, a photograph of which is in Figure 3.2. This is obviously an existing space, not one designed for the purpose, and it has features that would generate frowns among most acoustical cognoscenti. This was an example of a room in violation of several conventional design criteria: It was almost square, one side wall had a large opening to the rest of the house, the other side wall was mostly glass, there was a sloping ceiling and the front, and rear walls had large alcoves and depressions. To add insult to injury, the listeners were arranged for conversation, not for focusing solely on a screen, and they were placed toward the perimeter of the room, avoiding the "desirable" central area. It was not a good scenario. If this space could be made to work, anything might be possible.

The magnitude of the problem can be seen in Figure 13.18a, and the audible evidence was "abundant"; the bass changed dramatically from seat to seat. Applying SFM optimization to these two subwoofers greatly improved things, Figure 13.18b, but such an arrangement of subwoofers can exercise no control in the front-back dimension of the room. There was a significant front-row/back-row change in bass level, moving away from the screen. The "money" seat was in the back bass-deprived row, so this was obviously not the end of the exercise.

Adding another pair of subwoofers to the rear of the room resulted in the SFM-optimized results in Figure 13.18c. Not only were the seat-to-seat variations greatly reduced, but the need for global equalization was all but eliminated. Overall sound levels were significantly elevated, even with three of the subwoofers operating at reduced levels. How did it sound? Personally, it sounded superb then, and five years later, it still does. Again, the "corner" configuration has proved to be an excellent choice, and SFM optimization resulted in what appears from all perspectives to be a desirable performance in what seemed at the outset to be a hostile situation.

An interesting side story: the seat closest to the lower left corner subwoofer suffered from seriously excessive low bass in the two-subwoofer configuration. In fact, there was a tendency to localize bass to that rear corner of the room, partly because there was so much concentrated energy there that small noises

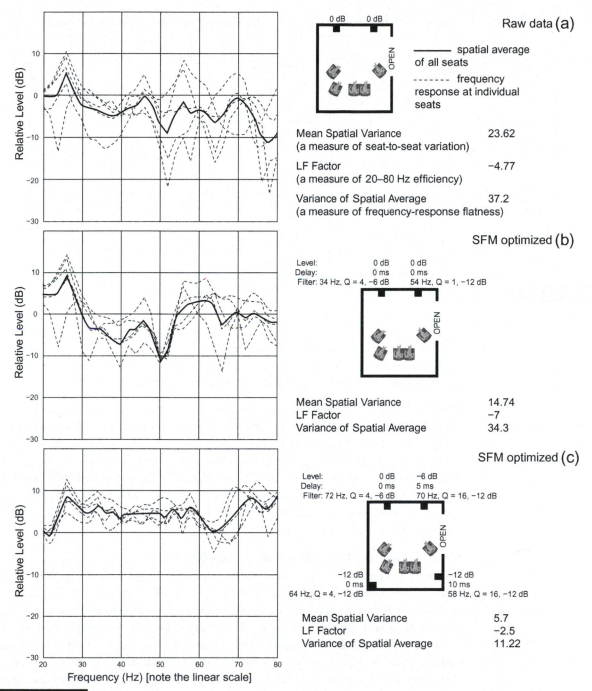

FIGURE 13.18 *Measurements made in the author's entertainment/family room. Subwoofers were located as space and visual considerations permitted, not by any acoustical rules. Listeners were arranged to allow verbal and visual communication with and without video. The room is 22 × 20 ft (6.7 × 6.1 m) with a ceiling that slopes from 8–12 ft (2.4–3.6 m). Data from Welti and Devantier, 2003; (a) their Figure 17, (b) their Figure 18, and (c) their Figure 19.*

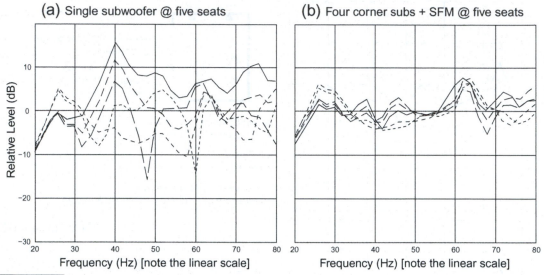

FIGURE 13.19 *Measurements pertaining to Example 3. From Toole, 2006, Figure 19.*

were emitted from vibrating structural elements and windows. There was no loudspeaker there. Adding the rear woofers eliminated the problem, and the woofers went unnoticed from a localization perspective.

Example 3. Figure 13.19 shows results of additional subwoofers and Sound Field Management in the nonrectangular domestic listening room of an audio journalist and reviewer. In (a) there are frequency responses at the five listening positions for a single subwoofer in the front-left corner. In (b) there are comparable frequency responses for four corner-located subwoofers, with SFM optimization. The curves speak for themselves. Seat-to-seat variations are all but gone, and little or no global equalization is called for. Obviously, this is another successful application of the technology.

13.3.7 Getting Good Bass in Small Rooms

Attentive readers will by now have concluded that there is nothing that can be done in advance of building and setting up a room that has a high degree of certainty of achieving good bass. Probabilities may be improved in some respects of acoustical performance, but not all respects. There is no "magic bullet."

In spite of claims of scrupulously "powerful, fast, tight, deep, and clean" performance, subwoofers are probably more accurately described as low-frequency energy sources. The room and loudspeaker and listener locations within it are the principal determinants of sound quality: the "punch and drive" at low frequencies. And together they determine that no two listeners in the same room will hear the same bass. Depending on circumstances, the differences may be small or large, but they are there. Looking at some of the frequency responses measured at listening positions, it is hard to imagine that the "tune" in a bass line will emerge unscathed, no matter how perfect the woofers or subwoofers are. Altogether, this is probably the least well-understood aspect of sound reproduction in audio, especially in the business of audio reviewing.

For a solitary listener there are solutions—simple solutions—here and now. A single competent subwoofer, combined with a competent measurement/equalization system, should be able to deliver respectable bass to a single listener. Other listeners in the room will hear different bass. Once equalization is introduced, the need for a "flat" response from a woofer or subwoofer is removed; it is an energy source and power output versus frequency is the dominant specification, with power (electrical clipping) and displacement (mechanical limiting) limits close behind, and distortion third (only because low-frequency distortions are generally difficult to hear).

Some equalization algorithms employ combinations of measurements made at several listening locations, including those that look for groups (clusters) of listeners that have similar sound. Equalization curves based on such data are attempting to find the best compromise. But it is a compromise because the seat-to-seat differences remain; they are in the physics of the room/loudspeaker/listener interface. In the end, global equalization in the overall signal path cannot solve the problem for multiple listeners.

To improve the situation, room modes must be attenuated; we must insinuate ourselves into the normal "physics" of the room modes. This can be done passively, using low-frequency absorbers—"bass traps"—and lots may be needed to achieve truly excellent results. Absorbers remove energy, and so more sound must be created to produce the original sound levels. Low-frequency absorption is always useful because no matter what else one may do, the problem is lessened.

Room modes can also be attenuated by active means using multiple subwoofers. Welti has provided some "statistical" guidance to the choice of subwoofer number and location for simple rectangular rooms. However, some of the solutions that reduce seat-to-seat variations leave us with unpleasant options for equalization and perhaps also the need for still more sound power to replace the lost energy. Still, it is an important move in the right direction.

Adding higher technology to the solution, transfer-function measurements between each loudspeaker and each listening location can be operated upon by optimization algorithms to yield great improvements in almost any situation. Performance is improved in all respects: smoother frequency response, reduced seat-to-seat variations, and, possibly, elevated sound level. It is an elegant solution, but it costs money, takes time and skill to implement, and adds more audio paraphernalia to a room. One example has been discussed here: Sound Field Management. There are others, one of which was evaluated in Welti and Devantier (2006). Celestinos and Nielsen (2005, 2006) have contributed another. Specific acoustical circumstances have an effect on final performance of any of the systems. Although casual listening indicates large improvements, we await more scientific psychoacoustic examination of the amounts of "non-ideal" aspects of low-frequency performance that are tolerable, or even audible. We also await widespread availability of any of these more exotic solutions in the marketplace.

Are there disadvantages to any of this? Nothing serious, it seems. As with any subwoofer system, the low-pass filtering must be such that the sound output is rapidly attenuated above the crossover frequency (≤80 Hz). Excessive output, distortion products, or noises at higher frequencies increase the risk that listeners will localize the subwoofers.

Correctly attenuating the acoustic output of subwoofers at crossover and achieving an optimum match with each of the satellite loudspeakers are matters that are currently not adequately handled except in rare custom installations. The idea that the normally supplied electronic high- and low-pass filters are sufficient is a dream. With acoustic performance in the region of the crossover frequency, typically 80 Hz, so much under the influence of adjacent boundary influences (Chapter 12), and the standing-wave factors discussed in this chapter, what is happening in any system is simply not known. Only acoustical transfer-function measurements in the room, at the listeners' head positions, can provide the necessary data to permit good subwoofer-satellite transitions to be achieved using additional electronic filtering.

13.3.8 Stereo Bass: Little Ado about Even Less

With apologies to William Shakespeare, this issue relates to the fact that for all the systems described above to function fully, the bass must be monophonic below the subwoofer crossover frequency. Most of the bass in common program material is highly correlated or monophonic to begin with, and bass-management systems are commonplace, but some have argued that it is necessary to preserve at least two-channel playback down to some very low frequency. It is alleged that this is necessary to deliver certain aspects of spatial effect.

Experimental evidence thus far has not been encouraging to supporters of this notion (Welti, 2004, and references therein). Audible differences appear to be near or below the threshold of detection, even when experienced listeners are exposed to isolated low-frequency sounds. The author has participated in a few comparisons, carefully set up and supervised by proponents of stereo bass, but each time the result has been inconclusive. With music and film sound tracks, differences in "spaciousness" were in the small to nonexistent category, but differences in "bass" were sometimes obvious, as the interaction of the two woofers and the room modes changed as they moved in and out of phase. These were simple frequency-response matters that are rarely compensated for in such evaluations. Even with contrived stereo signals, spatial differences were difficult to tie down. This is not a mass-market concern. In fact, some of the discussion revolved around the idea that one may need to undergo some training to hear the effects.

Another recent investigation concludes that the audible effects benefiting from channel separation relate to frequencies above about 80 Hz (Martens et al., 2004). In their conclusion, the authors identify a "cutoff-frequency boundary between 50 Hz and 63 Hz," these being the center frequencies of the octave bands of noise used as signals. However, when the upper-frequency limits of the bands are taken into account, the numbers change to about 71 Hz and 89 Hz, the average of which is 80 Hz. This means, in essence, that it is a "stereo *upper-bass*" issue, and the surround channels (which typically operate down to 80 Hz) are already "stereo" and placed at the sides for maximum benefit. Enough said.

13.4 LOOKING AT TIME AND FREQUENCY DOMAINS

Any reader of audio reviews knows that "tight" bass is a holy grail. It seems that no matter what product is being evaluated, from a power-line conditioner to amplifiers, wire, assorted tweaks, and, of course, loudspeakers, described improvements in sound frequently include "tighter bass," whether there is any possibility of the product impacting bass performance or not. It is one of those giveaway compliments like "better imaging." One can imagine a new control—a knob labeled "tightness." But then there would be discussions of *how* tight is right.

Perceived "tightness" is increased by attenuating low bass at frequencies that energize room resonances; it seems that listeners have a great desire to hear bass that is free from resonances. "Tight" implies good time-domain behavior, just as "boomy" describes bad behavior. But, orchestral bass drums actually do "boom," so we come to the inverse question: How much boom is right? Some amount of control of time-domain behavior is in order, but how much?

So far, discussions have revolved around frequency response data. It was discussed in Section 13.1 that room resonances at low frequencies behave essentially as minimum-phase phenomena, meaning that there is a relationship between the shape of the frequency response and behavior in the time domain. But let us not take it for granted; let's look at it. The following examples stem from situations in the author's previous home. Plato's phrase, "Necessity, the mother of invention" predates these events by roughly 2400 years, but it applies perfectly to what transpired. Motivated by personal dissatisfaction with bass performance in two rooms, the situations were examined and solutions were found, both of which, at the time, were not common, and would have been considered "highly suspect" in certain circles. Both were learning experiences, leading to some of the research described earlier in this chapter.

13.4.1 "Natural" Acoustical Equalization Versus Electronic Equalization

Figure 13.20 shows interesting measurements the author made in his previous home in 1990. The room was large and very irregular in shape, except for the single pair of parallel walls behind the loudspeakers and the listener. It was between these surfaces that the only resonances of note existed, and one of them was a monster. The room was used primarily for listening to classical music; it was intended to present a spacious illusion, and it did, nicely. However, the huge peak at 42 Hz, caused by mode 2,0,0, was seriously annoying in both frequency domain (organ pedal notes varied greatly in amplitude) and time domain (even an orchestral bass drum was excessively boomy); kick drums generated low-frequency drones.

Recognizing the cause of the problem, the immediate solution was to move the listening chair away from the wall, closer to the quarter-wavelength null, attenuating the resonance peak (Figure 13.20b). At a distance of about 1.5 m, everything sounded right; the problem was gone. The sequence of waterfall diagrams shows the sustained ringing tail of the resonance—the "boom"—in (c). Perceptive readers may be thinking that because of reciprocity, the same result could have been achieved by moving the loudspeakers into a similar location at the opposite end of the room. This is correct, but in this house, it was quite impossible due to pesky real-world restrictions. In (d), at a distance of 1.5 m where things sounded about right, the ringing is still visible, but it is much attenuated, starting with about a 12 dB rapid drop following the termination of the signal. In (e) the microphone was placed in the null, at the quarter-wavelength point, and the ringing is utterly gone. At this location, subjectively, there was not enough bass.

Before moving on, some things should be said about waterfall diagrams:

- They are highly decorative.

- They contain a lot of information.

- That information is compromised in both time domain and frequency domain axes, and the compromise can be manipulated to favor one or the other, but not both. In other words, one can have high resolution in the frequency domain and sacrifice resolution in the time domain, or the reverse. All of this is most relevant at low frequencies.

The steady-state frequency response curves shown in (b) indicate a peak at 42 Hz (which can be determined with some precision), rising at maximum almost 20 dB above the average spectrum level. This is visually very disturbing. However, there is no time-domain information. The curve at the back of (c) is the same data, but it looks very much subdued, being much flatter and rising a smaller amount. This is because to see any detail in the time domain, the measurement had to trade off frequency resolution. In this case, the frequency

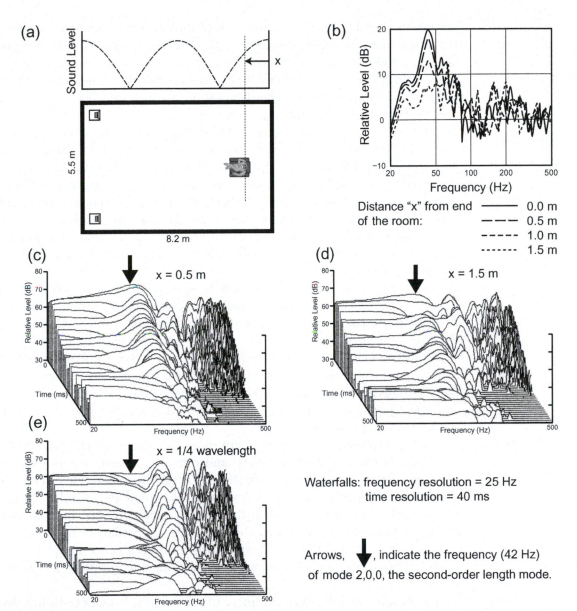

(a) Sound Level

x

5.5 m

8.2 m

(b) Relative Level (dB)

Frequency (Hz)

Distance "x" from end of the room: ———— 0.0 m
––––– 0.5 m
– – – – 1.0 m
········· 1.5 m

(c) x = 0.5 m
Relative Level (dB)
Time (ms)
Frequency (Hz)

(d) x = 1.5 m
Relative Level (dB)
Time (ms)
Frequency (Hz)

(e) x = 1/4 wavelength
Relative Level (dB)
Time (ms)
Frequency (Hz)

Waterfalls: frequency resolution = 25 Hz
time resolution = 40 ms

Arrows, ⬇, indicate the frequency (42 Hz)
of mode 2,0,0, the second-order length mode.

FIGURE 13.20 *Measurements made in a large (7770 ft³/220 m³) living/listening room. (a) The floor plan with the second-order length mode (2,0,0) displayed above it, and the distance x from the wall behind the listener, at which measurements were made. (b) The high-resolution frequency responses at each position. (c), (d), and (e) Waterfall diagrams for three of those positions. The parameters of the Techron TEF 12 were set to present frequency response with a 25 Hz resolution, which has a corresponding time resolution of 40 ms. (b) Adapted from Toole, 1990.*

resolution is 25 Hz (every point on each of the cascade of frequency responses is a weighted average of events over a 25 Hz bandwidth), and the time resolution is 40 ms, and the same can be said for each point in the time-domain decays. These values were chosen because they seemed to be able to reveal the necessary features of both dimensions. The steady-state curves in (b) show the frequency domain behavior in the highest possible resolution.

Getting back to the listening room problem, although moving the chair eliminated the objectionable resonance—but only for the single listener—there was a problem. The chair could not be left permanently in the middle of the living room floor, and moving it was leaving tracks in the carpet. What next?

Equalize, of course. Put the chair back where sensible room décor suggests it should be, closer to the wall, and attenuate the resonance with a single parametric filter tuned to 42 Hz, the appropriate Q and attenuation required to create a frequency response that looked like the one measured at the previously preferred listening location.

Figure 13.21 shows both results, side by side. On the left, (a) shows the frequency response and waterfall for "positional" equalization, with the listener at 1.5 m from the back wall, and (b) on the right shows the comparable results for the listener moved back to 0.5 m from the wall and electronic equalization engaged. The main point of the figure is to show how very similar the waterfall plots look. Most of the small differences in both the frequency responses and the waterfalls are the result of making the measurements at different locations, with the consequent different interactions with room standing waves in addition to the one being addressed.

Looking similar is one thing, but how did they sound? Over several months colleagues, audio journalists, and interested social visitors were subjected to simple A versus B comparisons of the two conditions. The overwhelming conclusion was that they sounded remarkably similar in every respect and very much better than the original condition. The most dramatic demonstration sound was a well-recorded kick drum that, before treatment, was amusingly fat and flabby and, after treatment, became an abrupt slam—as it should be. It was audibly obvious that the time-domain problem had been repaired.

Equalization had the huge advantage of allowing the listener to sit in a decorative location. Acoustically, there were advantages, too. With up to a 14 dB amplitude reduction around 42 Hz, the woofers no longer had to work so hard, distortion was lower, and they could play louder. There was also *much* less energy everywhere in the room at 42 Hz. This was noticeable as improved sound quality at other listening locations. This was a good solution to a personal problem, as well as a learning experience: The right kind of equalization sounds just fine, and electronics can provide an option equivalent to natural acoustical manipulations. Of course, it works best for a single seat.

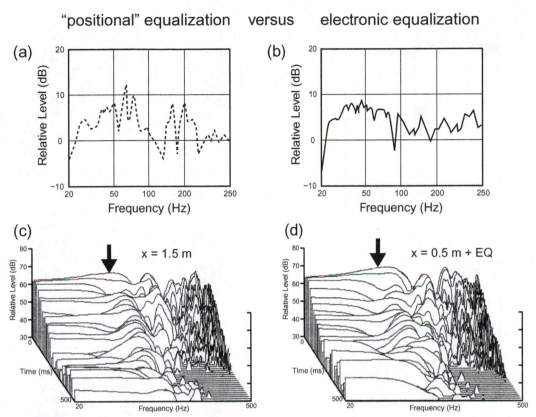

FIGURE 13.21 *A comparison of steady-state frequency responses (a) and (b) and waterfall diagrams (c) and (d) for both kinds of equalization.*

13.4.2 Another Room, Another Problem—A Very Different Solution

In the family room of the same house, there was a first-generation home theater, installed in 1987, with five channels and a subwoofer. It was all very exciting and impressive, but a case of "one-note" bass quickly became tiresome. After a brief experiment with equalization, which in this room worked for the prime listener and made things worse for some others, the author went back to the basics.

Figure 13.22 is well explained in the caption. It shows how two strategically located subwoofers can attenuate bothersome axial modes along both length and width dimensions. The result is that a pronounced peak in the frequency response is converted into a narrow destructive-interference dip, and an obvious energetic ringing in the time domain is substantially attenuated. The visibility of frequency shifting during the decay is very interesting. This undoubtedly happens in many rooms, and it cannot be a good thing. In this room, no listener

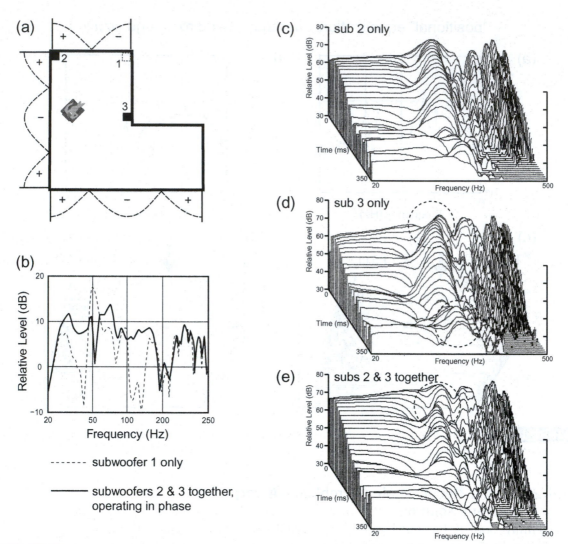

FIGURE 13.22 *A family room attached to a dinette and kitchen in an L-shaped space. (a) Three axial modes at similar frequencies, all of which were strongly energized by a single subwoofer located in the corner behind the video display—subwoofer 1. (b) (dotted curve) A powerful resonance around 50 Hz and large holes in the bass response above and below that frequency. Inspection of the polarities in the standing wave lobes suggested that a pair of subwoofers, one in a lobe of each polarity, should alleviate the problem. This was conveniently possible at locations 2 and 3, and the solid curve in (b) was the result, with no equalization. (c) The waterfall at the prime listening position for subwoofer 2 only. (d) One for subwoofer 3 only. (e) One for both subwoofers. (c) In addition to energetic ringing, it can be seen that the ringing skews downward in frequency. It is clear that the reason is that there are two, or possibly three, closely adjacent modes that are revealed as energy moves among the closely coupled resonances that have different decay times (amounts of damping). (d) See what appears to be a quite discrete downward frequency shift in the top circle, from a decaying mode to a more energetic lower-frequency one, and a stabilization in the bottom circle. (e) It is clear that with both subwoofers operating, the aggressive ringing is substantially attenuated; evidence of the frequency-shifting modal interaction can still be seen early in the decay. (a) and (b) Adapted from Toole, 1990, Figures 14 and 15.*

complained about pitch shifting per se, but in retrospect, it turns out that "one-note" bass was an incorrect description. Benjamin and Gannon (2000) reported hearing pitch shifts, dissonance, and beats when they focused on the interactions of music and low-frequency room resonances, so finding methods to control them is important. In this case, the mode-canceling solution worked wonders, and superb bass was the result. It is interesting to sit in a room and, at low frequencies, to not hear the room. It was addictive.

13.5 TIME AND FREQUENCY DOMAIN-MEASUREMENT RESOLUTION

Heisenberg's uncertainty principle, introduced to the world of physics in 1927, stated that in the context of a subatomic particle moving through space, "the more precisely the position is determined, the less precisely the momentum [mass × velocity] is known" (http://www.aip.org/history/heisenberg).

The notion that these two important descriptors of a physical event cannot be determined to equal precision simultaneously was a disruptive assertion. The physical parallel is not good, but there is a kind of parallel logic between this and the circumstances of certain acoustical measurements. Mention was made in Section 13.4.1 of the necessity to make choices of frequency and time resolutions when displaying waterfall data. Figure 13.22 illustrates some examples of the consequences of making the wrong choices.

Figure 13.23a shows a very high-resolution (about 2 Hz) steady-state frequency response. Many measuring devices, including some relatively inexpensive PC/laptop-based packages, can generate this kind of data with adequate resolution if the measurement parameters are set appropriately. It has wondrous detail in the frequency domain, but it reveals nothing *directly* about what is happening in the time domain. However, because we know that low-frequency room resonances generally behave in a minimum-phase manner, we know that if there are no prominent peaks protruding above the average spectrum level, there will not be prominent ringing in the time domain. It is this *indirect*, inferential knowledge that permits us to confidently use frequency responses as a primary source of information about room behavior at low frequencies.

Figure 13.23b, (c), and (d) show the *same* situation displayed in time and frequency, as a waterfall, but employing different resolutions. In (b) the frequency response at the back, near, time 0, looks a lot like the curve in (a) because a moderately narrow 7 Hz bandwidth was chosen. However, the successive curves comprising the waterfall all look very much like it. This is because in achieving detail—resolution—in the frequency domain, the time-domain resolution has been sacrificed; it is 142 ms. Within the total 500 ms decay that is shown, nothing much changes.

In (c) the frequency resolution has been reduced; now it has an effective resolution of about 14.2 Hz. The top frequency response is a little smoother,

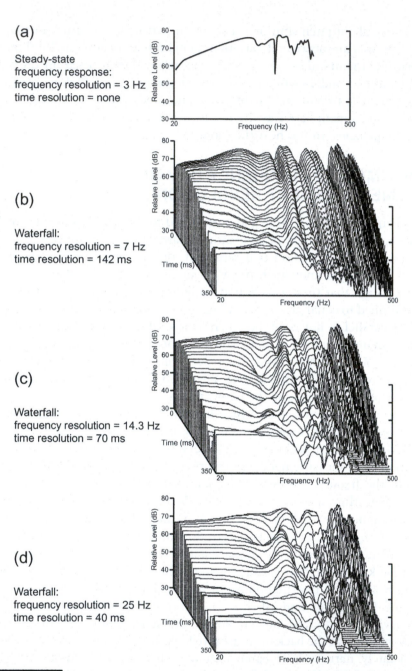

(a)

Steady-state
frequency response:
frequency resolution = 3 Hz
time resolution = none

(b)

Waterfall:
frequency resolution = 7 Hz
time resolution = 142 ms

(c)

Waterfall:
frequency resolution = 14.3 Hz
time resolution = 70 ms

(d)

Waterfall:
frequency resolution = 25 Hz
time resolution = 40 ms

FIGURE 13.23 *Examples of different resolutions in time and frequency domains for the same acoustical situation. (a) A very high-resolution steady-state frequency response. (b), (c), and (d) Waterfall diagrams with different frequency and time resolutions.*

revealing less detail, but we now can see some changes in the decay curves in the time domain. There is now clear evidence of a frequency shift around the bothersome room resonance in the early part of the decay, as seen in Figure 13.22e. At frequencies approaching 500 Hz, it can be seen that the output has dropped below the lowest measured level before 500 ms is reached.

In (d) is the same resolution used in Figure 13.22e, and the top frequency response, and all that follow it, are seriously compromised, being much smoothed by the reduced resolution, but because of this trade-off, it is possible to see more of what is happening in the time domain.

In informal audio literature and manufacturers' data, there have been numerous examples of erroneous conclusions being drawn from data of this kind. Usually the authors don't reveal the parameters of the measurement; in some cases, one suspects that they did not know or know that it mattered. Other mathematical methods exist that reveal different aspects of the time/frequency trade-off (e.g., Wavelets, Wigner-Ville, and Gabor distributions), but each requires skilled interpretation.

13.5.1 Practical Resolution Issues—How Some Reputations Get Tarnished

For at least half a century, acoustical measurements have been made using fractional-octave bandwidth analyzers. The most common has been the 1/3-octave version, partly because it somewhat resembled auditory critical bands over much of the audible bandwidth. Critical bands are not the final answer in terms of sound quality or timbre (more in Chapter 19).

In measurements of loudspeakers in rooms, the curve smoothing offered by the wider bandwidth was attractive; the ugly and not very informative "grass" went away, making everyone feel better. Massive numbers of 1/3-octave "real-time" analyzers invaded the acoustics field. Multifilter "graphic" equalizers of 1/3-octave band also flooded the market, providing a tidy match between the measured data and the "draw-a-curve" style of equalization. At frequencies above the transition frequency, there is still a modicum of utility in this approach. However, at low frequencies in small rooms, measurements of this kind can be greatly misleading. The big problem is that the limited bandwidth of the measurement system cannot reveal the true nature of resonances—amplitude, center frequency, and Q—and the corresponding limited nature of the matching equalizers cannot address the resonances with a matching filter. Adding to the deficiency are those measurement systems that employ fixed-frequency filters, presenting their data as histograms, bar graphs, or staircase line drawings, as shown here. A swept filter is preferable.

Figure 13.24a shows high-resolution, 1/20-octave measurements of a subwoofer in a badly resonating room before and after equalization using a parametric filter adjusted to accurately match the center frequency and Q (bandwidth) of the peak caused by the resonance. The time-domain data show that the

(a) 1/20–octave-resolution measurements and parametric equalization

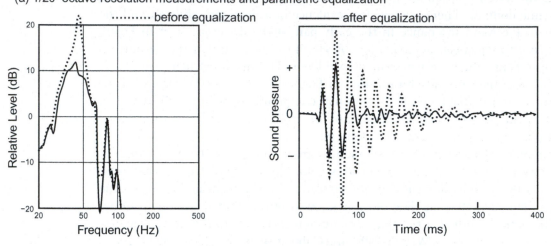

(b) 1/3–octave-resolution measurements and 1/3–octave multifilter equalization

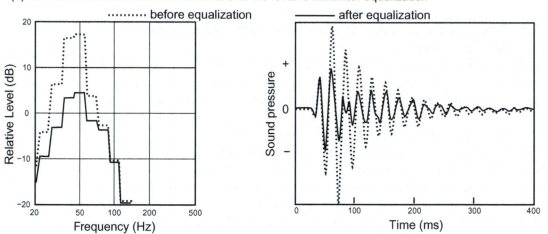

FIGURE 13.24 *Before and after measurements in frequency and time domains for two different measurement and equalization methods.*

resonance has been greatly damped by the process. The ringing is almost gone, and that is how it sounded.

In Figure 13.24b, the same situation is measured and equalized using 1/3-octave, fixed-center frequency measurements and equalization. First, it is seen that the measurements do not reveal small details in the data. Therefore, the measurement does not reveal the true nature of the problem, and the solution is capable only of modifying what is revealed. Thus, although the corrected frequency response looks good, the time domain data reveal that there is still some of the old problem left: it still rings, although at a lower level. With limitations of this kind, it is no wonder that room equalization acquired a poor reputation over the years.

CHAPTER 14

Summary of Part One: Looking for a Way Forward

The first part of this book was more than slightly esoteric, considering that our goal is simply to learn how to set up a home theater. However, those of you who have struggled through should understand that there is a reason for many of the things we do and also a reason to think it might be better if we did some things differently.

With no apologies, it is inevitable that summaries are biased by the author's priorities and interests. Here follows one such summary, based on quotations taken from the text:

CHAPTER 1 SOUND REPRODUCTION

If any sound is rewarding, better and more spatially complex sound may be more pleasurable.

Knowing that the production process will lead to a reproduction liberates a new level of artistic creativity. Capturing the total essence of a "live" event is no longer the only, or even the best, objective.

Sound reproduction has influenced music itself, especially jazz.

And so it will continue in the unending interplay between musical creation, reproduction technology, and listener expectations and preferences. . . . The technologies discussed in the rest of this book are part of the future of our audio industry, our music, and our movies. It will be seen that there are several ways to improve the process, ensuring a superior and more reliable delivery of the art we know. There are also opportunities to introduce new ingredients into the art.

CHAPTER 2 PRESERVING THE ART

In spite of its many capabilities, science cannot describe music. . . . The determination of what is aesthetically pleasing remains firmly based in subjectivity.

The evaluation of reproduced sound should be a matter of judging the extent to which any and all of these elements are accurately replicated or attractively reproduced. It is a matter of trying to describe the respects in which audio devices add to or subtract from the desired objective. A different vocabulary is needed.

In the audio industry, progress hinges on an ability to identify and quantify technical defects in recording and playback equipment while listening to an infinitely variable signal—music.

To a remarkable extent, we seem to be able to separate the evaluation of a reproduction technology from that of the program. It is not necessary to enjoy the program to be able to recognize that it is well reproduced.

The practical reality is that all recordings end up in a control room of some sort, where decisions are made about the blending of multiple micro- phone inputs, of sweetening with judicious equalization, of enrichment with a little electronically generated delayed sound. This is the second layer of art in recordings, added by some combination of recording engineers and perform- ing artists.

All recording professionals care about what is heard by their customers, but they differ enormously in how to estimate what it is.

Choosing a single or even a small number of "bad" loudspeakers cannot guarantee anything. Nobody in this massive industry seems to have undertaken a statistical study of what might be an "average" loudspeaker. The author's experience suggests that the performance target for almost all consumer loud- speakers is a more-or-less flat axial frequency response. Failure to achieve the target performance takes all possible forms. . . .

It remains a perplexing dilemma that there are no truly reliable technical standards for control room sound, making the reference a moving target.

Sounds exist in acoustical contexts. In live performances, we perceive sources at different locations, and at different distances, in rooms that can give us strong impressions of envelopment. A complete reproduction ought to convey the essence of these impressions.

Reproducing a persuasive illusion of realistic direction and space must entail multiple channels delivering sounds from many different angles surrounding a listener. The key questions are how many and where? The answer may not yet be completely known, but we are absolutely confident that it is more than one (mono) or two (stereo) loudspeakers, and very much less than some large numbers that have been proffered.

Auditory spatial illusions are no longer attached to visual correlates; they exist in the abstract, conceivably a different one for every instrument in a multi- miked studio composition. Traditionalists may complain about such manipula- tions, but most listeners seem to find it to be just another form of sensory stimulation.

All of this stands in stark contrast to the spatially deprived decades that audio has endured.

A coincidental influence was the development of the acoustical materials industry. In the 1930s, dozens of companies were manufacturing versions of resistive absorbers—fibrous fluff and panels—to absorb reflected sound and to contribute to acoustical isolation for bothersome noises. Acoustical treatment became synonymous with adding absorption. Dead acoustics were the cultural norm—the "modern" sound. . . .

This notion that reflections result in a corruption of "pure" music, and the apparent surprise in finding that musicians and ordinary listeners prefer "muddied up" versions, reappears in audio, even today.

In some ways, our problems with rooms, especially small rooms, began when we started to make measurements. Our eyes were offended by things seen in the measurements, but our ears and brain heard nothing wrong with the audible reality.

The audio industry has developed and prospered until now without any meaningful standards relating to the sound quality of loudspeakers used by professionals or in homes. The few standards that have been written for broadcast control and music listening rooms applied criteria that had no real chance of ensuring good, or even consistent, sound quality.

The penalty for this lack of standardization and control is that recordings vary, sometimes quite widely, in their sound quality, spectral balances, and imaging.

No matter how meticulously the playback equipment has been chosen and set up, and no matter how much money has been lavished on exotic acoustical treatments, what we hear in our homes and cars is, in spatial terms, a matter of chance.

How can we measure something that subjectively we react to as art? Measurements are supposed to be precise, reproducible, and meaningful. Perceptions are inherently subjective, evanescent, subject to various nonauditory influences within and surrounding the human organism. However, perceiving flaws in sound reproducing systems appears to be an activity that we are able substantially to separate from our critique of the art itself. We can detect flaws in the reproduction of music of which we have no prior knowledge and in which we find no pleasure.

The hope is that we may be able to "connect" with some of the key underlying perceptual dimensions. The fact that several chapters follow this one signals that there has been some success at doing so.

CHAPTER 3 SOUND IN ROOMS—MATTERS OF PERSPECTIVE

The science of room acoustics developed in large performance spaces: concert halls.

All acoustical measurements done in these spaces begin with a sound source intended to radiate sound equally in all directions. So the sound source is a "neutral" factor. . . .

Driven by habit and tradition, many of the same measurements have been carried over into small rooms used for recreational listening to music and movies.

Measurements made with directional loudspeakers [used in sound reproduction] are measurements of the loudspeaker and room in combination, a different thing. And it is the sounds radiated by the multichannel sound system (two, five, or more directional loudspeakers all aimed at the listeners, but from different directions) that deliver the impressions of direction and space, not the listening room itself—another different thing.

A listening room should be designed so as not to detract from a reproduced sound event. . . . In contrast, a concert hall is designed to be a substantial positive contributor to a live performance; without it, the performance suffers.

If we cannot reconstruct a specific sound field, what is it that we need to reconstruct?

At issue here is whether the record/reproduction system has the ability to capture and reproduce the principal perceptual variables that contribute to the spatial aspects of live listening experiences.

- *Direction—the ability to localize sound sources*

- *Distance*

- *Spaciousness or spatial impression—perceptions associated with listening in a space, especially a large space. It has two principal perceptual components: ASW (apparent source width) and LEV (listener envelopment).*

Spaciousness is level dependent, since the illusion requires that low-level reflected sounds be audible. The more of these sounds that are heard, the greater is the spatial impression. In live performances, profound spaciousness is a forte phenomenon. In reproduction, it is dependent on volume setting and background masking noises.

Listeners to the stereo and multichannel loudspeaker systems we are familiar with are responding to a sound field that is very different from that which they would experience at a live musical performance. Yet, from these very different physical sounds, we seem to be able to perceive what many of us judge to be very satisfying representations of our memories of concerts or other live experiences.

One of the common criticisms of recorded music is that it just doesn't quite sound like the "real thing." . . . What the microphones "hear" is only a small portion of the blended sounds that are delivered to the audience. . . . There is no possibility of anything like "waveform fidelity"—there simply is no single waveform that totally exemplifies the sound of real musical instruments.

Deliberately reflective recording studios spatially integrate the differing off-axis sounds into a pleasing whole. Still, microphone placement is a factor, and some amount of equalization in a microphone channel is a common thing; the portion of the sound field sampled by the microphone often exhibits spectral biases that are not in the overall integrated sound of the instrument.

A great deal of research underlies our understanding of concert halls, but that is not where most of us spend most of our listening time. This book is an attempt to adjust the balance, just a little, toward the circumstances where we mostly live and listen.

CHAPTER 4 SOUND FIELDS IN ROOMS

Physical measures of the sound fields in rooms are important because, through them, we hope to understand the perceptual dimensions of speech, movies, and musical performances that we enjoy in those rooms.

It is important to note that in the calculations of reverberation time, it is assumed that the acoustical activity occurs on the room boundaries and that the volume of the room is empty.

In reality, the level of the reverberant sound field gradually falls with increasing distance as energy is dissipated.

In addition, individual voices and instruments do not obey the simplifying assumption of omnidirectionality, so the sound field at different listening positions will be different for different instruments. . . . We hear this at concerts in the contrast between the penetrating clarity of brasses that deliver a higher proportion of direct sound compared to the open and airy strings that radiate their collective energy more widely.

Because reverberation, in effect, prolongs all acoustical events, reverberation time is an obvious influence on how we hear speech and music.

In terms of speech intelligibility in large spaces, it has long been recognized that the early reflections that are a component of the early portion of reverberation are important aids to speech intelligibility.

Increased early reflection energy has the same effect on speech intelligibility scores as an equal increase in the direct sound energy.

Most of us think of live performances in good concert halls as "reference" experiences, not only greatly enjoyable but an opportunity to recalibrate our perceptual scales. That is because there are no technical devices to get in the way, no microphones, recordings, and loudspeakers. But that does not guarantee unimpaired sound transmission. There are acoustical phenomena, one of which has come to be known as the "seat-dip effect."

[When we examine what happens as listening spaces shrink from concert hall size to small listening rooms] When the floor area shrinks from office/factory

to domestic dimensions, it seems probable that this behavior will continue because key features of the commercial spaces are present. Large portions of one or more surfaces have significant absorption in the form of carpet, drapery, and, perhaps, acoustical ceilings. There are also sound-absorbing and -scattering objects, such as sofas, chairs, tables, cabinets, and vertically stepped arrangements of bulky leather chairs in custom home theaters, all of which are large relative to the ceiling height in typical homes.

The acoustical explanation is the dominance of relatively isolated room modes and standing waves at low frequencies and of a complex collection of overlapping modes and reflected sounds at high frequencies. . . . In between the orderly low-frequency room resonances and the disorderly higher-frequency acoustical behavior is a transition zone.

No matter how it is identified, or what it is called, the transition region is real, and it is necessary to take different approaches to dealing with acoustical phenomena above and below it.

Considering the distances at which we listen in our entertainment spaces and control rooms, it is clear that we are in the transitional region, where the direct and early-reflected sounds dominate, and late-reflected sounds are subdued, and progressively attenuated with distance. The sound field is not diffuse, and there is no critical distance, as classically defined. If we were to speculate at this early stage about loudspeaker performance in these rooms, it would seem that a combination of direct and early-reflected sounds would figure prominently in their potential sound quality and that sound power would not be the dominant factor.

Diffusion is a property of a sound field. . . . Perceptually, a diffuse sound field sounds spacious and enveloping. However, a diffuse sound field is not a requirement for the perception of spaciousness and envelopment. Much simpler sound fields work, too, especially if multichannel sound reproduction is involved, because then it is possible to deliver sounds to the ears that are perceived to have those qualities—with or without a room.

A highly diffuse sound field may be a worthy objective for performance spaces and recording studios, where the uniform blending of multiple sound sources and the reflected sounds from those multidirectional sources is desired. However, it is conceivable—indeed, probable—that such a sound field may not be a requirement for sound reproduction.

Diffuse-field theory may not apply perfectly to concert halls, but it applies even less well to other kinds of rooms. In the acoustical transition from a large performance space to a "small" room, it seems that the significant factors are a reduced ceiling height (relative to length and width), significant areas of absorption on one or more of the boundary surfaces, and proportionally large absorbing and scattering objects distributed throughout the floor area.

What we hear in a small listening room is dominated by the directional characteristics of the loudspeakers and the acoustic behavior of the room bound-

aries at the locations of the strong early reflections. RT reveals nothing of this. As a measure, it is not incorrect; it is just not useful as an indicator of how reproduced music or films will sound. Nevertheless, excessive reflected sound is undesirable, and an RT measurement can tell us that we are in the ballpark, but so can our ears, or an "acoustically aware" visual inspection.

The transitional sound field appears to extend over the entire range of listening distances we commonly employ in small rooms. It is therefore necessary to conclude that the large-room concept of critical distance is also irrelevant in small rooms.

The numbers produced by traditional acoustical predictions and by measuring instruments, while not totally irrelevant, are simply not direct answers to the important questions in small-room acoustics. What, then, are the important questions? The accumulating evidence suggests that they have to do with reflections but not in a bulk, statistical, sense. We need to find, in Schultz's words, a "practical theory that is usable in real rooms" to explain what we are hearing.

CHAPTER 5 THE MANY EFFECTS OF REFLECTIONS

As the story develops, it will be evident that measurements are indeed relevant, but some measurements are much more useful than others. It will also become clear that humans are wonderfully adaptable, able to compensate for some things that we can measure, and for some things that can be heard while moving, or while they are changing, but that fade away once stability is established. It is as if when we walk into a room, we hear all of the reflections that give us a great deal of information about the acoustical nature of the space. Then, when we sit down, within a very short time the perceptual effects of the reflections are attenuated, some more than others, and we settle in to listen to the sound sources, whatever they may be.

Within some range of "normal" rooms, we seem to have a built-in ability to "listen through" a room to attend to even minute details of the sound source.

We know that in real rooms, there are multiple reflections. However, to understand the influence of many, it is useful to begin by understanding the influence of a few, even one. It also makes experiments practical and controllable. As will be seen, there is a logical progression of effects from a single to multiple reflections, giving us, in the end, a better insight into the perceptual mechanisms at play.

CHAPTER 6 REFLECTIONS, IMAGES, AND THE PRECEDENCE EFFECT

The precedence effect describes the well-known phenomenon wherein the first arrived sound, normally the direct sound from a source, dominates our sense of where sound is coming from. Within a time interval often called the "fusion zone," we are not aware of reflected sounds arriving from other directions as separate spatial events. All of the sound appears to come from the direction of the first arrival. Sounds that arrive later than the fusion interval may be perceived as spatially separated auditory images, coexisting with the direct sound, but the direct sound is still perceptually dominant.

Haas described this as an "echo suppression effect." Some people have mistakenly taken this to mean that the delayed sound is masked, but it isn't. Within the precedence effect fusion interval, there is no masking; all of the reflected (delayed) sounds are audible, making their contributions to timbre and loudness, but the early reflections simply are not heard as spatially separate events. They are perceived as coming from the direction of the first sound; this, and only this, is the essence of the "fusion." The widely held belief that there is a "Haas fusion zone," approximately the first 20 ms after the direct sound, within which everything gets innocently combined, is simply untrue.

Haas noted audible effects having nothing to do with localization. First, the addition of a second sound source increased loudness. There were some changes to sound quality, "liveliness" and "body," and a "pleasant broadening of the primary sound source."

This graphical display is very different from most discussions of the precedence-effect fusion phenomenon, in which it tends to be stated as a delay interval: a single number. This is simply wrong, as it presumes certain conditions that may or may not exist.

Individual reflections in normal small rooms cannot generate multiple images from speech produced by a person or reproduced by a loudspeaker (the directivity of a human talker is within the range of directivities for conventional forward-firing cone/dome loudspeakers. . . . In small rooms, the precedence effect is undoubtedly the dominant factor in the localization of speech.

The thresholds for the side wall and the ceiling reflections are almost identical. This is counterintuitive because one would expect a lateral reflection to be much more strongly identified by the binaural discrimination mechanism. . . .

It is remarkable that a vertically displaced reflection, with no apparent binaural (between the ears) differences, can be detected as well as a reflection arriving from the side, generating large, binaural differences. Not only are the auditory effects at threshold different—timbre versus spaciousness—the perceptual mechanisms required for their detection are also different.

In explanation, Rakerd et al. (2000) agreed with other referenced researchers that there may be an "echo suppression mechanism mediated by higher auditory centers where binaural and spectral cues to location are combined."

A long-standing belief in the area of control room design is that early reflections from monitor loudspeakers must be attenuated to allow those in the recordings to be audible. . . . In conclusion, it seems that the basic audible effects of early reflections in recordings are remarkably well preserved in the reflective *sound fields of ordinary rooms. There might be reasons to attenuate early reflections within listening rooms, but the ability to hear individual reflections in a reflective listening environment, it seems, is not one of them.*

When detection threshold and image-shift threshold determinations were done first with real and then with phantom center images, in the presence of an asymmetrical single lateral reflection, the differences were insignificantly small. It appears that concerns about the fragility of a phantom center image are misplaced.

The message is that if we believed the impulse response measurements, we might have concluded that by breaking up the large reflecting surface, we had reduced the audible effects. This is one of the persistent problems of psychoacoustics: human perception is usually nonlinear and technical measurements are remarkably linear.

So now there are both subjective and objective perspectives indicating that breaking up reflective surfaces may not yield results that align with our intuitions.

In measurements of reflections, we need to measure the spectrum level of reflections to be able to gauge their relative audible effects. This can be done using time-domain representations, like ETC or impulse responses, but it must be done using a method that equates the spectra in all of the spikes in the display—for example, bandpass filtering. Examining the "slices" of a waterfall would also be to the point, as would performing FFTs on individual reflections isolated by time windowing of an impulse response.

All of this is especially relevant in room acoustics because acoustical materials, absorbers and diffusers, routinely modify the spectra of reflected sounds. Whenever the direct and reflected sounds have different spectra the simple broadband ETC or impulse responses are not trustworthy indicators of audible effects.

CHAPTER 7 IMPRESSIONS OF SPACE

Impressions of space are the paramount audible factors distinguishing good spaces for live performances. They contribute much of the interest and identity to all large reverberant spaces we encounter. . . .

It is not necessary to replicate the sound field of a real space in a listening room; it is sufficient only to provide key cues in order to elicit a recollection or an emotion.

With good two-channel stereo recordings, one can get impressions of these kinds. With multichannel audio, such illusions can be delivered in any amount— including excess.

Envelopment requires multiple loudspeakers, delivering recorded sounds containing the appropriately delayed sounds from the appropriate directions. It is possible that reflections within the listening room may assist in impressions of envelopment by adding repetitions, but they must be initiated by recorded sounds having the large initial delays.

Hearing a change (a threshold detection) tells us nothing about whether the change is good, bad, or neutral. What happens when listeners are allowed to choose the level of a single reflection, based on what they perceive as a sense of pleasantness—a preference?

When listening to speech, the preferred levels just avoid the "second image" curve, indicating that the preferred reflections were all within the precedence effect fusion zone, thereby not generating distracting second images. The inevitable conclusion is that in listening to live speech, and for a single loudspeaker reproducing speech, individual room reflections are not problems. In fact, they are not loud enough.

When listening to music, all preferred levels are far above the natural reflections provided by small rooms. . . . They are simply not consequential factors in this matter. . . . Obviously, listeners were willing to add reflections having sound levels and delays that would cause strong image shift and broadening, and perhaps second images, and still indicate a preference for listening to music in that state.

For maximum "preference," it seems that reflections from about 30° to 90° are most effective. When IACC is measured, a broad minimum is seen around 60°, corresponding to a maximum in the preference ratings. Preference, therefore, is associated with low interaural cross correlation.

The most important message is that "preference" is associated with a strong "spatial impression." Technically, it seems to be possible to find correlation with both a measure of the sounds arriving at the ears (a low IACC) and a measure of the physical sound field in which the listener is immersed (a high proportion of lateral vs. frontal sound in the room).

Complete listener gratification is likely to require reflections that are higher in level and later in time than those naturally occurring in small listening rooms. This is where multichannel sound reproduction systems enter the picture.

CHAPTER 8 IMAGING AND SPATIAL EFFECTS IN SOUND REPRODUCTION

Localization is not perfect; there is localization "blur," a region of uncertainty, the size of which depends strongly on direction. In live performances, we have visual information to substantiate localization (the ventriloquism effect), and generations of audiences have voted in favor, not of pinpoint localizations of musicians but of spatially embellished sound images, called apparent source width (ASW). Stating this again, we know where the sound is coming from, and we derive pleasure from having the auditory directional information corrupted!

Absorbing the first reflections has a powerful effect on the diffusivity, the IACC, and thereby the perceived spaciousness of sound in a room.

When listening tests were done in the two versions of the room, it was found that the condition with absorbing side walls was preferred for monitoring of the recording process and examining audio products, whereas reflective side walls (which reduced IACC) were preferred when listeners were simply "enjoying the music." As might be expected, reflective side walls resulted in a "broadening of the sound image."

Musicians judge reflections to be about seven times greater than ordinary listeners, meaning that they derive a satisfying amount of spaciousness from reflections at a much lower sound level than ordinary folk. . . . It is logical to think that this might apply to recording professionals as well, perhaps even more so because they create artificial reflections electronically and manipulate them at will while listening to the effects. There can be no better opportunity for training and/or adaptation. This is a caution to all of us who work in the field of audio and acoustics. Our preferences may reflect accumulated biases and not be the same as those of our customers.

It is tempting to speculate that the direct sounds from the stereo loudspeakers combined with all of the reflections remaining in a room after the first lateral reflections are removed appear to have about the same potential to generate ASW/image-broadening as a single, well-aimed, lateral reflection. . . .

When we look at the situation leading to a phantom center image, the picture is much more complex. . . . We hear a "phantom" image. But it is a phantom image with a spatial effect associated with it because of the reflected sound field. . . . The common impression is that the left and right panned sounds appear to originate in the loudspeakers themselves, while the intermediate images appear to originate further back, in a more spacious setting, and sometimes elevated. Instead of a soundstage extending across a line between the loudspeakers, the center images tend to drift backward. Since the impression of distance is dependent on early reflections this is a plausible perception.

Listeners comparing a discrete center channel with a phantom center image generated by a stereo pair in a normal room consistently rated the phantom image higher in perceptual dimensions of width, elevation, spaciousness, envelopment, and naturalness. In a situation where the discrete center sound was unsupported by any sounds from other loudspeakers, this is consistent with expectations.

However, it is the task of the recording engineer to augment the spaciousness of a discrete center channel by using appropriately delayed and level-adjusted sounds sent to the left and right front channels and surround channels. If a phantom center is thought to have audible advantages, a real center channel, used in proper collaboration with processed signals delivered through other channels, has the potential to be better in every respect and much more flexible. It is a matter of having the necessary signal processing tools during the mixing process and the knowledge of how to use them.

For the surround channels, the dominant impression of spaciousness will likely be delivered by the direct sound from the surround channels; they arrive at the listener from useful angles, but it is up to the recording engineer to optimize the amplitudes and delays relative to the front channels. Additional reflections contributed by the room will embellish the sense of space, making it more complex than that possible with a pair of surround loudspeakers. Here, the opposite wall reflection is likely to be a contributor (good angle and delay). Reflections from the front and rear walls arrive from relatively unproductive directions and will contribute less to the effect.

It is reasonable to think that encouraging first-order side-to-side reflections from the side-located surround loudspeakers may be advantageous to the creation of envelopment when the number of channels and loudspeakers is limited. Obviously, one does not wish to create conditions for flutter echoes between the side walls.

In mono versus stereo loudspeaker comparisons, spatial quality and sound quality ratings were obviously not independent; one tracks the other. Is it possible that listeners cannot separate them even though, consciously, most were confident that they could. If indeed they are separable factors, it is fair to consider which one is leading. In monophonic tests listeners reported large differences in both sound quality and spatial quality, and, if anything, there were stronger differentiations in the spatial quality ratings. This was definitely not anticipated, but these listeners had little doubt that there were substantial differences in both rating categories. However, in stereo listening, most of the differences between the loudspeakers disappeared.

The principal conclusion is that recording technique is often the prime determinant of spatial impressions perceived in sound reproduction. The directivity of the loudspeakers is a factor, as is the reflectivity of the surfaces involved in the first lateral reflections, especially in recordings incorporating left- or right-hand panned sounds.

The provocative suggestion is that the two domains (sound and spatial quality) are interrelated and that the spatial component is greatly influential. Listeners appeared to prefer the sound from wide-dispersion loudspeakers with somewhat colored off-axis behavior to the sound from a narrow-dispersion loudspeaker with less colored off-axis behavior. In the years since then, it has been shown that improving the smoothness of the off-axis radiated sound pushes the subjective ratings even further up, so it is something not to be neglected.

The implication is that in multichannel recordings where all channels are generously used for spatial enhancement, the nature of the loudspeaker off-axis behavior or listening room acoustics may be perceptually even less important.

In summary, it is clear that the establishment of a subjective preference for the sound from a loudspeaker incorporates aspects of both sound quality and spatial quality, and there are situations when one may debate which is more important. The results discussed here all point in the same direction, that wide dispersion loudspeakers, used in rooms that allow for early lateral reflections, are liked by listeners especially, but not exclusively, for recreational listening. There appear to be no notable sacrifices in the "imaging" qualities of stereo reproduction, indeed there are several comments about excellent image stability and sensations of depth in the soundstage.

The industry mantra for decades has been to absorb, diffuse, or deflect early lateral reflections. The so-called "reflection-free zone" in control room design, in which early reflections are attenuated, seems therefore to be a means of reducing the impression of spaciousness in the playback environment. As shown in Section 6.2, it certainly is not to permit early reflections in the recording to be audible; they are not masked by these sounds. The apparent preference by many recording professionals for reduced spaciousness may have nothing to do with their abilities to hear subtleties in recordings, so much as to address an unusual sensitivity to it—a side effect of the profession.

We are left, though, with a problem: how to explain why the often-mentioned comb filtering engendered by early reflections is not a problem. None of these listeners heard it, or at least they didn't comment on it. If it is reflected in their subjective ratings, it appears to have had a positive effect. There is an explanation.

CHAPTER 9 THE EFFECTS OF REFLECTIONS ON SOUND QUALITY/TIMBRE

In perceptual terms, timbre is what is left after we have accounted for pitch and loudness. It is that quality of a sound that allows us to recognize different voices and musical instruments and what allows us to distinguish the intonations of a superb musician from those of a learner.

When we talk of timbre change as a result of reflections, or anything else for that matter, the natural tendency is to think that any audible change is a negative thing—a degradation. However, as we will see, in some circumstances, judgments can go either way—better or worse.

There appear to be two primary mechanisms for timbre change as a result of reflections:

- *acoustical interference . . . when the direct and reflected sounds combine at the ears*

- *repetition, the audible effect of the same sound being repeated many times at the ears of listeners.*

The worst situation for the audibility of comb filtering is when the summation occurs in the electrical signal path. In the listening room the direct sound and all reflected versions of it contain the same interference pattern.

Another difficult situation is one in which there is only a single dominant reflection arriving from close to the same direction as the direct sound. In a control-room context, this could be a console reflection in an otherwise dead room.

Interestingly, humans seem to cope with lateral reflections in rooms very well because the spectrum we perceive is a combination of those existing at both ears. It is a "central spectrum" that is decided at a higher level of brain function.

If there are many reflections, from many directions, the coloration may disappear altogether, a conclusion we can all verify through our experiences listening in the elaborate comb filters called concert halls.

The spectral smoothing from multiple reflections occurs even when the delayed sounds are at levels 30–40 dB below the direct sound. This remarkable finding helps further to explain why sound in rooms is so pleasant. . . .

Superimposed on all of this is a cognitive learning effect, a form of "spectral compensation" wherein listeners appear to be able to adapt to these situations and to hear "through and around" reflections to perceive the true nature of the sound source.

The upshot is that in any normal room, audible comb filtering is highly improbable.

The reflections causing comb filtering are the same reflections that result in the almost entirely pleasant, pleasurable, and preferable impressions of spaciousness discussed in the previous two chapters.

Acoustical crosstalk associated with the phantom center image generates an important one-toothed comb—a fundamental flaw in stereo.

All of this should provide reasons to employ a real center channel in recordings, another point made by Augspurger (1990), who notes how very different, timbrally and spatially, a phantom image sounds in comparison to a discrete center sound source. "But no matter what kind of loudspeakers are used in what kind of acoustical space, conventional two-channel stereo cannot produce a center image that sounds the same as that from a discrete center channel, even if it is stable and well defined."

Resonances are the "building blocks" of most of the sounds that interest, entertain, and inform us. Very high-Q resonances define pitches; they play the notes. Medium- and low-Q resonances add complexity, defining the character of voices or musical instruments. We learn to recognize patterns of resonances, including their relative amplitudes.

The crux of the matter is repetition of a transient sound by reflections in the listening environment or by electronic regeneration of the signal makes low-Q resonances more audible and the threshold is lowered.

Do we hear the spectral bump or the temporal ringing?

At frequencies above 200 Hz at least, the detection process for resonances employs spectral information, not temporal cues. It seems that we are responding to the "bump" in the frequency response, an energy concentration, not ringing in the time domain. Repetitions, whether they are in the signal itself because of its temporal structure or added by the environment, are obviously well used by the perceptual process in improved detection of medium- and low-Q resonances.

The most distinctive timbral cues in the sounds of many musical instruments have been found to be in the onset transients, not in the harmonic structure or vibrato of sustained portions. . . . This being so, it is reasonable that repetitions of these transient onsets give the auditory system more opportunities to "look" at them and to extract more information.

Summarizing this chapter, on the topic of the role of reflections in the corruption or enhancement of timbre—sound quality—it is now evident that in normal listening rooms there is little risk of corruption (by comb filtering) and substantial evidence that resonances will be rendered more audible. If those resonances are in the program material, it is highly probable that the added tonal richness and timbral subtleties will be welcomed. If those resonances are in loudspeakers, it is highly probable that their enhanced audibility will not be welcomed.

CHAPTER 10 REFLECTIONS AND SPEECH INTELLIGIBILITY

In the audio community, it is almost ritualistic to claim that reflected sounds within small listening rooms contribute to degraded speech intelligibility. The concept has an instinctive logic and "rightness," and it has probably been good for the fiberglass and acoustic-foam industries. However, as with several perceptual phenomena, when they are rigorously examined, the results are not quite as expected. This is another such case.

Summarizing the evidence from these studies, it seems clear that in small listening rooms, some individual reflections have a negligible effect on speech intelligibility, and others improve it, with the improvement increasing as the delay is reduced.

Early reflections (<50 ms) had the same desirable effect on speech intelligibility as increasing the level of the direct sound.

Signal-to-noise ratio is important, but the noise levels at which significant degradation occurs far exceed anything that would occur, much less be acceptable, in any home situation.

To achieve high percentages in speech intelligibility, a signal-to-noise ratio of 5 dB is good, and 15–20 dB nearly perfect. Noise, in this context, is everything other than the speech. In music it is the sound of the band with which a vocalist is singing. In movies it is everything else in a soundtrack occurring at the same time as the dialogue.

Summarizing the results of experiments comparing different center channel levels, relative to left and right front levels, listeners with normal hearing find themselves conflicted. In terms of "dialogue clarity," things improved as the L&R channels were progressively attenuated, and even turned off. In terms of "enjoyment," they thought that 3 or 6 dB attenuation of the L&R channels was an improvement, suggesting that they put substantial value in dialogue clarity, but here they voted not to turn them off. However, in terms of "overall sound quality," 3 dB attenuation of the L&R channels was acceptable, but more than that was rejected. Overall, these normal hearing listeners would seem to be better satisfied with a system in which the L&R channels were not running at reference levels, but perhaps 3 dB lower. Listeners with impaired hearing were utterly predictable. Anything less than mono was a degradation. This does not mean that they dislike the sound of a multichannel presentation but that they place a higher priority on the clarity of dialogue.

It is clear that "listening difficulty" ratings are more sensitive indicators of problems than conventional intelligibility/word recognition scores, and would seem to be more relevant to the assessment of entertainment content and reproduction systems.

It is a convenient fact that the directivity of human talkers is not very different from those of conventional cone-and-dome loudspeakers. . . . The consequence of this is that if casual conversation is highly intelligible with one person in the location of the loudspeaker and another in the audience area, then it is probable that loudspeaker reproduction of close-miked vocals will be comparably intelligible.

CHAPTER 11 ADAPTATION

In the contexts of precedence effect (angular localization), distance perception, and spectral compensation (timbre), humans can track complex reflective patterns in rooms and adjust our processes to compensate for much that they might otherwise disrupt in our perceptions of where sounds come from, and of the true timbral signature of sound sources. In fact, out of the complexity of reflected sounds we extract useful information about the listening space, and apply it to sounds we will hear in the future. We are able, it seems, to separate acoustical aspects of a reproduced musical or theatrical performance from those of the room within which the reproduction takes place.

Under these circumstances, where the component can be aurally "tracked," it is highly probable that it can be heard at levels below those at which it is likely to be audible when listening normally to the completed mix. Thus, sounds that may be gratifying to the mixing or mastering engineer may be insufficient to reward a normal listener, or worse, simply not heard at all.

If it is necessary to absorb, attenuate, scatter, or redirect reflections, the acoustical devices should be similarly effective over the entire spectrum above the transition frequency (say, 300 Hz), not part of it, so the sounds arrive spectrally intact at the listeners' ears.

It seems safe to take away from this a message that listeners in comparative evaluations of loudspeakers in a listening room are able to "neutralize" audible effects of the room to a considerable extent.

There remains one compelling result: When given a chance to compare, listeners sat down in different rooms and reliably rated loudspeakers in terms of sound quality. Now we need to understand what it is about those loudspeakers that caused some to be preferred to others. If that is possible, it suggests that by building those properties into a loudspeaker, one may have ensured that it will sound good in a wide variety of rooms; a dream come true.

CHAPTER 12 ADJACENT-BOUNDARY AND LOUDSPEAKER-MOUNTING EFFECTS

Where a loudspeaker is placed in a room has a major effect on how it sounds, most especially at low frequencies.

By averaging several room curves, measured at different locations—a spatial average—the effects of the position-dependent variations are reduced, and evidence of the underlying adjacent-boundary effects is clearly seen.

Correcting for the adjacent-boundary errors involves choosing the position of the loudspeaker with respect to the boundaries in a manner that minimizes the variations in frequency response at the listening locations. Equalization, changing the frequency response of the loudspeaker, is another one.

After looking at the common placement options, it can be concluded that there are really only two locations in which a loudspeaker has the potential of performing at its best: free standing or flush-mounted in a wall (or ceiling). All other options involve compromises of some sort. The on-wall placement of this generic bookshelf loudspeaker is flawed but, as we will see, loudspeakers can be specifically designed to perform extremely well as on-wall products.

The idea of a universally applicable, one-type-does-all, loudspeaker is a "steam-era" concept, but it is the basis of most of today's designs. It would seem that there is an "opportunity" for something different.

In-wall, flush-mounting is excellent, but with good design, on-wall configurations work very well. Many surround loudspeakers are designed in this fashion, a welcome trend. Ironically, it is the front loudspeakers, arguably the most important ones, that routinely are designed with little or no regard for the acoustical settings into which they will be placed.

A final thought: There is a somewhat reciprocal effect, imperfect, but significant, for the location of a listener's head with respect to adjacent boundaries.

CHAPTER 13 MAKING (BASS) WAVES—BELOW THE TRANSITION FREQUENCY

Real-world experience is that bass reproduction in small rooms is a game of chance. Rooms are different from one another. Different listening positions in the same room can be quite different. . . . A strategy is needed that can ensure the delivery of similarly good bass to all listeners in all rooms.

In conclusion, it is not that the idea of optimum room ratios is wrong; it is simply that, as originally conceived, it is irrelevant in our business of sound reproduction.

The summary of this section could be that in understanding the communication of sound from loudspeaker(s) to listener(s) in small rooms:

- *Everything matters.*

- *There are no generalized "cookbook" solutions, no magic-bullet room dimensions.*

- *Without your own acoustical measurements, you are "flying blind."*

- *Without high-resolution measurements, you are myopic.*

- *With good acoustical measurements and some mathematical predictive capability, you are in a strong position to identify and to explain major problems.*

- *There are indications that some combination of low-frequency acoustical treatment, multiple subwoofers, and equalization will be helpful*

- *The idea of optimizing room dimensions has not been abandoned, but future investigations must take into account where the loudspeakers and listeners are located.*

- *There are two methods to vary the amount of energy transferred from loudspeakers to modes:*
  —Locate the subwoofer at or near a pressure minimum in the offending standing wave.
  —Use two or more subwoofers to drive the standing waves constructively or destructively.

These will be described in the following sections as a progression of sophistication, and success, in delivering predictably good bass to several listeners.

1. *Recommendations of subwoofer arrangements that offer increased probability of similar bass in seats within a defined listening area in rectangular rooms of any shape.*

2. *As number 1, but we pay attention to the end result in terms of equalization needs and power output requirements.*

3. *Selecting optimum room length-to-width ratios to maximize the benefits of multiple subwoofer installations.*

4. *Using measurements, an optimization algorithm, and signal processing to deliver predictably similar bass within rooms of arbitrary shape, rectangular or non-rectangular, with listeners and multiple subwoofers in arbitrary locations. Or using the same process in an iterative process to maximize performance in a given situation.*

All of these, especially number 4, are greatly different from traditional methods in that we are using increasing amounts of control over the extent and manner with which both the subwoofers and the listeners couple to room modes. The primary goal is to reduce seat-to-seat variations. However it is found that in doing so, the modes are attenuated to such an extent that the need for global equalization is substantially reduced.

Designing Listening Experiences

In this portion of the book, the scientific background knowledge is combined with technical ingenuity to create equipment that delivers rewarding listening experiences from a variety of signal sources, in many listening environments, considering all of the normal constraints of budget, complexity, visual clutter, and so on.

The audio industry has come a long way in spite of occasional ill-conceived ventures and a tendency to resist change when change is precisely what is needed. Audiophiles have much to choose from, and there are many exciting listening opportunities for those who wish to explore. The mass market has benefited enormously from scientific and technical progress. The expensive high end is not the only place where excellent sound can be found, but high prices can buy exotic wood cabinetry, adventurous industrial design, and, of course, exclusivity.

High prices can also purchase loudspeakers capable of playing loud without stress, something that those who enjoy blockbuster movies and rock and roll music may find greatly satisfying. As I peruse magazines showing elaborate, obviously expensive and photogenic, home theater installations, it is disappointing to see how many of them use loudspeakers and amplifiers that are simply not capable of delivering the full audio experience. The automotive parlance would be: lots of "show" but not enough "go." Realistic dynamic range, which includes "loud," can be beautiful when it is not compromised by power compression, and the distressed sounds of amplifiers and loudspeakers operating at or beyond their output limits.

The biggest change in audio has been the integration of digital processing into virtually everything . . . except loudspeakers. With only rare exceptions, this product has stubbornly resisted the logical "next step" of integrating amplifiers and DSP into a wired and wireless, plug-and-play, automatically configurable environment. This is a case where computer-like technology can actually be advantageous to both good sound and simplicity of setup. Perhaps it will happen when the current generation grows up. In the meantime, there is plenty for do-it-yourselfers, consultants, and contract installers to do.

CHAPTER 15

Multichannel Options for Music and Movies

In the beginning, there was mono. Everything we heard was stored in and reproduced from a single channel. In those early days, listeners enthused, and critics applauded the efforts of Edison, Berliner, and others as being the closest possible to reality. They were wrong, but a revolution in home entertainment had taken place.

It would be 50 years before stereo, the minimalist multichannel system, would emerge. With two channels came dramatic improvements in the impressions of direction and space. Once we got past the exaggerated "ping-pong," "hole-in-the-middle" problems of many early recordings, listeners enthused, and critics applauded the efforts of artists and recording engineers as being the closest possible to reality. They were wrong again, but clearly *another* revolution in home entertainment had taken place.

Now we demand still more . . . more realism, more dramatic effects, and more listeners to share the auditory experiences. After another 50 years, multichannel audio is a reality. Is this finally the solution that we have been searching for?

15.1 A FEW DEFINITIONS

Monaural Listening through one ear. This term is widely misused, as in "monaural" power amplifier, a single-channel amplifier that of course can be listened to binaurally or, with a finger in one ear, monaurally.

Binaural Listening through two ears. Natural hearing is binaural. When the ears are exposed to the sounds in a room, we can enjoy any number of channels binaurally. However, there is another audio interpretation of the word, and that narrowly applies to "binaural" recordings made with an anatomically correct dummy head, a mannequin, that captures the sounds arriving at each ear

271

location so subsequently these two signals can be reproduced at each of the two ears. This is most commonly done through headphones, which offer excellent separation of the sounds at each ear, or through two loudspeakers, using a technique called *acoustical crosstalk cancellation*. The idea is that the listener hears what the dummy head heard.

Monophonic   Reproduction through a single channel.

Stereophonic   *Stereo*, as a word, has the basic meaning of "solid, three-dimensional." It seems that in the early days of our industry, some influential people thought that two channels were enough to generate a three-dimensional illusion. Now, stereophonic, or just stereo, is firmly entrenched as describing two-channel sound recording and reproduction. In its original incarnation, the intent was that stereo recording would be reproduced through two loudspeakers symmetrically arrayed in front of a single listener. Nowadays, stereo recordings are enjoyed by multitudes through headphones. What is heard, though, is not stereo; it is mostly inside the head spanning the distance between the ears, with the featured artist placed just behind and maybe slightly above the nose. There may be a kind of "halo" of ambience in some recordings. This is sound reproduction without standards, but the melodies, rhythms, and lyrics get through.

Multichannel   An ambiguous descriptor because it applies to two-channel stereo as well as to systems of any higher number of channels. At the present time, in the mass market, that number is 5, plus a limited-bandwidth channel reserved for low-frequency special effects in movies. Together they are known by the descriptor "5.1." Already this has evolved into 6.1 and 7.1 versions, although programs encoded for those playback configurations are rare, but new program delivery schemes offer hope. There are other systems with higher channel counts—for example, 10.2 (Holman, 1996, 2001) and 22.2 (Hamasaki et al., 2004), used for special exhibits and presenting arguments to expand the system further.

Bass management   A signal processing option in surround processors with which it is possible to combine the low frequencies in any or all of the five channels, add them to the low-frequency effects (LFE) channel, and deliver the combination signal to a subwoofer output. The normal crossover frequency at which this is done is 80 Hz (this can often be changed), the frequency below which it is difficult or impossible to localize the source. Bass sounds will be stripped from all channels in which the loudspeakers have been identified as "small" and will be reproduced through those identified as "large," as well, of course, as the subwoofer(s). Holman (1998) gives a good history.

Downmixer, downwards conversion, down-converter   An algorithm that combines the components of a multichannel signal, making it suitable for reproduction through a smaller number of channels. It is also widely used to store the multichannel signal in a smaller number of channels. This is not a "discrete" process; there is inevitable cross-channel leakage when the signal is subsequently upmixed. Dolby Surround/Dolby Stereo is an example of a specific kind of down-mixer, processing four channels for storage in two, as Lt + Rt. Dolby Digital, a 5.1 channel signal, can be downmixed by the Dolby Digital decoder into mono, stereo, or Lt + Rt outputs. Lt + Rt can then be upmixed to 5.1, 6.1, or 7.1 channels by Dolby ProLogic or any of several other competing algorithms.

Upmixer, upwards conversion, up-converter (a.k.a. surround processor) An algorithm or a device that processes a signal and makes it suitable for reproduction through a larger number of channels. Two-channel signals can be upmixed for reproduction through five or more channels, or five channels can be upmixed for reproduction through six, seven, or more. In one common application, multichannel recordings are downmixed, encoded, with a specific form of upmix decoding in mind, as in the case of Dolby Surround (which generates Lt + Rt composite signals) and Dolby ProLogic (which upmixes those signals into 5.1 channels). Other upmixers are designed to operate on these same encoded two-channel signals, arguing that they have a superior strategy to generate a multichannel result for listeners. Finally, upmixers may also be optimized to convert standard stereo music recordings into multichannel versions. These are known as "blind" upmixers because the stereo recordings were not made with this processing in mind, and, consequently, there is no way of predicting the result. Some recordings inevitably will work better than others. In the uncontrolled reality of life, it is highly probable that any two-channel signal will find itself being upmixed/decoded/surround-processed by any of the preceding options.

15.2 THE BIRTH OF MULTICHANNEL AUDIO

Monophonic reproduction conveys most of the musically important dimensions: melody, harmony, timbre, tempo, and reverberation, but no sense sound-stage width, depth, or spatial envelopment—of being there. In the 1930s, the essential principles by which the missing directional and spatial elements could be communicated were understood, but there were technical and cost limitations to what was practical. It is humbling to read the wisdom embodied in the Blumlein-EMI patent (Blumlein, 1933), applied for in 1931, describing two-channel stereo techniques that would wait 25 years before being popularized. It is especially interesting to learn that the motivation for stereo came in part from a desire to improve cinema sound (Alexander, 1999, pp. 60, 80). Cinema played a central role in the development of present-day incarnations of multichannel audio. Then there are the insights of Steinberg and Snow (1934) at the Bell Telephone Laboratories who, when considering the reproduction of auditory perspective, concluded there were two alternative reproduction methods that would work: binaural and multichannel.

Binaural (dummy head) recording and headphone reproduction is the only justification for the "we have two ears, therefore we need two channels"

BINAURAL RECORDING AND REPRODUCTION

The binaural recording and reproduction system—a true encode/decode system—has been a great tease to the audio industry. Timbral, directional, and spatial cues are "encoded" into the two recorded channels, one for each ear, by the sounds approaching the dummy head from different directions, arriving at the ears at different times and with different amplitudes because of the acoustical interaction with the head and torso. All of this is captured at the entrances to the mannequin's ear canals, and if the same sounds can be delivered to the entrances of a listener's ears, one should hear what would have been heard by a human listener seated where the recording mannequin was placed. The "decoding" is done by the ears and brain of the listener.

Does it work? In the beginning it really never had a chance. Microphones were large, noisy, cumbersome things—impossible to fit into a modeled ear canal. Headphones were rudimentary transducers designed for decoding morse code or for basic voice communication but not for reconstructing a Beethoven concert. Things eventually improved, of course, but a lingering bias against being iso-lated under a pair of headphones posed resistance in the marketplace. Thinking more about it, perhaps it wasn't so much the headphones themselves as it was the headphone cord, the tether tying the listener to the static playback hardware. The current "portable music" generation has no such problem and headphone listening is widespread. But there was another negative bias: the fact that most of the time, for most listeners, the sound was perceived to be inside or very close to the head. Externalization was a problem.

For every problem there is a solution. Now, with the benefit of exceptionally accurate and sensitive small microphones and superb headphones, the sound quality can be impressive. Employing head-position tracking devices and appropriately modifying the signals at the ears in real time, it is possible to substantially overcome the front-back reversal and in-head localization problems. With, or even without, customization for individual ears, the result is an amazingly accurate reconstruction of a three-dimensional acoustical event.

argument. Two-channel stereo as we have known it is the simplest form of multichannel reproduction; it is not binaural.

Multichannel reproduction is more obvious, since each channel with its associated loudspeaker creates an independently localizable sound source, and interactions between multiple loudspeakers create opportunities for "phantom" sources. Inevitably, the question "How many channels are necessary?" must be answered. Bell Labs scientists assumed that a great many channels would be necessary to capture and reproduce the directional and spatial complexities of a musical front soundstage—not even attempting to recreate a surrounding sense of envelopment. Their goal was to capture a performance in one hall using a row of microphones across the front of a stage and then reproduce that "wavefront" in another hall. One loudspeaker would be used for every microphone channel in a similar position arrayed across the front of the performance stage. There was no need to capture ambient sounds, as the playback hall had its own reverberation. Being practical people, they investigated the possibilities of simplification, and they concluded that although two channels could yield acceptable results for a solitary listener, three channels (left, center, and right) would be a workable minimum to establish the illusion of a stable front soundstage for a group of listeners (Steinberg and Snow, 1934).

By 1953, ideas were more developed, and in a paper entitled "Basic Principles of Stereophonic Sound," Snow (1953) describes a stereophonic system as one having two *or more* channels and loudspeakers. He says, "The number of channels will depend upon the size of the stage and listening rooms, and the precision in localization required. . . . For a use such as rendition of music in the home, where economy is required and accurate placement of sources is not of great importance if the feeling of separation of sources is preserved, two-channel reproduction is of real importance."

So two channels were understood to be a compromise, "good enough for the home," or words to that effect, and that is exactly what we ended up with. The choice had nothing to do with scientific ideals but with technical reality that at the time stereo was commercialized, nobody knew how to store more than two channels in the groove of an LP disc.

Vermeulen (1956) had a superb understanding of what stereo could and could not do:

Although stereophonic reproduction can give a sufficiently accurate imitation of an orchestra, it is necessary to imitate also the wall reflections of the concert hall, in order that the reproduction may be musically satisfactory. This can be done by means of several loudspeakers, distributed over the listening room, to which the signal is fed with different time-lags. The diffused character of the artificial reverberation thus obtained seems to be even more important than the reverberation time.

The host of spatial enhancers over the subsequent years, up to and including contemporary stereo-to-multichannel upmix algorithms, absolutely support his insight.

Around that same time, the film industry managed to succeed where the music side of the audio industry failed, and several major films were released with multichannel surround sound to accompany their panoramic images. These were discrete channels recorded on magnetic stripes added to the film.

Although they were very successful from the artistic point of view, the technology suffered because of the high costs of production and duplication. Films reverted to monophonic optical sound tracks, at least until the development of the "dual bilateral light valve." This allowed each side of the optical sound track to be independently modulated, and two channels were possible. As we will see, it didn't stay that way for long, and, ironically, it has been the film industry, not the audio industry or audiophiles, that has driven the introduction of multichannel sound reproduction in homes.

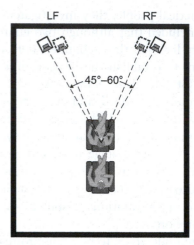

FIGURE 15.1 *Two-channel stereo. The ±30° arrangement is a widespread standard for music recording and reproduction, although many setups employ a smaller separation, especially those associated with video playback. To hear the phantom center image, and any other panned images between the loudspeakers correctly located, listeners must be on the symmetrical axis between the loudspeakers. Away from the symmetrical axis, as in cars, and through headphones, we don't hear real stereo; we hear a spatially distorted, but still entertaining, rendering.*

15.3 STEREO—AN IMPORTANT BEGINNING

In contrast with binaural audio, stereophony is not endowed with an underlying encode/decode system. It is merely a two-channel delivery mechanism. Yes, there have always been some generally understood rules about setting up the playback loudspeakers and about sitting in the symmetrical "stereo seat." But everybody knows that these simple rules are routinely violated. In professional recording control rooms, there was an attempt to adhere to standard playback geometry (loudspeakers at ±30° or so), but otherwise, it was simply wide open for creativity. There were no rules about microphone selection or placement, mixing, and signal manipulation. Nor, in the beginning, was there much in the way of a scientific foundation of knowledge to guide the creative process. It was a period of trial and error, and in many ways, it still is.

Over the years, the struggle to capture, store, and reproduce realistic senses of direction and space from two channels and loudspeakers has been a mighty one. There has been no single perfectly satisfactory solution, even after all these years. Professional audio engineers have experimented with many variations of microphone types and techniques, trying to capture the directional and spatial essence of live musical events. Several coexist, each with its adherents. For pop music, the analog and digital signal processors used to expand the soundstage are countless. Even simplified binaural crosstalk-cancellation processing has been used to place sounds outside the span of the loudspeaker pair.

At the playback end, multitudes of loudspeaker designs have come and gone, all attempting to present a more gratifying sense of space and envelopment. What does one say about a system that accommodates loudspeakers that have directional characteristics ranging from omnidirectional, through bidirectional in-phase (so-called "bipole"), bidirectional out-of-phase (dipole), predominantly backward firing, and predominantly forward firing? The nature of the direct and reflected sounds arriving at the listeners' ears from these different designs runs the entire gamut of possibilities. From this perspective, stereo seems less like a system and more like a foundation for individual experimentation. Older audiophiles may remember the rudimentary "four-dimensional system" that used four loudspeakers, sold by Dynaco as the QD-1 Quadapter. It delivered a sum of both channels to a center-front loudspeaker and a difference signal to a center-rear loudspeaker. David Hafler (1970), the inventor, proposed a quadraphonic multichannel recording system to complete the package.

Taking a different tack, the "Sonic Hologram" (Carver, 1982) was a simplified approach to binaural crosstalk cancellation in the electronic signal path, whereas the Polk SDA-1 loudspeakers tried it at the sound production end of the chain. Lexicon's "panorama" mode, being digital, allowed for individual setup adjustments to cater to different loudspeaker/listener geometries. The goal of all of these was to expand the soundstage beyond the stereo loudspeakers, potentially out to ±90°. None were doing anything that was intended by the recording artists, but all were attempting to reward listeners with an expanded, more enveloping listening experience. A host of digital "hall" and other artificial-reverberation effects came along in this period; they came to be known as "DSP" effects. Most were not very good, which by association gave digital processing an undeservedly bad name. The reputation, however, did not last.

The most recent, and the most ambitious, attempt to extract the maximum from legacy stereo recordings is Ambiophonics (Glasgal, 2001, 2003; www.ambiophonics.org). It has gone through several phases of evolution, incorporating binaural techniques as well as complex synthesis of spatial effects to provide optimum sound delivery.

Added to these fundamental issues is the inconvenience of the "stereo seat." Because of the stereo seat, two-channel stereo is an antisocial system: Only one listener can hear it the way it was created. If one leans a little to the left or right, the featured artist flops into the left or right loudspeaker, and the soundstage distorts. When we sit up straight, the featured artist floats as a phantom image between the loudspeakers, often perceived to be a little too far back and with a sense of spaciousness that is different from the images in the left and right loudspeakers (see Figure 8.4 and the associated discussion).

This puts the sound image more or less where it belongs in space, but then there is another problem: the sound quality is altered because of the acoustical crosstalk, as described graphically in Figure 9.7. As shown in that figure, the audibility of this significant dip in the frequency response depends entirely on how much reflected sound there is to dilute the effect. In the reflection-controlled environment of a typical recording control room, it is likely to be very audible. If the recording or mastering engineer attempts to compensate for this effect with equalization, another problem is created. When such recordings are played through an upmix algorithm, and the featured artist is sent to a center channel loudspeaker, the sound will be too bright. The fault may be attributed to the center channel, but the problem is in the recording.

In fairness, however, it must be said that after over 50 years of experimentation, the best two-channel stereo recordings reproduced over the right set of loudspeakers in the right room can be very satisfying indeed. Sadly, only a fraction of our listening experiences fall into that category. The music, and enthusiasm for it, survived well in spite of it all. Meanwhile, in cinemas, audiences since the 1950s had occasionally been enjoying four to six *discrete* channels of magnetically recorded audio on 70 mm prints. Naturally, there was a center channel.

15.4 QUADRAPHONICS—STEREO TIMES TWO

In the 1970s, we broke the two-channel doldrums with a misadventure into four-channel, called quadraphonics. The intentions were laudable: to deliver an enriched sense of direction and space. The key to achieving this was in the ability to store four channels of information in the existing two channels, on LPs at that time, and then to recover them.

There were two categories of systems in use at the time: matrixed and discrete. The matrixed systems crammed four signals into the bandwidth normally used for two channels. In doing this, something has to be compromised, and as a result, all of the channels did not have equal channel separation. In other words, information that was supposed to be only in one channel would appear in smaller quantities in some or all of the other channels. The result of this "crosstalk" was confusion about where the sound was coming from and an inordinate sensitivity to listener position; leaning left, right, forward, or back caused the entire sound panorama to exhibit a bias in that direction.

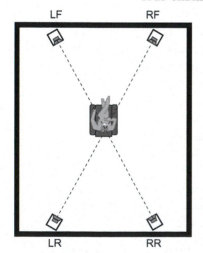

FIGURE 15.2 *A quadraphonic listening arrangement showing the side-to-side/front-to-back restricted seating caused, mainly, by acoustical crosstalk, signal leakage among the channels.*

Various forms of signal-adaptive "steering" were devised to assist the directional illusions during the playback process. The "alphabet soup" is memorable: SQ from CBS, QS from Sansui, E-V from ElectroVoice, and others. Peter Scheiber, a musician with a technological bent, figures prominently as a pioneer in matrix design, with his patented encoder and decoder ideas being incorporated into many of the systems. The best of these systems were remarkably good in creating the illusion of completely separate, or discrete, channels when an image was panned around the room. However, this clear separation breaks down when there is a demand for several simultaneously occurring discrete images. In the limit, the steering ceases, and we listen through the raw matrix with its generous crosstalk.

Ultimately, what was really needed were four discrete channels. However, achieving this on the vinyl LPs required that the recorded bandwidth be extended to about 50 kHz—quite a challenge. Nevertheless, it was accomplished, as CD-4 from JVC, and although this quadraphonic format was short lived, the technology necessary to achieve the expanded bandwidth had a lasting benefit on the quality of conventional two-channel LPs. Half-speed cutting processes, better pressings and playback cartridges with high compliance, low-moving mass, and exotically shaped styli combined to yield wider bandwidth and reduced tracing and tracking distortions. All of these had a continuing positive influence on the industry.

Discrete multichannel tape recordings were available, but open-reel tape was a nuisance to say the least, and high-quality packaged tape formats (e.g., cassettes) were not yet ready for true high-fidelity multichannel sound.

Years passed, with the industry unable to agree on a single standard, which was an intolerable situation from a business perspective. There were issues with mono, stereo, and broadcast compatibility (Crompton, 1974), and eventually, the whole thing dissolved (Torick, 1998). The industry lost money and credibility, and customers were justifiably disconcerted.

Looking back on this unfortunate episode in the history of audio, one can see another reason for failure: The system was not psychoacoustically well founded. Lacking an underlying encode/decode rationale, the problems of two-channel stereo were simply compounded. There were even notions of "panning" images front to back using conventional amplitude-panning techniques, something that Ratliff (1974) and others have found problems with. The quadraphonic square array of left and right, front and rear, was still an antisocial system with even stricter rules. The sweet spot now was constrained in the front-back as well as the left-right direction.

Perhaps most important, there was no center channel—a basic requirement needed to eliminate the stereo seat. Placing the additional channels symmetrically behind the listener is now known not to be optimum for generating envelopment and a sense of spaciousness. Placement more to the side is better. Sounds that arrive from the rear are extremely rare in the standard repertoire of music, but the need for a credible spatial impression is common. Ironically, a 1971 paper entitled "Subjective Assessment of Multichannel Reproduction" (Nakayama et al., 1971) reported that listeners preferred surround loudspeakers positioned to the sides compared to those placed behind, awarding subjective rating scores that were two to four times higher. It seems that nobody with any influence read it. Fortunately, much of the clever technical innovation that went into quadraphonics was not wasted; it went to the movies.

15.5 MULTICHANNEL AUDIO—CINEMA TO THE RESCUE!

The key technological ideas underlying quadraphonics were (1) four audio channels stored in two channels and (2) the ability to reconstruct them with good separation by using adaptive matrices—electronically enhanced steering. Dolby Laboratories Inc. was well connected to the real multichannel pioneers, the moviemakers, in the application of its noise-reduction system to stereo optical sound tracks. Putting the pieces together, Dolby rearranged the channel configuration to one better suited to film use: left, center, and right across the front, and a single surround channel that was used to drive several loudspeakers arranged beside and behind the audience. All of this was stored in two audio-bandwidth channels. With the appropriate adjustments to the encode matrix and to the steering algorithm in the active decoding matrix, in 1976 they came up with the system that has become almost universal in films and cinemas: Dolby Surround, or as it is also known in the movie business, Dolby Stereo.

This system was subject to some basic rules that have set a standard for multichannel film sound: well-placed dialogue in the center of the screen and music and sound effects across the front and in the surround channel. Reverberation and other ambience sounds are steered into the surround channel, as are various sound effects. At times, the audience can be enveloped in sound (as if in a football game), transported to a giant reverberant cave or gymnasium, inside the confines of a car engaged in a dramatic chase, or treated to an intimately whispered conversation between lovers where the impression is that of being embarrassingly close. Because the optical film sound track was relatively noisy, even with Dolby noise reduction, and relatively distorted, occasional "splatters" of vocal sibilants would leak into the surround channel and be radiated by the surround loudspeakers, causing them to be localized. Consequently the surround channel was attenuated above about 7 kHz, eliminating the annoying misbehavior but also degrading the overall spectral balance.

To achieve this dynamic range of spatial experiences requires a flexible multichannel system, controlled-directivity loudspeakers, as well as a degree of control over the acoustics of the playback environment. When it is done well, it is remarkably entertaining . . . and it is not antisocial! There are still better and worse seats in the house, but there are multiple acceptable good seats.

It is important to note that the characteristics of the encoding matrix—the active decoding matrix; the spectral, directional, and temporal properties of the loudspeakers; and the room—are all integral parts of the functioning of these systems. Fortunately, the film industry recognized the need for standardization, and so for many years it has tried to ensure that sound-dubbing stages, where film sound tracks are assembled, resemble cinemas, where audiences are to enjoy the results. Although the industry standards provided a basis, there were still inconsistencies. The good cinemas were superb, but many fell far short of expectations. This left a need, and an opportunity, for THX, then a Lucasfilm subsidiary, to establish a program to certify the audio performance of cinemas so audiences would have an even greater assurance of quality. George Lucas, having authored films that made extravagant use of multichannel effects and wide-bandwidth dynamic sounds, had a special interest in seeing that the customers were properly served by cinema audio systems.

15.6 MULTICHANNEL AUDIO COMES HOME

With the popularity of watching movies at home, it was natural that Dolby Surround made its way there on videotape, laserdisc, television, and all the other delivery formats that have followed and that continue to be created. Adapting it to the smaller environment required only minor adjustments to the playback apparatus. Reducing the number of surround loudspeakers to two ensured greater consumer acceptance, and recommending the placement of these loudspeakers to the sides of the listeners ensured that they would be most effective in creating the required illusions of space and envelopment (see Figure 8.6 and associated discussion). Delaying the sounds to the surround speakers used the precedence effect to ensure that even in a small room the surround sounds would be perceptually separated from those in the front channels.

At the outset, a simple fixed-matrix version was available in entry-level consumer systems. The fixed-matrix systems exhibited so much crosstalk among the channels that listeners were surrounded by sound most of the time, even when it was inappropriate.

If memory serves, it was Fosgate and Shure HTS who brought the first active-matrix decoders to the home theater market. Julstrom (1987) describes the HTS device, which had an innovative feature, an "image-spreading technique . . . to diffuse the rear image and discourage localization at the closer surround loudspeaker." It also avoided the monophonic "in-head" localization effect that could be heard by listeners seated on the center line of the room. This was done with a complementary-comb-filter technique to introduce differences between the left- and right-side surround loudspeakers, decorrelating the otherwise monophonic surround channel. Decorrelation of the surrounds was included as a feature of Home THX a few years later. These were discrete-component products, and they came at premium prices. When low-cost silicon chips incorporating the active-matrix Dolby ProLogic decoder hit the market, home entertainment entered a new era.

Having enjoyed the spatial illusions in movies, it was inevitable that listeners would play conventional stereo recordings through a Dolby ProLogic processor. The results were spotty; some recordings worked quite well, and others didn't.

The translation of a phantom center to a dedicated center loudspeaker did not always work well. The high-frequency rolloff in the surround channel was also noticeable as dullness in the surround sound field. The active matrix steering could sometimes be caught manipulating the music. Recordings made specifically for Dolby Surround were better, but they failed to establish a significant following in the music recording industry. There was work yet to be done.

15.6.1 THX Embellishments

In a natural succession to their THX program for certifying cinema sound systems, around 1990 Lucasfilm established a licensing scheme for certain features intended to enhance, or in certain ways ensure, the performance of home theater systems based on Dolby ProLogic decoders. Home THX, as it was called, added features to a basic ProLogic processor and to the loudspeakers used in home theater systems, and it set some minimum performance standards for the electronics and loudspeakers. At a time when the market was being inundated with small, inexpensive, add-on center and surround loudspeakers and amplifiers, THX made a clear statement that this was unacceptable. All channels had to meet high standards. Tomlinson Holman deserves credit for assembling this amalgam of existing and novel features into what became an early benchmark for consumer home theater. Arguably the most positive lasting contribution to the industry has been the certification program for components that ensures that their specifications are adequate for satisfactory real-world home theaters, that their performances live up to those specifications, and that all of the functions in the evermore complex surround processors actually work as they should.

The first generation of THX certified components also embodied some features unique to the THX program. Much has changed since those early days, so not all of the original THX embellishments continue to be relevant, and some have been phased out.

The following are the THX embellishments of relevance to this discussion. It is important to spend a little time discussing them because their influence is still felt within the home theater industry. Comments have been added that attempt to put them into a present-day context:

1. *High- and low-pass filters in the surround processor to approximate a proper crossover between the subwoofer and satellite loudspeakers.* A glaring omission from previous systems was any consideration of how the outputs of subwoofers and satellite loudspeakers merged within the crossover region. It was a matter left entirely to chance, and inevitably there were many examples of really bad upper-bass sound. However, preset electronic filters cannot do it all because the loudspeakers and the room greatly influence the final result, and these effects cannot be predicted in advance. The high-pass

characteristics of loudspeakers vary substantially (although THX-approved loudspeakers are somewhat controlled). More important, the room is part of the system; as we discussed in Chapters 12 and 13, the listening room is powerfully influential in this (80 Hz) frequency region. The subwoofer and satellite loudspeakers—and the listener, of course—are in different locations, so there can be no assurance that the low-pass and high-pass slopes will add as intended. Only in situ acoustical measurements and equalization can do that. At the time, this was impractical to consider, but the application of *any* high-pass filter to the satellite loudspeakers will prevent them from trying to duplicate the job of the subwoofer, and the entire system should be able to play louder, with less distortion—so the basic idea was constructive.

2. ***Electronic decorrelation between the left and right surround signals.*** These were the days of a single surround channel. Reducing the number of surround speakers to two and putting them in a small room eliminated much of the acoustical decorrelation (randomization of the sounds arriving at the listeners' left and right ears) that multiple speakers in a large cinema accomplished. Picking up the Shure HTS idea (Julstrom, 1987), THX recommended decorrelation of the signals supplied to the left and right surround loudspeakers. A pitch-shifting algorithm was suggested, but in fact they approved other forms as well. This was a very useful feature at the time, but now it is needed only for playback of older films. Today's upmixers incorporate decorrelation or at least delays in the process of subdividing the surround signal(s) into multiple channels. With discrete formats, it is up to the recording engineers to decide how much decorrelation is appropriate for the surround channels; no further processing is needed.

3. ***"Timbre matching" of the surround channels to the L, C, R (front) channels is a feature worthy of discussion.*** Holman (1991) describes some experiments in which the timbre of the surround array of loudspeakers was compared to that of a front loudspeaker. He found that in a cinema, such a match was impossible, ostensibly because there were 22 surround loudspeakers and only a single front loudspeaker (see Figure 15.3). In a home version of the test, involving either two or four surround loudspeakers, a reasonable timbre match

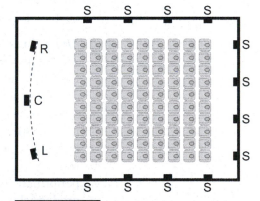

FIGURE 15.3 *Dolby stereo as it was developed in 1976: three channels across the front and a single surround channel delivered identically to loudspeakers surrounding the audience. This is the foundation from which the home theater and multichannel music technologies we know today evolved.*

was found to be possible only with two surrounds. Obviously, multiple-source surround channels do not match with a single (front) source. This is consistent with the listeners being in the complex near field of a spatially distributed collection of loudspeakers radiating correlated sounds. There is no location in the cinema or in the domestic room where listeners are in a stable far field. Because of multiple delays in the propagation paths, sounds arriving at listening locations is strongly nonminimum phase, timbre will be distinctive, and equalization cannot provide a match to a simple single-loudspeaker source. The ultimate impossibility of this venture is explained in more detail in Section 18.4.5, which includes the additional complication of dipole surrounds. Because home systems are now expected to reproduce momentary discrete left, right, front, and back sounds in movies, and sustained vocal and instrumental sounds in music, the entire topic is put to rest: all channels need to be identical in performance.

4. *"Re-equalization" of the sound track.* This is a compensation for excessive treble that is sometimes found in film soundtracks as a result of the sound systems used in large cinemas and the manner in which they are calibrated. A single correction curve was chosen. This was a useful feature in those early days, but it needed to be an adjustable tone control, since sound tracks varied in treble balance; some needed more treble cut, some less, and many seem to be fine with no correction. One reason for the variability is in the inconsistent calibrations of film production facilities and cinemas (see discussion of the "X" curve in Section 18.2.6).

5. *Limited vertical dispersion of front (L, C, R) loudspeakers.* The argument here was that it resulted in improved dialogue clarity and localization of sounds in the front soundstage. Holman (1991) describes an experiment (done around 1981) using undocumented loud-speakers in a large 10 500 ft^3 (297 m^3) room. Whatever the intrinsic performances of the loudspeakers might have been, they were modified by equalizing them to match the film-industry-traditional room curve, the X curve. We now know that this is not the way to calibrate the performance of loudspeakers above the transition frequency. The product claiming to have narrower dispersion was said by listeners to produce a greater " 'clarity' of dialog and better localization of individual sounds in a complex stereo sound field. . . ." Numerous scientific studies of speech intelligibility and listening difficulty, some done before this event and others since, do not support this finding (see Chapter 10). Chapter 8 discusses in detail the effects of early reflections on imaging. It seems that this directivity requirement has now been removed from the THX specifications.

6. ***Dipolar radiation pattern for the surround loudspeakers.*** Dipoles are normally thought of as single-panel diaphragms that are allowed to radiate freely in both directions (Figure 8.7). Most (all?) of the loudspeakers promoted as meeting the THX specification, though, have been designed as bidirectional out-of-phase systems consisting of two sets of drivers mounted on different sides of the same box, one facing forward in the room and one facing back. The null in the radiation pattern was intended to face the listening location. Examples of the genre will be shown in Chapter 18, and we will see that timbral misbehavior is inevitable. As for the ability of this design to generate spaciousness or envelopment, as shown in Figure 8.6, they appear to be aimed in the wrong directions to have the greatest beneficial effect (sounds need to arrive at listening locations from the sides). A survey of current offerings from several prominent manufacturers seems to indicate a trend toward simple wide-dispersion designs achieved using bidirectional in-phase (bipole) or monopole configurations, sometimes with selectable directivity.

Obviously, if all of this were done today, things would probably have been done differently. THX has also evolved, it is under new ownership, and not all of these requirements exist or are promoted as they were originally. It has also expanded into the certification of many kinds of audio and video products.

15.7 MULTICHANNEL AUDIO—THE AMBISONICS ALTERNATIVE

The Ambisonics premise has two parts. The first is that with the appropriate design of microphone, it is possible to capture (record in some number of channels) the three-dimensional sound field existing at a point. The second part is that with the appropriate electronic processing, it should be possible to reconstruct a facsimile of that sound field at a specified point within a square or circular arrangement of four or more loudspeakers. Therefore, this system distinguishes itself in that it is based on a specific encode/decode rationale. Several names are associated with the technology. The basic idea for this form of surround sound was patented first by Duane Cooper (Cooper and Shiga, 1972). Patents were also granted to Peter Fellgett and Michael Gerzon (1983), who were working simultaneously and independently in England. Peter Craven contributed to the microphone design and, with the support of the NRDC, the U.K. group commercialized the Ambisonics record/reproduction system. See also http://www.ambisonic.net/ for historical perspectives and enthusiastic support.

It is an enticing idea, and the spatial algebra tells us that it should work. And it does, up to a point, at a point in space. Ambisonics remains a niche player in the surround-sound industry. Most people know little to nothing about it, although there are some encoded recordings (http://www.ambisonic.net/), and the Soundfield microphone continues to be used in some recordings. The scarcity of playback decoders is a clear problem. However, there are other considerations that may be significant.

The author has heard the system several times in different places, including a precise setup in the NRCC anechoic chamber in which he participated. There, theoretically, it should have worked perfectly, since there were no room reflections to contaminate the delivery of sounds to the ears. In general, Ambisonics seems to be most advantageous with large, spacious classical works, with which the system creates an attractively enveloping illusion for a listener with the discipline to find and stay in the small sweet spot. It tolerates a certain amount of moving around, but leaning too far forward results in a front bias, leaning too far backward creates a rear bias, leaning too far left . . . well, you get the idea. Big, spatially ambiguous, reverberant recordings are more tolerant of listener movements, of course. All of this should be no surprise in a system in which the mathematical solution applies only at a point in space, and then only if the setup is absolutely precise in its geometry and the loudspeakers are closely matched in both amplitude and phase response. Room reflections absolutely corrupt the theory. So what did it sound like in the anechoic chamber? It sounded like an enormous headphone; the sound was *inside* the head. When the setup was moved to a nearby conventional listening room, the sound externalized, and all previous comments apply.

To reconstruct the directional sound-intensity vectors at the center of the loudspeaker array, some amount of sound may be required to be delivered by many, or all, of the loudspeakers simultaneously; that is the way the system works. A practical problem then arises because we listen through two ears, each at different points in space, and both attached to a significant acoustical obstacle: the head. If a head is inserted at the summation point, then it is not possible for sounds from the right loudspeakers to reach the left ear without timing errors and large head-shadowing effects, and vice versa. The system breaks down, and we hear something other than what was intended. What is heard can still be highly entertaining, but it is not a "reconstruction" of the original acoustical event. For that, one would need to generate individual sound fields at two points, one for each of the ears, binaural audio.

There are numerous ways to encode and store the Ambisonics signals and even more ways to process the signals into forms suitable for reproduction from different numbers of loudspeakers in different arrangements. Higher-order Ambisonics is held by some as the real solution, but this means more channels, more paraphernalia, and more cost. Ambisonics may yet have a role to play. Certainly, having multiple digital discrete channels within which to store data can only be an advantage.

15.8 UPMIXER MANIPULATIONS: CREATIVE AND ENTREPRENEURIAL INSTINCTS AT WORK

Some of the criticisms of the first-generation Dolby ProLogic stemmed from it having been designed to deal with the imperfections of optical sound tracks on films. When delivered to homes on formats not having these problems, the limitations were obvious, especially when the upmixer was used for music. The surrounds were dull, and there was too much emphasis on the center channel; the stereo soundstage seemed to shrink in width.

Recognizing an opportunity to improve on a good thing, inventors have found great satisfaction manipulating the parameters of the matrixes, with delays and with steering algorithms, all in attempts to finesse the multichannel decoders either to be more impressive when playing movies or to be more compatible with stereo music or both. Most of them allowed for full bandwidth surround channels, and the more adventurous ones augmented the system with additional loudspeakers behind the listeners.

It must be emphasized that when playing stereo music through these algorithms, we are hearing "ambience extraction," not reverberation synthesis. All of the reflected and reverberated sound that is reproduced in the surround channels was in the recording. It is just redirected to the side and/or rear loudspeakers rather than being reproduced exclusively through the front channels. Consequently, it sounds more natural. It can also sound exaggerated with some recordings because the stereo recordings were not designed for this form of reproduction. To get a sense of spaciousness in stereo reproduction through two loudspeakers, more "ambiance" (decorrelated sound) is often recorded than would have been required if surround channels had been anticipated. The solution: use the remote control and turn the surrounds down.

Willcocks (1983) provides a good overview of surround decoder developments in the fruitful period of the 1980s. The following examples are those that the author had some contact with in more recent years.

15.8.1 The Fosgate 6-Axis Algorithm

A veteran of the quadraphonic wars, Jim Fosgate developed ways to decode Dolby Surround sound tracks in ways that many people found to be preferable to more mainstream means. Part of the improvement had to do with the responsiveness of the steering logic, and part of it had to do with providing some amount of left-right distinction in the full-bandwidth surround channel. Since there is no such separation in the encoded program, the "art" has been in judging how much and when left and right front information should be directed to the surrounds, with what spectral modifications (if any), and with what delay.

Fosgate practiced his art well, and over the years he generated several well-received designs optimized for films and for different kinds of music, all in the analogue domain. An interesting feature was the provision for separately driving forward- and rear-firing drivers in the surround loudspeakers to allow for more directional enrichment. His designs were found in products bearing his own name, as well as some older models from Harman/Kardon, Citation, and JBL Synthesis (the Fosgate company had been acquired by Harman International Industries, Inc.). The name 6-Axis came about because, in addition to the basic five steered channels, there was an optional sixth behind the listener to complete the surround effect.

15.8.2 The Harman/Lexicon Logic 7 Algorithm

Working independently, and in the digital domain, David Griesinger from Lexicon (another Harman International company) did similar things to move beyond the basic ProLogic process. Driven by an intense interest in the acoustics and psychoacoustics of concert halls, Griesinger's efforts in surround-sound decoding and multichannel synthesis benefited from years of these studies (Griesinger, 1989). The challenge in all of these exercises is to accentuate the desirable aspects of complex multidimensional sound fields while avoiding undesirable artifacts.

The result was a suite of film and music playback algorithms initially embodied in Lexicon digital surround processors. The product was called Logic 7 because it provided for two additional channels and loudspeakers behind the listener. Using clever detection, enhancement, attenuation, and steering process, these rear loudspeakers are supplied with strongly uncorrelated sounds, such as reverberation, applause, and crowd sounds, or sounds that are strongly directed to move from front to surround, or vice versa. Thus, multiple listeners were treated to an enveloping sense of ambiance and to occasional sounds that swept dramatically forward or backward, even with appropriate left or right biases. An important focus in the continuing development of Logic 7 was the quest for compatibility in multichannel reproduction of film sound tracks and music, as well as that between two-channel and multichannel reproduction of stereo music mixed for two channels. Logic 7 decoders now exist in several variations and in many products from Harman International brands, including versions optimized for automotive audio systems.

15.8.3 "Surround-Sound" Upmixing

Holding on to algorithmic exclusivity is difficult. It is, after all, signal processing of a type that relies greatly on the ability to adjust any of many parameters of a complex signal processor to produce specific audible effects from nonstandard recordings. The basic principles are widely understood, but proprietary processing and know-how are sometimes the principal determinants of audible differences. Only lawyers seeking expensive lawsuits may be able to determine ownership of any portion of any specific algorithm. The battles may or may not be worth fighting. Consequently, since the early days just described, many alternative upmix processes and processors have emerged, all claiming to do wondrous things with two-channel inputs. Some work better than others, but they all seem to be able to provide entertainment.

Rumsey (1999) conducted controlled listening tests on some upmix alternatives, comparing all of them to the stereo original. There were substantial variations among the upmixers, strongly depending on program, which would be expected, and significantly on the listeners and their accumulated listening experience (all were either students in a sound recording program or active in the recording/broadcasting industry). In general, the upmixers were judged to have degraded the front soundstage. These expert listeners (presumably having grown up and worked with stereo) preferred the stereo original to any upmixer. Nevertheless, the best upmixers were given only slight demerits. Opinions about spatial impression were different, with some listeners giving upmixed versions substantial bonus points, whereas others thought the opposite. In the end, some of the "expert" listeners wanted to be left with their stereo systems, but others thought the new formats had some interesting and engaging things to offer. Clearly, there is a cultural component to tests of this kind, and, as Rumsey points out, it would be interesting to conduct similar kinds of tests with a wider population of listeners, asking more basic, "preference"-oriented, questions.

Choisel and Wickelmaier (2007) compared mono, stereo, wide stereo, three matrix upmixers, and five-channel discrete playback formats. The results are worth looking at, as they indicate that a good upmixer can occasionally be even more rewarding than a five-channel original. But there are also some unrewarding upmixers. Stereo appears to provide much of what listeners want, but it loses in terms of perceptions like width, spaciousness, and envelopment, very much as one would expect. Stereo also suffered in terms of sound quality (brightness) of the phantom center image due to acoustical crosstalk (see Figure 9.7).

As shown in Figure 15.4, there is almost a limitless supply of two-channel program material and numerous options for their playback. Perhaps the most consequential development of upmixers is that they are everywhere, including in many automobiles. In fact, it may be car audio that exposes the current "portable audio" generation to the pleasures of surround sound for music more than anything in homes. Adapting surround sound to a car environment is a special challenge because of the proximity of the loudspeakers and the off-center locations of listeners. But this is not new; all of us for years have listened to stereo program in cars and for those same reasons have never heard it as intended. Still, it has been enjoyable. Multichannel audio seems like a format better adapted to the automobile cabin, as long as there is a center channel. The only downside is that background noise at highway speeds greatly diminishes the subtle envelopment illusion.

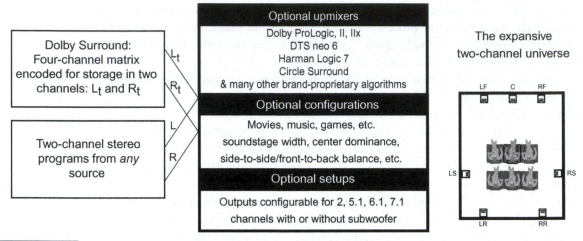

FIGURE 15.4 *A summary of processing options for two-channel audio signals.*

15.9 MULTICHANNEL AUDIO GOES DIGITAL AND DISCRETE

The few samples of discrete multichannel recordings from the quadraphonics era were sufficient to generate a lasting interest, if not an outright lust, to develop a viable format that did not suffer from crosstalk among the channels. Today we are experiencing several versions of that dream. There are the expected pro and con arguments about which ones sound better, but under the bluster and ballyhoo, all of the systems, so far, have sufficient sonic integrity that our entertainment is unlikely to be compromised. Purist audiophiles have pushed for systems of such bandwidth and dynamic range that even the most fastidious superhumans, dogs, and extraterrestrials will be pleased. Professional audio needs extra "space" in recordings to cope with inevitable artist excesses, mistakes, multiple overdubs, and so forth, but at the point of delivery to consumers, we are more than adequately served by several of the popular media. One of the joys of digital encoding and decoding is that it is all available for a price—bandwidth or data rate.

Data rate, in audio terms at least, is abundant and inexpensive but not limitless. There are situations where it is restricted, as when Dolby squeezed packages of digital data between the sprocket holes in movie film. The optical sound tracks were left intact, for cinemas not digitally equipped, and as a fallback in the event of digital crashes. The consequence was that there was only a certain amount of digital data space into which 5.1 channels could be stored. It was not enough to allow uncompressed audio, so Dolby incorporated an audio codec, a perceptual encode/decode scheme that devoted less data space to recording those sounds or components of sounds that were estimated not to be audible due to masking. The codec they developed, AC-3, has proved to be highly durable as the basis for Dolby Digital, now the nearly universal standard.

15.9.1 Comments on Codecs

All audio codecs are scalable, meaning that the amount of data compression can be adjusted according to bandwidth/data-rate availability, so when evaluating the performance of any codec, it is important to know which version is being used. In the case of Dolby Digital, film sound operates at 320 kbit/s. On DVD, that can expand, but it is limited by standards to 448 kbit/s. On the current Blu-Ray high-definition video discs, it can stretch to its maximum: 640 kbit/s.

There is a point of diminishing returns in the performance of all codecs; as the data rate is reduced, each codec has a different data rate at which it begins to exhibit audible problems—artifacts. And to make things more complex, different codecs exhibit different kinds of artifacts. It depends on the strategy used to identify components of the sound that are to be either encoded more simply or discarded. *All* codecs can be made to misbehave, but the best of them misbehave in a manner that is revealed as a momentary "difference" in sound, not a gross distortion or lapse in information. Most of the time, competent codecs are transparent. Those people involved with the subjective evaluation of codecs have found it necessary to train listeners to know what to listen for. It is not common for these problems to be discovered in casual listening. Obviously, if the data rate is drastically reduced, no codec can perform flawlessly, and there is ample evidence of this in the now-popular Internet audio and the often highly compressed portable audio devices. Check the data rate. Years of experience suggest that Dolby Digital is operating acceptably at the current data rates.

Audiophile paranoia suggests that all perceptually encoded systems are fatally flawed, alluding to the discarded musical information. Well, it is only lost if it could have been heard. Auditory masking is a natural perceptual phenomenon, operating in live concert situations just as it does in sound reproduction. It has assisted our musical enjoyment by suppressing audience noises during live performances and, over several decades, by rendering LPs more pleasurable. If we talk here about compressing data, it would be fair to say that LPs perform "data expansion," adding unmusical information in the form of crosstalk, noise, and distortions of many kinds. More comes off of the LP than was in the original master tape. However, because of those very same masking phenomena that allow perceptual data reduction systems to work, the noises and distortions are perceptually attenuated. So successful is this perceptual noise and distortion reduction, that good LPs played on good systems can still sound impressive. Chapter 19 offers some explanations.

Serious subjective evaluations by experienced and trained listeners have been involved with the optimization of these encode/decode algorithms to ensure that critical data are not deleted. These are in tests where listeners can repeat musical phrases and sounds as often as necessary for them to be certain of their opinions.

Having participated in comparative listening tests of some of these systems, the author can state categorically that the differences among the good systems at issue here are not "obvious." Even in some of the aggressive data reduction configurations, audible effects were quite infrequent and limited to certain kinds of sounds only. It was helpful to know what to listen for. And even then the effects were not always describable as better or worse; sometimes they could only be identified as being "not quite the same."

A rival to Dolby in the marketplace is DTS, Digital Theater Systems, which in cinemas operates with an outboard audio playback CD-ROM that is synchronized with the film. The DTS perceptual encoding algorithm also discards data, but it operates at a higher data rate than Dolby Digital; movie sound appears to be encoded at rates in the 768 to 1103 kbit/s range. This may or may not ensure superior performance because it all comes down to whether the higher data rate is necessary to yield acceptable performance with this compression strategy or whether the strategy is comparable to that of Dolby Digital and the result is fewer audible artifacts. There have been many heated arguments about which system is better, accusations of manipulating sound levels in the channels (meaning that in a comparison the encoded signals were subtly different), and so on. It seems that both of these systems are good enough not to interfere with our entertainment, but it is something where customers are free to exercise a choice.

Finally, there is lossless compression, in which nothing is discarded. Taking advantage of redundancy, quiet moments, and so on, the data rate can be reduced, but the reconstructed data are complete; there is no loss. In formats and delivery systems with adequate bandwidth, such systems are attractive in that they simply eliminate all discussions about possible degradations.

15.10 FINDING THE OPTIMUM CHANNEL/ LOUDSPEAKER ARRANGEMENT

All of these discrete systems are really transparent multichannel transport media; none of them incorporates or is based on an underlying method for encoding and decoding spatial information. All the matrix systems discussed up to now put serious constraints on the creative process because of the cross-channel leakage and steering artifacts; these characteristics operated as part of the multichannel encode/decode process. Discrete systems have no such limitations, and in fact, recording engineers have had to learn new techniques and need new production tools to recreate some of the illusions with which we have become familiar in the matrix systems. Without interchannel leakage, a "hard pan" to a single channel in a discrete system is "harder" than it is in a matrix system. In short, we have entered a realm of multichannel entertainment wherein what we hear will be almost entirely the result of individual creative artistry in the recording process and how this interacts with the particulars of the playback systems. Since there are no rigid standards, we can expect considerable variety in the results.

The audio system in cinemas was developed within constraints imposed by screen size and practical room sizes and shapes, all blended with judgment and common sense. At the time, there was no real science to guide progress. When video delivery into homes became a reality, Dolby, THX, and others considered what alterations, additions, or adaptations were necessary to ensure a satisfying experience in small rooms. Early thoughts slavishly followed the lead set by cinemas. Apart from being logical, there were decades of satisfied audiences saying that it was pleasurable.

Figure 15.5 shows the comparison, and it is easy to see that the important sounds are localized where they need to be: with the picture. Subsidiary sounds, directional and spatial, are served by the two surround channels. The success of this depends on specific circumstances, none of which, yet, considered the needs of music as a stand-alone experience, although music is a critical component in most movies. Occasionally, it is part of the foreground entertainment, but more often, it operates as a supporting device, setting moods, atmosphere, and so on. Some of the most spatially engaging multichannel music that the author has heard has been in film sound tracks. This should come as no surprise, as the film industry has had decades of experience with multichannel formats. The question to be asked at this point is, if there were an opportunity to do so, would we wish to change anything? Can we improve on the familiar 5.1-channel format?

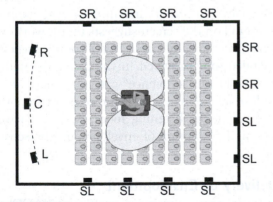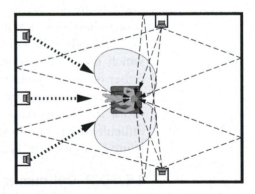

FIGURE 15.5 *A schematic comparison of five-channel audio playback in cinemas and in homes. The comparison includes the "spatial effect balloon," introduced in Chapter 7 and discussed in Chapter 8, showing that the cinema situation is especially well arranged to encourage perceptions of spaciousness and envelopment from the array of surround loudspeakers for listeners in most of the seats. By comparison, the home situation is less complex but is also well configured to create basically similar directional and spatial effects, at least for well-located listeners. As we will see, the addition of more loudspeakers and channels improves things.*

15.10.1 Scientific Investigations Look at the Options

Let us begin by defining the duties of a surround-sound system.

- *Localization.* The perception of direction: where the sound is coming from. The minimum number of locations would be the number of discrete or steered channels in the system. Beyond that, we rely on phantom images floating between pairs of loudspeakers, those across the front being familiar because of stereo. With a center channel these are even more stable. In multichannel systems, other opportunities exist—for example, between the front and sides. These are rarely used except to convey a brief sense of movement because these capricious illusions move around depending on where one is sitting relative to the active loudspeakers. Anyone seated away from the sweet spot is likely to hear a distorted panorama of phantom sound images.

- *Distance.* By the appropriate addition of delayed, reflected, sounds in the recordings, it is possible to create impressions of distance, moving the apparent locations well beyond the loudspeakers themselves.

- *Spaciousness and envelopment.* The sense of being in a different space, surrounded by ambiguously localized sound. This is a very important function.

The possibilities for improving localization are limited by the practical number of channels the industry feels customers will buy and install in their homes. Anything beyond the present number seems like a difficult sell. For sounds associated with on-screen action, there is the powerful "ventriloquism" effect to help, and the hours spent by moviegoers listening to a monophonic center channel as action moves around on the screen suggests that it works very well. Sounds originating off-screen are usually momentary sound effects for which no real precision is demanded (nor delivered in the cinema situation). Other off-screen sounds fall into the broad "ambience" category where, if anything, ambiguity of location is desirable. A sense of distance is an important factor in "transporting" listeners out of the listening room, but this factor is difficult to separate from the essential perception of envelopment, the sense of being in a different, larger space.

15.10.2 Optimizing the Delivery of Envelopment

As discussed in Chapter 7, interaural cross-correlation coefficient (IACC) is a strong correlate of a perception of ASW, image broadening, spaciousness, and envelopment (Figure 7.4). The more different the sounds are at the two ears, at certain frequencies and delays, the greater the sense of these spatial descriptors. The locations of the ears then determine that sounds arriving from different directions generate different amounts of IACC and perceived ASW (Figure 7.5). Sounds from the sides are most effective, and those from front and back are least effective. It is also known that diffusion in a sound field is a contributing factor, but that diffusion—or at least directional diversity in many reflections—is not a requirement for the perception of spaciousness (Figure 8.2).

At this point it is necessary, again, to emphasize that reflections occuring in small rooms cannot *alone* generate a sense of true envelopment. Envelopment requires delays (more than about 80 ms) that can only be supplied by recorded signals reproduced through multiple loudspeakers. Additional room reflections of those greatly delayed signals may enhance the impression, but the initial delay and the appropriate directions must be provided in the recorded sound and an arrangement of playback loudspeakers. How many channels do we need, and where do we put the loudspeakers? The following three studies merge nicely to provide significantly useful illumination of this topic.

Tohyama and Suzuki (1989) looked at a few arrangements of two and four loudspeakers, comparing measured IACCs to those found in a truly diffuse sound field—an all but unachievable goal in the real world. Results shown in Figure 15.6 indicate, not surprisingly, that two-channel stereo did not come close to replicating the diffuse-field IACC. The real news, though, is that doubling the number of channels by adding a pair of loudspeakers behind the listener at the same angular separation did not really change anything. The solid (two channels) and open (four channels) dots in Figure 15.6 are very similarly distributed, and neither matches the target curve. This was the original layout for quadraphonic sound—obviously not an optimum concept.

When the extra pair of loudspeakers was deployed at several angles between 45° and 75°, as shown in Figure 15.7, the IACC results all move closer to the target curve. Two things have changed: All four loudspeakers are in front of the listener, and they are at different horizontal angles. Which is responsible for the improvement? Front-back symmetry in hearing suggests that it may be the different angles, but more direct data are needed. At this stage, it is possible only to make a tentative statement that avoiding equal front and surround-channel angles may be a good idea.

Hiyama, Komiyama, and Hamasaki (2002) conducted subjective evaluations of how closely the sound of a reference diffuse sound field, generated by a circular array of 24 loudspeakers, could be approached by arrays of smaller

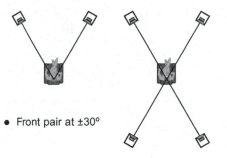

- Front pair at ±30°

o Front & rear pairs at ±30°

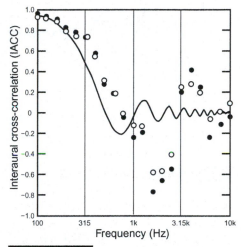

FIGURE 15.6 *A comparison of IACC measured for a stereo pair of loudspeakers at ±30° and with a second pair of loudspeakers added at the same angles behind the listener. Signals delivered by the individual loudspeakers were narrow bands of uncorrelated noise, as might be recorded in a perfectly diffuse sound field. The target curve, the solid line, is the IACC of a perfectly diffuse sound field. Adapted from Tohyama and Suzuki, 1989.*

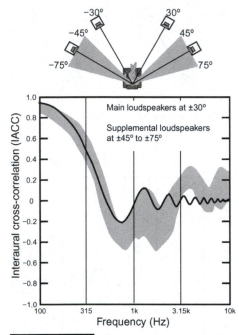

FIGURE 15.7 *The same as Figure 15.6, but with the second loudspeaker pair located at different angles within the range ±45° to ±75°. All of the measurements fall within the shaded area. Adapted from Tohyama and Suzuki, 1989.*

numbers of loudspeakers. Listeners were required to judge the degree of impairment in perceived envelopment (LEV, defined in Section 4.1.3) for each of the loudspeaker configurations, when compared to the 24-loudspeaker reference array. All loudspeakers radiated uncorrelated noise.

The first experiment they conducted examined the performance of different numbers of loudspeakers *equally spaced* in circular arrays. The results indicated that arrays of 12, 8, and 6 loudspeakers did well at imitating the perceived envelopment of the 24-loudspeaker array. Arrangements of 4 and 3 loudspeakers did poorly, however. It was a very clear delineation, suggesting that equally spaced arrays of 4 and 3 loudspeakers should be avoided.

Figure 15.8 shows some excerpts from the results of their numerous experiments in which two to five loudspeakers in different arrangements attempted to imitate the perceived envelopment of the 24-loudspeaker reference array (score = 0). The closer the "difference grade" is to 0, the better the subjective performance of the array. There are much more data in the paper; these selections related most closely to multichannel sound reproduction options. Results for three test signals are shown: 100 Hz to 1.8 kHz noise (the frequency range over which there appears to be the strongest correlation with envelopment), and dry recordings of cello and violin to which have been added convolved and simulated early and late reflections from the appropriate directions for a concert hall. The results are interesting.

- Two-channel stereo does not fare well (a), and neither does the "quad" arrangement (f), which performed very similarly, strongly confirming the results of Tohyama and Suzuki.

- Symmetrical front-back arrays, it seems, contribute nothing to envelopment but only add two more locations for special-effects sounds in movies and voices or instruments in music.

- A center-rear loudspeaker is worse (g).

- All combinations of a pair of loudspeakers at ±30° and another pair of loudspeakers at angles from ±60° to ±135° perform superbly (b), (c), (d), and (e). Avoid ±150° (f), or whatever angle identifies the spread of the front loudspeakers.

- Four loudspeakers behind the listener (h) do not perform as well as four in front, at the same reflected angles (b).

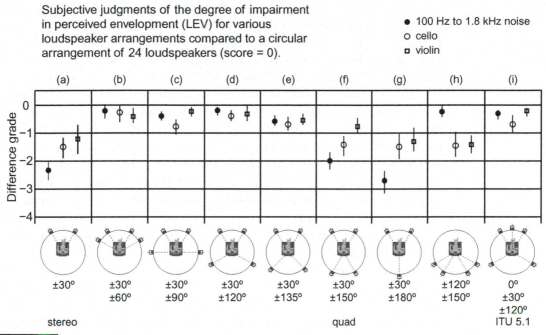

FIGURE 15.8 *Results of subjective comparisons in which listeners judged the degree of impairment in perceived envelopment (LEV) when an array of a smaller number of loudspeakers attempted to imitate the performance of a circular array of 24 loudspeakers. Adapted from the data in Hiyama et al., 2002.*

■ The five-channel arrangement described in ITU-R BS.775-2, shown in (i), performed about as well as any other configuration.

This last statement is obviously very good news, as a lot of energy has been invested in promoting the scheme by many persons and industry groups.

Muraoka and Nakazato (2007) used a measurement of frequency-dependent interaural cross-correlation (FIACC) as a measure of the successful "sound-field recomposition." The idea was simple: FIACC was measured in four large spaces: a large lecture hall and three concert halls. An omnidirectional loudspeaker on the stage was the source. Recordings were made at the measurement locations using a circular array of 12 equally spaced microphones. These recordings were reproduced through different numbers of these channels, using loudspeakers placed at 2 m from a measurement mannequin in an anechoic chamber. The FIACC of the sound field reproduced by each of the loudspeaker configurations was measured, and a "square error" metric was computed, describing the degree of difference between the FIACC at the original location and that reproduced by the test arrangement of loudspeakers. The difference was computed over a "full" bandwidth (100 Hz to 20 kHz) and over a "fundamental" bandwidth (100 Hz to 1 kHz), the frequency range believed to be most related to the perception of envelopment.

Figure 15.9 shows selected data from the experiments. With the exception of (e), the arrangements shown in the top row are the same as those shown in Figure 15.8. That particular arrangement was not tested in this study, so the space (e) has been filled with results for what *should* be the best possible configuration: 12 channels at 30° intervals.

This is a very different kind of experiment from that of Hiyama et al., and yet the conclusions are almost identical. There are some minor differences, as might be expected, because here the goal was to replicate the sound field of real rooms, not a mathematical ideal or a synthesized approximation.

Of special note is the great similarity between (e), the 12-channel system, and (i) the widely used five-channel "home theater" arrangement. All of the combinations of a front pair of loudspeakers at ±30° and another pair of loudspeakers at angles from ±60° to ±120° performed reasonably well, as did the front or rear combinations of four channels (b) and (h). Stereo (a) performed poorly, as did the front-back symmetrical "quad" arrangement (f).

The lower row of results show the effect of adding a center channel to (b), (c), (d), and (f), creating optional five-channel configurations, including a repetition of the ITU arrangement as (l). The already good performance of the four-channel versions is improved, with (j), (k), and (l) all exhibiting highly attractive results. The front-back symmetrical arrangement (f) is slightly improved by the addition of the center channel (m), but it is still not an attractive option. The lesson: Avoid symmetrical front-back arrangements of left-right loudspeakers.

15.10.3 Summary

There is *very* good news. Large numbers of channels are not necessary to provide excellent facsimiles or reconstructions of enveloping sound fields. This is true whether the evaluating metric is subjective or objective. The optimal selection

A CHANNEL-NUMBERING SCHEME

To keep track of how many loudspeakers are in a multichannel system and to help in understanding where they are located, the industry has adopted a simple designation. It consists of two numbers: the first number is the number of front channels, and the second number is the number of surround/side/rear channels. Therefore, 2/0 is stereo; 3/1 is the original Dolby Stereo/Surround system with a single surround channel; 3/2 is conventional five-channel (5.1) surround, with L, C, and R across the front, and two surround channels; and 3/4 has four surround channels, which is called 7.1 in the consumer world. However, 7.1 in the cinema world is more likely to be interpreted as 5/2, which is Sony's SDDS system with five channels across the front and two surround channels. The numbers 3/3 indicate any of the systems that have two side surrounds and a single rear channel.

Comparisons of frequency-dependent interaural cross-correlation (FIACC) measurements in four large halls (A,B,C,D) with FIACC measurements of reproductions of those spaces through different multichannel loudspeaker arrangements in an anechoic space. The results are expressed as a "square error" computed over the frequency range.

▬▬▬ "Full" bandwidth: 100 Hz to 20 kHz

▬▬▬ "Fundamental" bandwidth: 100 Hz to 1 kHz, the frequency range most related to the perception of spaciousness.

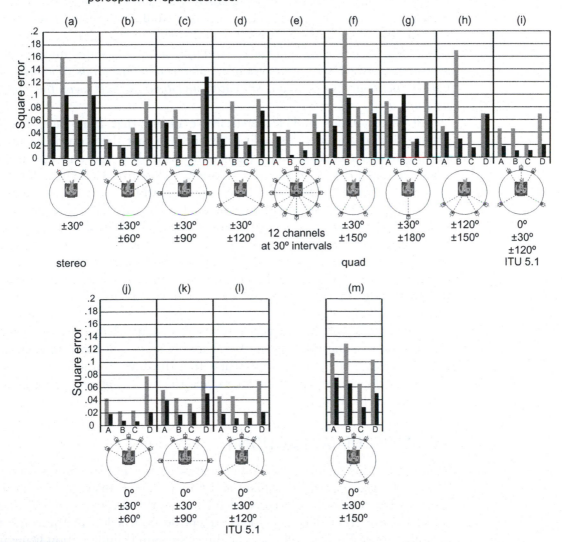

FIGURE 15.9 *Selected data from Muraoka and Nakazato (2007) showing how closely various configurations of reproduction channels and loudspeakers, (a) through (m), can replicate the FIACC measured in four large venues: a large lecture hall "A" and three concert halls "B," "C," and "D." The shorter the vertical bars, the better the reconstruction of the original sound field. It is probable that the black bars are more meaningful.*

of four or five channels and loudspeakers can provide performance very similar to circular arrays of 12 or 24 loudspeakers.

Two elaborate studies, one subjective and one objective, concluded that the existing popular standard arrangement of L, C, R loudspeakers spanning a 60° arc across the front, combined with two surround loudspeakers at ±120° (i) and (l), performed superbly. However, it is also clear that there are other five-channel options, (j) and (k), that work comparably well. Only if one wants discretely panned localizable images at other locations would more than five channels surrounding a listener be necessary. *(NOTE: The previous sentence said "a listener," but if there is an audience of multiple listeners located away from the symmetrical sweet spot, more channels and/or different kinds of loudspeakers will be necessary to generate similarly enveloping effects over the enlarged listening area.)*

All three studies provide persuasive evidence that front-back, left-right symmetrical arrangements should be avoided, in that they contribute little or nothing to the perception of envelopment over simple stereo, only adding directional options for panned sounds. Morimoto (1997) did some experiments in which he concluded that sounds that originated behind listeners are important to LEV; in situations exhibiting similar IACC, listener envelopment was improved when a greater proportion of the sound arrived from the rear. Unfortunately, the loudspeaker arrangement used in the experiments was symmetrical left-right, front-back, a situation already disadvantaged in the generation of convincing LEV. Since all of the practical loudspeaker arrangements for home theater require rear loudspeakers for localized sounds, this is an issue that is settled automatically.

A caution: all of these experiments were conducted in anechoic listening circumstances. Reflections within the listening room will have some effect on the conclusions, but it is highly probable that the direct sounds from the loudspeakers will have the dominant effects, and the more channels, the greater the dominance is likely to be.

15.11 RECOMMENDATIONS

The choice of the number of surround loudspeakers is in part a matter of intent for the playback system. Nakahara and Omoto (2003) discuss it in terms of direct surround (two loudspeakers, each aimed at the sweet spot) or diffuse surround (two or more surround loudspeakers on each side). In the context of recording control rooms, they see the direct surround configuration being used for "high-grade musical contents such as DVD-Audio or SACD," and diffused surround for "general purpose" program material such as DVD-Video, games, and so forth. Although not part of their thoughtful investigation, it is clear that without any additional considerations, home entertainment systems would use more than two surround loudspeakers.

All of these scientific investigations just discussed focused on a listener in the sweet spot, the "money seat." As important as this is, additional channels, discrete or upmixed, are needed to deliver similarly persuasive spatial illusions to listeners seated away from the prime location. The studies were also based on anechoic listening. If more channels are able to improve circumstances for listeners seated away from the prime location, it is also probable that certain combinations of room reflections may be able to do the same. It is a situation ripe for research. It is time to focus research efforts on finding optimum solutions for real-world listening situations. If there is a problem with all of these schemes, it is that listeners close to the perimeter of the room are likely to localize the surround loudspeaker closest to them. It is a topic that will come up in the next two chapters, and possible solutions are proposed.

As we will see in Chapter 16, video adds its own requirements to the choice of optimum seating arrangements, so it is well to consider the number of loudspeakers and the *ranges* of angles at which they may be deployed that yield desirable audible effects. In the real world, it is essential to have options.

3/2 (5.1) Channel Considerations. Looking at the evidence just discussed, a front soundstage of 0° center and left and right loudspeakers at ±30° is a solid beginning. To this, one could add side channels with the prime task of generating envelopment and providing occasional directed sound effects. For many of the sound effects in movies, the specific angles at which the effects are perceived are not critical. However, if the system is a 3/2 (5.1) channel configuration, these loudspeakers must also provide occasional persuasive rearward localizations. Consequently, a compromise is usually reached, which places the loudspeakers to the sides of the prime listener and slightly behind, somewhere in the angular range of 110° to 120°. If there are multiple rows of listeners, it may be necessary to place additional pairs of side/rear loudspeakers, simply connected in parallel with the others. Although it is possible to do this with some impunity in large rooms, in the close confines of a small listening room it is a situation that might generate occasional audible artifacts because the multiple sources are not adequately decorrelated from each other. Adding a delay to the supplemental loudspeaker signal paths is the least that can be done to improve circumstances, which is precisely why 3/4 (7.1) channel systems exist.

3/4 (7.1) Channel Considerations. Having the luxury of four surround loudspeakers opens some options. It is possible, for example, to position two of the loudspeakers slightly forward of the listening area for maximum envelopment. Locations in the range ±60° to ±90° seem to be good choices, leaving the rear clear for additional channels. These additional channels cannot simply be parallel-wired versions of the side channels; there must be at least a delay and possibly a spectrum change to provide additional decorrelation between the surround channels. This is what happens in well-designed upmixers. As for locations, to assist in localizing sound effects to the rear, a definite rear bias is appropriate. In 3/2 systems it is advisable to avoid surround-loudspeaker angles that are symmetrical with the front left and right loudspeakers. In 3/4 systems, the existence of the side loudspeakers would most likely eliminate this as a factor for the rear loudspeakers.

15.11.1 The ITU Perspectives

On the international scene, there is little doubt that the International Telecommunication Union document, Recommendation ITU-R BS.775-2, updated in 2006, is the most influential. Figure 15.10a shows the well-known five-channel recommended layout, and (b) shows a broad angular range within which four *or more* side/rear loudspeakers can be positioned. It would seem that writers of this standard were aware of the scientific data discussed in Section 15.10 because they have expanded the permissible locations for side/rear loudspeakers to include those forward of the listening position.

If more than two loudspeakers per side are to be used, it is suggested that they be distributed at equal intervals on the sides. They also state that the signals to the additional side/rear loudspeakers may need to be delayed or otherwise decorrelated. This is most likely to reduce the risk that listeners may localize the loudspeakers, to further enhance the sense of envelopment, and to eliminate the possibility of acoustical interference effects. In 7.1-channel surround processors, this should be a standard feature. The angular range includes the symmetrical front-back situation, something that has been shown to be less than ideal in 3/2 systems (see Section 15.10). However, the existence of an additional pair of side-located loudspeakers in 3/4 or higher-order surround configurations would likely alleviate this problem.

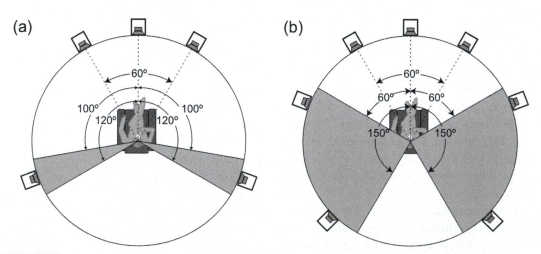

FIGURE 15.10 *The ITU-R BS.775-2 recommendations for (a) five channels (3/2) and (b) seven channels (3/4). Reproduced with the permission of the International Telecommunication Union.*

15.11.2 Other Perspectives

Looking across the industry for guidance, Dolby is a logical starting point. According to www.dolby.com, the recommendations for 7.1-channel configurations is as shown in Figure 15.11a. The spacing of the front left and right loudspeakers is allowed to creep inward; a 60° separation is the most common recommendation for music reproduction, but for movies, a separation of 45° is an option introduced by viewing-angle considerations, as will be explained in Chapter 16. The side and rear loudspeakers are well positioned. In a home system intended for multiple purposes, including music and games, the 45° separation of the front L and R loudspeakers might be considered too narrow.

Figure 15.11b shows the author's composite recommendation, combining the Dolby frontal recommendations with the ITU side/rear recommendation. This has the advantage of allowing side loudspeakers to be placed ahead of the listeners, while additional side and/or rear loudspeakers occupy the more traditional locations to the rear of the listener. Of course, if there are multiple listeners, more surround loudspeakers are clearly advantageous.

For interest, the "spatial-effect balloon" introduced in Chapter 7 is included in (c), indicating that the suggested locations are well chosen. In (d) it is seen that the arrangement is totally compatible with home theater installations. Adjustments to delays in the signal paths will compensate for distance discrepancies. Note that in this example, the "rear" loudspeakers are not on the rear wall. If the rear wall happens to be closer to the seating area, the rear loudspeakers would simply migrate around the corner to that wall, with the only stipulation that locations in or very close to the corners be avoided to minimize coloration (see Figure 12.2). With steady progress in surround processor developments, it is probable that high-end models will one day provide processed outputs for more than four surround channels, making it easier to cater to several rows of listeners. In the meantime, there are several outboard multichannel digital delay/equalization/crossover processors available in the professional audio domain that are extremely useful in these applications.

15.12 ASSIGNING THE CHANNELS AND THE CENTER-REAR OPTION

Some readers may be wondering why we have not discussed center-rear loudspeakers more, since there are processors and delivery formats that support such an option. Indeed, it is an option available for choice, but even the proponents often recommend using a pair of loudspeakers rather than a single centrally located unit.

Why? I can only offer an opinion based on limited experience. Experience is limited because there are very few films containing encoded rear-directed sounds

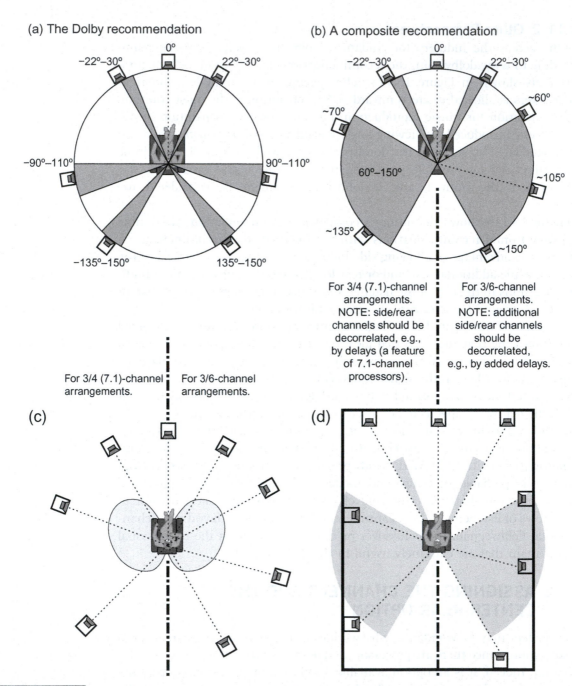

(a) The Dolby recommendation

(b) A composite recommendation

For 3/4 (7.1)-channel arrangements. NOTE: side/rear channels should be decorrelated, e.g., by delays (a feature of 7.1-channel processors).

For 3/6-channel arrangements. NOTE: additional side/rear channels should be decorrelated, e.g., by added delays.

For 3/4 (7.1)-channel arrangements.

For 3/6-channel arrangements.

(c)

(d)

FIGURE 15.11   *(a) The 7.1-channel recommendations from Dolby. (b) The author's composite of the Dolby angular range for front left and right loudspeakers, the ITU angular range for 3/4, and higher-order surround configurations. (c) A superimposition of the "spatial-effect balloon" from Chapter 7, showing that all of the loudspeakers are well positioned to create the desired impressions. (d) One way that the recommendations may translate to a rectangular home theater installation. (b), (c), and (d) are divided left and right to show arrangements for 3/4 and 3/6 multichannel configurations.*

for this form of playback. THX and Dolby collaborated in the development of THX Surround EX/Dolby Digital Surround EX, which is a matrix-encoded monophonic signal included in the left and right surround channel signals. Surround upmixers configured to drive four surround channels extract this signal and deliver it to the rear loudspeaker or equally to a pair of rear loudspeakers. The rest of the time, depending on the processing chosen by the designers of the upmixers, the rear channel(s) are supplied with manipulated versions of the side signals. DTS-ES Discrete 6.1 is currently the only system capable of driving a rear channel with its own discrete signal.

In general, a solo center-rear loudspeaker is not an attractive idea because movies are already center dominated by the relentless front-center channel. Some form of lateral expansion, spaciousness, or envelopment is a welcome relief. Then there are the front-back or back-front reversals that happen with sounds originating on or close to the median plane (the vertical plane running front to back through the head). It is normal, and it happens occasionally in normal life, most often with unfamiliar sounds. We reflexively rotate the head slightly to resolve such ambiguities, but with brief sounds that is not possible— and a lot of movie sound effects are brief. We make these mistakes most often with tonal, or narrow-band, sounds. It is a problem without a solution, and in movies it is aggravated by the presence of the picture, a logical place for directional ambiguities to find a default location. So back-to-front reversals are much more likely than front-to-back reversals.

Using a pair of loudspeakers is a good option because, for the vast majority of the time, they will add noticeably to an improved sense of envelopment. Even when they are called upon to do duty as a true "rear" channel, they are not compromised. Because the signal is monophonic, listeners on the center line will hear a phantom image in the middle of the back wall. Listeners to the left or right will hear an image to the left or right of center in the back wall. In all cases, the image, if it is a flyover, will simply proceed to pan to the front of the room as it should. All of these sounds tend to be brief because moviemakers don't like sounds that dwell at off-screen positions. Part of this may be because in cinemas, where surround channels consist of a collection of small loudspeakers, the sound quality is greatly different from and usually disappointing compared to the front channels.

CHAPTER 16

Putting Theory into Practice: Designing a Listening Experience

This chapter attempts to distill the scientific guidance developed in the earlier chapters and apply it to the design of home theaters and entertainment spaces. It is taken in stages: the room, video considerations, seating, loudspeaker placement and the multichannel options, and, finally, identifying the basic requirements for loudspeakers to make it all work. The intent is to provide an understanding of how things should be done and why. Ideally, this should enable designers and installers to cope with real-world circumstances—to be adaptable and to be able to find alternatives, knowing what the rigid requirements are and what aspects of design are flexible. Because rooms and loudspeakers are interdependent, we will be discussing which factors matter most, what is less important, and those things that are optional. Fortunately, this is not a matter that demands micrometer precision; some decisions are merely differences, not matters of better or worse, right or wrong. There are opportunities for choice on the part of the designer or to allow interested customers to participate in the design.

As discussed in Chapter 11, humans are remarkably adaptive. Listeners can adjust to many circumstances that fall far short of any rational ideals, deriving great pleasure from many forms of audio/video entertainment in spite of insults to sound and video quality that some of us might consider to be intolerable. A "happy customer" may, therefore, not be evidence of a job well done in acoustical terms, especially given the other dimensions to a fully integrated home entertainment installation. In fact, one of the great frustrations of this industry is the extent to which consumers continue to support, sometimes at great expense, ill-conceived and poorly-designed products and entertainment room installations. Marketing is clearly as important as substance, appearance as important as performance, and, to an acoustically unsophisticated clientele, mediocrity often passes for excellence. Acoustical consultants frequently measure their success by the number of installations they have done. The real questions are: How many of them were truly good? How many of them could have been better?

In reality, with a savvy choice of components and some acoustical guidance, excellence can be achieved at modest cost. The bad news is that it is possible to end up with sonic mediocrity, having mortgaged the house and postponed the purchase of a car.

16.1 THE ROOM

The first notion that needs to be put to rest is that there is no uniquely good set of room dimensions. In most cases, installations must fit into existing spaces, so reality sets a limit on screen size, viewing distances, number of seats, and so on. If there is the luxury of choice, the size of the room is dictated first by what needs to fit within it. The issue related to the dimensional proportions of a room is how it behaves at low frequencies—only. In small rooms, because of the acoustical influences of seating, furnishing, and decoration, the mathematical predictions of room resonances and standing waves cease to be meaningful above the transition frequency region, around 200–300 Hz, as explained in Chapter 13. There it was also explained that bass performance depends on how many woofers there are, where they are placed, where the listeners are located, and the room itself. So it is a multifactored equation. All solutions that offer improved sound equality among several seats require two or more subwoofers.

If the room is perfectly rectangular, with all the walls of the same construction, then you can roll the dice and simply choose a benevolent arrangement, as shown in Figure 13.12. Or one can go the next step and take advantage of the fact that for each of several arrangements of multiple subwoofers, there is a range of dimensions that would have advantages so far as minimizing seat-to-seat variations (Figure 13.16). Note, however, that all of these predictions apply to the specified seating areas. Figure 22.4 and the associated discussion provide additional insight into the fine tuning of room dimensions and/or seating locations.

If all that matters is that a single listener is happy, then put a subwoofer wherever it is convenient, preferably in a corner, and equalize it to eliminate prominent peaks in the frequency response at that seat. Other listeners in the room must take their chances. Algorithms that involve equalization to satisfy measurements made at several locations are seeking a compromise in which the fewest listeners are offended, but none may be completely gratified.

The ultimate form of intervention in controlling the sound field in a room is the type of process exemplified by Sound Field Management (Section 13.3.5). It is an optimization algorithm that begins with measurements in the room and ends with a description for processing of the signals fed to each of multiple subwoofers. Here we are actually manipulating the room resonances, the standing waves, to minimize variations among some number of designated listening positions. High-resolution measurements and parametric equalization at low frequencies are essential ingredients of any of these systems. Acquiring and learning to use one of the laptop-based measurement systems is hugely advantageous to anyone planning to make a career in this industry. In the following chapters, there will be many examples of these measurements, and the clarity they bring to a situation is profound.

The number of audience members, combined with the available space, will influence the style of seats. Massive reclining chairs take up a lot of space, and it is necessary to allow for people to come and go during a show. If fully reclined viewers are not to be disturbed, the front-back space occupied by a row can be considerable. This is a matter to be discussed with the customer at an early stage. In less-formal entertainment/media rooms, consider having some light-weight chairs that can be easily moved into rows for movies and then arranged differently for conversation and normal living. One can see from Figure 3.2 that the author has a bias for "sociable" seating arrangements and moveable seating, the principle being that row-upon-row seating is acceptable for groups of strangers, but among friends it is nice to share body language and facial expressions during the entertainment. With large audiences, there is obviously no choice.

The acoustical treatment of the room interior is an important matter and a small industry has developed to cater to it. Purveyors of acoustical materials and consulting services both actively promote the importance of a properly "designed" listening space. The design usually begins with traditional reverberation-time (RT) measurements. For small rooms, the name *reverberation time* is sometimes changed to *reflection-decay time* to acknowledge the reality that there is no sustained reverberant sound field in small, relatively dead listening rooms and that the sound fields are not very well diffused. It is an accurate description of what is measured, but whatever you choose to call it, the measurement process is the same (see Chapter 4). In this book, the traditional term *reverberation time* will be used out of respect for over 100 years of history and the fact that all instruments that measure it are so labeled.

Obviously, we would look for a value of RT that provides satisfactory listening circumstances. Such optimum values exist for large performance spaces, and there is a rich literature tracking developments in architectural acoustics over many decades. However, as was pointed out in Chapter 4, sound reproduction is fundamentally different. There is not a large, omnidirectional sound source (the orchestra) radiating into a highly reflective space (the hall) entertaining hundreds or thousands of listeners. Instead, we have two to seven moderately directional loudspeakers surrounding and aimed at a few people in the central area of a small room. The reverberation time of any practical listening room is typically much shorter than that in the program, so it is simply not a factor in any normal sense. The only reason to measure RT in a small room is to be certain that it is not excessively high (over about 0.5 s) to preserve high speech intelligibility or low (under about 0.2 s) to avoid oppressive "deadness."

Fortunately, the nuisance is easily overcome because if a room has normal domestic furnishings—carpet, drapes, upholstered chairs, bookcases, fireplace, lamps, side tables, and so on—it is probable that it falls within the broadly accepted midfrequency (around 500 Hz) reverberation time of 0.3–0.5 s. Rooms that are overly damped are not so much a problem for multichannel listening as they are for the inevitable pre- and postperformance conversation. Acoustically dead rooms are simply not very pleasant spaces to converse in. Vocal effort must be elevated, and face-to-face communication improves clarity. As has been explained in Chapter 10, room reflections improve speech intelligibility. Unfortunately, in custom home theaters, overzealous consultants and installers sometimes sell too much fabric and fiberglass.

Figure 16.1a shows the acoustical evolution of a listening room; in this case, it was the prototype IEC listening room, and these measurements were made in 1975. It shows the very high RT of an empty room being significantly tamed by the addition of wall-to-wall carpeting. The effect is dramatic. However, the

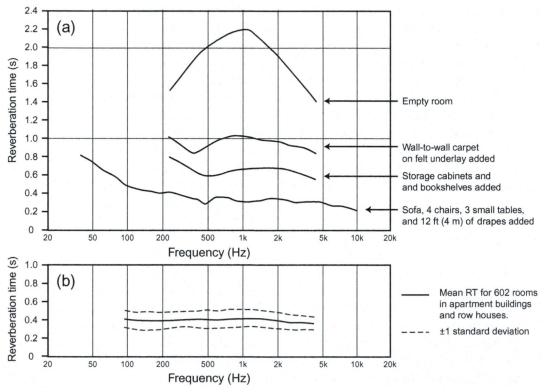

FIGURE 16.1 *(a) Reverberation time measurements made at different stages in the furnishing of a typical room. The only special materials employed in this example were some custom-built low-frequency absorbers used to compensate for the fact that the floor and two walls were of masonry construction and therefore contributed no low-frequency absorption. These had effect only below about 100 Hz. (b) The results of a large survey of domestic rooms in Canada. From Bradley, 1986.*

room was still far too reflective, a character being especially noticed in flutter echoes between and among the walls. Adding some functional storage and display cabinet and bookcases (no significant absorption but a lot of scattering) dropped the RT and eliminated most of the "empty" sound in the room. The scattering devices were redirecting more of the sounds into the carpet, making it work harder. Bringing in the rest of the furniture and some drapes finished the task. The drapes were chosen to be acoustically effective: heavy cotton with lining, pleated to less than one-half fabric length, and hung 4 in. (100 mm) from the wall so they would function at lower frequencies. The room sounded utterly "normal"; conversation was very comfortable, and reproduced sound, then in stereo, was excellent.

It was in this room that experience was gained in understanding the role of first reflections from the side walls. The drapes were on tracks, permitting them to easily be brought forward toward the listening area so listeners could compare impressions with natural and attenuated lateral reflections (see Figures 4.10a and 8.8). In stereo listening, the effect would be considered by most as being subtle, but to the extent that there was a preference in terms of sound and imaging quality, the votes favored having the side walls left in a reflective state. In mono listening, the voting definitely favored having the side walls reflective. See the discussions in Chapter 8, and Figures 8.1 and 8.2, which show that attenuating first reflections seriously compromises the diffusivity of the sound field and the sense of ASW/image broadening. One of the problems with both music and movies is that sounds that in real life occupy substantial space—multiple musicians or crowds of people, for example—end up being delivered through a single loudspeaker—a tiny, highly localizable source. The precision of the localization is the problem. Most of what we hear in movies and television is monophonic, delivered by the center channel, so a certain amount of locally added room sound may be beneficial; this is definitely a case where a personal opinion is permitted.

Figure 16.1b shows the results of a large survey of domestic rooms in Canada, conducted by Bradley (1986), that indicates a remarkable constancy among the rooms we live in. A few other, much smaller studies exist that were done in Europe and the United States, and those data fall comfortably within this set. The only result that distinguished itself was one from Sweden where the Scandinavian style of sparse furnishing resulted in significantly higher RTs. One may presume that different trends exist in other regions of the world as well. The basic message here is that target range of midfrequency RT should not be difficult to achieve in any well-furnished room.

In custom-constructed listening spaces and home theaters, the only required furnishings are the chairs, so additional acoustical materials must be added to achieve the reverberation time of a normal space. A custom listening room is therefore the ideal opportunity to optimize the loudspeaker/room system, and the topic of where to selectively place absorbing, reflecting, and scattering surfaces and devices will reappear in the following sections. It is probably safe to assume that the floor will have carpet (best if it is a clipped-pile, jute-backed, acoustically porous type), installed over a thick felt cushion so that the combination behaves like a broadband absorber (see Figure 21.3). It should be in a location that provides some attenuation of the floor bounce from the front L, C, and R loudspeakers.

In the following sections, it will be assumed that the loudspeakers are well behaved both on- and off-axis (see Chapters 17 and 18 for examples of a sometimes horrible reality). It will also be assumed that if one wishes to reflect (change the direction of), absorb (attenuate), or scatter (redirect to many different directions) sounds, any devices chosen to perform the function will be similarly effective at all frequencies above the transition frequency: about 200–300 Hz. As we will see in Chapter 21, this requires materials and devices that are much more substantial (i.e., thicker) than many of those in the marketplace.

16.2 BASIC VIDEO

16.2.1 The Cinema Reference

In a cinema, all the seats do not allow for equally gratifying experiences. Arrive late for a movie, and you will fumble your way in semidarkness to the front of the cinema, where, for the remainder of your stay, you will stare upward into an exaggerated, geometrically distorted, fuzzy, jiggly picture. It is not a great experience, which is why most people like to sit farther back. The industry believes there is something special about being two-thirds of the way back from the screen because that is where they have chosen to specify that a horizontal viewing angle of 45° to 50° is optimal for widescreen (1:1.85) presentations (Robinson, 2005). This is important because it seems that about 80% of Hollywood films are issued in this format. If screen height is maintained, this becomes 56° or more for 1:2.35 Cinemascope blockbusters. For those who find themselves in the back row, the recommendation is that the viewing angle be not less than 30°, presumably for widescreen images.

Some interesting statistics about a true reference cinema, the Samuel Goldwyn Theater in the Academy of Motion Picture Arts and Sciences, Hollywood, come from an article by Bishop (2007). According to this, the Goldwyn Theater has 24 rows of seating, and the screen is 54 ft wide. In the front row, the viewing angle is over 100°. Rows eight through ten are highly rated, where the viewing angle is 60° to 75°. Of course, not all presentations use the full screen width, so viewing angles are often smaller. It undoubtedly helps that this cinema gets prime film stock to run and has a meticulously maintained projector and screen. The author has watched several films at this facility, sitting at different seats each time, and can confirm that the very big image certainly can be beautiful. The Goldwyn also sounds impressive. The loudspeaker system is shown in Figure 16.2 and described technically in Eargle et al. (1997).

On the video resolution side, 35 mm film was the medium to beat when home theater became a serious business. No realistic person thought that 70 mm and IMAX experiences could be replicated on the domestic "big" screen. In the beginning, it seemed that even 35 mm resolution might be unobtainable in the home, with some very large resolution numbers being talked about. Indeed, a fine-grained negative can be equivalent to about 12 megapixels. However, that is not the distribution medium. When the degradations of conversion to a positive, then to an internegative, and then to the positive "release prints" are accumulated, it is estimated that horizontal resolution has dropped to about 2000 pixels (Koebel, 2007). To this must be added optical flaws and limitations in camera and projection lenses, and jitter introduced by the film gate. All considered, many real-life film experiences are no better than we can now achieve in our homes. Koebel concludes, "A two-million-pixel display system is more than sufficient for showing 35 mm-originated material better

FIGURE 16.2 *The Samuel Goldwyn Theater in the Academy of Motion Picture Arts and Sciences, Hollywood, California, with the screen down and showing the five front channels and subwoofers, all JBL loudspeakers. Photo courtesy of JBL Professional; "Academy Award" and "Oscar" image © AMPAS®. THX® Lucasfilm, LTD.*

than most public movie theaters (those that don't use digital projectors!)." Two megapixels is approximately the 1920 × 1080 pixel 16:9 HDTV display, and its excellence is assured only if there are no encode/decode artifacts added by the video delivery system.

16.2.2 Transferring the Video Experience to Homes

The implication of this is that the acceptable viewing angles for 1080p images will be similar to those for cinema displays, and indeed they are. However, video in the home has evolved using sources and display devices with substantially less resolution. Figure 16.3 shows viewing angles considered to be acceptable in cinemas and, by inference, for 1080p displays. The chart shown along the left side of the figure indicates the distances from this screen at which viewers should be able to appreciate the full benefits of several popular resolutions (Bale, 2006; Koebel, 2007; Ranada, 2006).

It is obvious that only recently has there been a mass-market delivery system with resolution sufficient to justify large screens viewed at normal distances. Over many years, video enthusiasts have supported the notion that a large, "fuzzy" image is better than a small, "sharp" image. Thus begins the discussion of "immersion" in the film versus loss of resolution and the visibility of artifacts in the projected image. From Figure 16.3 it is clear that, on this size of screen and at these viewing angles (and distances), standard DVDs with 480p resolution should be unacceptable. Undoubtedly, higher-resolution images are preferable, and a switch to HDTV is an obvious improvement. Still, there is a repertoire of legacy material that will be with us forever, so designers must consider options.

A very practical option that caters to both 480p and 1080p material is to provide two displays:

- A smaller, direct-view or rear-projection display for daytime viewing and for viewing standard-resolution images—about 50 in. (1.27 m) diagonal would be appropriate at these viewing distances.

- A large, front-projection system, perhaps with a motorized screen, for cinema-like experiences—123 in. (3.12 m) diagonal in this example.

Sometimes the decision about which display to watch has nothing to do with resolution but with content. Watching "talking heads" on a huge screen is disconcerting. Such TV news and entertainment programs were not created for cinema-screen sized presentation.

Let us now move on to see how to fit an audience and an audio system around this display in this room. Figure 16.4 shows two alternative seating arrangements in elevation. Obviously, vertically staggered seating is much preferable from the perspective of sight lines. It also changes the room dimensions in a way that will disrupt the simplicity of the vertical standing waves, reducing their strength and shifting the null locations.

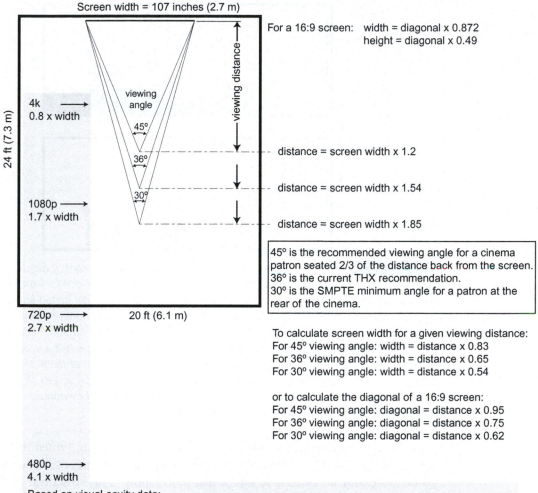

Screen width = 107 inches (2.7 m)

24 ft (7.3 m)

For a 16:9 screen: width = diagonal x 0.872
                     height = diagonal x 0.49

viewing distance

4k →
0.8 x width

viewing angle

45°

distance = screen width x 1.2

36°

distance = screen width x 1.54

30°

distance = screen width x 1.85

1080p →
1.7 x width

45° is the recommended viewing angle for a cinema patron seated 2/3 of the distance back from the screen.
36° is the current THX recommendation.
30° is the SMPTE minimum angle for a patron at the rear of the cinema.

720p →
2.7 x width

20 ft (6.1 m)

To calculate screen width for a given viewing distance:
For 45° viewing angle: width = distance x 0.83
For 36° viewing angle: width = distance x 0.65
For 30° viewing angle: width = distance x 0.54

or to calculate the diagonal of a 16:9 screen:
For 45° viewing angle: diagonal = distance x 0.95
For 36° viewing angle: diagonal = distance x 0.75
For 30° viewing angle: diagonal = distance x 0.62

480p →
4.1 x width

Based on visual acuity data:
Arrows show distances from this screen at which viewers can see the full benefit of each resolution.
Persons seated closer to the screen will see a progressive loss of resolution.
Persons seated farther away will not be able to see all of the advantages of the stated resolution.

FIGURE 16.3 *In the 20 ft × 24 ft × 9 ft room used in several previous examples, a 123 in. diagonal (107 in. wide × 60 in. high) 16 × 9 (1: 1.78) screen serves as display. This figure shows some of the relevant dimensions: viewing angles considered to be standard references and viewing distances at which persons with normal 20/20 vision are able to appreciate the detail in displays having different popular resolutions.*

The structure supporting the seats can be part of the experience. In a good audio system, there is enough low-frequency energy to stimulate vibrations in the floor. This is pleasant if the floor surface responds about equally to all bass frequencies. For this to be assured, it may be necessary to laminate the floor surface with gypsum board or other forms of added mass and elastomeric damping to prevent single-frequency mechanical resonances. The author can

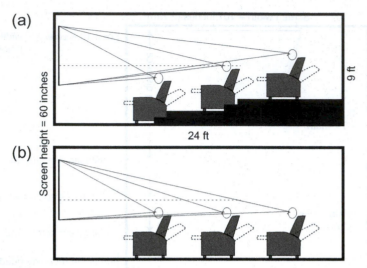

FIGURE 16.4 *Examining sight lines as they might be in such a room, it can be seen that there are two options. (a) A possible vertical staggering of head locations that yields a clear view of the entire screen for all in the audience. The horizontal dotted line is a popular objective: the horizontal sight line from the prime location should be about one-third of the distance up the screen. (b) Everyone at floor level. This situation requires, at a minimum, that successive rows of viewers be shifted sideways by one-half a seat width so viewers can see the screen between the heads of the people in front of them. If this is done, try to arrange for the sweet spot to be on the central axis of the screen. A second consideration is to lift the screen, which forces everyone to look slightly upward. The latter can be aided by tilting the reclining seats slightly backward.*

recall being in one very expensive home theater where, during a musical number, it became quickly obvious that the floor was responding energetically to one bass guitar note and no others. It was a real distraction. The structural surface of the staged seating was a single layer of plywood.

Another option, of course, is to drive the floor structure or entire chairs with shakers. If this is overdone, the result is silly and totally unreal (except for gaming, where almost anything goes, it seems). Done with subtlety, it can be another sensory embellishment for over-the-top blockbusters, but one must first consider the option of simply supplying more and/or better-quality low bass from the loudspeakers. The small, middle-of-the-back vibrators in chairs are amusing, not impressive.

16.3 MERGING AUDIO AND VIDEO

After the discussions in Chapter 13, where we saw the challenges of delivering good bass to multiple listeners in small rooms, it must be evident that bass management (combining the low frequencies from all channels and adding the level-corrected LFE) and a system of subwoofers is compulsory. The some-times-promoted "purist" notion of five or more full-range, floor-standing loudspeakers has historical attachments to the origins of audio, but knowing what we do now, it is simply not a good solution (see Chapter 13). So *all of the following discussion applies to signals above the subwoofer crossover frequency— normally 80 Hz*.

Let us now take a close look at the direct and reflected sounds from the loudspeakers in the optional multichannel arrangements. The first consideration is to provide a superlative listening experience for the prime listening location. Beyond that, it is necessary to avoid excessive degradation as the listening locations move toward the perimeter of the room. Perfection for *all* listeners is impossible.

All loudspeakers must deliver a strong, high-quality direct sound to all listeners; otherwise, simple localization cannot occur, the precedence effect will be compromised, and any ambient/enveloping effects will be degraded. The next step is to look at first reflections because, being the second-loudest sounds to arrive at listeners, they can be organized so they contribute to the sensations of ASW/image broadening and envelopment.

It is especially important to understand what contributes to the illusion of envelopment. This is arguably the greatest benefit of multichannel audio, and sound tracks are designed to provide it most of the time, if not in the main story line, then in the accompanying background music. This perception has nothing to do with noticing that sounds are coming from different directions. Those are "special effect" gunshots, bullet ricochets, flyovers, and so forth. Envelopment is the feeling of being in a different, larger space than the listening room. In movie sound tracks, it is generously used as a spatial accompaniment to front channel sounds and almost always as a component of atmospheric music. Chapters 7 and 8 discuss these effects in detail, and Chapter 15 discusses some of the multichannel options for achieving it.

We know that this perception is closely associated with interaural cross-correlation coefficient, IACC, a measure of the *difference* in the sounds arriving at the two ears. Obviously, sounds must arrive at the listening locations from both sides—originating in the side surround loudspeakers—but it is a well-known experience that the sensation of envelopment deteriorates as one moves away from the center of the room, ending up with a localization of the sound emerging from the nearest side-surround loudspeaker. This is a problem of listening in small rooms, and the root cause is physics: propagation loss, the inverse-square law. The desired low IACC is possible only when the sounds arriving from both sides are similar in amplitude, which with conventional loudspeakers occurs only in the central portion of the room. The solution to this is not to radically change the directivity of the surround loudspeakers, hoping for the improbable creation of a "diffuse" sound field. There are other options, and in a later section we will consider some solutions that have not had much, or any, promotion. In the meantime, designers should attempt to place as many listeners as possible close to the left-right axis of symmetry of the loudspeaker arrangement.

Starting with the video requirements, Figure 16.5a shows a seating arrangement that would work in this room. It places the prime location in the center of the second row (see also Figure 16.4a), with a viewing angle of 36°. The front row is a bit close, and the rear row is a bit far away. An option: move the rows closer together if the seats permit.

NOTE: All rows have a seat on the central axis of the room. To some extent for video and most certainly for audio, there is a sweet spot. Sitting there is not as uniquely rewarding in multichannel audio as it is in two-channel stereo, but there are significant benefits. You can be sure that during the final audio mix or in its transfer to a distribution medium, any person passing judgment would be sitting in the sweet spot. It would be a shame if nobody in a home theater gets to have that experience. Yet, sadly, it often happens.

In this example, the center of the front row is almost perfect for listening to stereo through LF and RF loudspeakers. If this were important to the customer, a small positional adjustment to the seating or loudspeakers would yield the ideal ±30° listening angle.

If all seats are at floor level, as in Figure 16.4b, it is recommended to start with the middle row as it is, placing the prime listener in the sweet spot. Then the front and rear rows of seats could have four seats each and each shifted half a seat width from the middle row. This allows all viewers to look between the heads of those in front, but there is no symmetrical seat in two of the rows.

Figure 16.5b shows the composite recommendation from Figure 15.11b superimposed on the room layout. Placing the LF and RF loudspeakers just slightly beyond the screen yields a listening angle of ±22.5° for the prime listener. They could be farther apart, up to the 30° recommended limit, at the customer's option. The only argument for crowding the screen, or for violating the listening angle recommendation and placing them behind an acoustically translucent screen, is to keep the sound aligned with the on-screen action. The problem with this argument is that it is an uncommon event in movies for the sound to actually track the on-screen action, and when it does, it is usually for dramatic effect (it costs money to do it), and it happens quickly, like a car driving by. Most of the time, it is the ventriloquism effect that allows our minds to think that sound emerging from a loudspeaker (usually the center front channel) is actually emerging from the mouths or other visible sound sources somewhere on the wide screen. Localization errors are so routine in movies that a sound that actually follows an image has a kind of startle effect because of its rarity. It is not likely that anyone will notice or care whether the side front loudspeakers are perfectly aligned with the edge of the screen or not, and this is the opinion of a critical movie watcher.

So use your judgment. A larger separation of LF and RF may contribute to a more engaging audio experience, but too extreme a separation may be bothersome; in situations involving small direct-view screens, the decision is an agonizing one. Placing the loudspeakers near the screen destroys any sense of

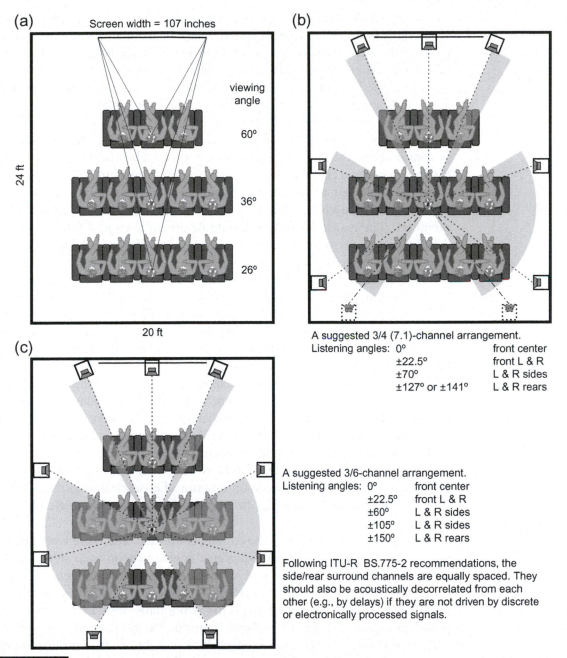

(a)

Screen width = 107 inches

viewing
angle

60°

36°

26°

24 ft

20 ft

A suggested 3/4 (7.1)-channel arrangement.
Listening angles: 0°   front center
                  ±22.5°   front L & R
                  ±70°   L & R sides
                  ±127° or ±141°   L & R rears

(b)

(c)

A suggested 3/6-channel arrangement.
Listening angles: 0°   front center
                  ±22.5°   front L & R
                  ±60°   L & R sides
                  ±105°   L & R sides
                  ±150°   L & R rears

Following ITU-R BS.775-2 recommendations, the
side/rear surround channels are equally spaced. They
should also be acoustically decorrelated from each
other (e.g., by delays) if they are not driven by discrete
or electronically processed signals.

FIGURE 16.5 *(a) One possible seating arrangement for this screen in this room. Row-to-row spacing is generous, following a manufacturer's guidance, allowing for easy access and egress during a show. This is an option to discuss with the customer, with reference to the viewing angles. (b) A 7.1-channel (3/4) arrangement, with an option to place the rear loudspeakers on the side or rear walls. (c) A 3/6-channel arrangement.*

soundstage width and prevents them from contributing usefully to a sense of space. This is a decision that should involve the customer because with a small picture, the perception of source location is not a big issue, but hearing the benefits of an elaborate multichannel audio system may be. In fact, in some situations the L, C, R separation is so small that it approximates a mono illusion. This author's bias: try to follow the listening-angle guidelines. A big soundstage can make a small picture seem bigger.

If music video concerts, multichannel audio playback (discrete or upmixed), or games are on the customer's entertainment agenda, a wide-front soundstage is likely to be appreciated with any size of screen. Continuing with Figure 16.5b, it can be seen that a four-channel surround configuration was chosen in an attempt to provide a better surround illusion—envelopment—for more audience members. The rear loudspeakers should be driven from a (3 front/4 surround) 7.1-channel processor or receiver, not just connected in parallel with the sides. The suggested angle for the rear loudspeakers was dangerously close to the corner, a location to be avoided for anything except a subwoofer. Consequently, two options are given: side-wall and rear-wall locations. There is little to cause one to be preferred over the other. Each one is so close to the corner that there will be some desirable image broadening from the strong acoustic reflection from the adjacent right-angled surface in any event (assuming that the relevant area has not been covered with absorbing material).

Figure 16.5c proposes a further elaboration: six surround loudspeakers. In large rooms, it is not uncommon to use this many loudspeakers, but normally two pairs of them are connected in parallel. Here, to preserve the impact of rear directed sounds, that would be the two side surrounds. For maximum envelopment, it would be the loudspeakers spanning the rear corners. The proximity of strong reflections in the vicinity of the two sets of loudspeakers at the side/rear of the room would generate a certain amount of acoustical decorrelation of the sounds emerging from each pair because of the nearby right-angled boundary reflections.

Ideally, none of them would be parallel connected, and there would be some form of electronic decorrelation between the channels on each side/rear of the room to enhance the sense of envelopment and to reduce the tendency of off-center listeners to localize the surround loudspeakers. Delaying the channels from one another is a good option. This is normally done in 7.1-channel surround processors for the four-channel surround configuration. An additional delay is necessary for the extra loudspeakers in a 3/6 configuration, something that must be added externally. Expensive custom installations now use multichannel digital signal processors from the professional audio side of the business for equalization, level, and delay duties. These have many channels, two of which might be made available for this purpose. A 10 ms time differential would be a good starting point, but some experimentation is in order to cater to any peculiarities of a particular room. A small amount of high-frequency roll-off might also be appropriate to simulate realistic reflections, much delayed, from the rear.

Because localization is dominated by the direct sound arriving at the listening location, attenuation of high frequencies would also reduce the attraction of these loudspeakers as localizable sources. This was a trick employed years ago in the original Dolby ProLogic system, when imperfect analog film sound tracks caused unwanted leakage of center-channel voice sibilants into the surround channels. The surround channel was low-pass filtered around 7 kHz. The reason why "dipole" surround loudspeakers are less easily localized is because they greatly attenuate the high frequencies in the direct sounds that arrive at listening locations (see Figure 18.20). The effect is that of a very imperfect low-pass shelving filter above about 500 Hz; the negative impact on sound quality is apparent.

None of these measures ensures that listeners seated near the perimeter of the room will get a full sense of envelopment. They may also occasionally localize a surround loudspeaker that is close to them. Imperfections of this kind also happen in cinemas and concert halls; not all seats are equally good.

16.4 DIRECTIVITY REQUIREMENTS FOR THE LOUDSPEAKERS

Moving into the details of acoustical interactions among loudspeakers, listeners, and the room, let us examine the paths taken by direct sounds and first-order reflections as they travel from each of the loudspeakers to the sweet spot, and to listening locations farthest from the sweet spot. The idea is to determine what kind of dispersion pattern is required for each of the loudspeakers in the multichannel array in order, first to create strong senses of location for directed sounds, and second, a strong sense of envelopment when it is required.

16.4.1 Delivery of the Direct Sounds: Localization

The left half of the room layout in Figure 16.6a shows angular dispersion extremes required of each of the loudspeakers so they can deliver strong direct sounds to all parts of the audience. If they cannot do this, any "directed" sounds in a sound track may not be localized correctly. Symmetry allows us to learn what is necessary from looking at one-half of the room.

The answer is simple for the L, C, R front loudspeakers: a nondemanding ±30° if the loudspeakers are aimed at the sweet spot. If they are flat against the wall, the requirement expands to about ±50°, at which angle most loudspeakers will not be performing at their best. Lesson: angle the left and right front loudspeakers toward the audience.

For the side and rear surround loudspeakers, the angular span for delivering direct sounds extends to ±70° to ±75° (considering both sides of the room); this

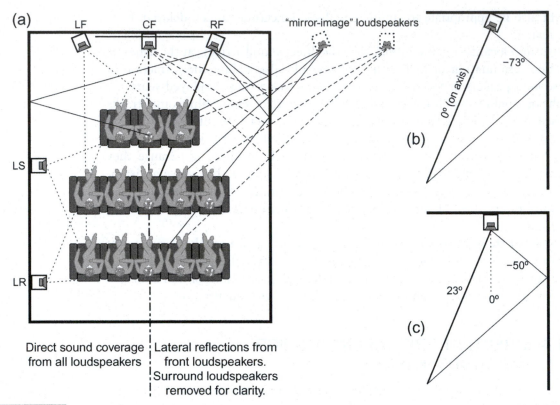

FIGURE 16.6 *(a) The left half of the room layout drawing shows the angular spread required from all the loudspeakers to deliver direct sounds to all of the listeners. The right half shows first laterally reflected sounds delivered by the front L, C, and R loudspeakers to center seats in each of the rows. (b) and (c) The consequences of not orienting the L and R front loudspeakers toward the sweet spot.*

will require some thoughtful loudspeaker design if all listeners within that angular range are to receive similarly loud, similarly good sound. A thought: from this perspective only, aiming the surround loudspeakers at the audience could significantly reduce this angular requirement.

16.4.2 L, C, R First Lateral Reflections

The right half of the room in Figure 16.6a shows first lateral reflections from the front loudspeakers being delivered to the center seats in each row. If the LF and RF loudspeakers are aimed toward the sweet spot, the lateral reflections are created by sounds radiated at angles from about 60° to almost 90° off axis. This is significant for the following reasons:

- First, the sound from normal forward-firing loudspeakers will be less energetic at such large off-axis angles, thus reducing whatever effects they may have.

■ Second, few loudspeakers are well behaved at such large off-axis angles, so any reflections of those sounds might not spectrally resemble the direct sound and thus degrade perceived sound quality (see Figure 8.10, loudspeakers AA and E).

The former will diminish the effect on image broadening/ASW. The second may have an impact on sound quality/timbre. The subjective loudspeaker evaluations discussed in detail in Section 8.2 showed that side-wall reflections of this kind were regarded as contributing positively to the perceived sound and spatial qualities of loudspeakers with normally wide (cone/dome) dispersions. A dipole loudspeaker that, due to its directivity, lacked these reflections was considered to be less desirable. All of this was in spite of some severe off-axis frequency response misbehavior on the part of the preferred cone/dome loudspeakers. As has been discussed several times, it is obvious that listeners *like* the effects of reflected sounds. The issue of sound quality is more complex. In these experiments, done in both mono and stereo, listeners appeared not to be disturbed by less than ideal sound quality contributed by the reflected sounds themselves. In stereo phantom center images, there is the matter of the 2 kHz dip created by the crosstalk. This significant coloration is reduced by reflected sounds, as shown in Figure 9.7.

It is reasonable to assume that absorbing the side-wall reflections would give the same result as using a loudspeaker that does not radiate sound toward the side walls. In this case, the important observations are that wide dispersion loudspeakers and, by inference, strong side-wall reflections were preferred (1) when listening to a single loudspeaker (think center channel), and (2) for hard-panned—that is, monophonic—sounds that originate in the L&R loudspeakers. In stereo, when both channels were actively producing combinations of direct sounds and recorded reflections of those sounds (i.e., decorrelated sounds), there were no observable effects of the lateral—side-wall—reflections in either the sound quality or spatial ratings. So when audible effects due to side-wall reflections exist, they range from negligible to positive, depending on the nature of the program material. The positive effects were that loudspeakers that operated alone were judged to sound better.

Many movie soundtracks deliver essentially stereo music from left and right front channels much of the time (with leakage and recorded reflections in the surrounds), whereas dialogue and on-screen sounds emerge from the center channel. Noting that the center channel spends much of its time operating alone, as a monophonic source, allowing lateral reflections of it to exist may be doing it a favor. However, the effect is likely to be small and not at all related, as is sometimes claimed, to problems with speech intelligibility (Chapter 10). In fact, as discussed in Section 10.3, the greatest degradation to dialogue intelligibility and clarity would seem to be competition from other sounds in the sound tracks themselves, especially for listeners with hearing disabilities (an increasing percentage of our customers). The message from this work is that reducing the levels of all other channels, emphasizing the center, is what works. For them—and, it seems, for most others, too—the audible effects of naturally occurring lateral reflections would be utterly negligible.

The findings of Kishinaga et al. (1979), shown in Figure 8.2, and several other authors mentioned in the associated text all suggest leaving the side-wall reflections intact and/or using wide-dispersion loudspeakers. Only for the special task of mixing recordings do listeners show any consistency in preferring lateral reflections that are attenuated. In any event, as shown in Figures 7.2 and 7.3, reflected sounds that matter enough to be judged to be a seriously positive contribution to the listening experience are at levels so high that they must be in the recordings. Sound levels of naturally occurring first reflections in small rooms are much lower than those at which listeners express opinions of "preference." Nakayama et al. (1971) and Olive and Martens (2007) both noted that the effects of a normally furnished listening room (or the acoustical equivalent) on subjective opinions of multichannel sound quality are small to nonexistent.

The key point in that last statement is "normally furnished." It must be obvious that a certain amount of absorption and scattering of sound is required to eliminate the "empty-room" sound, with its excessive reflected sound field and flutter echoes. However, eliminating these is not difficult; no "heroic" measures are required. The issue is not whether one needs a certain amount of absorbing material or sound scattering-furnishing or devices but how much and where to put them. In the case of the first-order side wall reflections from the front L, C, R loudspeakers, it is difficult to make a strong case for either side. Leaving them appears to contribute a small positive effect for monophonic components of the sound field, where listeners have a chance to hear a direct sound and to correlate early room reflections with it. The center channel is the most likely beneficiary of this. On the other hand, when there are recorded reflections accompanying the direct sounds, the natural room reflections are swamped by the much louder and greatly rewarding recorded reflections.

Lesson: absorbing side-wall reflections of L, C, and R loudspeakers is an option. If the loudspeakers have good off-axis performance, and especially if the customer likes to listen to stereo music, leave some blank wall at the locations of the first lateral reflections from the front loudspeakers. An area with a minimum dimension of about 4 ft (1.2 m) centered on the reflection path is sufficient. It need not be the entire wall height. If the customer only watches movies, it probably doesn't matter. Nevertheless, there are some who will insist on eliminating those pesky reflections as a matter of ritual. The ritual had its origins in recording control rooms—listening in stereo—justified by alarmist cautions of comb filtering (see Chapter 9) or degraded speech intelligibility (see Chapter 10). These are not problems. The real factor appears to be spaciousness and the possibility that recording engineers, like musicians, are many times more sensitive to it and the reflections causing it than ordinary folk (see Section 8.1). They appear to feel that their work is impeded by lateral reflections, but many (most?) of them prefer to have them in place for recreational listening.

All of this may be modified when readers see Figure 21.9, the modifications to the frequency response of sounds reflected from normal 2-in. (50 mm) absorbers. Ask yourselves: is it better to leave a natural lateral reflection that has no known negative effects and some possible positive effects or to grossly distort the reflected spectrum of otherwise excellent loudspeakers using this kind of material? These distortions can be avoided by using *much* thicker absorbers (about 4 in., 100 mm) or even thicker diffusers (8–12 in., 0.2–0.3 m), options that are practical but perhaps not aesthetically suitable for all interior décors.

Sounds reflected from the rear wall are not shown, but it is obvious that they originate from small angles off axis of the loudspeakers and that they arrive at most listening locations from small angles off the front-back axis, meaning that they contribute little to the spatial effect. This would seem to be a useful place to put absorption or a combination of absorbing and scattering surfaces.

Figures 16.6b and (c) describe the difference between the two common orientations of LF and RF loudspeakers. In (b), with the loudspeaker aimed at the prime listening location, the direct sound is the pristine on axis output, and the reflected sound originates far off axis, at 73°. When flat against the wall, aimed forward, the direct sound originates 23° off axis and the reflection at a more energetic 50°. Based on normal behavior of forward-firing cone/dome loudspeakers, (c) is likely to generate more ASW, image broadening, and frontal spaciousness; (b) is likely to present a more compact soundstage, with a reduced sense of soundstage width. In stereo, listeners have been known to prefer each of these for different reasons. For purposes of this multichannel audio discussion, it is not a matter of great consequence because there are the surround channels with nothing much to do except to generate envelopment and create spacious "atmospheres." Home theater is not the single-listener experience that stereo is, so it is important to ensure good coverage of the audience. Therefore, aim the loudspeakers at the audience.

16.4.3 The Surround Loudspeakers—Horizontal Dispersion Requirements

Figure 16.7 looks at requirements for the surround loudspeakers. Don't be put off by the complexity of diagram (a); follow the reflected pathways if you like, but the important information is summarized in (b) and (c). In (b) it can be seen that uniform dispersion over a huge horizontal angle—almost ±90°—is required to deliver similarly good sound in the direct and reflected pathways to all of the listeners, not just the sweet spot in the center of the audience.

Figure 16.7c summarizes what this is all about: delivering a sense of envelopment to the listener in the prime location—the center of the second row. Here, direct and reflected paths from surround loudspeakers on both sides of the room

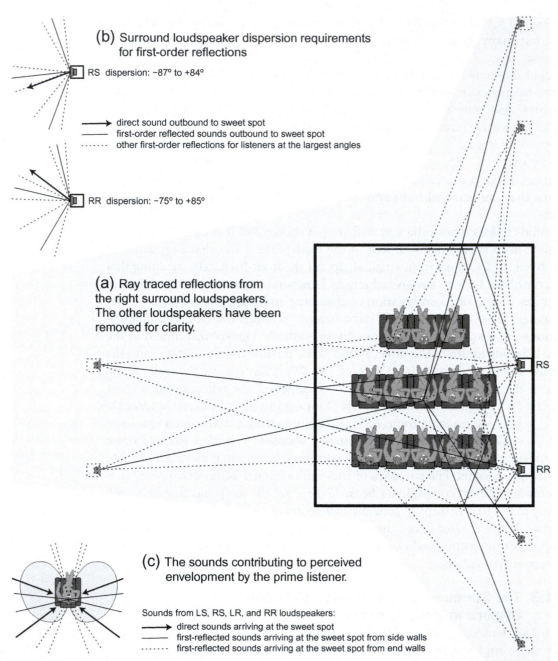

(b) Surround loudspeaker dispersion requirements
    for first-order reflections

RS dispersion: −87° to +84°

→ direct sound outbound to sweet spot
── first-order reflected sounds outbound to sweet spot
······ other first-order reflections for listeners at the largest angles

RR dispersion: −75° to +85°

(a) Ray traced reflections from
the right surround loudspeakers.
The other loudspeakers have been
removed for clarity.

RS

RR

(c) The sounds contributing to perceived
    envelopment by the prime listener.

Sounds from LS, RS, LR, and RR loudspeakers:
→ direct sounds arriving at the sweet spot
── first-reflected sounds arriving at the sweet spot from side walls
······ first-reflected sounds arriving at the sweet spot from end walls

FIGURE 16.7 *(a) For the surround loudspeakers, a horizontal-plane sound-ray diagram showing direct-sound and reflected-sound routes to the sweet spot and to extreme listening locations in the audience area. (b) From (a) some of the important outbound sounds. (c) The sounds that most contribute to a perception of envelopment, employing the "spatial-effect balloon" from Chapter 7.*

are combined. Also shown is the "spatial-effect balloon" from Chapter 7, indicating which of the incoming sounds is most productive at creating the desired impression of envelopment. It is evident that the main contributions are sounds that arrive directly from the surround loudspeakers and those that arrive after a single reflection from the opposite side walls. Reflected sounds arriving from the front and back walls fall within the angular range that does not contribute to much in the way of interaural differences, and therefore they contribute little to the desired spatial illusions. Bidirectional out-of-phase surround loudspeakers ("dipoles") are intended to use the end-wall reflections and therefore are not optimal designs.

As a side note, it needs to be mentioned that if the screen is acoustically transparent, it is normally recommended that the wall behind the screen be treated with absorbing material. If so, the ray diagram shows that several of the potential front-wall reflections will pass through the screen, encounter the absorbing material, and disappear. In fact, as a general rule, it is a good idea to put absorbing material on the front wall. As noted in Section 8.1, Kishinaga et al. (1979) observed that it improved image localization and reduced coloration. It is of no real consequence to the operation of the surround loudspeakers, as the perception of envelopment is created by the direct sounds from the surround loudspeakers and the lateral reflections from the opposite side walls.

As shown in Figure 8.6, and discussed there, the surrounds perform two jobs:

■ *They provide momentary localizable sound effects in films and stationary localizable effects in music.* For these tasks, natural reflections may or may not be the only spatial accompaniment to the sound. This is of more consequence for the sustained sounds of music than for the momentary sounds in movies. Recording engineers should provide deliberate reflected and/or reverberant sounds in the other channels to provide impressions of distance and/or space for these directed sounds. When this is not done the effect is of a sound originating in a nearby loudspeaker. The result is that listeners are surrounded by sounds, but not placed in a spatial context.

■ *They reproduce delayed versions of sounds that originate in the front soundstage (reflections in the recording).* In this, which is a primary function of the surround channels, reflections from the opposite wall can assist the envelopment illusion. Be very careful to avoid creating an opportunity for flutter echoes between the opposing walls. Because it is lower frequencies—approximately 1 kHz down to 100 Hz—that generate envelopment (see Figure 7.1), broadband diffusers alone or in combination with reflecting surfaces would be suitable. Reflecting surfaces would generate the most energetic reflections, but at the same time they address a restricted listening area and they create the highest risk of flutter echoes. A secondary function of these strong early reflections is to provide more directional distraction from the surround loudspeakers, perhaps making them less localizable by listeners seated near the perimeter of the room.

16.4.4 Outside the Sweet Spot: The Effect of Propagation Loss

Up to this point in this book, and in most of the research that has been done, a critically important assumption has been made: what the listener in the sweet spot hears is most important. Few investigations have probed the changes in perceptions and preferences as the listeners move away from the point or axis of symmetry in the loudspeaker array. We know from decades of experience with stereo that the front soundstage deteriorates very quickly, with lateral movement away from the symmetrical axis. In multichannel audio systems, there is the center channel to stabilize the position of certain sounds, but others are susceptible to lateral movement, and the balance between front soundstage and surround effects is altered by forward-back movements. When systems are calibrated, they are calibrated in sound level and timing from the various channels for only one seat: the sweet spot.

Sitting in the right location makes a difference, and we routinely gravitate to that location if given the opportunity. Recording and mastering engineers belly up to the center of the console for any critical listening; others hover with their heads just behind and overhead to share the experience. Yet cinemas and home theaters are sometimes filled with as many seats as they can physically hold. We know that all of those seats are not equally good, but we try not to talk about it.

Large rooms have advantages. As shown in Figure 4.12, the transition frequency falls with increasing size until, in large cinemas and concert halls, it becomes so low that there are no consequential standing-wave, room-resonance effects. Bass is very similar everywhere. All of the discussions in Chapter 13 apply to small rooms and the mighty struggle to deliver similarly good bass to all listeners. There, solutions were found. Now we turn our attention to delivering the rest of the spectrum and its directional and spatial effects to all listeners.

Figure 4.13 shows the decline in steady-state sound level as a function of distance from several different sound sources in several different rooms. They all decline at about −3 dB per double-distance. This is steady-state sound, a combination of direct and all reflected sounds from all directions at all delays. It represents something about what we hear but by no means all of it. It mainly tells us that the sound field is not diffuse, that there is a persistent source-to-sink loss; energy is being depleted as we move away from the source. It is being lost to absorbing materials, accelerated by large scattering objects in the space. This is not bad; it is merely a fact.

Also a fact is that the direct sound component radiated by loudspeakers is attenuated with distance. Direct sounds are important because they determine localization for all of the real and phantom sound images in the artistic panorama. They also precondition the precedence effect, the perceptual phenomenon that allows us to localize with precision in acoustically reflective rooms. They hold a special place in the perception of sound quality but not an exclusive place. Reflected sounds are also important. In establishing the perceptions of space (image broadening, ASW, and envelopment), it has been shown in experiments done in anechoic chambers that direct sounds alone are sufficient to generate convincing impressions. The envelopment of a concert hall can be persuasively simulated by direct sounds from as few as five loudspeakers located appropriately (see Figures 15.8 and 15.9). But, again, the experiments addressed the needs only of a single listener in the sweet spot; the sounds arriving at the ears from left and right directions (it is the lateral sounds that matter most) had the same amplitudes. If all listeners in a room are to have the same perceptions, it is necessary that they hear the same sounds. Propagation times to different listening locations prevent that from being absolutely possible. However, for the sustained, repetitive, decorrelated sounds that create impressions of envelopment, the amplitude decline with distance (propagation loss), is possibly a more significant factor in achieving the desired low interaural cross-correlation (IACC) for off-center listeners.

The introduction to Chapter 18 explains these phenomena, but for now, it is sufficient to say that propagation loss refers to the rate at which the direct sound is reduced in level as we move away from a source. The energy is not lost in the sense that it dissipates in the air. That happens, but it is a tiny effect in small rooms. The reduction in level we are talking about has to do with the rate at which the area of the sound wavefront expands as it propagates away from the source. For a source that is small compared to the measuring or listening distance, called a point source, the attenuation rate is about 6 dB per double-distance. The key point is that the sound level drops very rapidly close to the source but less rapidly at greater distances. For example, moving from 4 ft to 8 ft, the level drops by 6 dB, which is the same amount it falls when moving from 20 ft to 40 ft. Live human voices are approximate point sources. So are most conventional cone/dome loudspeakers used in front and surround-channel applications, even multidirectional versions.

In contrast, a closely packed line of small transducers running from floor to ceiling would, along with the acoustical reflections at the floor and ceiling, approximate a line source (see Section 18.1.2). This kind of source radiates a direct sound that declines at 3 dB per double-distance. Sometimes vertical arrangements of a few transducers are called line sources. They aren't; they are simply tall conventional loudspeakers. Real-world loudspeakers and sound sources tend to fall somewhere within this range of 3 to 6 dB per double-distance, and to make things worse, they sometimes exhibit different attenuation rates at different frequencies.

Figure 16.8 shows how this affects the sense of envelopment in small rooms. In (a) we look at five seats in the middle of a large cinema. At such large distances from the left and right surround loudspeakers, the sound levels change little from one seat to the next. When the differences between sound arriving from left and right are considered they also are not great. Of course, there is no difference in the center seat—the sweet spot. All of these listeners probably get a decent impression of envelopment.

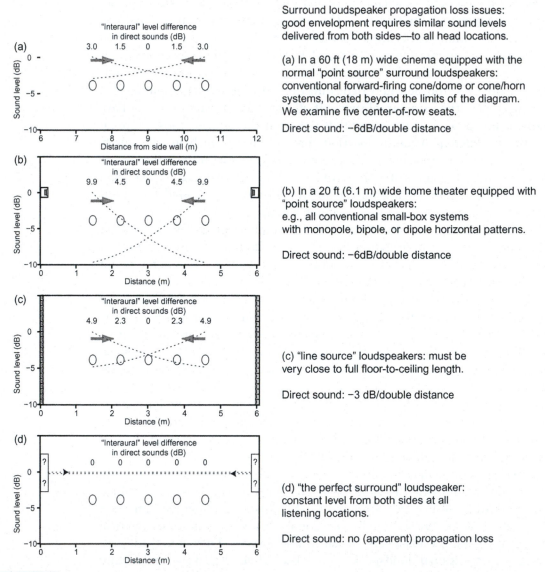

Surround loudspeaker propagation loss issues: good envelopment requires similar sound levels delivered from both sides—to all head locations.

(a) In a 60 ft (18 m) wide cinema equipped with the normal "point source" surround loudspeakers: conventional forward-firing cone/dome or cone/horn systems, located beyond the limits of the diagram. We examine five center-of-row seats.

Direct sound: −6dB/double distance

(b) In a 20 ft (6.1 m) wide home theater equipped with "point source" loudspeakers: e.g., all conventional small-box systems with monopole, bipole, or dipole horizontal patterns.

Direct sound: −6dB/double distance

(c) "line source" loudspeakers: must be very close to full floor-to-ceiling length.

Direct sound: −3 dB/double distance

(d) "the perfect surround" loudspeaker: constant level from both sides at all listening locations.

Direct sound: no (apparent) propagation loss

FIGURE 16.8   (a) The anticipated sound levels arriving at five listening locations halfway across the width of a cinema, from conventional surround loudspeakers to the left and right. (b) The comparable situation in a home theater. (c) The effect of changing the surround loudspeakers to floor-to-ceiling line sources. (d) What we really want.

In Figure 16.8b, the only thing that has changed is that we have moved to a normal-sized home theater. Because of the proximity of listeners to the loudspeakers, the "interaural" left-right differences are now much larger. It is still zero for the lucky person in the sweet spot, but it is very large for those forced to sit close to the perimeter of the room. For them, unfortunately, the nearly 10 dB left-right difference will result in little sense of envelopment, and the sound will be localized to the nearby loudspeaker. In the days of a monophonic surround channel, this was a great affliction, leading to the introduction of electronic decorrelation in the signals fed to the left and right surround loudspeakers in Shure HTS surround processors, and later in THX licensed processors "to diffuse the rear image and discourage localization at the closer surround loudspeaker" (Julstrom, 1987). With discrete multichannel audio, this decorrelation is now—or should be—incorporated into the master recording through the use of electronic processing or multiple microphone pickups in large acoustical spaces.

This is the situation that led to the promotion of multidirectional surround loudspeakers as a means of sending more of the sound in other directions, enhancing the "diffuse" sound field. The problem is that the diffuse sound field in small normally furnished rooms is not strong; there is too much absorption, and sounds reflected from the front and rear walls arrive from the wrong directions to be effective at generating impressions of space (see Figure 8.6). If the design follows the well-known dipole surround loudspeaker configuration (more correctly called a bidirectional out-of-phase loudspeaker), then there is the additional consideration that the null is oriented in the direction that is most productive for creating envelopment. This has the effect of turning down the level of the useful (lateral) direct sounds and replacing them with less-useful (front-back) indirect sounds. Because the null, such as it is, exists only at frequencies above about 500 Hz, the effect is reduced treble. As a result, the loudspeaker may be less easily localized, but the desirable spatial effect, even for the sweet spot, has been compromised. Figure 18.20 illustrates that there is another problem with this loudspeaker configuration: sound quality.

Thinking creatively, there are other solutions. It seems clear that what we need is a means of delivering sound across the width of a room without the dramatic attenuation rates of point sources. The line source comes to mind, and this option is shown in Figure 16.8c. The situation is much improved—almost as good as in the middle of the cinema. Those of you who have heard line source loudspeakers will no doubt recall the remarkable experience of slowly approaching one from a distance and noting that it doesn't really sound "loud" when you are close to it. That is because the sound energy is radiated over the entire length of the unit, so no one portion ever gets very loud. This is the kind of solution we are seeking, but line sources are expensive to build because they require a tightly packed vertical array of small transducers. Line loudspeakers that are less than floor-to-ceiling in height, called truncated line sources, behave in a much less orderly manner, unless they are curved or "shaded" (the output is gradually reduced toward the ends). Figure 18.3 shows examples of this. In common parlance, vertical arrangements of several drivers in a tall box are also called line sources. They aren't. They are simply tall, ordinary, loudspeakers with untidy directional properties.

Figure 16.8d shows what would be ideal: a loudspeaker design capable of delivering a constant sound level across the whole width of the room. The following chapter suggests what might get us close to this ideal. It has yet to reach the marketplace, but such a design seems to be plausible. It is likely to be more expensive than simple box loudspeakers, even multidirectional ones, but it is reasonable to think that fewer would be necessary. As stated before, it is not necessary to surround listeners with many loudspeakers to create the perception of being enveloped by sound. It is important only that the right sounds arrive at the two ears. The number of loudspeakers is really dictated by how many independent directions sounds will be steered to in the program material. Right now that is a total of five or six.

And now, what about the front L, C, and R loudspeakers? Surely a reduction in the sound attenuation rate from front to back in a room would also be desirable so all listeners in all rows receive a similar sound level and maintain a similar front-to-surround balance. Indeed, it is, and these design options apply there as well. Figure 16.9 shows sound levels for point-source front loudspeakers and the decline in steady-state sound level for a point source, including all reflections; it is identical to the decline of the direct sound from a line source. In a normal program, what is perceptually important would likely be a combination of the two.

The attenuation rate of the direct sound from a line source is clearly an improvement over that of the point source. Other array options may be equally rewarding. Of concern, though, is that the nature of the front soundstage may be changed by the vertically extended sources having very wide horizontal dispersion. Add the low (as low as −3 dB/dd) propagation loss, and reflected sounds will be significantly louder. There is anecdotal evidence that such loudspeakers are beneficial to stereo but in a multichannel system, experimentation may be required to determine whether they are suitable for the front L, C, R channels.

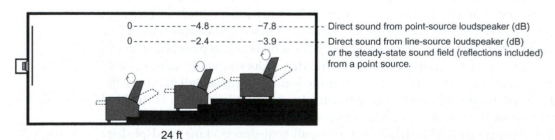

FIGURE 16.9 *Sound levels at listening locations for front loudspeakers of point- and line-source configurations. The 3 dB/double-distance attenuation rate for the line source is also the same as is normal for the steady-state sound field from a point source in a typical listening room (see Figure 4.13).*

The greatest challenge is likely to be the center channel, from which the bulk of the dialogue originates. A point source loudspeaker should be a better imitator of a point-source human voice, but, in the context of a movie, it may be perceived differently. Nevertheless, the evidence suggests nothing but benefits for surround channel applications. It is tantalizing to think that at this stage in the mature audio industry, there may be novel loudspeaker designs delivering performance advantages.

16.5 A SUMMARY OF LOUDSPEAKERS AND THE ACOUSTICAL TREATMENTS IN ROOMS

16.5.1 LF, CF, and RF Loudspeakers

These exist primarily to create the front soundstage and to deliver dialogue. Direct sounds are delivered by sounds radiated within a horizontal angular range ±30°. As summarized in Figure 16.10a, this is easily achieved by conventional cone/dome or horn designs, although some large-panel designs will have problems (see Figure 18.14, speaker M). The common simple MTM horizontal-center designs present a serious problem. See Figure 18.18 for a description of the problem and a solution; such loudspeakers *need* a centrally located midrange driver. Side-wall reflections from the CF loudspeaker originate in the angular range from about ±48° to about ±70°, which may contribute moderately to ASW/image broadening. Since this channel operates alone a substantial part of the time, one could argue that this is desirable. The LF and RF loudspeakers produce side-wall reflections from very large angles off axis, especially when, as they should be, they are aimed at the prime listening position. This means that, due to the forward-biased directivity of typical cone/dome forward-firing loudspeakers, these sounds will be significantly attenuated at middle and high frequencies. There will be a contribution to spatial effect and, surprisingly, also to perceived sound quality when these loudspeakers are fed hard-panned sounds. In professional environments, like recording control rooms, it has been common practice to absorb these side-wall reflections from the front loudspeakers. As discussed in Chapter 8, most recreational listeners have voted that they enjoy them in stereo reproduction. In a multichannel context, the matter is open for discussion. If the surround channels are active, it is probable that the modest spatial contributions of these front-channel reflections will be masked. If only the front channels, especially the center channel, are active, it is possible that a small spatial effect may be beneficial. In the grand scheme of things, these are factors but not the dominant factors. This opens up some options:

- Use wide-dispersion loudspeakers with relatively constant directivity that can supplement ASW/image broadening without damaging sound quality.

- Absorb the first sidewall reflections. According to the ray diagrams in Figure 16.6a, this could require covering most of the front half of the

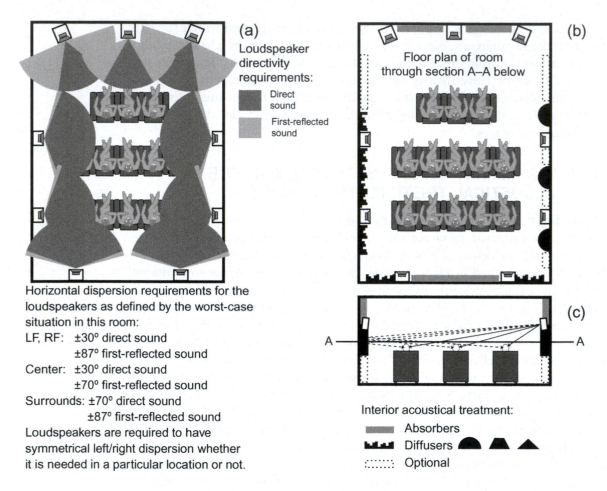

(a)

Loudspeaker directivity requirements:

■ Direct sound

■ First-reflected sound

Horizontal dispersion requirements for the loudspeakers as defined by the worst-case situation in this room:

LF, RF: ±30° direct sound
±87° first-reflected sound

Center: ±30° direct sound
±70° first-reflected sound

Surrounds: ±70° direct sound
±87° first-reflected sound

Loudspeakers are required to have symmetrical left/right dispersion whether it is needed in a particular location or not.

(b)

Floor plan of room through section A–A below

(c)

A ——— A

Interior acoustical treatment:

▬ Absorbers

▬▬▬ Diffusers ◖ ◣ ▲

∷∷∷∷ Optional

sidewalls with resistive absorbers at least 3 to 4 in. (76 to 102 mm) thick. Thinner absorbers absorb only a part of the spectrum and degrade sound quality. Only an area centered at ear level needs treatment, having a smallest dimension of about 4 ft (1.2 m).

■ Use directional loudspeakers, such as large horn units, which can reduce the amount of sound sent in the direction of the side-wall reflection points. Panel dipole loudspeakers could have their nulls oriented in the correct direction, so these are an option, but remember that 50% of the sound from dipole loudspeakers is radiated away from the listeners and will reflect off the front wall unless that is absorbed using materials as previously described, for the same reasons. Remember also that large panels, unless they are progressively subdivided into smaller areas at higher frequencies, can be very directional.

■ Never use simple midrange-tweeter-midrange horizontal center loudspeakers because of their atrocious off-axis performance. Those with

FIGURE 16.10 *(a) A summary of the horizontal-plane angular dispersions required of the loudspeakers to deliver direct sounds of comparable quality and level to all listeners and to deliver sounds to the wall surfaces from which the first reflections occur. The propagation loss due to the inverse square law will inevitably cause differences in level at different distances. The criterion of excellence for direct sounds (the darker shaded angular range) is that they should all be as similar as possible to the on-axis performance of the loudspeaker. This is obviously a challenge for the surround loudspeakers because of the very large, almost 180° dispersion, required of these units. Requirements for performance within the expanded angular range embracing first reflections can probably be relaxed, but it is not known by how much.*

(b) The horizontal plan for room acoustical treatment based on concepts discussed earlier in the book. The materials described here apply to a horizontal band around the middle of the room, around and above seated ear height. The front side walls are "optional" territory, meaning that one can do nothing (reflect), absorb, or diffuse. Evidence indicates that these reflections have a positive effect on subjective impressions of loudspeakers in mono or stereo. However, when multiple channels are simultaneously active, these natural reflected sounds may be masked and their effects diminished. The absorbers on the front and rear walls avoid reflections within the angular ranges that contribute little to the perception of envelopment. The diffusers along the side walls provide reflections of sounds from surround loudspeakers on the opposite walls, from directions that aid in the perception of envelopment. Envelopment is most influenced by sounds in the 100 Hz to 1 kHz frequency range—the lower frequencies. Therefore, these diffusers should be about 12 in. (0.3 m) deep if they are geometric shapes (as illustrated on the right wall) and about 8 in. (0.2 m) for well-engineered surfaces (as illustrated on the left wall). The corners are available for low-frequency absorbers if they are needed.

(c) The room in elevation, with the diffusers placed at the appropriate elevation to redirect sound to the listening locations (determined by geometry or mirrors). A height of 36 to 48 in. (0.9 to 1.2 m) should be sufficient for this treatment. The space above can have patches of absorption. The space below is "optional" because sounds reflected from this portion of the wall will likely be captured by the carpet or the seating.

midrange units to accompany the tweeter in the center can be excellent. See Figure 18.18.

It is difficult to ignore the small benefits (without apparent disbenefits) of using normal forward-firing loudspeakers with wide dispersion, good off-axis behavior, and allowing the relevant areas of side walls to reflect. However, in a multi-channel context, this is an issue where the customer and/or the consultant can express some free will.

16.5.2 The Surround Loudspeakers

These channels provide impressions of both direction and space. In movies, sounds are momentarily localized to off-screen locations. The locations are noncritical; general left-right, front-back directions are sufficient, which is all that is possible in cinemas. In music, they may be called upon to provide stable images of vocalists, musical instruments, and so on, and again, great precision is not a requirement. Most of the time, these loudspeakers provide ambient sounds, atmospheric mood-setting music, and other sounds that are intended to float ambiguously in space. This illusion is associated with the impressions of spaciousness or envelopment that are experienced in concert halls, and the same physical and psychoacoustic rules apply.

The notion that envelopment can be perceived in a highly diffused sound field like a concert hall has led to a belief that a diffuse sound field is a requirement. It isn't. An impression of spaciousness or envelopment exists when the sounds arriving at the listeners' ears have random differences in amplitude and time—that is, they are uncorrelated, a condition that is encouraged by sounds arriving from the sides and not by sounds arriving from angles close to the front-back axis (see Chapters 7, 8, and 15).

In the present example, to deliver the direct sounds, the surround loudspeakers require well-behaved horizontal dispersion out to ±75°, and delivering first reflections may require dispersion out to ±87°. Looking at the ray diagrams in Figure 16.6b, it is clear that the dispersion needs to be uniform over the entire angular range. A simple technical description of such a loudspeaker would be a horizontally omnidirectional (*very* wide dispersion) on-wall or in-wall loudspeaker.

16.5.3 Propagation Loss

It has been shown that conventional "small box" loudspeakers are incapable of delivering similar surround experiences to all members of an audience in a small room. There is the challenge of covering the angular range just described, which is addressed with well-executed existing designs. There is also the problem that the direct sounds from the left and right surround loudspeakers are so attenuated in traversing the width of the room that, for listeners away from the central axis, the envelopment illusion is seriously diminished. And finally there is the fact that for these off-center listeners the adjacent surround loudspeakers may be localizable sources. At the present time, multichannel audio systems are optimized for listeners in the middle of the room. The degradations one hears in moving away from the sweet spot are not as dramatic as they are in stereo, but they are real.

There are solutions based on loudspeaker designs capable of delivering more similar sounds to more listeners. A conventional floor-to-ceiling line source is one. In the next chapter we will discuss other possibilities. All of them seem like significant improvements on the present situation.

In the evolving story thus far, we have defined the essential directional (sound dispersion) requirements of loudspeakers for front and surround duties in a multichannel system. We have not yet discussed the core issue: the fundamental requirement for *any* loudspeaker—that it should *sound* good. The following chapter opens with that discussion and leads on in the following chapter to merging the requirements for good sound with those of directivity, sensitivity, power handling, and other important characteristics.

CHAPTER 17

Loudspeakers I: Subjective Evaluations

In a book about sound reproduction, it may seem strange that it has taken 16 chapters to get to the topic of loudspeakers. Well, the reason is it took that long to understand what it is that loudspeakers need to do; they are a means to an end. Loudspeaker systems and rooms have traditionally been designed in isolation. As if by magic, the loudspeaker is expected to sound good in a small reflective space of uncertain size and acoustical properties, with it and listeners positioned with minimal understanding of the acoustical consequences.

Of course, people know that the room is part of the process, but the details of the relationship have been hard to tie down. In the trial-and-error process of traditional loudspeaker design, countless hours of concentrated listening have been devoted to "voicing" the loudspeaker. The result? It might sound "good" for the program material favored by that listener in that particular room, but change the room, and parts of the experience change. Change the recording, and still other parts of the experience change. It has been a great problem, causing much confusion about what is responsible for the sound quality we are judging. Historically, the loudspeaker has carried most of the burden. We now know that above the transition frequency the responsibility is justified. But in the very important bass region, the room and the arrangement of listeners and loudspeakers within it dominate, so no loudspeaker design has any chance of sounding the same at all frequencies in all rooms.

Add to these uncertainties the notion that what we like—taste—is a personal thing. One man's meat might be another man's poison, so the saying goes. Ask people at an audio show which loudspeakers they think are best, and you will get many different answers. Are our preferences in sound quality as distinctive as our preferences in "wine, persons and song," to paraphrase another saying?

While listening to products in stores and the homes of friendly audiophiles, people are participating in informal experiments. The experimental variables—loudspeaker, room, music, mood, price, size, brand, and whatever verbal chatter is happening at the time—are mingling and merging. Whatever opinion emerges from the amalgam of influences might have a relationship to the inherent capabilities of the loudspeaker . . . or it might not. It is impossible to know. These exercises set out to assess personal preference in sound but end up being influenced by personal susceptibility to many nonauditory influences. No malicious intent need be assumed; it is just what happens. Any competent audio salesperson knows that most customers can be "persuaded" by the right kind of presentation and a well-orchestrated demonstration.

Along with the personal preference argument is the one asserting that people who live in different parts of the world have distinctive needs. It extends also to beliefs that different kinds of music may need different spectral balances. If these were so, there would be additional controls on audio components or loudspeakers, as shown in Figure 17.1.

As discussed in Chapter 1, the word *reproduction* implies that somewhere there was an "original," and it is the task of the sound reproducing system to emulate that original. It is well understood that a perfect three-dimensional acoustical replica of a live performance is simply not feasible, and recordings have gone on to create their own artistic and abstract interpretations of reality. As discussed in Chapter 2, our real goal is to "connect with some of the key underlying perceptual dimensions" so artists and listeners can go beyond the limitations of small playback spaces. In the intervening chapters, it has become clear that most of us find considerable pleasure in impressions of direction, space, and envelopment. This much we have in common. Let us now examine the matter of sound quality.

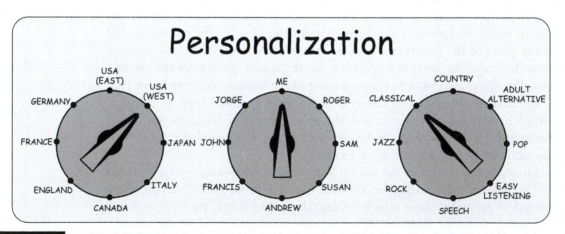

FIGURE 17.1 *Certain beliefs would have controls of this kind on loudspeakers. One is as silly as the other.*

17.1 THE GENESIS OF A LIFE'S WORK

This is where the story becomes personal because much of the content that follows derives from research done by the author and his colleagues over the past 40 years. Acoustical measurements, and the devices to do them, were well developed by the time I became a research scientist at the National Research Council of Canada, in Ottawa, as a fresh PhD EE graduate. It was April 1965. My PhD research had been on the topic of sound localization and, in particular, the manner in which sounds at the two ears were processed by the brain to yield perceptions of direction (Sayers and Toole, 1964; Toole and Sayers, 1965a, 1965b). All of the experiments to that point had been done with headphones, which allowed signals to each ear to be controlled independently.

A thrilling prospect of the new job was that there was an excellent anechoic chamber, within which the research could be extended to include listening under natural circumstances, starting in a reflection-free environment. For this, loudspeakers were needed. When anechoic measurements were made on some highly-rated audiophile loudspeakers of the time, the results were depressing. Most of the frequency responses were far from flat, and these were simple on-axis anechoic measurements made for the purpose of performing anechoic listening tests. Up to this point, the author had only seen "specifications" for frequency response, and if it were not for the unimpeachable pedigree of the measurement circumstances, it would have been possible to think that there had been a tragic error in making the measurements. Suddenly, claims that the loudspeaker was the "weakest link" in the audio chain rang true. But could these products really sound as bad as some of the curves looked?

A logical "Friday afternoon" experiment was to do a simple comparison listening test in one of the laboratory rooms. Having learned the basics of experimental psychology for the thesis work, it was obvious that this test had to be somewhat controlled. So cotton sheeting was hung up to render the experiment "blind." The sounds being compared in the monophonic A/B/C/D comparisons were adjusted to be equally loud. There was no statistical imperative for listening in groups of four; it just seemed convenient. Interestingly, four-way multiple comparisons have remained the norm in our subjective evaluations ever since. A supportive technician built a simple relay switch box. After that, I and a few interested colleagues took turns sitting in, forming opinions and making notes. A "Gestalt" impression, a summarized overall rating, was required: a number on a scale of 10.

The results surprised all of us. The audible differences were absolutely enormous, but there was general agreement about which ones seemed to sound good. It remained a topic of discussion for days. The need for loudspeakers for my anechoic sound localization experiments remained, and the winner of this simple test showed promise. It was dismantled and some improvements were made. The experiments proceeded.

Many months passed before another listening test was staged in early 1966. By then I had learned that bed sheeting is not acoustically transparent and that music passages needed to be short and repeated, which meant that we needed a disc jockey (remember this was the LP era). That would be the agreeable technician. Word had spread and audio enthusiasts from within the organization lined up to participate and in some cases to bring their personal loudspeakers to be evaluated. This test went on for several days and yielded enough subjective data to warrant rudimentary statistical analysis. Again, there was good agreement about the products that were preferred and those that were not. The winning loudspeaker was the redesigned unit that was being used in the anechoic chamber tests. It also had the best-looking set of measured data, assuming one puts any value in smooth and flat frequency responses on- and off-axis (see Figure 17.3b).

Figures 17.2 and 17.3 show several loudspeakers that were used in listening tests in that period of time. My archives still have some of the handwritten listener notes from these tests, done 42 years ago! It is interesting to look them over and, with the benefit of hindsight, to comment briefly on these products. First, it is obvious that they sounded *very* different from one another. These were among the first loudspeaker measurements I ever made, and a standard format had not yet been established, so some of the measurements go to 45° off axis and others to 60°.

I discovered KEF at an audio show in London (Figure 17.2a) while I was there as a student. I liked what I heard and purchased two kit versions of the KEF Concord to be the nucleus of my first postgraduation stereo system. In these tests it was judged to have good overall balance but some midrange coloration. This was found to be caused by flexure along the long dimension of the woofer diaphragm (a breakup mode) above 1 kHz, occurring in the crossover region to the tweeter. The large (1.5-in. diameter) tweeter became directional above about 4 kHz, and this was noticeable as a slight dullness.

The Acoustic Research AR-3 (Figure 17.2b) was famous for its novel acoustic suspension woofer, and it came to be one of the reference loudspeakers of that generation. Its acoustic performance was well documented in the literature (e.g., Allison and Berkovitz, 1972), which was a great credit to the company. A major design goal was to achieve constant directivity, and they did well; the only minor exceptions are the woofer beginning to beam as it approaches the crossover to the midrange and the tweeter at very high frequencies. The essential issue with this product is its frequency response, which significantly rolls off toward both low and high frequencies. Low-frequency output would be aided by boundaries (see Figure 12.5 for boundary interaction data). Almost all listeners found it to be slightly "dull" sounding, and some identified coloration around 1 kHz.

The dual-concentric Tannoy (Figure 17.2c) had a strong following, attracted by the logic of all of the sound emerging from a single point. However, this example indicates that there were problems putting the theory into practice. The horn portion of the system, above about 2 kHz, had serious dispersion issues. Directivity increased progressively, with a good deal of acoustical interference, and possibly some resonances. The result was that listeners complained about shrill highs, sibilant voices, "metallic" strings, and the like. The woofer in the large enclosure exhibited underdamped behavior (the bump around 100 Hz), but this was not commented on by listeners.

The Quad Electrostatic, Mark 1 ESL (Figure 17.2d) was designed in 1957. It must have been a revelation at the time. Its on-axis performance was, and is, exemplary. Many audiophiles found great pleasure listening to it in circumstances where the direct sound field was prominent. However, electrostatic loudspeakers have limited diaphragm displacement. Large diaphragm areas are necessary to compensate for this, and directional behavior therefore tends to be problematic. In a typical reflective room, this loudspeaker did not sound as good as the on-axis curve suggested. Its successor, the ESL-63, had much improved dispersion, as can be seen in Figure 8.10 (loudspeaker "BB").

The Wharfedale W-90 (Figure 17.3a) was a good example of a "big box" loudspeaker: multiple largish drivers (leading to acoustical interactions and interference) and high sensitivity (then often meaning sacrificing low bass). The fashion faded quickly, and with good cause, as these curves suggest. The directional problems extend down to below 500 Hz, with evidence of resonances and massive acoustical interference above that. Words cannot describe the difference in sound between this and the loudspeaker that follows.

The author's research loudspeakers, the redesigned KEF system, are shown in Figure 17.3b. Although optimized for anechoic tests—that is, on-axis performance—they were clearly able to sound good in reflective spaces. This

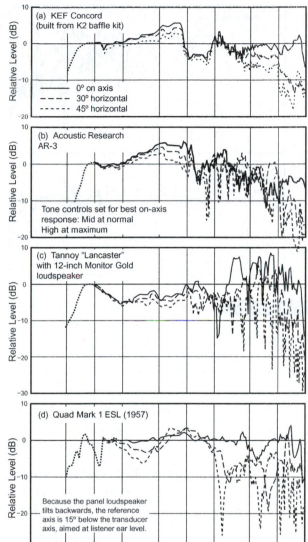

FIGURE 17.2 *On- and off-axis anechoic measurements on four loudspeakers that were highly respected in the mid-1960s. Dotted curves at low frequencies indicate uncertain accuracy due to anechoic chamber errors.*

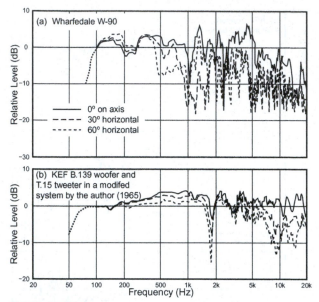

FIGURE 17.3 *On- and off-axis anechoic measurements on a mid-1960s loudspeaker and on the redesigned unit that was created for use in anechoic chamber experiments. Dotted curves at low frequencies indicate uncertain accuracy due to anechoic chamber errors.*

loudspeaker was preferred over all the preceding designs at the time. The implication is that smooth flat axial frequency responses and well-behaved off-axis performance combine to yield reproduced sounds that many listeners like. That is a conclusion that has only been reinforced in subsequent years.

This happened in the late 1960s. It remained a background activity for several years. I read what I could find on the topic (not much), and a lot of that was more anecdotal than scientific. The BBC Research Department had done some fine insightful work—for example, Shorter (1958)—but budget cuts slowed further progress. Professional psychoacousticians, most of them in universities, nibbled around the edges of the real problems in sound reproduction, but none chose to engage the topic directly. Phone calls and visits to some prominent loudspeaker designers found that all were very cooperative and candid. There were more questions than answers and more opinions than facts. This was a field that could use serious scientific investigation, but the industry was too fragmented to be able to mount such an effort on its own.

On the measurement side, there were three distinct camps:

1. The "on-axis" school of design, based on the thought that the first sound to arrive at the ears had a dominant role not only in localization but also timbre. The important information was thought to reside in the on-axis frequency response curve. Reflected sounds in rooms, according to this idea, would be perceptually suppressed.

2. The "sound-power" philosophy took the opposite view, making the assumption that the listening room is a highly reflective space with a sound field that is sufficiently diffuse that all of the sound radiated from a loudspeaker in any direction is fully integrated. According to this idea, an integration of all sound radiated in all directions, a single curve, would be the perceptually dominant factor.

3. The "room curve" viewpoint took into account the frequency response and directivity of the loudspeaker, as well as the reflective properties of the room. The logical argument was that this combination was what we heard, and therefore this single curve must be the principal metric.

The problem with this argument is that two ears and a brain respond differently to a complex sound field than does a microphone, which simply adds all sounds together without regard for the direction from which they arrive or how much time has passed since the direct sound. This philosophy disregards notions of precedence; binaural discrimination; the spatial effects of early, late, lateral, and vertical reflections; and temporal integration phenomena.

Such is the problem of seeking a single curve that embodies all of the truths; the real situation is much more complex. An associated problem was that measurements were often done using 1/3-octave filter sets (filters with fixed-center frequencies producing a staircase, not a continuous, curve). The poor frequency resolution limited the ability to see narrow-band irregularities, and the histogram form of displaying spectral data did not lend itself to revealing more than the crudest trends. Toole (1986) has a detailed review of this historical background.

Looking back, the data that were used in those days could not be reliable indicators of sound quality in rooms. This very likely was the origin of the belief that "we cannot measure what we can hear." Indeed, if the measurements lack adequate frequency resolution and not enough measurements are made, the belief is fulfilled.

As a result, a lot of listening went into the design process. Listening tests tended to be fairly casual affairs, with participants almost always knowing what was being auditioned. Great weight was placed on the opinions of musicians or regular concertgoers (in those days, this meant classical, acoustic concerts). The source material was usually LPs, themselves subject to significant imperfections and variability. In such a climate of uncertainty, it could not be surprising that many loudspeakers of indifferent quality were being produced.

In the mid-1970s, the project became a foreground activity for the author, time-shared with other tasks. This was a fruitful period. Around 1986 the Athena project was established, a partnership between the National Research Council of Canada and a nonprofit consortium of five Canadian audio manufacturers, to help fund the effort. The mission was to explore the interface among loudspeakers, rooms, and listeners. That project wound up around the time, in 1991, when I joined Harman International Industries, Inc., where I established a corporate research group that has continued to add to the base of scientific knowledge. (I retired in 2007. It has been a very gratifying career.)

The following discussion of loudspeakers will be presented in two parts: subjective evaluations (Chapter 17) and objective evaluations (Chapter 18). Along the way, there will be overlap and interaction because of the final objective: to identify the measurable quantities that correlate with listener opinions. Finally, all of the knowledge is put to the test, to find a means of predicting listener opinions from an analysis of measured data (Chapter 20). It is an interesting story.

17.2 SUBJECTIVE MEASUREMENTS OF LOUDSPEAKERS—TURNING OPINION INTO FACT

It may seem oxymoronic to place the words *subjective* and *measurements* in such close proximity. However, when it is possible to generate numerical data from listening tests, and those numbers exhibit relatively small variations and are highly repeatable, the description seems to fit. It has been this more than anything else that has allowed the exploration of correlations between technical measurements and subjective opinions. Technical measurements, after all, don't change, but however accurate and repeatable they may be, they are useless without a method of interpreting them or a way to process them so they relate more closely to perceptions. Subjective measurements provide the entry point to understanding the psychoacoustic relationships. The key to getting useful data from listening tests is in controlling or eliminating all factors that can influence opinions other than the sound itself.

In the early days, most of us thought that listeners were recognizing excellence and rejecting inferiority when judging sound quality. As logical as this seemed, it was soon thrown into question when listeners in the tests showed that they could rate products just as well with studio-created popular music as they could with classical music painstakingly captured with simple microphone setups, sometimes even better. How could this be possible? None of us had any idea what the studio creations should sound like, with all of the multitrack, close-miked, pan-potted image building and signal processing that went into them. The explanation was in the comments written by the listeners. They commented extravagantly on the problems in the poorer products, heaping scorn rich in adjectives on things that were not right about the sound. In contrast, high-scoring products received only a few words of simple praise. People seemed to be able to separate what the loudspeakers were doing to the sound from the sound itself. The fact that from the beginning, all the tests were of the "multiple-comparison" type may have been responsible. Listeners were able to freely switch the signal among three or four different products while listening to the music. Thus, the "personalities" of the loudspeakers were revealed through the ways the program changed. In a single-stimulus, take-it-home-and-listen-to-it kind of test, this would not be nearly so obvious.

Humans are remarkably observant creatures, and we use all our sensory inputs to remain in control in a world of everchanging circumstances. So when asked how a loudspeaker sounds, it is reasonable that we instinctively grasp for any relevant information to put ourselves in a position of strength. In an extreme example, an audio-savvy person could look at the loudspeaker, recognize the brand and perhaps even the model, remember hearing it on a previous occasion and the opinion formed at that time, perhaps recall a review in an audio magazine, and, of course, would have at least an approximate idea of the cost. Who among us has the self-control to ignore all of that and to form a new opinion simply based on the sound?

It is not a mystery that knowledge of the products being evaluated is a powerful source of psychological bias. In comparison tests of many kinds, especially in wine tasting and drug testing, considerable effort is expended to ensure the anonymity of the devices or substances being evaluated. If the mind thinks that something is real, the appropriate perceptions or bodily reactions can follow. In audio, many otherwise serious people persist in the belief that they can ignore such nonauditory factors as price, size, brand, and so on.

This is especially true in the few "great debate" issues, where it is not so much a question of how large a difference there is but whether there *is* a difference (Clark, 1981, 1991; Lipshitz, 1990; Lipshitz and Vanderkooy, 1981; Nousaine, 1990; Self, 1988). In controlled listening tests and in measurements, electronic devices in general, speaker wire, and audio-frequency interconnection cables are found to exhibit small to nonexistent differences. Yet, some reviewers are able to write pages of descriptive text about audible qualities in detailed, absolute terms. The evaluations reported on were usually done without controls because it is believed that disguising the product identity *prevents* listeners from hearing differences.

That debate gives no indications of slowing down, with periodic editorial assaults from the subjective reviewing side and some animated debates in Internet discussion groups. This is a segment of the audio industry that is aptly described as "faith-based." If you believe something, there is a possibility that you will hear it, and if you hear it, nothing can persuade you that, in a fully sighted evaluation, you might have been mistaken. If there is to be a resolution to this matter, it may require the recruitment of resources outside the domains of physics and engineering.

Science is routinely set up as a "straw man," with conjured images of wrongheaded, lab-coated nerds who would rather look at graphs than listen to music. Disputes between "subjectivists" and "objectivists" are not new. London (1963) reports the following:

In 1852, as Helmholtz began his research, the sciences and the arts were at loggerheads. The cold rationalism seemingly preached by Darwin's biology, Faraday's new atomic chemistry, and Gustav Magnus' physics was being strenuously opposed by a hostile school of arch-Romanticists, who believed that exposure of the roots of art as mere extensions of matter and energy meant the complete destruction of all beauty.

Fortunately, then and now, there are those who bridge the gap, who fully acknowledge that the subjective experience is "what it is all about," and who use their scientific and technical skills to find ways to deliver rewarding experiences to more people in more places.

In the category of loudspeakers and rooms, however, there is no doubt that differences exist and are clearly audible. Because of this, most reviewers and loudspeaker designers feel that it is not necessary to go to the additional trouble of setting up blind evaluations of loudspeakers. They believe that their professionalism can overcome any biases from nonauditory inputs. This attitude will be tested.

17.3 CONTROLLING THE EXPERIMENTAL VARIABLES

Any measurement requires controls on variables that can influence the outcome. Some can be completely eliminated, but others can only be controlled in the sense that they are limited to a few options (such as loudspeaker and listener positions) and therefore can be randomized in repeated tests, or they can be held as constant factors. Much has been written on the topic (Toole, 1982, 1990; Toole and Olive, 2001; Bech and Zacharov, 2006). The following is a summary.

17.3.1 Controlling the Physical Variables

The listening room. The discussions of Chapter 13 show graphically and dramatically how much a room can change what can be heard at low frequencies. If comparing loudspeakers, listen to all in the same room. As pointed out in Chapter 11, listeners have the ability to adapt to many aspects of room acoustics as long as they are not extreme problems.

Loudspeaker position. Listen in the same room, with the loudspeakers each brought to the same position. If that is not possible, ensure that all loudspeakers are auditioned in each of a set of standard locations and the results averaged.

Listener position. Listen in the same room, using a single listener in the same location. If there are multiple listeners, on successive evaluations it is necessary to rotate listeners through all of the listening positions.

Relative loudness. Perceived loudness depends on both sound level and frequency, as seen in the well-known equal-loudness contours (Figure 19.3). Consequently, something as basic as perceived spectral balance is different at different playback levels. In comparing the sound from audio components, loudness levels must be very closely matched. If the frequency responses of the devices being compared are identical—that is, flat—as in most electronic devices, it is a task easily accomplished with a simple signal like a pure tone and a voltmeter. Loudspeakers are generally not flat, and individually they are not flat in many different ways. They also radiate a three-dimensional sound field into a reflective space, meaning that it is probably impossible to achieve a perfect loudness equality for all possible elements (e.g., transient and sustained) of a musical program. There has been a long quest for a perfect "loudness" meter. Some of the offerings have been exceptionally complicated, expensive, and cumbersome to use, requiring narrow-band spectral analysis and computer-based loudness-summing software. A few years ago, when Aarts (1992) suggested that B-weighted sound-level measurements were adequate, many of us were greatly relieved. More recently, that option has been challenged; additional research suggests an even better solution (Soulodre, 2004; Soulodre and Norcross, 2003). Fortunately, it also is simple to implement: a high-pass characteristic somewhere between that of B and C weighting but with no high-frequency roll off. Figure 17.4 shows the standard A, B, and C weighting curves, along with the new proposal, the RLB (Revised Low-frequency B) curve. Nevertheless, the more different the spectra of the program material being compared, the greater will be the difficulty in achieving a loudness match. There is no set of sound level adjustments that could ensure equal perceived loudness for the loudspeakers shown in Figures 17.2 and 17.3; a balance achieved with one kind of signal would not apply to a signal with a different spectrum. Fortunately, as loudspeakers have improved and are now more similar, the problem has lessened, although not disappeared entirely. This principle, of course, also applies to the loudness balancing of the channels in a multichannel audio system. The typical method of using a midfrequency band-limited noise cannot be reliable, but, fortunately, perfection is not necessary in that application. In all of these cases, if in doubt, turn the instruments off and listen; a subjective test is the final authority.

Absolute loudness/signal-to-noise ratio. Spectral balance is affected by playback sound levels, and so are several other perceptual dimensions: fullness,

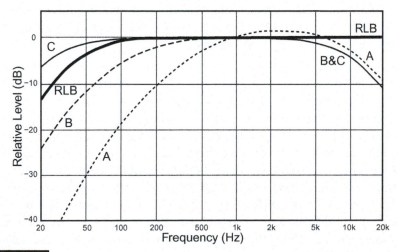

FIGURE 17.4 *Frequency-weighting curves that are available in sound-level meters (A, B, and C) and a new RLB curve proposed by Soulodre and Norcross (2003) for providing an objective measure of the perceived loudness of typical audio signals.*

FREQUENCY WEIGHTING

Frequency-weighting curves began with the notion that a single-number measurement might be able to correlate with overall perceptions of loudness. The A-weighting curve, for example, resembles the shape of an inverted 40-phon loudness contour (Figure 19.3a) and thereby was thought to be appropriate for measuring low-level sounds. B- and C-weighting curves were thought to be more representative of progressively higher sound levels. It didn't work, and over the years, C weighting came to be used as an approximation to flat but with some discrimination against very low and very high frequencies. A weighting has become widely accepted as a general-purpose measure where low frequencies are not an issue, including assessment of hearing damage risk. B weighting has been all but forgotten and is no longer a standard feature in sound-level meters. For balancing the loudness of program material, in which spectrum and level vary continuously and in which there are transient and sustained components, a single, simple measure of loudness has long been sought. The most recent evaluation of options suggests that a simple high-pass filter, with rolloff characteristics somewhere between B and C weighting, works as well as anything. (See Figure 17.4.) However, any of these measures are useful only as estimates of long-term, average loudness. They fail on a moment-to-moment basis and will exhibit differing amounts of error, depending on the spectrum and temporal structure of the program.

spaciousness, softness, nearness, and the audibility of distortions and extraneous sounds, according to Gabrielsson et al. (1991). Higher sound levels permit more of the lower-level sounds to be heard. Therefore, in repeated sessions of the same program material, the sound levels must be the same. Allowing listeners to find their own "comfortable" playback level for each listening session may be democratic, but it is bad science. Of course, background noises can mask lower-level sounds. This can be a matter of concern in automotive audio, where perceptual effects related to timbre and space can change dramatically in going from the parking lot to highway speeds.

Program material. The ability to hear differences in sound quality is greatly influenced by the choice of program material. As discussed in Chapter 9, the audibility of resonances is affected by the repetitive nature of the signal, including reflections and reverberation in the recording and in the playback environment. The poor frequency response of some well-known microphones is enough to disqualify recordings made with them, whatever the merits of the program itself (Olive, 1990; Olive and Toole, 1989). Some of the popular large diaphragm units have frequency responses at high frequencies that would be unacceptable in inexpensive tweeters. Olive (1994) shows how, in training listeners to hear differences in loudspeakers, it is possible to identify the programs that are revealing of differences and those that are merely entertaining. In general, solo instruments and voices are not very helpful, however comfortable we may feel with them. A program that is good for demonstrations (e.g., selling) can be different from that which is needed for evaluation (e.g., buying).

Electronic imperfections. The maturity of this technology means that problems with linear or nonlinear distortions in the signal path are normally not expected. However, they happen, so some simple tests are in order. It is essential to confirm that the power amplifier(s) being used are operating within their safe operating ranges. It is wise to measure the frequency-dependent impedances of loudspeakers under test to ensure that they do not drop below values that the amplifiers can safely drive. Since specifications may not be adequate to determine this, nothing is better than an oscilloscope to monitor the amplifier output during a preliminary trial.

Electroacoustic imperfections. In the subjective evaluation of electronic components, one makes huge assumptions about the performance of loudspeakers and rooms or of headphones, all of which are well known to introduce audible colorations and distortions that are orders of magnitude larger than those found in electronics and, even more, audio-frequency wires and interconnects. All that can be said is that, in such tests, it may be possible to detect *differences* introduced by such components, but it would be a high-risk venture to make the next step and to pronounce a *preference*.

Mono, stereo, multichannel—the listening configuration. Chapter 8 discusses in some detail the matter of stereo versus mono listening when establishing subjective opinions about loudspeakers (Toole, 1986). Monophonic tests turn out to be more revealing of the essential nature of the loudspeakers. Winners of monophonic tests win stereo tests. However, losers of monophonic tests are routinely scored higher in stereo tests in which the program material contains substantial interchannel decorrelation. Program material with hard-panned signals (i.e., monophonic left or right signals) may be judged as if they were monophonic sounds. In multichannel systems, this is especially true for the center channel in film and TV sound, which operates alone, monophonically, much of the time. Monophonic reproduction may have been superseded by stereo in the 1950s, but much of what we hear today has strong monophonic components. In general, single-loudspeaker comparisons are the recommended starting point. For program material, some people insist on using one of a stereo pair of channels, but summing the channels is possible with many stereo programs; listen, and find those that are mono compatible. Since complexity of the program is a positive attribute of test music, abandoning a channel is counterproductive.

17.3.2 Controlling the Psychological Variables

Knowledge of the products. This is the primary reason for blind tests. An acoustically transparent screen is a remarkable device for revealing truths.

Familiarity with the program. The efficiency and effectiveness of listening tests improves as listeners learn which aspects of performance to listen for during different portions of familiar programs. Select a number of short excerpts from music known to be revealing, and use them repeatedly in randomized sequences. It is not entertaining, but it can be informative.

Familiarity with the room. Chapter 11 discusses the importance of listening in a constant environment. We adapt to the space and appear to be able to "listen through" it to discern qualities about the source that would otherwise be masked. It takes time to acquire that ability, so schedule a warm-up session before serious listening begins.

Familiarity with the task. For most people, critical listening is something out of the ordinary. It will take some time before new listeners can relax and focus on the essential task, without devoting time and energy dealing with procedural matters, the mechanism of registering a response, and so on. For many beginners, it is also an intimidating experience. Seeing some preliminary test results showing that their opinions are not random is very confidence inspiring. Plan to discard the results of the first few test sessions. Experienced listeners are operational immediately.

Judgment ability or aptitude. Not all of us are good listeners, just as not all of us can dance or sing well. If one is establishing a population of listeners from which to draw over the long term, it will be necessary to monitor their decisions, looking for those who (a) exhibit small variations in repeated judgments and (b) differentiate their opinions of products by using a large numerical range of ratings (Olive, 2001, 2003; Toole, 1985). An interesting side note to this is that the interests, experience, and, indeed, occupations of listeners are factors. Musicians have long been assumed to have superior abilities to judge sound quality. Certainly, they know music, and they tend to be able to articulate opinions about sound. But what about the opinions themselves? Does living "in the band" develop an ability to judge sound from the audience's perspective? Does understanding the structure of music and how it should be played enable a superior analysis of sound quality? When put to the test, Gabrielsson et al. (1979) found that the listeners who were the most reliable and also the most observant of differences between test sounds were persons he identified as hi-fi enthusiasts, a population that also included some musicians. The worst were those who had no hi-fi interests. In the middle, were musicians who were not hi-fi oriented. This corresponds with the author's own observations over many years. Perhaps the most detailed analysis of listener abilities was in an EIA-J (Electronic Industries Association of Japan) document (EIA-J, 1979) that is shown in Figure 17.5. It is not known how much statistical rigor was applied to the analysis of listener performance, but it was the result of several years of observation and a very large number of listening tests (personal communication from a committee member).

In a different kind of test, Olive (2003) analyzed the opinions of 268 listeners who participated in the same loudspeaker evaluation. Twelve of those listeners had been selected and trained and had participated in numerous double-blind listening tests. The others were visitors to the Harman International research lab, as a part of dealer or distributor training or in promotional tours. One interest in the results was to see if opinions of a large number of visiting listeners agreed with the subjective ratings of the 12 internal listeners. They did, but there were differences of the kind being described here. Listeners from the normal population exhibited greater variability in their opinions and tended not to differentiate their ratings as strongly as the experienced listeners. This is measurable as the F_L statistic, and Figure 17.6 shows the relative performances of listeners having different audio experience and expertise. A high number indicates that the person gives very repeatable ratings in repeated tests and that the ratings of good and less good loudspeakers are strongly differentiated. The trained, veteran listeners distinguished themselves by having the highest rating by far, but obviously years of experience selling and listening on store floors has had a positive effect on the retail sales personnel. Olive (1994) describes the listener training ritual, which is also a part of the selection process; those lacking the aptitude do not improve beyond a moderate level.

Hearing ability. It is inconceivable that a person with defective hearing would perform well in listening tests, and it was no surprise that this was found to be a profound factor in the results of listening tests. So interesting are aspects of this matter that it will be discussed separately in the following section.

	Categories of listeners/audio	Sound analytical ability	Knowledge of reproduced sound	Attitude toward reproduced sound	Application objective	Reliability
1	General public	mediocre	mediocre	biased	various general investigations	
2	General public interested in audio equipment	mediocre	mediocre	biased	various general investigations	
3	General public strongly interested in audio equipment	fairly	some	strong self-assertion tendency	semi-specialized studies and measurements by specific group	low
4	Audio equipment engineers	high	sufficient	most correct	specialized studies and high precision measurements	high
5	Experienced acoustic experts	adequately high	excellent	correct	various studies and high precision measurements	high
6	Musicians, including students	adequately high	some	over-rigorous or non-interest	valuable opinions but unsuited for measurement	
7	General public strongly interested in music	high	some	mostly uninterested	unsuited for study & measurements (opinions are valuable)	
8	General public interested in music	somewhat high	mediocre	biased	various studies as representative of the general public	
9	General public strongly interested in recorded music	fairly high	fair	roughly correct	various studies and measurements	fairly high

FIGURE 17.5 *A chart describing several aspects of listener capabilities and their suitability for certain kinds of subjective audio evaluations. From EIA-J, 1979.*

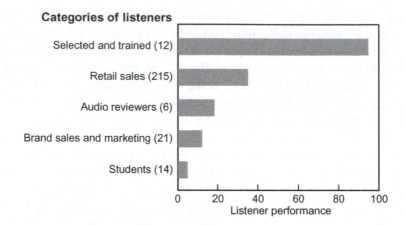

FIGURE 17.6
A performance metric of listener performance, related to the consistency of repeated judgments as well as the strength of rating differentiation of products having different sound quality levels. From Olive, 2003.

Listener interaction. Listeners in groups are sometimes observed to vote as a group, following the body language, subtle noises, or verbalizations of one of the number who is judged to be the most likely to "know." For this reason, controlled tests should be done with a single listener, but this is not always possible. In such cases, firm instructions must be issued to avoid all verbal and nonverbal communication.

Recognition. The premise of controlled listening tests is that each new sound presentation will be judged in the same context as all others. This theoretical ideal sometimes goes astray when one or more products in a group being evaluated exhibit features that are distinctive enough to be recognized. This may or may not alter the rating given to the product, but it surely will affect the variations in that rating that would normally occur. If listeners are aware of the products in the test group, even though they are presented "blind," it is almost inevitable that they will attempt to guess the identities. The test is no longer a fair and balanced assessment. For this reason it is good practice to add one or more "unknowns" to the test population and to let that fact be known in an attempt to preclude a distracting guessing game.

17.3.3 Controlling the Experimental Variables

Method. The method most preferred by product reviewers is the "take-it-home-and-listen-to-it," or single-stimulus, method. In addition to being "sighted," and therefore subject to all manner of nonauditory influences, it allows for adaptation. As noted in Chapter 11, humans are superb at normalizing many otherwise audible factors. This means that characteristics that might be perceived as flaws in a different kind of test can go unnoticed. On the other hand, in living with the product and auditioning many, many different programs, it is possible that a flaw might be found that would not be noticed in a more abbreviated controlled test that has fewer program selections. The scientific method prefers more controls, fewer opportunities for nonauditory factors to

BREAKING IN

In parts of the audio industry, there is a belief that all components from wires to electronics to loudspeakers need to "break in." Out of the box, it is assumed that they will not be performing at their best. Proponents vehemently deny that this process has anything to do with adaptation, writing extensively about changes in performance that they claim are easily audible in several aspects of device performance. Yet, the author is not aware of any *controlled* test in which any *consequential* audible differences were found, even in loudspeakers, where there would seem to be some opportunities for material changes. A few years ago, to satisfy a determined marketing person, the research group performed a test using samples of a loudspeaker that was claimed to benefit from "breaking in." Measurements before and after the recommended break-in showed no differences in frequency response, except a very tiny change around 30–40 Hz in the one area where break-in effects could be expected: woofer compliance. Careful listening tests revealed no audible differences. None of this was surprising to the engineering staff. It is not clear whether the marketing person was satisfied by the finding. To all of us, this has to be very reassuring because it means that the performance of loudspeakers is stable, except for the known small change in woofer compliance caused by exercising the suspension and the deterioration—breaking *down*—of foam surrounds and some diaphragm materials with time, moisture, and atmospheric pollutants. It is fascinating to note that "breaking-in" seems always to result in an improvement in performance. Why? Do all mechanical and electrical devices and materials acquire a musical aptitude that is missing in their virgin state? Why is it never reversed, getting worse with use? The reality is that engineers seek out materials, components, and construction methods that do *not* change with time. Suppose that the sound did improve over time as something broke in. What then? Would it eventually decline, just as wine goes "over the hill"? One can imagine an advertisement for a vintage loudspeaker: "An audiophile dream. Model XX, manufactured 2004, broken in with Mozart, Schubert, and acoustic jazz. Has never played anything more aggressive than the Beatles. Originally $1700/pair. Now at their performance peak—a steal at $3200!"

influence an auditory decision. What is often ignored by critics is that a time limit is not a requirement for a controlled test. It could go on for days, months, even years. All that is required is that the listeners remain ignorant of the identity of the products. A simple A versus B comparison is a start. Several of them randomized would be even better. A multiple comparison with three or four products available for comparison is arguably best. The problem with a solitary paired comparison is that problems shared by both products are not likely to be noticed. In addition, there are important issues of what questions to ask the listeners, scaling the results, statistical analysis, and so on. It has become a science unto itself. (See Toole and Olive, 2001, and Bech and Zacharov, 2006, for more discussion of experimental procedures.) However, as will be seen, even rudimentary experimental controls and elementary statistics can take one a long way.

17.4 HEARING PERFORMANCE IN LISTENING TESTS

The first clear evidence of the issue with hearing came in 1982 during an extensive set of loudspeaker evaluations conducted for the Canadian Broadcasting

(a)

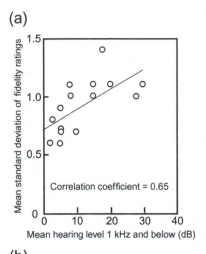

(b)

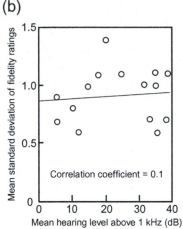

FIGURE 17.7 *Variability in fidelity rating judgments (mean standard deviation) as a function of hearing level (0 dB indicates normal hearing, and levels above that are indications of reduced hearing ability—that is, elevated hearing threshold). Audiometric measurements were made at the standard frequencies of 250, 500, 1k, 2k, 4k, and 8k Hz. (a) Hearing levels at 1 kHz and below were averaged. (b) Those above 1 kHz were averaged. From Toole, 1985, Figure 8.*

Corporation (CBC), in which many of the participants were audio professionals: recording engineers, producers, musicians. Sadly for them, hearing loss is an occupational hazard. During analysis of the data, it was clear that some listeners delivered remarkably consistent overall sound-quality ratings (then called fidelity ratings, reported on a scale of 10) over numerous repeated presentations of the same loudspeaker. Others were less good, and still others were extremely variable, liking a product in the morning and disliking it in the afternoon, for example. The explanation was not hard to find. Separating listeners according to their audiometric performances, it was apparent that those listeners with hearing levels closest to the norm (0 dB) had the smallest variations in their judgments. Figure 17.7 shows examples of the results. Surprisingly, it was not the high-frequency hearing level that correlated with the judgment variability but that at frequencies at or below 1 kHz. Figure 19.4 shows typical hearing threshold measurements for some of these listeners.

Noise-induced hearing loss is characterized by elevated thresholds around 4 kHz. Presbycusis, the hearing deterioration that occurs with age, starts at the highest frequencies and progresses downward. These data showed that by itself, high-frequency hearing loss did not correlate with trends in judgment variability (see Figure 17.7b). Instead, it was hearing level at lower frequencies that showed the correlation (Figure 17.7a). Some listeners with high-frequency loss had normal hearing at lower frequencies, but all listeners with low-frequency hearing loss also had loss at high frequencies—in other words, it was a broadband problem.

Hearing loss can occur as a result of age itself and as a result of accumulated abuse over the years. Whatever the underlying cause, Figure 17.8 shows that in terms of our ability to make reliable judgments of sound quality, we do not age gracefully. It certainly is not that we don't have opinions or the ability to articulate them in great detail; it is that the opinions themselves are less consistent and possibly not of much use to anyone but ourselves. In my younger years, I was an excellent listener, one of the best in fact. However, listening tests as they are done now track not only the performance of loudspeakers but of listeners—the metric shown in Figure 17.6. About age 60, it was clear that it was time to retire from the active listening panel. Variability had climbed, and, frankly, it was a noticeably more difficult task. It is a younger person's pursuit. Music is still a great pleasure, but my opinions are now my own. When graybeards expound on the relative merits of audio

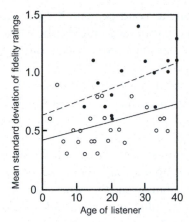

FIGURE 17.8   *Variations in fidelity ratings versus the age of the listener. The results are shown for two sets of experiments. One, indicated by black dots and the dashed regression line, was an early test in which all controls were not in place. The second series of tests (open circles and the solid regression line) was better controlled and therefore showed lower levels of variability. However, both sets of results show the same trend with age. As we get older, our judgments of sound quality tend to be less consistent. From Toole, 1985, Figure 9.*

products, they may or may not be relevant. But be polite—the egos are still intact. Figure 19.4 shows relevant data on hearing performance with age, including my own.

The effects of elevated hearing thresholds were also found in the opinions themselves. Looking for evidence of bias—a change of opinion—related to hearing level, regression lines were calculated for distributions of judgments as a function of low-frequency hearing level (see Figure 17.9). At the time this was done, listeners had been screened to eliminate those with more than about 20 dB mean hearing level over this frequency range, so the horizontal axis spans only 20 dB, all very much within the range of hearing that your neighborhood audiologist would classify as "normal." However, to an audiologist and to OSHA and other health-regulating agencies, normal hearing is evaluated on the basis of understanding the spoken word. Here we are asking listeners to perform a *much* more demanding task, and the consequences are very clear.

For all of the loudspeakers shown, there are shifts in fidelity ratings as a function of hearing level, some

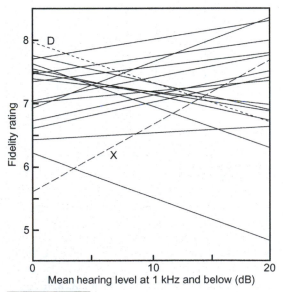

FIGURE 17.9   *Regression lines showing shifts in fidelity ratings for several loudspeakers as a function of low-frequency hearing level. From Toole, 1985, Figure 14.*

slight and some not so slight. Loudspeaker "X," for example, is heartily disliked by listeners with normal hearing, who rated it below 6. Other listeners found it to be highly pleasurable, scoring it over 7. It is interesting to look at some raw data to see what is happening.

Figure 17.10 shows results for individual listeners who evaluated four loudspeakers in a monophonic multiple-comparison test. Each dot is the mean of several judgments on each of the products, made by one listener. To illustrate the effect of hearing level on ratings, listeners were grouped according to the variability in their judgments: those who exhibited standard deviations below 0.5 scale unit and those above that. The low-variability listener results are shown in the white areas and the high-variability results are in the shaded areas. Looking back to Figure 17.9, it is seen that loudspeaker D (a PSB Passif II conventional forward-firing 8-in. (200 mm) woofer and 1-in. (25 mm) tweeter design) has a mean fidelity rating that falls with increasing hearing level, and loudspeaker X (a Quad ESL-63 full-range electrostatic; see Figure 8.10, "BB" for measurements) has a mean fidelity rating that rises dramatically.

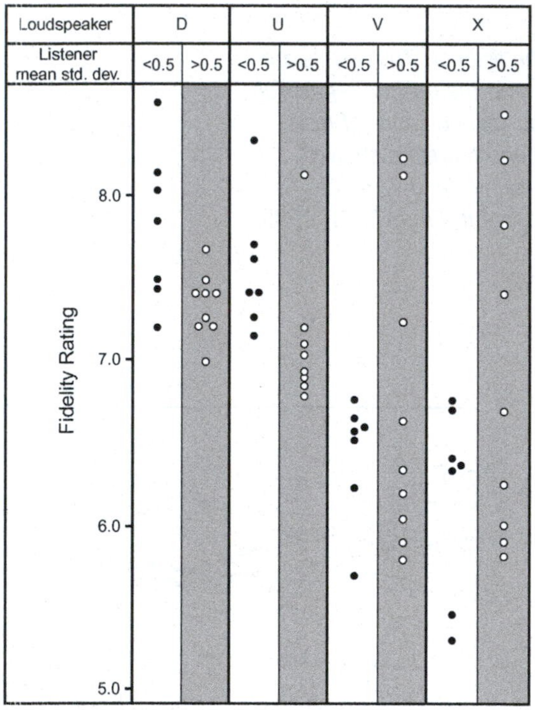

In Figure 17.10, we can see that fidelity ratings for loudspeakers D and U are closely grouped by both sets of listeners, but the ratings in the shaded areas are simply lower. Things change for loudspeakers V and X, where the close groupings of the low-variability listeners change to widely dispersed ratings by the high-variability listeners. This is a case where an averaged rating does not reveal what is happening. Listener ratings simply dispersed to cover the available range of values; some listeners thought it was not very good (fidelity ratings below 6), whereas others thought it was among the best loudspeakers they had ever heard (fidelity ratings above 8). This is not a consensus rating or multimodal but is nonmodal. Listeners simply exhibited strongly individualistic opinions. That they most likely have some form of hearing loss is a probable explanation, but recall that this amount of hearing-level elevation is comfortably within the conventional audiometric "normal" range. Both of these loudspeakers exhibit relatively smooth and flat on- and off-axis frequency responses, but they differ greatly in directivity. "D" is a conventional wide-dispersion cone/dome system, and "X" is a full-range electrostatic with a relatively narrow front-oriented

FIGURE 17.10 *Averaged rating judgments for several listeners who were classified according to the variability of their ratings. Those with mean standard deviations below 0.5 rating unit were placed in one category (vertical white bars), and those with variations above 0.5 unit were placed in the other (vertical shaded bars). Four loudspeakers were evaluated, two of which, D and X, were included in the data of Figure 17.9. For interest: D was a PSB Passif II, U was a Luxman LX-105, V was an Acoustic Research AR58S, and X was a Quad ESL-63. From Toole, 1985, Figure 16.*

dispersion pattern (the rear lobe is deliberately attenuated). This was the topic of discussion in Chapter 8 in the context of spaciousness and envelopment; it seems that there may be a connection to hearing level as well.

Finally, let us return to the stereo versus mono issue. Chapter 8 discussed this at length, concluding that results were strongly program dependent and that monophonic listening evaluations were more likely to reveal the true performance of loudspeakers. The data of Figure 17.11 add more strength to that argument, showing that only those listeners with hearing levels very close to the statistical normal level (0 dB) are able to perform similarly well in both kinds of tests. Even a modest elevation in hearing level causes judgment variations in stereo tests to rise dramatically, yielding less-trustworthy data.

All of this emphasis on normal hearing seems to imply that a criterion excluding listeners with greater than 15–20 dB hearing level may be elitist. According to USPHS data, about 75% of the adult population should qualify. However, there is some concern that the upcoming generation may not fare so well because of widespread exposure to high sound levels through headphones and other noisy recreational activities.

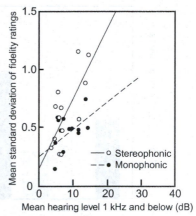

FIGURE 17.11   *A comparison of judgment variability as a function of hearing level for tests done in mono and stereo.*

17.5 BIAS FROM NONAUDITORY FACTORS

A widespread belief among audio professionals is that they are immune to the influences of brand, price, appearance, and so on. They persist in conducting listening evaluations with the contending products in full view. This applies to persons in the recording industry, audio journalists/reviewers, and loudspeaker engineers. As this is being written, the 45th anniversary issue of *Stereophile* magazine arrived (November 2007). In John Atkinson's editorial, he interviewed J. Gordon Holt, the man who created the magazine. Holt commented as follows:

As far as the real world is concerned, high-end audio lost its credibility during the 1980s, when it flatly refused to submit to the kind of basic honesty controls (double-blind testing, for example) that had legitimized every other serious scientific endeavor since Pascal. [This refusal] is a source of endless derisive amusement among rational people and of perpetual embarrassment for me, because I am associated by so many people with the mess my disciples made of spreading my gospel.

When I joined Harman International, listening tests were casual affairs, usually sighted. At a certain point it seemed appropriate to conduct a test, a demonstration that there was a problem. It would be based on two listening evaluations that were identical, except one was blind and one was sighted (Toole and Olive, 1994).

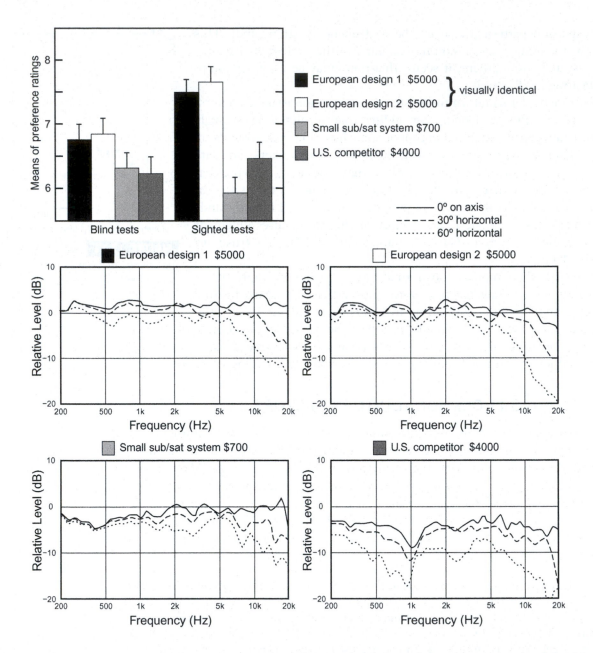

Forty listeners participated in a test of their abilities to maintain objectivity in the face of visible information about products. All were Harman employees, so brand loyalty would be a bias in the sighted tests. They were about equally divided between experienced listeners, those who had previously participated in controlled listening tests, and inexperienced, those who had not.

Figure 17.12 shows that in the blind tests, there were two pairs of statistically indistinguishable loudspeakers: the two European "voicings" of the same

The results of the subjective evaluations of four loudspeakers and anechoic data on the products. The anechoic data were unreliable below 200 Hz. Two of the loudspeakers were visually identical, large floor-standing units, representing alternative crossover network designs from different sales/marketing regions in Europe thought to cater to special regional tastes in sound. The third product was a recently introduced, inexpensive subwoofer satellite system with sound-quality performance that belied its small size and low cost. This was to be the honesty check-in sighted tests. The fourth product was a respected high-end product, a large floor-standing unit, from a competitor. One review of it claimed sound quality "equal to products twice its price." Another allowed that there were "a few $10 000 speakers that come close." Because this test was an evaluation of sound quality, not dynamic capabilities, care was taken not to drive the small system into overload. Loudness levels were equalized as well as possible, using a combination of measurements and listening. They remained unchanged throughout the test. The small bars on top of the large verticals are 95% confidence error bars, an indication of the difference between the ratings required for the difference not to be attributable to random factors. From Toole and Olive, 1994.

hardware and the other two products. In the sighted version of the test, loyal employees gave the big attractive Harman products even higher scores. However, the little inexpensive sub/sat system dropped in the ratings; apparently its unprepossessing demeanor overcame employee loyalty. Obviously, something small and made of plastic cannot compete with something large and stylishly crafted of highly polished wood. The large, attractive competitor improved its rating but not enough to win out over the local product. It all seemed very predictable. From the Harman perspective, the good news was that two products were absolutely not necessary for the European marketing regions. (So much for intense arguments that such a sound could not possibly be sold in [pick a country].) In general, though, what listeners saw changed what (they thought) they heard.

Dissecting the data and looking at results for listeners of different genders and levels of experience, Figure 17.13 shows that experienced males (there were no females who had participated in previous tests) distinguished themselves by delivering lower scores for all of the loudspeakers. This is a common trend among experienced listeners. Otherwise, the pattern of the ratings was very similar to those provided by inexperienced males and females. Over the years, female listeners have consistently done well in listening tests, one reason being that they tend to have closer to normal hearing than males. Lack of experience in both sexes shows up mainly in elevated levels of variability in responses (note the longer error bars), but the responses themselves, when averaged, reveal patterns similar to those of more experienced listeners. With experienced listeners, statistically reliable data can be obtained in less time.

The effects of room position at low frequencies have been well documented in Chapters 12 and 13. It would be remarkable if these did not reveal themselves in subjective evaluations. This was tested in a second experiment where the loudspeakers were auditioned in two locations that would yield quite different sound signatures. Figure 17.14 shows that listeners responded to the differences

FIGURE 17.13

The results shown in Figure 17.12 for three groups of listeners. From Toole and Olive, 1994.

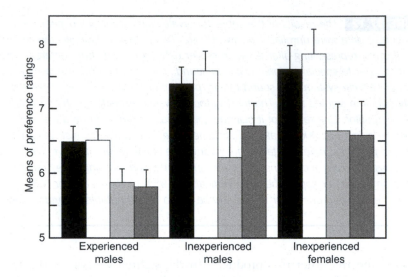

FIGURE 17.14 *Results of experiment 2 in which the same listeners auditioned the same loudspeakers located at two different positions in the room, in blind, and then in sighted evaluations. These are the same loudspeakers that were used in the previous test. Compare the heights of adjacent narrow bars of the same color to see the effects of changing loudspeaker position. In the blind tests, listeners heard big differences. In the sighted tests, they thought there were almost no differences. From Toole and Olive, 1994.*

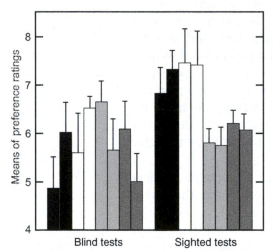

in blind evaluations: adjacent bars of the same color have different heights, showing the different ratings when the loudspeaker was in position 1 or 2. In contrast, in the sighted tests, things are very different. First, the ratings assume the same pattern that was evident in the first experiment; listeners obviously recognized the loudspeakers and recalled the ratings they had been given in the first experiment. Second, they did not respond to the previously audible differences attributable to position in the room; adjacent bars have closely similar heights. Third, some of the error bars are quite short; no thought is required when you know what you are listening to. Interestingly, the error bars for the two visually identical "European" models (on the left) were longer because the eyes did not convey all of the necessary information.

It is normal to expect an interaction between the preference ratings and individual programs. In Figure 17.15, this is seen in the results of the blind tests, where the data have been arranged to show declining scores for programs toward the right of the display. In the sighted versions of the tests, there are no such changes. Again, it seems that the listeners had their minds made up by what they saw and were not in a mood to change, even if the sound required it. The effect is not subtle.

Summarizing, it is clear that knowing the identities of the loudspeakers under test can change subjective ratings.

- They can change the ratings to correspond to presumed capabilities of the product, based on price, size, or reputation.

- So strong is that attachment of "perceived" sound quality to the identity of the product that in sighted tests, listeners substantially ignored easily audible problems associated with loudspeaker location in the room and interactions with different programs.

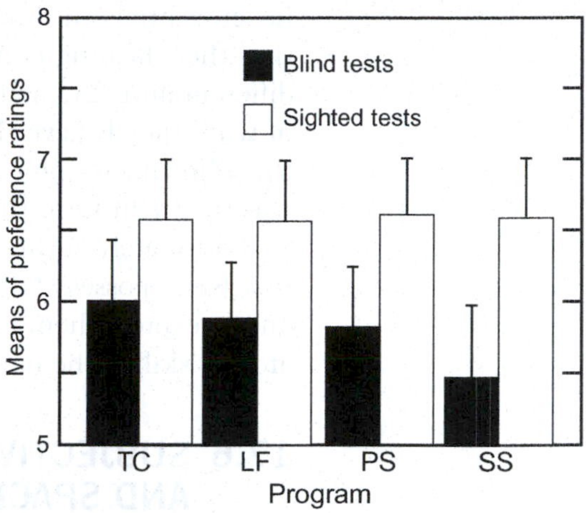

These findings mean that if one wishes to obtain candid opinions about how a loudspeaker sounds, the tests *must* be done blind. The good news is that if the appropriate controls are in place, experienced and inexperienced listeners of both genders are able to deliver useful opinions. Inexperienced listeners simply take longer, more repetitions, to produce the same confidence levels in their ratings.

FIGURE 17.15 *A mean of all ratings for all loudspeakers and positions, shown as a function of program, for both blind and sighted tests. From Toole and Olive, 1994.*

Other investigations agree. Bech (1992) observed that hearing levels of listeners should not exceed 15 dB at any audiometric frequency and that training is essential. He noted that most subjects reached a plateau of performance after only four training sessions. At that point, the test statistic F_L should be used to identify the best listeners. Olive (2003), some of whose results are shown in Figure 17.6, compiled data on 268 listeners and found no important differences between the ratings of carefully selected and trained listeners and those from several other backgrounds, some in audio, some not, some with listening experience, some with none. There were, as shown in Figure 17.6, huge differences in the variability and scaling of the ratings, so selection and training have substantial benefits in time savings. Rumsey et al. (2005) also found strong similarities in ratings of audio quality between naive and experienced listeners, anticipating only a 10% error in predicting ratings of naïve listeners from those of experienced listeners.

In the end, the best news for the audio industry is that if something is done well, ordinary customers may actually recognize it. The pity is that there is no source of such unbiased listening test data for customers to go to for help in making purchasing decisions.

It is paradoxical that opinions of reviewers are held in special esteem. Why are these people in positions of such trust? The listening tests they perform violate the most basic rules of good practice for eliminating bias. They offer us no credentials, no proofs of performance, not even an audiogram to tell us that their hearing is not impaired. Perhaps it is the gift of literacy that is the differentiator, the ability to convey in a colorful turn of phrase some aspects of what they believe they hear. Adding insult to injury, as will be discussed in the following chapter, most reviews offer no meaningful measurements so that readers might form their own impressions.

Fortunately, it turns out that in the right circumstances most of us, including reviewers, possess "the gift"—the ability to form useful opinions about sound and to express them in ways that have real meaning. All that is needed to liberate the skill is the opportunity to listen in an unbiased frame of mind.

17.6 SUBJECTIVE EVALUATIONS OF DIRECTION AND SPACE—AND MORE

Multichannel audio is relatively new, certainly in the context of music. There has been a lot of research focused on the perceptual dimensions of direction, apparent source width and envelopment, some of which was discussed in detail in Chapter 15. However, what happens when listeners are placed in a situation of judging multichannel audio in its entirety? Most people would simply not know where to start in analyzing such a complex sound picture. In fact, it is only now that systematic investigations of the perceptual dimensions and subjective reporting and rating schemes are being mounted (AES Staff Writer, 2004, is a good summary).

Rumsey et al. (2005a, 2005b) found that in multichannel evaluations, naive listeners placed more emphasis on surround effects. Experienced listeners placed more emphasis on the front soundstage. Is this because in gaining their experience, those listeners were immersed in a two-channel stereo world? It was interesting that ratings of both groups were dominated by timbral fidelity—sound quality. If it doesn't sound good, direction and space don't matter much.

17.7 CREATING A LISTENING ENVIRONMENT FOR LOUDSPEAKER EVALUATIONS

If one is conscientious about controlling the many nuisance variables just listed, a ritualized procedure is necessary, and a dedicated listening room is not an unreasonable request. This is not something that can be put together on short notice. Establishing a pool of selected, trained, and well-practiced listeners is a long-term commitment, requiring constant attention.

The problems of loudspeaker and listener positions are most commonly ignored. Figure 8.8 shows a solution to the problem used in the early 1980s by the author. In those tests, a person was required to rotate the turntables between sound examples.

Motorizing the turntables would be a step forward. However, that would limit the possible loudspeaker arrangements that could be tested. Figure 17.16 shows a schematic of the system installed in the Harman International research laboratory—the "shuffler." The loudspeakers are mounted on pallets that are able to move, individually, forward (toward the listener), or back (toward the wall). The assembly of nine moveable pallets itself is able to move left and right. So with position controllable on both *x* and *y* axes, it is possible to bring some number of loudspeakers to the same location in front of the listener. Figure 17.17 shows a stereo comparison setup.

In this case, the range of movement permits the comparison of up to four single loudspeakers for monophonic comparisons, four stereo pairs, or three sets of front L, C, R loudspeakers. The platform and the pallets are activated by pneumatic rams under computer control, so there are many positional options that can be explored. The rate of movement is adjustable, but typical exchanges take about 3 s. The shuffler has been an absolutely revolutionary component of the listening test program. No additional repetitions and averaging of responses in different locations were necessary. The tests were *much* shorter and the subjective data more consistent. With a few trained listeners, useful results could be obtained in hours, not days. With wireless computer data collection, statistically processed, graphically illustrated results are quickly available. See Olive et al. (1998) for more details.

More recently, a second room has been commissioned, with a different kind of positional substitution device for a different purpose. With the market moving to more in- and on-wall and ceiling loudspeakers, there was a need for a convenient method of comparing such products. Figure 17.18 shows a plan view of the device: a three-sided structure, each side of which can accommodate an in-wall or on-wall loudspeaker. This one is electrically motivated and changes positions in approximately one second. Of course, a free-standing loudspeaker can be placed below and in front, so there are several options for comparisons.

In the following chapters, all of the subjective data referred to was gathered in these facilities, using trained listeners that were selected for aptitude and normal hearing.

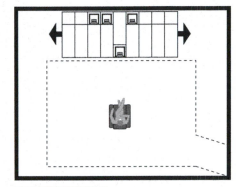

FIGURE 17.16 *The loudspeaker "shuffler" in the Harman International research department. The dashed line is a floor-to-ceiling acoustically transparent screen.*

FIGURE 17.17 *A photograph of a stereo loudspeaker comparison, showing two products in the forward, active locations and two others parked quietly at the wall.*

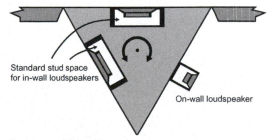

FIGURE 17.18 *Installed in the wall of a listening room this device allows the comparison of any combination of three in- or on-wall loudspeakers.*

CHAPTER 18

Loudspeakers II: Objective Evaluations

By now it is absolutely obvious that in sound reproduction the room and the loudspeaker are inseparable; they operate as a system. Yet, loudspeakers must be designed as separate devices, so it is necessary to develop measurements and interpretations for those measurements that allow us to anticipate how the loudspeakers are likely to sound in normally reflective rooms.

Chapters 12 and 13 demonstrated several ways in which the room imposes its considerable will at low frequencies. Still, woofers and subwoofers must be measured, just to give us a starting point. The wavelengths at the frequencies of interest are long compared to the dimensions of conventional box loudspeakers and subwoofers, so they can be considered to be approximately omnidirectional radiators. This means that a single curve can describe the frequency response in the range below about 100 Hz. The fact that the room resonances and standing waves wreak havoc with the sounds arriving at different listening locations is a separate challenge, but Chapter 13 shows that it can be managed. However, this can only be done after the system is set up in the room because all rooms are different, and the locations of subwoofers and listeners are crucially important.

At higher frequencies, we must be concerned with the nature of sounds radiating in different directions from the loudspeaker because they constitute the direct and reflected sounds arriving at listeners' ears. This means that many measurements must be made, and a system for organizing, processing, and displaying the data must be developed to allow it to be usefully interpreted. No single curve will be sufficient to describe the complex interface between the loudspeaker and all of the reflecting surfaces in a room.

18.1 TWO SIMPLE SOURCE CONFIGURATIONS

At the end of Chapter 16, the topic of propagation loss was introduced as a significant factor in what is heard in multichannel audio systems in small rooms. It was suggested that anything that could be done to deliver more uniform sound levels as a function of distance from the loudspeakers would be beneficial. It is convenient, therefore, to start this chapter with a description of the two basic radiation patterns of sound sources.

18.1.1 Point Sources: Spherical Spreading, Near- and Far-Field Designations

Figure 18.1a shows an ideal point source that, as a function of distance, experiences a rapid increase in the surface area over which the sound energy is distributed. Because the energy per unit area (sound intensity) is inversely proportional to the square of the distance from the source, this phenomenon has come to be called the *inverse-square law*. The sound level correspondingly falls rapidly, at a rate of −6 dB/dd (dd = double-distance). This happens only in the far field of the source. Beranek (1986) suggests that the far field begins at a distance of 3 to 10 times the largest dimension of the sound source. At this distance, the source is small compared to the distance, and a second criterion is normally satisfied: $distance^2 = wavelength^2/36$

In the near field, as shown in Figure 18.1b, the sound level at any frequency is uncertain. Figure 18.1c shows estimated distances at which far-field conditions should prevail for a loudspeaker system and for its components. This would be the minimum distance at which a microphone should be placed for measurements and at which listeners should sit to have a predictable experience.

In a room, closely adjacent reflecting surfaces must be considered to be part of the source. This means that the far field for the combination (loudspeaker plus a very early reflection) can be very far away. Diffusers behave as secondary sources of sound, and they can cover significant areas of room surfaces. Cox and D'Antonio (2004, p. 37) point out that listeners should be placed as far from scattering surfaces as possible, at least three wavelengths away. For devices that are effective to 300–500 Hz, this is a minimum distance of about 10 ft (3 m). As they realistically point out, "In some situations, this distance may have to be compromised."

So what is heard while standing close to the loudspeaker and its immediate environs can be very different from what is heard farther away, especially if one is moving around and by doing so enhancing the audibility of any near-field lobing or acoustical interference. Such effects are especially audible with stable broadband sounds like pink noise. Back in the listening area, sitting down, listening to music or movies, the audible result will be very different and much more pleasant.

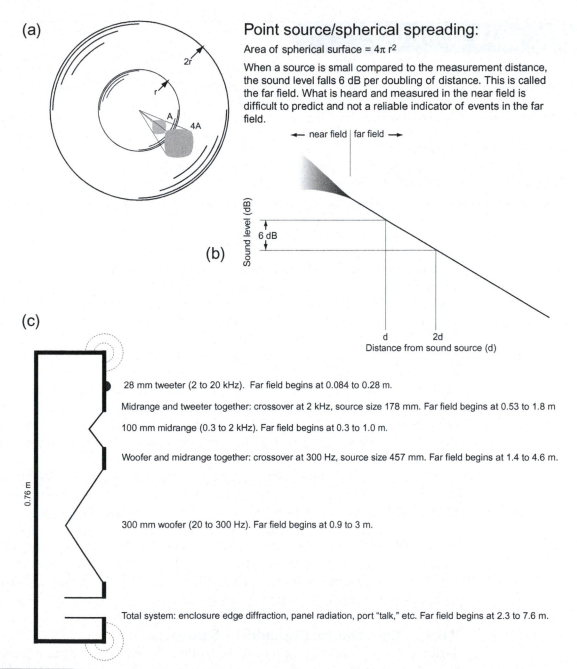

(a)

Point source/spherical spreading:

Area of spherical surface = $4\pi r^2$

When a source is small compared to the measurement distance, the sound level falls 6 dB per doubling of distance. This is called the far field. What is heard and measured in the near field is difficult to predict and not a reliable indicator of events in the far field.

← near field | far field →

(b)

Sound level (dB)

6 dB

d   2d
Distance from sound source (d)

(c)

0.76 m

28 mm tweeter (2 to 20 kHz). Far field begins at 0.084 to 0.28 m.

Midrange and tweeter together: crossover at 2 kHz, source size 178 mm. Far field begins at 0.53 to 1.8 m

100 mm midrange (0.3 to 2 kHz). Far field begins at 0.3 to 1.0 m.

Woofer and midrange together: crossover at 300 Hz, source size 457 mm. Far field begins at 1.4 to 4.6 m.

300 mm woofer (20 to 300 Hz). Far field begins at 0.9 to 3 m.

Total system: enclosure edge diffraction, panel radiation, port "talk," etc. Far field begins at 2.3 to 7.6 m.

FIGURE 18.1 *(a) The classic illustration of spherical spreading, originating with a point source. In the far field, the sound level falls at a rate of −6 dB per double-distance. (b) A graphic illustration showing the disorderly near field and the predictable far field behavior of a source. (c) Estimates of the distances at which far-field conditions are established for a three-way loudspeaker system and for its components, singly and in combination.*

NEAR-FIELD MONITORS

In recording control rooms, it is common to place small loudspeakers on the meter bridge at the rear of the recording console. These are called near-field or close-field monitors because they are not far from the listeners. As shown in Figure 18.1c, the near field of a small two-way loudspeaker (the midrange and tweeter of the example system) extends to somewhere in the range 21 in. to almost 6 ft (0.53 to 1.8 m). Including the reflection from the console under the loudspeaker greatly extends that distance. There is no doubt, then, that the recording engineer is listening in the acoustical near field, and that what is heard will depend on where the ears are located in distance, as well as laterally and in height. The propagating wavefront has not stabilized, and as a result this is not a desirable sound field in which to do precision listening, but as they say, perhaps it is "good enough for rock-and-roll."

Some of these far-field distances are much greater than the 1 m distance universally used for specifying loudspeaker sensitivity (e.g., 89 dB @ 2.83 v @ 1 m). There is no problem here because in the standards that specify the rituals of loudspeaker measurements, it is stated that the measurement should be made in the far field, whatever that may be, and then the sound level that would be expected from a point source at 1 m should be calculated. For example, if a measurement is made at 2 m, 6 dB should be added to arrive at the sound level at the reference distance, even though 1 m may be within the near field of that particular loudspeaker. The 1 m standard distance is therefore a convenience, not a directive that a microphone should be placed at that distance. *Many* people have misunderstood the intent of the standard distance, including some major players in the loudspeaker business.

If it is necessary to make measurements within the near field, useful data can still be obtained by spatial averaging: making several measurements at the same distance but at several different angular orientations with respect to the loudspeaker and averaging them. This is another of those uncertainty principle situations. By spatial averaging we have a better idea of the true frequency response, but we don't know the axis to which it applies. If we measure at a single point within the near field, we know the axis precisely, but we don't have a good measure of the frequency response.

18.1.2 Line Sources: Cylindrical Spreading

Figure 18.2 shows another extreme—the "infinite" line source—that , if it could be realized, would radiate a perfectly cylindrical sound wave, the area of which expands linearly with the radius. As a result, the sound level falls at the lower rate of −3 dB per double-distance. Practical line sources have finite lengths, so the critical issue becomes one of keeping listeners within the near field of the line, where the desirable −3 dB/dd (dd = double distance) relationship holds and out of the far field where even line sources revert to −6 dB/dd.

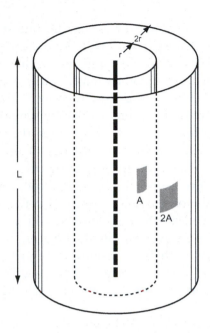

Line source/cylindrical spreading:

Area of cylindrical surface = $2\pi rL$

When a source is long compared to the measurement distance, the sound level falls 3 dB per doubling of distance. For a line loudspeaker this requires that it run from floor to ceiling, using "image" reflections from those surfaces to extend the effective length of the line. Most practical line loudspeakers are truncated (shortened) lines and they behave differently.

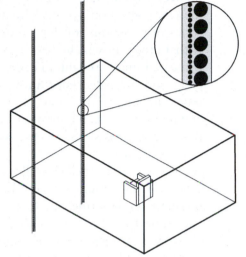

A stereo pair of line sources in a room, showing "images."

FIGURE 18.2 *An illustration of a theoretical infinite line source and of a practical approximation.*

Obviously the distance at which the near- far-field transition occurs is a function of frequency and the length of the line. Figure 18.2 shows a stereo pair of full-height lines, taking advantage of the ceiling and floor reflected images to make them appear to be even longer. A portion of one line has been expanded to show that it is a two-way system using conventional cone and/or dome loudspeaker drivers, densely packed (ideally spaced by less than about 1/2 wavelength of the highest reproduced frequency) to simulate a continuous sound source.

It is possible to use less than a full-height floor-to-ceiling array if one understands the variables and how they can be traded off. Lipshitz and Vanderkooy (1986) provide a thorough theoretical background to the behavior of "finite length" (not full height), truncated, line sources and they point out a number of problems, ultimately concluding that "there is little to recommend the use of line sources as acoustic radiators." They did grant that full-height lines had potential if the –3 dB/octave tilt in the frequency response is corrected.

There are advantages to collections of drivers: They share the workload and therefore can play loud without distress. However, most of the products casually referred to as "line sources" or "truncated line sources" in the industry are simply vertical arrangements of drivers that are too short to be useful even as truncated lines and with the drivers too far apart to be any kind of line. These loudspeaker systems obey the rules of collected point sources, with the disadvantage that, due to their size, the far field is a long distance away.

Griffin (2003) gives a comprehensive and comprehensible presentation of what is involved in designing practical line sources that approach the performance of full-height lines using less hardware. Smith (1997) describes a commercial realization and explains why it does what it does. Keele culminates a series of papers on constant-beamwidth transducers (CBTs) in a collaboration with Button, in which they examine the performance of several variations of truncated lines: straight and curved, "shaded" (drive power reduced toward the end), and unshaded (all transducers driven equally), all standing on a plane-reflecting surface (Keele and Button, 2005). It is a masterpiece of predictions and measurements that provide many answers and suggest many more possibilities. Figure 18.3 shows a small sample of the informative sound field simulations in the paper.

It is rare to see such clear illustrations of what is right and wrong with certain aspects of sound reproduction. In Chapter 12, we looked at adjacent boundary interactions, pointing out that the immediate surroundings of loudspeakers affect how they function and that some of the effects are not subtle. Figure 18.3a shows how just a single reflecting surface, the floor, disrupts an omnidirectional point source. Instead of tidy expanding circular contour plots, we see an example of gross acoustical interference with alternating lobes of high and low sound levels. The constant directivity of the source, indicated on the right, means that this problem exists at all frequencies, but the patterns will be different because of differing wavelengths. Additional boundaries—ceiling, side walls—add more of the same, of course, and the merged combination usually ends up being more satisfactory than this single-dimensional perspective suggests. This is, after all, another perspective on comb filtering, discussed in Chapter 9.

Chapter 12 finished with examples of loudspeakers designed to interface with room boundaries. Illustration 18.3b and those that follow show how much better things can be if a boundary is considered as part of the loudspeaker design. Figure 18.3b shows that a simple truncated line seems to be an improvement over the elevated point source, but note that uniform directivity has been sacrificed. The directivity index has a sharply rising character, indicating high-frequency beaming.

Figure 18.3c shows that shading the output, reducing the drive delivered to the transducers closer to the top of the line according to a Hann contour, greatly simplifies the pattern, but it still beams at high frequencies. We are not there yet.

Curving the line, as shown in (d), seems to be a step in the right direction. The contour lines are not yet smooth, but there is an underlying desirable order

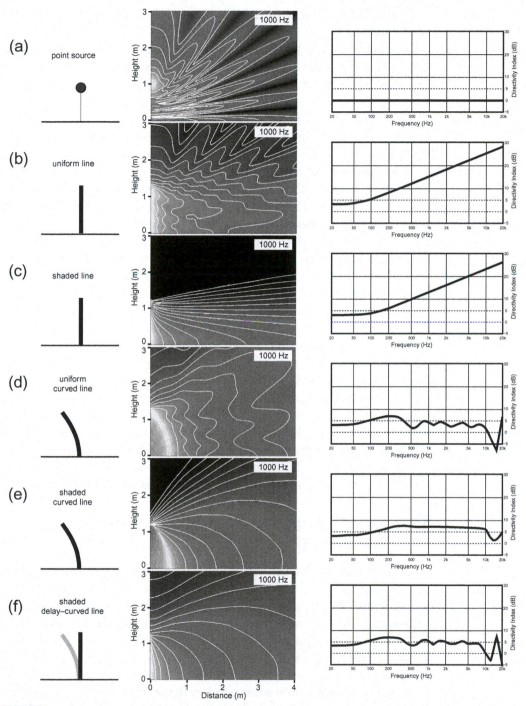

FIGURE 18.3 *Illustrations of the near-sound fields generated above a ground plane by several sound sources. The shading gets darker as sound levels drop; adjacent contour lines represent sound levels that differ by 3 dB. The original paper displays results for several frequencies; all of those shown are for 1 kHz. The words and graphics on the left explain the sources. On the right are far-field directivity indexes. Data from Keele and Button (2005).*

to them. The constancy of the directivity index tells us that it applies over a wide bandwidth.

Shading the curved line using the Legendre contour yields a set of plots that have a sense of order and beauty, (e). The constant directivity index indicates that it will be similar at most frequencies. This is the kind of thing we like to see.

If the marketing department thinks that the customers might prefer a straight line, applying the right delays to the drive signals can, in effect, contour the line (f). When shaded, the result is very similar to (e)—and good.

Scanning from (a) to (e) and (f), it is easy to see that there are improvements that can be made in the delivery of sounds from loudspeakers, through rooms, to listeners. This is a two-dimensional example of what is possible. Interfacing the source with the floor benevolently uses that reflection, and directivity control reduces the effect of the ceiling reflection. Line sources, by their nature, have a narrow frontal aspect, so horizontal dispersion can be wide and uniform.

How did (e) and (f) sound? Excellent—at least that is the author's opinion from a biased, sighted test. It was distinctive in how little the sound level and timbre appeared to change with location in the room and how the loudspeaker did not get "loud" as one walked up to it. Note that the sound level contours around ear height (just under 2 m) are only gently sloped.

Any of these line radiators can be positioned at the ceiling interface—for example, as surround loudspeakers—or positioned between floor and ceiling. In the latter situation, they lose the boundary reflection and will need to be physically lengthened to regain comparable radiation performance. The shaded versions would have the lower half inverted so the acoustical output would decline toward both ends, top and bottom. So as we move into the detailed characterization of loudspeaker performance, it is important to keep in mind that directivity and propagation characteristics are important parts of the data set.

18.2 MEASURING THE ESSENTIAL PROPERTIES OF LOUDSPEAKERS

Frequency response is the single most important aspect of the performance of any audio device. If it is wrong, nothing else matters. That is a statement without proof at this point in the book, but that will come. It is interesting to consider that for as long as anyone in audio can remember, all electronic devices had basically flat frequency responses. No manufacturer of an amplifying device, a storage device, or a music or film distribution medium would even momentarily consider a frequency response specification that was far from what could be drawn with a ruler from some very low frequency to some very high frequency. Yet, when we come to loudspeakers, it is as though we threw away the rule book and suddenly tolerances of ±3 dB or more are considered acceptable. The measurements in Figures 17.2 and 17.3 show a few loudspeakers from the 1960s.

Some of these needed all of that tolerance, and more, to embrace even the on-axis digressions from flat. Yet, two of them, over substantial portions of the frequency range, behaved quite well. It could be done. Still, bad habits are hard to shake off, and the industry is still burdened with that embarrassingly inadequate descriptor for the most important specification. 20 Hz to 20 kHz ±3 dB is meaningless without seeing the curve that it describes. It could be a horizontal straight line that simply falls off sharply at the upper and lower frequency limits (perfection), or it could be a line that undulates randomly between +3 dB and −3 dB, a 6 dB range (absolute rubbish).

But even worse than the uselessness of that description of frequency response is the fact that it is often assumed that a single curve is sufficient to describe the performance of a device that radiates sound in three dimensions—in all directions—into a room. When a manufacturer shows a specification for a loudspeaker frequency response in numerical form only, and the tolerance is more than about ±0.5 dB, ignore it. If a curve is shown, but there is only one, it might be correct, but by itself, it is not enough data.

18.2.1 What Do We Need to Know?

Figure 18.4 illustrates three distinct sound fields in small listening rooms:

- The direct sound from the source to the listener, normally represented by the on-axis behavior of the loudspeaker.

- The early reflections, sounds that have been reflected only once on their way to the listener. These would be represented by measurements made at the appropriate off-axis angles, taking into account the positions of the room boundaries and the arrangement of loudspeaker and listener within them.

- The late reflections, arriving after multiple reflections from all directions. This would be called the reverberant sound in large performance spaces. This is the sum of all other sounds radiated by the loudspeaker in all possible directions and is described by the sound power.

It is believed that some combination of these is sufficient to describe much of what listeners hear from loudspeakers in small rooms.

This much was evident in the early 1980s when the author set up a semiautomated data-gathering and -processing system at the National Research Council, in Ottawa. A computer-

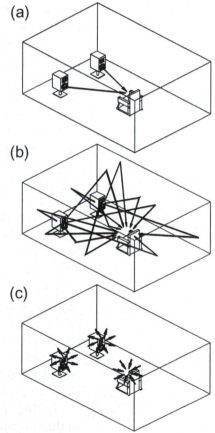

(a)

(b)

(c)

FIGURE 18.4 *The three sound fields in small rooms: (a) Direct sound. (b) Early reflections. (c) Late reflections.*

controlled oscillator stepped through a logarithmically spaced set of 200 frequencies from 20 Hz to 20 kHz (20 measurement frequencies per octave). The anechoic chamber was calibrated at low frequencies by fixing the locations of the rotational center of the loudspeaker (the center of the front baffle) and the microphone, and deriving a correction curve based on true free-field measurements made on a 10 m tower outdoors (an alternative method of generating a reference curve is to employ the ground-plane technique; Gander, 1982). It is believed that the resulting frequency response curves were accurate within ±1 dB down to 30 Hz. Of course, the correction applied only to monopole woofers. Even then, some had sufficient interference from port radiation that the sound power measurement became the definitive measure at low frequencies. Measurements were made at 2 m, on both vertical and horizontal orbits, at angular increments of 15° in the frontal hemisphere and 30° in the rear hemisphere. The data were postprocessed by the computer to yield several individual and spatially averaged frequency responses, sound power, directivity index, and phase (Toole, Part 2, 1986).

Figure 18.5a shows some of the basic data on a loudspeaker designed to have a flat and smooth on-axis response, but clearly off-axis performance was not given comparable attention (from Figure 8.10, loudspeaker "E"). The raw data were then manipulated using dimensions from the prototype IEC listening room (Figure 4.10a) to generate a picture of the sequence of sounds that would arrive at the listener's ears in the room. Specular (mirror-like) reflections were assumed. Figure 18.5b shows those predictions.

Let us pause at this point and ask the question: If we want to measure a loudspeaker, and from those measurements try to anticipate how it might sound in a room, what should we measure? The answer at low frequencies is sound power; it is the highest curve. However, we know that if the woofer radiation is omnidirectional, the shape of the curve at frequencies up to about 100 Hz will be the same in all of the curves, which it appears to be for this loudspeaker (for loudspeakers with collections of woofers, or a physically separated port, the sound power will be the true measure). At the highest frequencies, it is the on-axis frequency response that is dominant; it is the highest curve. Over the rest of the frequency range, which includes voices and most musical instruments, the curves weave among each other and are never far apart. So the global answer to the question is that we must measure *everything*. No single curve tells the complete story. Performing an energy summation of the data, we get a curve that is a prediction of what might be measured in a room.

Taking the loudspeaker into the real room and measuring its performance in three locations, averaged over the listening area, yields the three curves shown in Figure 18.5c. As pointed out in Chapter 4, where these data are also shown, the standing waves in the listening room cause huge fluctuations at low frequencies. However, at frequencies above about 300–400 Hz, the curves become quite similar to each other and also to the predicted room curve, which is superim-

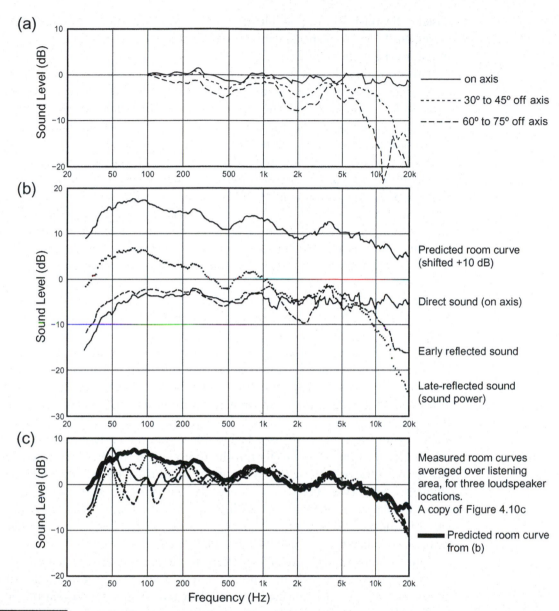

FIGURE 18.5 *(a) Anechoic data for a three-way loudspeaker with uneven directivity versus frequency. (b) Sounds arriving at the listening position predicted from anechoic measurements: direct sound; the early reflected sounds from floor, ceiling, and two side walls; and late reflected sounds. These are summed to show a predicted room curve, elevated by 10 dB for clarity. (c) The same loudspeaker measured at three locations in a real room, compared with the prediction in (b). From Toole, Part 2, 1986.*

posed. Obviously, it is possible to use anechoic chamber measurements to anticipate how a loudspeaker might sound in a room at frequencies above the transition frequency.

However, this observation is much more important than may appear. Some have argued that it is the shape of the room curve that determines how a loudspeaker sounds. If so, then one could ignore a loudspeaker's anechoic behavior and equalize the room curve to have the desired shape. Then the peculiarities of both the loudspeaker and the room are accommodated in the one action. What is missing from this perspective is that two ears and a brain are far more analytical than an omnidirectional microphone and an analyzer. The measurement system simply adds up all sounds, from all directions, at all times, and renders a single curve. A loudspeaker turned to face the wall, after equalization, should sound like that same loudspeaker facing the listeners. It doesn't. There is insufficient data to describe the source and thereby how it interacts with the room boundaries. Humans have a remarkable ability to separate a source from the room it is in and offer up detailed descriptions of how it sounds, even when the room is changed (see Chapter 11). We need measurements that describe the nature of the source *and* that provide insight into what happens in a room.

The loudspeaker in Figure 18.5 is an example of a common problem. Engineering convention dictates that the on-axis frequency response be as flat and smooth as possible. This, after all, is likely to appear, perhaps artistically enhanced, in the brochure. However, this loudspeaker exhibits different directivity at different frequencies, and it is this pattern of directivity variation that shows up in the room curve (see Figure 18.5a and the early reflection curve in (b)). If equalization is used to flatten the room curve, the pristine on-axis curve will be lost—the only thing that was correct about the loudspeaker. An equalizer changes frequency response, not directivity. The cure for this room curve is a better loudspeaker, one with better directional consistency. Many loudspeakers suffer from combinations of both problems—faulty axial frequency response and inconsistent directivity—which makes life complicated.

Once a loudspeaker is in a room, there are no measurements that will enable us to separate—with high measurement resolution and accuracy—the direct, early-reflected and late-reflected sounds. Without detailed information on the loudspeaker, equalization within the room is a game of chance. However, if one has sufficiently detailed information on a loudspeaker, it may be possible to predict what may happen in a room. If a loudspeaker is properly designed, and strong early reflections are not spectrally corrupted, equalization might not be necessary above certain frequencies. This is a worthy objective.

18.2.2 Improved Data Gathering and Processing
The system that generated the curves for a specific room in Figure 18.5 was an important beginning, but improvements were possible. What was missing was

a statistical perspective on many rooms, so we might develop a similar kind of measure that suggested performance in a "typical" room.

Figure 18.6 describes the data-gathering system at Harman International, Northridge, California. It incorporates a computer-controlled rotating platform upon which the loudspeaker is placed on its bottom to measure the horizontal orbit at 10° intervals and then on its side to measure the vertical orbit. The height of the platform is adjusted to bring the reference axis to the same point. The data for the 70 frequency response curves have a frequency resolution of 2 Hz, the curves are 1/20-octave smoothed, and the anechoic chamber is anechoic

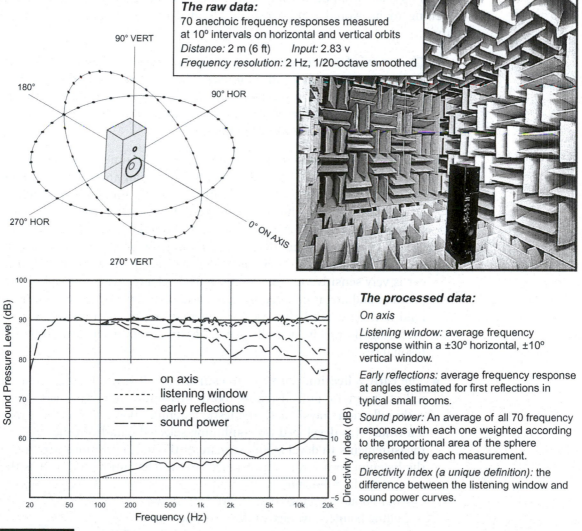

FIGURE 18.6 *The data collecting system used at Harman International and the basic set of processed data that it yields.*

(±0.5 dB, 1/20-octave) from 60 Hz to beyond 20 kHz and has been calibrated to be accurate (±0.5 dB, 1/10-octave) from 20 to 60 Hz (Devantier, 2002). The vertical scale has been adjusted to show the sound level corrected to the standard 1 m distance so sensitivity can be read directly.

The family of curves shown in the lower half of the figure is the set of data calculated to describe sounds that might arrive at a listener's ears in an average room. All of the data are based on the selection of a *reference axis*, the axis along which the on-axis curve is measured. Normally this has a point of origin between the tweeter and midrange drivers, and it extends perpendicularly outward from the front baffle. It is possible for a manufacturer to specify any axis as its reference, but logically it would be the line that, if extended into the listening room, would come close to a seated listener's ears. These are the curves:

- The **on-axis frequency response** is the universal starting point, and in many situations it is a fair representation of the first sound to arrive. However, as shown in the Devantier (2002) survey, over half of those investigated had the prime listening position 10° to 20° off axis. Hence, a justification for the following measure.

- The **listening window** is a spatial average of the nine frequency responses in the ±10° vertical and ±30° horizontal angular range. This embraces those listeners who sit within a typical home theater audience, as well as those who disregard the normal rules when listening alone. Because it is a spatial average, this curve attenuates small fluctuations caused by acoustical interference, something far more offensive to the eye than to the ear, and reveals evidence of resonances, something the ear is very sensitive to: interference effects change with microphone position and are attenuated by the spatial averaging, whereas resonances tend to radiate similarly over large angular ranges and remain after averaging. Bumps in spatially averaged curves tend to be caused by resonances.

- The **early reflections** curve is an estimate of all single-bounce, first reflections in a typical listening room. Measurements were made of early reflection "rays" in 15 domestic listening rooms. From these data, a formula was developed for combining selected data from the 70 measurements to develop an estimate of the first reflections arriving at the listening location in an "average" room (Devantier, 2002). It is the average of the following:
  —Floor bounce: average of 20°, 30°, 40° down
  —Ceiling bounce: average of 40°, 50°, 60° up
  —Front wall bounce: average of 0°, ±10°, ±20°, ±30° horizontal

—Side wall bounces: average of ±40°, ±50°, ±60°, ±70°, ±80° horizontal
—Rear wall bounces: average of 180°, ±90° horizontal

The number of "averages" mentioned in that description may make it seem as though anything useful would be lost in statistics. However, this turns out to be a very useful metric. Being a substantial spatial average, a bump that appears in this curve, and in other curves is clear evidence of a resonance. It is also, as will be seen, the basis for a good prediction of what is measured in rooms.

- **Sound power** is intended to represent all the sounds arriving at the listening position. It is the weighted average of all 70 measurements, with individual measurements weighted according to the portion of the spherical surface that they represent. Sound power is a measure of the total acoustical energy radiating through an imaginary spherical surface with the radius equal to the measurement distance. Thus, the on-axis curve has very low weighting because it is in the middle of other, closely adjacent measurement points (see the perspective sketch at the top of the figure), and measurements further off axis have higher weighting because of the larger surface area that is represented by each of those measurements. Ideally, such a measurement would be made at equally spaced points on the entire surface of the sphere, but this simplified spatial-sampling process turns out to be a very good approximation. The result could be expressed in acoustic watts, the true measure of sound power, but here it is left as a sound level, a frequency response curve having the same shape. This serves the present purposes more directly. Any bump that shows up in the other curves and persists through to this ultimate spatial average is a noteworthy resonance.

- **Directivity index (DI)** is defined as the difference between the on-axis curve and the sound-power curve. It is thus a measure of the degree of forward bias—directivity—in the sound radiated by the loudspeaker. It was decided to depart from this convention because it is often found that because of symmetry in the layout of transducers on baffles, the on-axis frequency response contains acoustical interference artifacts, due to diffraction, that do not appear in any other measurement. It seems fundamentally wrong to burden the directivity index with irregularities that can have no consequential effects in real listening circumstances. Therefore, the DI has been redefined as the difference between the listening window curve and the sound power. In most loudspeakers, the effect of this choice is negligible, but in highly directional systems it is significant because the listening window curve is lower than the on-axis curve. In any event, for the curious, the raw evidence is there to inspect.

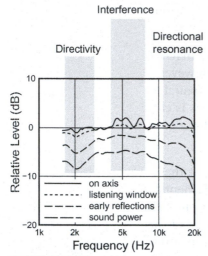

FIGURE 18.7 *An examination of different forms of misbehavior, as revealed in the family of frequency-response curves.*

Obviously, a DI of 0 dB indicates omnidirectional radiation. The larger the DI, the more directional the loudspeaker in the direction of the reference axis.

18.2.3 Interpreting the Data: Exercises in Detection

Having single-axis data along with data representing progressive increases in spatial averaging is highly useful. For example, Figure 18.7 illustrates the identification of three distinctive forms of misbehavior in an early prototype of the example loudspeaker shown in Figure 18.6.

- Off-axis misbehavior leading to a change in directivity. This is revealed in the fact that around 2 kHz the on-axis curve is quite flat, but as the data embraces more off-axis measurements into the spatial average, a shallow dip develops. In Figure 18.6, it can be seen that the DI shows a small bump at 2 kHz.

- On-axis misbehavior: ripples in the frequency response caused by enclosure diffraction. Proof that they are not resonances is that they are much attenuated by the moderate spatial averaging incorporated in the listening window curve, and have all but vanished in the increasing spatial averaging of the lower curves. It is highly improbable that this would be audible in a room, but it has a threatening appearance in the on-axis curve. Figure 18.6 shows that the problem was eliminated in the final product. This is the type of circumstance that led to a redefinition of DI, as just discussed.

- A low-Q (i.e., well-damped) resonance at the upper limit of the tweeter. It is visible in the top three curves, but there is little evidence of it in the sound power, meaning that it has a forward directional bias. Figure 18.6 shows that for the final product, the tweeter was improved, extending the frequency range and, in the process, eliminating even this innocuous resonance.

For perspective, this now-discontinued loudspeaker was included in numerous double-blind listening tests over several years. It always was a front-runner, either winning or being in a statistical tie with the best competitors. So although there are imperfections, one may conclude that this family of curves describes a highly commendable standard of performance.

Figure 18.8a shows an active/equalized loudspeaker at an early stage in its development. The well-mannered directivity seen here is characteristic of good constant-directivity horns. If directional control is required, as in an acoustically "live" room that cannot be altered, horns are an excellent solution, especially since horizontal and vertical directivity can be independently manipulated.

Belief that horns and waveguides are inherently colored is an idea that good modern designs have put to rest.

Cone/dome systems are best suited to wide-dispersion applications, and there the challenge is to maintain a relatively constant, or a least smoothly changing directivity, as a function of frequency. In achieving that objective, it is increasingly common to add shallow horns, often called waveguides to mid- and high-frequency cone/dome drivers to subtly manipulate directivity so that they better integrate into the entire system.

This design is well optimized for small rooms in that above the transition region—200–300 Hz—the directivity is quite constant. The small undulations in the curves are very consistent from the on-axis curve, down to the sound power curve. Normally, bumps that are seen in several spatially averaged curves are evidence of resonances. Here there are a lot, more than might have been expected from the transducers and enclosure. The question is: where did they come from? Looking closely, it is seen that the bump around 700 Hz gets larger as spatial averaging increases and the DI drops, indicating that most of the energy is radiated off axis. This behaves like radiation from an enclosure resonance, possibly the large rear panel. But the other frequency response peaks and dips look much the same in all of the curves, and they do not show up in the DI curve. This is unusual.

These directionally independent bumps can be fixed by equalization. But it turned out that some of the undulations were in fact *caused* by a preliminary casual equalization that had been done earlier, outside the anechoic

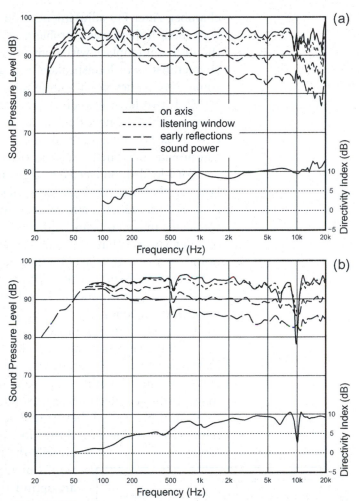

FIGURE 18.8 *(a) A large professional audio monitor loudspeaker: two 15-in. (380 mm) woofers vertically flanking a constant-directivity horn. It was active, with a dedicated equalizer. These data were taken at an early stage in the development and illustrate the challenges in separating the factors responsible for visible features in the family of curves. This is discussed in the text. The production version was significantly smoother than this representation (it is now discontinued). Note the high sensitivity of this and the following loudspeaker, about 95 dB at 1 m for 2.83 v input. (b) A large high-end loudspeaker with a 15-in. (380 mm) woofer, a midhigh horn, and a high-frequency horn. The benefits of a dedicated high-frequency horn are apparent compared to (a). This totally passive design is admirably free of resonant colorations, with superbly controlled directivity.*

chamber. After EQ adjustments, the loudspeaker sounded as it (finally) looked: very good. If there was a problem, it was a tendency to play it very much louder than is commonplace with consumer loudspeakers. That is one of the seductive characteristics of loudspeakers that do not power compress or distort at high sound levels; they don't sound loud until they are dangerously loud.

With two large woofers separated by a horn, the measurement distance of 2 m is within the near field, and there is evidence of directivity at bass frequencies. This is yet another advantage of spatially averaged measurements: they can provide meaningful data in the near field.

Horn-loaded loudspeakers are very well suited to large home theater installations; they deliver high-level crescendos effortlessly, and their directional control minimizes the amount of sound converted into heat in absorbers—which translates into significantly higher overall efficiency. Figure 18.8b is an example of a loudspeaker that is admired equally by audiophile stereo traditionalists and home theater enthusiasts (with deep pockets!).

There is a lesson to carry away from the example in (a). Equalization changes the intrinsic performance of a loudspeaker, and this can be good (if it is needed to repair the frequency response) or bad (if it was not needed). Transducers, within their normal operating ranges, behave as minimum-phase devices (the misbehavior of the large horn above 10 kHz is evidence that it is outside of the predictable operating range). Parametric equalization of resonances in transducers is an effective solution within this frequency range. But the measurements must be made without reflections. Measurements done in a reflective space are non-minimum-phase, and they cannot be trusted to reveal accurate evidence of resonances in transducers and therefore of the corrective equalization appropriate to remedy them. Therefore, notions that "room equalization" can address the problems of inferior loudspeakers are optimistic. Equalization cannot alter directivity, and steady-state measurements in rooms cannot reliably identify resonances. Such equalization is most useful for subtle adjustments to already well-behaved, loudspeakers, and then most likely at the lower frequencies.

All of this assumes that the measured data have sufficient resolution to reveal what can be heard. The next example begins with a prototype of an inexpensive loudspeaker that exhibited an easily audible problem, but there was no evidence in the measurements to explain it. At this time, it had been common for engineers to do time-windowed FFT measurements in their listening rooms. The dotted curve in Figure 18.9 shows what these people were looking at.

The problem was revealed when a female vocalist sustained a certain note. The loudspeaker "howled."

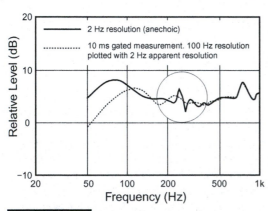

FIGURE 18.9 *Measurements of a loudspeaker with an audible problem at 280 Hz. One measurement used a 10 ms time-windowed FFT, as done in a normal room (dotted), and the other measurement was performed in an anechoic chamber using 2 Hz resolution.*

From the listening experience one could deduce that a very frequency-specific, high-Q resonance was the problem. A quick pitch match using an oscillator revealed that the problem was around 280 Hz. The time gating used for the measurement was 10 ms, chosen to eliminate room reflections from the data. This window yields data with 100 Hz frequency resolution (resolution = 1/ window duration), so it was clear that any high-Q events around 280 Hz would simply not be visible. The loudspeaker was then measured in an anechoic chamber where high resolution at low frequencies is possible; the solid curve in Figure 18.9 shows clearly that there was indeed a resonance. Gated measurements are very useful, but their limitations need to be kept firmly in mind. Howard (2005) discusses some interesting measurement options for those without access to anechoic chambers.

18.2.4 The Relationship Between Anechoic Data and Room Curves

It may seem absurd to use anechoic data to predict what happens in real rooms, but the connection, as was seen in Figure 18.5, is not a loose one. That was a demonstration using a single specific room. The examples in Figure 18.10 will use the statistical room data. The loudspeakers used are seen in Figure 18.14, along with subjective ratings: loudspeaker "I" is shown in (a) and loudspearker "B" in (b). "I" wins. Explanations are in the captions.

An obvious message from Figure 18.10 is that measuring a steady-state room curve and equalizing it cannot guarantee excellent sound. Equalization can only change frequency response. In loudspeaker "B," the dominant problem is non-uniform directivity as a function of frequency. The correction for this problem is back in the engineering department.

The real world has an infinite variety of room shapes, sizes, and acoustical treatments, and each one may require a different "blend" of curves from the anechoic data set to perfect the match to a measured room curve. Doing this in each of the infinite number of real-world examples is obviously unproductive, and unnecessary, because it is evident that human listeners have abilities that surpass those of a microphone and spectrum analyzer; the room curve is not the definitive answer. The message of real value here is that the right set of anechoic data, presented in the right fashion, does a creditable job of not just predicting room curves but, more important, permitting us to examine in detail the temporal sequence of sounds: direct, early reflections, and late reflections that listeners hear.

18.2.5 Sound-Absorbing Materials and Sound-Scattering Devices

The dominant factor in the shape of the room curves is the off-axis sound, the first reflections paramount among them. This observation sends a clear message that to preserve whatever excellence exists in the loudspeakers, the room boundaries must not change the spectrum of the reflected sound. This means that

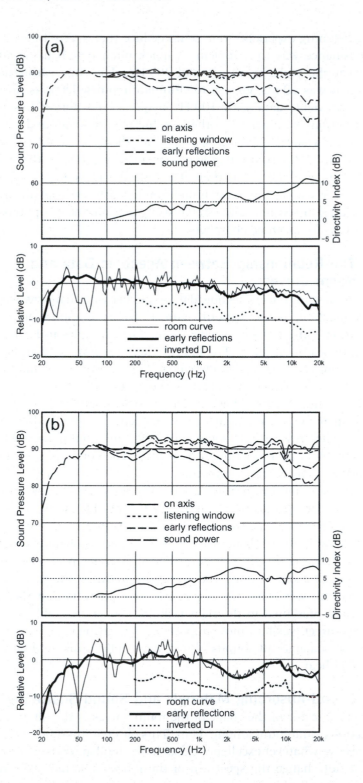

(a) Loudspeaker "I." Top: the full anechoic data. Bottom: in-room measurements with the loudspeaker in the front-left location averaged over several head locations within a typical listening area in a typical rectangular room. As a means of understanding which sounds from the loudspeaker contribute to this measurement, the "early-reflections" curve from the anechoic data set at the top has been superimposed. It is not a perfect match for the room curve, but it is not far off. Obviously, it is the off-axis performance of the loudspeaker that is the dominant factor in determining the sound energy at the listening location, and, in the several comparisons that have been done, the early-reflections curve seems to be a better fit than the sound power curve. As noted in Figure 18.5, the on-axis curve is the dominant factor at high frequencies. If some of the high-frequency portion of the "on-axis" data were added to the "early-reflections" data, the resulting curve would be an even better fit to the measured room curve. The inverted directivity index (DI) curve is included just to add support to the concept that the directivity of the loudspeaker is a factor not to be ignored, although in common audio discourse, it routinely is. As is expected, the standing waves in the room take control at low frequencies and the prediction fails. (b) Loudspeaker "B." This loudspeaker has directivity problems. In the DI curve, one can see the directivity of the woofer rising with frequency and then falling in the crossover around 350 Hz to a rather large (6-in., 150 mm) midrange driver that exhibits increasing directivity up to around 2 kHz before crossing over to an unbaffled tweeter with wide dispersion, which by 5 kHz has taken over. This behavior is clearly seen in the family of frequency responses, even at small-angles. There is a significant difference between the on-axis and the listening-window curves. In the lower box of curves, the pronounced mid- to upper-frequency undulations seen in the room curve are clearly associated with the off-axis behavior of this loudspeaker system. The shape of the room curve is clearly signaled in the shapes of both the "early-reflections" curve and the inverted DI. As in (a), the addition of some "on-axis" to the "early-reflections" curve will improve the match at very high frequencies. However, the on-axis frequency response by itself is not a useful indicator of how this loudspeaker will perform in a room.

absorbing and scattering/diffusing surfaces must have constant performance over the frequency range above the transition frequency, below which the room resonances become the dominant factor. As we will see in Chapter 20, such devices are significantly larger and thicker than some of those that have been used in the past.

The next step is to try to understand where these metrics stand in the hierarchy of measurements that are usefully revealing of perceived sound quality in rooms. But first, a short digression into an important part of the industry and how its standardization practices compare with what has just been discussed.

18.2.6 The "X" Curve—The Standard of the Motion Picture Industry

Sound reproduction in homes began with the belief that it is necessary to understand what the loudspeaker is doing. It hasn't been very difficult—the loudspeakers are smallish, the far field is not very far away, and anechoic chambers can be built for, admittedly large, but affordable sums of money. We have just seen that with enough anechoic measurements, it is possible to get impressively close to predicting the shape of room curves, even in nonspecific, typically furnished, rooms—except, of course, at low frequencies.

In professional audio, things started off quite differently. Sound reinforcement loudspeakers for auditoriums are large, heavy, and used in arrays, aimed in different directions to cover a widely distributed audience. Measuring them is a physical and acoustical challenge; the far field of an array is a long distance away. Consequently, room curves (or "house curves" as they are known in pro audio) were really all that could be measured once a system was assembled, usually at several locations throughout the audience. Early instruments permitted only steady-state measurements, using warbled tones and, later, bandpass-filtered pink noise.

Loudspeakers that were quite flat on axis, when measured in the audience area of an auditorium, exhibited substantial high-frequency roll-off. This was the result of several influences:

- Increasing directivity of the loudspeakers at high frequencies (early horns did not have constant directivity), reducing the high-frequency energy radiated into the room.

- Acoustical absorption in the rooms, typically more effective at high frequencies.

- Atmospheric absorption of the high frequencies because of the long propagation distances. This would apply not only to the direct sounds but also to reflections.

- The measurement microphones, early versions of which became quite directional at high frequencies.

- In cinemas, a high-frequency screen loss would be added to this list.

There was an instinctive belief that a loudspeaker with a flat axial frequency response and good directivity was probably a good objective, but the measured room curve that represented such a loudspeaker could not be flat; it had to roll off at high frequencies. Various "target" curves evolved, undoubtedly influenced by inconsistencies in loudspeakers of the day, and by inconsistencies in shape, size, and acoustical treatment of the auditoriums. Proof that it was a flawed measurement existed in the fact that almost always the final equalization of a large system was done subjectively, by listening. Using a room/house curve as a means of measuring the loudspeaker simply cannot be reliable, and experience suggests that, by itself, it is not a reliable indicator of sound quality. More and better information is needed.

Audio professionals knew that sufficient knowledge of the loudspeaker as a sound source should allow an estimate of what might arrive at a listener's ears to be computed. The predictive process will be different from small rooms where, as we have found, the boundary reflections are mostly benign. Sound reinforcement often takes place in acoustically hostile venues, with large reflecting surfaces that inject long-delayed disruptive echoes into the audio scene so loudspeakers tend to be much more directional, putting the listeners in a proportionately stronger direct sound field. All of this is becoming increasingly predictable, and elaborate acoustical modeling programs exist exclusively for that purpose. Using these programs, it is possible to predict the directional radiation properties of large loudspeaker arrays, knowing the detailed acoustical performance of the individual modules and the geometry of the array. Measurements in the audience have not been abandoned, but new instruments permit time-windowed measurements so different intervals—early, middle, late, or everything combined—can be independently inspected. Foreman (2002) discusses large-venue system equalization, suggesting that more attention should be paid to the direct sound and less to the reflected sound (p. 1189), a point made again in Eargle and Foreman (2002).

Starting back in the 1930s, the motion picture industry commendably attempted to standardize the sound in cinemas and in production facilities. There were few loudspeaker options, and the physical arrangements were somewhat uniform, which allowed the industry, in the late 1930s, to settle on a standard target curve for the power amplifier output that seemed to result in good sound from signal sources of the day, through some loudspeakers of the day, in some cinemas of the day. The story of events between then and now is documented by Allen (2006). The historical account is interesting, but there is not enough technical detail to understand why some things were decided, and, sadly, the story does not have what this author considers to be a happy ending. Since about 1970, cinema sound-reproducing systems (B-chains) have been calibrated using steady-state, low-resolution room curves. The culmination of decades of effort is a target curve, called the "X" curve (ISO 2969, 1987, and SMPTE 202M-1998), shown in Figure 18.11a.

The "X" in the name hints that there may be an element of the traditional mathematical "unknown quantity" in these measures. Room curves measured in the relevant motion-picture dubbing/mixing theaters, review rooms, and cinemas include all of the frequency response and directivity foibles of nonstandardized loudspeakers, both of which are altered by a perforated screen. They include the inconsistencies of nonstandardized rooms that vary in acoustical treatment, shape, and size. The measurements are made using 1/3- or 1/1-octave filters that yield histogram, bar-graph images of frequency response. Nothing of any subtlety can be deduced from such data; no narrow-band phenomena can be revealed. Finally, as if admitting that this whole exercise is only a crude estimation, a generous ±3 dB tolerance is blended into the mix.

To this expansive tableau of uncertainties must be added the great differences in volume between a large cinema and a much smaller dubbing or mix room; sound tracks are expected to sound the same in all locations. Acknowledging the influence of room size, Figure 18.11b shows variations in the high-frequency roll-off slope that are suggested for rooms capable of seating different numbers of listeners. Can it really be as simple as this, adjusting a treble control? What

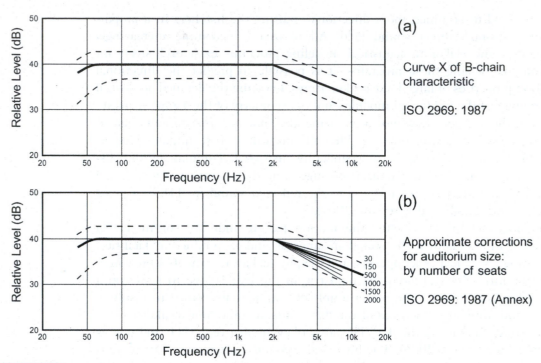

FIGURE 18.11 *(a) The "X" curve according to the ISO 2969 standard. The SMPTE version extends the high-frequency target to16 kHz, with a slope increased to –6 dB/octave above 10 kHz. (b) Both standards suggest these corrections for rooms of different seating capacities. © ISO. This material is reproduced from ISO 2969 with permission of the American National Standards Institute (ANSI) on behalf of the International Organization for Standardization (ISO). No part of this material may be copied or reproduced in any form, electronic retrieval system or otherwise or made available on the Internet, a public network, by satellite or otherwise without the prior written consent of the ANSI. Copies of this standard may be purchased from the ANSI, 25 West 43rd Street, New York, NY 10036, (212) 642-4900. http://webstore.ansi.org.*

about reflections arriving at large delays in large rooms and at small delays in small rooms? This is obviously another simplistic estimate. Using a head count as a measure of room acoustical properties sends a poor message, but the recommendation is in the right direction in that the curves become flatter as the rooms get smaller. Interestingly, the ±3 dB tolerance in the reference curve embraces most of the suggested variations.

Of course, with no anechoic data on the loudspeakers, it is impossible to interpret what they might have contributed to the shape of the room curves. Without information about the acoustical properties of the rooms, it is not possible to infer what the room might have done to the sound from the loudspeaker. Without higher resolution in the measurements, it is not possible to see all of the problems that may be audible. Without controlled, double-blind, subjective evaluations, the true merit of any equalizations is in doubt. It seems that no real science has been done.

A white paper by JBL Professional (2003) describes the present situation, with examples of their own products. Having anechoic data on the loudspeakers is a great advantage. Obviously, improvements to the standardized cinema calibrations are possible, and many within the industry realize it (see Allen, 2001, 2006; Engebretson and Eargle, 1982; Holman, 1993, 2007), and even a "future work" finale to the SMPTE 202M-1998 standard. It would be interesting to expand the science of sound reproduction in small rooms, as displayed in this book, into larger venues. That would make it necessary to take into account the radiation characteristics of the loudspeaker, some of the elemental room-boundary interactions, and some of the perceptual characteristics of the listener, elements totally ignored by or incomprehensible to an omnidirectional microphone and a spectrum analyzer.

Engebretson and Eargle (1982) conclude that a loudspeaker with "flat power response, in conjunction with constant directivity-frequency characteristics, will yield the desired results with no further adjustments required, other than compensation for through-the-screen losses." Flat power combined with constant directivity means that the loudspeaker has a flat on-axis frequency response. Such loudspeakers, compensated for the directional and spectral effects of perforated screens (Eargle et al., 1985), would be the "defined" sound source that current understanding of loudspeakers in rooms suggests is an essential starting point. This is eminently possible in new installations. The practical problem for the motion picture industry is that there is an enormous base of existing installations, especially cinemas, but also production facilities. Not all of these may meet the new requirements, and the industry is not any longer prospering as it has in the past.

Until something changes, it must be concluded that at this time the reference sound quality within the motion picture industry is somewhat "approximate," even within its own hierarchy of listening spaces. However, once an audience is into the thrills, spills, and drama of a film, it seems that the "willing suspension of disbelief" extends also to an acceptance of audio that is often less than excellent. When the film product is transported into the home, it is occasionally apparent from the sound quality that "this is a movie." It seems like a positive trend for home theater that some film-to-video transfers are now taking place in small rooms set up in a manner similar to home theaters and using the same kinds of loudspeakers. This is the only form of "mastering" that is appropriate for the noncinema audience, the dominant source of revenue for the industry. If anyone chooses to follow the guidance in this book in such mastering situations and in home theaters, we will have made an important step toward preserving the motion picture arts from producer to recipient.

18.2.7 Trouble in Paradise—The Pros Must Set an Example
However, even more important, it is long past the time in this mature industry that manufacturers of loudspeakers, especially those aimed at the professional

market, need to provide comprehensive anechoic data on their products. A few already do, and that is commendable. For loudspeakers intended for smaller listening spaces, there is no real challenge to acquiring meaningful data. A reasonable description of how the product performs needs to be the "price of entry" to this marketplace. It should not be up to the customer to discover information that should be publicly available. During the design of the product, this information was presumably available to the engineers who designed the product. If such data were *not* available to those engineers, then one is left to contemplate the competence of the source of the product. The descriptions of acoustical performance offered by many of the significant players in the loudspeaker business are simply insulting in their inadequacy.

18.3 COMPARING THE SUBJECTIVE AND OBJECTIVE DOMAINS

Figures 17.2 and 17.3 showed that even primitive measurements and listening tests reveal a subjective preference for loudspeakers with frequency responses that are flat and smooth. These results were not published at the time; they were not really in the category of scientific data, but they provided a stimulus to do more. Interest was not lost, and activity continued, but it was many years before the vagaries of life and work conspired to produce the circumstances for the next big step.

18.3.1 Measurements

It has been said before that measurements don't change. If they are done properly, they can be repeated many times in many places and the answer is closely similar. Opinions are different. Not only opinions from different persons, but from the same person at different times and places. In audio, it has been long regarded as a "given" that personal taste in sound quality is variable, not really to be trusted in any generalized sense.

In 1986, almost 20 years after the amateurish first tests described in Chapter 17, the author performed an elaborate subjective-objective investigation. It led, as discussed in Chapter 17, to a selection process for listeners. Of the 42 listeners who started the tests, only those results from the 28 most consistent are included in the following results. They all had hearing threshold levels within 10 dB of ISO audiometric zero at frequencies below 1 kHz and within 20 dB up to 6 kHz (Toole, 1986, Part 2). All of the listening tests were double-blind, of course; care was taken to avoid biasing the listeners.

Listeners auditioned the loudspeakers in mono in groups of four, presented in different randomized combinations, until each listener heard each loudspeaker three to five times. After completing a questionnaire of subjective qualities, listeners were required to provide an overall "fidelity rating" on a scale of 10, where 10 describes the *best* imaginable and 0 represents the *worst* imaginable reproduction. No reference sound was provided, and there was no formalized training of the listeners. Most had no prior experience in structured listening tests, although all had experience in critical listening either in their professions or audio hobby.

Figure 18.12a shows some results from the tests, sorted into columns according to the overall fidelity rating, and into rows according to the combined vertical

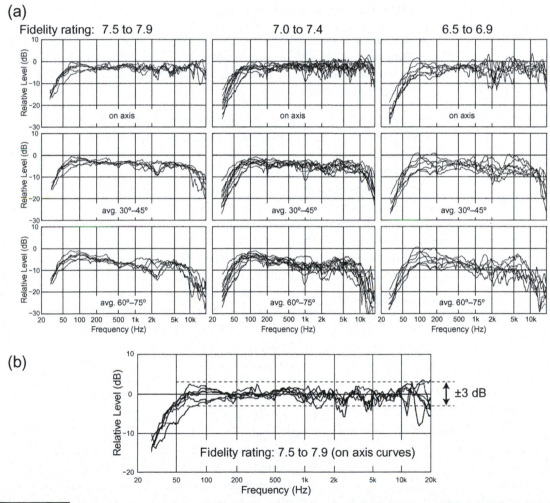

FIGURE 18.12 *(a) A sample of results, showing loudspeakers grouped according to subjective fidelity ratings in three categories. There were 6 loudspeakers awarded ratings between 7.5 and 7.9, 11 loudspeakers in the range 7.0 to 7.4, and 7 loudspeakers in the range 6.5 to 6.9. The original data include a fourth, lower category. The measurements are unsmoothed, 200-point, log-spaced stepped-tone anechoic measurements. To eliminate the effects of loudspeaker sensitivity, the vertical positions of the curves were normalized to the mean sound level in the 300–3000 Hz band. This same frequency band was used to normalize listening levels. (b) An enlargement of the upper left graph in (a). From Toole, 1986, Figure 7.*

and horizontal angular range embraced by the measurements. It is not difficult to see a progressive degradation in the smoothness of the curves as the fidelity rating decreases. The paper includes a lower category that is even less regular. It is important to note two trends here:

■ There is an underlying "flat" trend in these clusters of curves. The variations, even the larger ones, seem to be fluctuations around a horizontal line for the on-axis groups and around quite straight gently-sloping lines for the off-axis groups.

■ The average bass extension—the low-cutoff frequency—progressively decreases as the fidelity rating increases. The listeners liked low bass— not *more* bass, in the sense that it is boosted, but bass extended to lower frequencies.

Figure 18.12b shows an enlarged version of the top-left group of curves, the on-axis measurements of the highest rated loudspeakers. They have been vertically shifted to be symmetrical around 0 dB, and a ±3 dB tolerance band is shown. Clearly a ±3 dB numerical description does not do these loudspeakers justice. It seems evident that smooth and flat was the design objective for all of these loudspeakers. Deviations from this target are seen at woofer frequencies (below 150 Hz), in the woofer/midrange to tweeter crossover region (1 to 5 kHz), and in the tweeter diaphragm breakup region above 10 kHz. The variations seen among these six loudspeakers from different designers and manufacturers, and even different countries, are smaller than the production tolerances allowed by some manufacturers for a single model. Over 20 years later, these loudspeakers would not be embarrassed if compared to products in today's marketplace.

The apparent preference for extended bass motivated Figure 18.13, where the low-cutoff frequencies for all of the loudspeakers were determined at two levels relative to the reference 300–3000 Hz band. The normal −3 dB "half-power" level was also tried, but it yielded no relationship, which was anticipated because of the substantial effects of solid angle gains (Chapter 12) and bass resonances (Chapter 13) in the room in which the listening tests took place (or indeed any room). Therefore, it was not a total surprise to find that the relationship between the fidelity rating and the low-frequency cutoff reached a maximum correlation for cutoff frequencies determined at the −10 dB level. Bearing in mind that this is a correlation achieved with all other factors

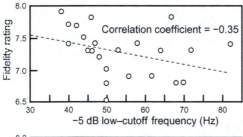

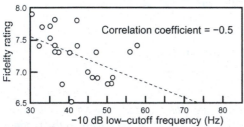

FIGURE 18.13 *Plots of the low cutoff frequencies determined at −5 dB and −10 dB relative to the average sound level over the 300–3000 Hz band.*

varying indicates that bass extension is a *very* important factor in overall sound quality evaluations.

18.3.2 A Contemporary Test

In the 40 years since those first tests described in Chapter 17 some things have changed, and some things have not. Loudspeakers, in general, are much better, and quite acceptable sound quality is available at eminently affordable price levels if you do a little shopping. None of it happened by accident; it is the result of better measurement systems being widely available and relatively inexpensive, and designers developing a healthy respect for good technical performance. The widespread availability of excellent raw transducers has made the design of good sounding loudspeaker systems a lot easier than it once was.

Yet, in other parts of the audio world, there are few signs of progress. Reviewers continue to ignore the scientific method, and a few even disparage those who follow it. Measurements are not a requirement for product reviews, and those that are seen cover a wide range from almost useless to quite impressive. Sometimes there are amusing examples of evasive writing when the technical data suggest something that runs contrary to the subjective component of the review. In one memorable case, the technical reviewer commented that he thought his measurements may be faulty—the data described an inferior product, the subjective reviewer really liked it. The measurements were correct. Listening tests continue to be of the "take-it-home-and-listen-to-it" kind, so many important variables are not controlled, and adaptation and bias are both factors. This was the basis of comments by pioneering audio journalist J. Gordon Holt in Section 17.5, lamenting the lack of scientific controls in subjective evaluations.

The following evaluation involved four high-end loudspeakers in the $8000 to $11 000 range per pair. All had been applauded by the audiophile press, with accolades like "Editor's choice," "Class A," "Product of the Year," and so on. Readers of these publications were led to believe that any one of these products would be a superb choice, and the prices only enhance such a belief.

Figure 18.14 is another example of the subjective and objective domains exhibiting harmony. The results of double-blind listening tests indicate that two loudspeakers tied for first place—one was a clear second and one brought up the rear. In looking at the corresponding measured data, it is not difficult to see the progressive decay in performance. There would seem to be no great mystery here, but two key elements underlying this display are very rare in the audio business: properly conducted double-blind listening tests and accurate, comprehensive measurements.

This test was a precursor to a much more ambitious test conducted by Olive (2003) who used some of these same loudspeakers, and others, in a test that, by the time of his paper, had involved 268 listeners and, by the time it was

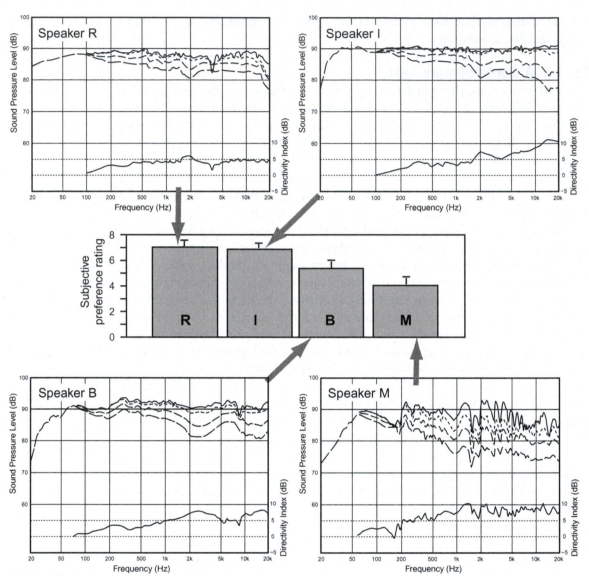

FIGURE 18.14 *Subjective ratings and objective data on four high-end loudspeakers. The error bars on the top of each histogram bar show 95% confidence levels, confirming that loudspeakers "R" and "I" are in a statistical tie for first place. The other differences are highly significant. Loudspeakers "R," "I," and "B" are forward-firing cone-dome systems. "M" is a hybrid system with an electrostatic mid-high-frequency driver mated with a conventional woofer. Loudspeakers "R," "I," and "B" are free from obvious resonances and differ only in the constancy of their frequency responses and directivities. Loudspeaker "M" is very directional, so directional in fact that listeners seated across the width of a sofa will experience different spectra. It also shows evidence of multiple resonances, bumps that appear in all of the curves including the sound power.*

disbanded, over 350 listeners. As reported in the paper, no group of these listeners, most of whom were visitors to the facility, differed in the order of their product ratings. Those tests were not as well controlled as the normal tests, in that time constraints allowed for only two rounds of listening and the listeners were in small groups, not individuals. It does show, though, that if listening tests are blind, with multiple comparisons among products, with revealing program and equal loudness levels, then it is possible to get remarkably consistent opinions from ordinary listeners. As shown in Figure 17.6, the consistent performance of selected and trained listeners is an advantage when time is of the essence, but listeners with an interest in audio and some attentive listening experience can yield the same answers if they are given enough time for repetitions.

Looking at Figure 18.14, it appears that recognizing true excellence or true inferiority in the measurements would not require any special training. Interpreting the relative merits of the vast middle ground of loudspeakers that don't do everything well is more difficult. Here the decisions come down to trading off problems and virtues, the audibility of which are likely to be dependent on program material. A listening test may settle such a debate. However, this is a debate over the relative merits of flawed products, and the winner of this kind of test is still not an example of excellence. It is common in listening tests for there to be little disagreement about the winners and losers, products with the highest and lowest ratings, but less certainty about the ratings of the intermediate products.

18.4 THE REAL WORLD OF CONSUMER LOUDSPEAKERS

It is important to have a perspective on where we stand at the present time; what kinds of products are in the marketplace? Measurements of the kind being shown here are almost never on public display, so the following is a brief tour of some real-world products, good and not-so-good. To be fair, no brands are mentioned because things can and do change. Past performances are not always reliable indicators of future performances, and it can go either way: up or down. The purpose is merely to provide the readers with enough visual data to allow them to see that there are recognizable patterns in the measured performances of loudspeakers that are awarded high ratings in double-blind listening tests.

One can only hope that one day data of this kind will be readily available to consumers, to help them in choosing products, and to reviewers, to help them explain some of the things that they hear. Arguments that families of curves like this are too complicated for common consumers to understand have merit, but does this justify displaying specifications data so simplified that all meaning is lost? Understanding and interpreting specifications is the duty of salespeople, and, as can be seen, no advanced degrees are required. If this kind of comprehensive data were suddenly available, the industry would undoubtedly suffer some angst as manufacturers adjust to the new, more level playing field. There may even be rebellion in some quarters, but such are the consequences of change. The automobile industry is a good example of one that has come to face the reality that their products will be subject to highly technical analysis as well as to demanding subjective evaluations. Sometimes the numbers and the opinions correlate better than others, but there is always a correlation. Being interested in cars, I have observed that some enthusiast magazines are seriously exploring new telemetric measures that do better at describing what is happening—lateral G force, slip angle, steering angle, and so on—at 100 mph in a decreasing-radius left-hand turn. Compared to this, showing a collection of frequency response curves on a simple product, decent examples of which are in the price range of tires and wheels, seems utterly trivial.

What is not trivial is the matter of production quality control. It is one thing to build a superb prototype, but it is a very different matter to duplicate that performance in mass production. A few manufacturers have impressive measurement capabilities on production lines—better perhaps than others have in their engineering labs. Sadly, this is the one reason why even honest specifications can mislead, because consumers have no option but to trust that all products with the same brand and model are the same. In some cases, they are amazingly similar, but in others . . . .

18.4.1 Examples of Freestanding L, C, R Loudspeakers

These are the loudspeakers commonly used as left and right front loudspeakers, and occasionally as center loudspeakers as well. Some are small bookshelf-sized products, needing a subwoofer to complete the spectrum, and others are full-range floor-standing units that may need no such help.

In Figure 18.15 we start with examples of poor loudspeakers, or loudspeakers that fail to live up to expensive expectations. That there are loudspeakers with problems is no surprise. That they are rewarded with excellent reviews is harder to swallow. The evidence of many years of experience is that all these reviewers, in an unbiased listening situation, would very likely have recognized the problems for what they were. Real measurements would have settled any debates, and the manufacturer would, in a fair world, get a reminder to pay closer attention to the first task of a loudspeaker: to be an accurate reproducer. Instead, the message is perpetuated that just about anything goes. The fact that truly good loudspeakers also get favorable reviews (as was exemplified in Figure 18.14) indicates that the existing reviewing process has a large tolerance range.

One could go on exploring the infinite ways of failing to be good, but instead let us look at what is being done by other manufacturers at affordable prices (see Figure 18.16). At low prices there are no fancy hand-polished wood finishes, no high-toned industrial designs, and the ability to play *very* loud may have been sacrificed, but in terms of sound quality, these are all excellent loudspeak-

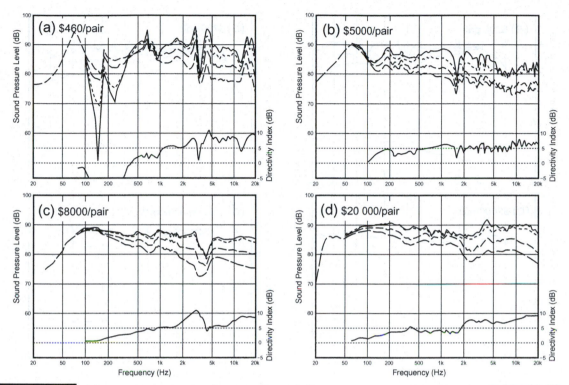

(a) An attractive-looking floor-standing unit from an internationally highly-regarded brand. The sound is not good. (It is what I call an "ecologically irresponsible" design, since it is truly a waste of raw materials.) If an engineer was involved, one has to suspect that the design was "phoned in." The cynicism of this brand is made all the more striking because the parent company has competent loudspeaker design capabilities. This product was farmed out to a local supplier to meet a "marketing" need in another country. (The world apparently needed another example of bad sound.) The loudspeaker in (b) is ten times more expensive but unfortunately not ten times better sounding. It is evident that this combination of an electrostatic mid-high-frequency unit and a conventional woofer does not go far enough in the sound-quality department. The DI of around 5 dB is about what might be expected of a dipole radiator (4.8 dB), but the frequency response seriously sags over the top two octaves, and the crossover around 100 Hz to the "one-note" bass unit could be improved. (c) An expensive two-way cone/ dome bookshelf with "exotic" ingredients. However, any virtues they may have contributed are swamped by frequency-response and directivity problems. It had no low-bass output, but the excessive upper bass made it sound notably flat, dull, and moderately colored. It got flattering reviews and was the chosen "reference" speaker of one magazine editor. (d) A very expensive product claiming the advantages of exotic diaphragm materials. At this price, one would have reason to expect something better than this performance, which exhibits both frequency-response and directivity anomalies. There also appears to be a significant resonance around 4 kHz, a bump that is present in all of the curves. The reviews have been fulsome.

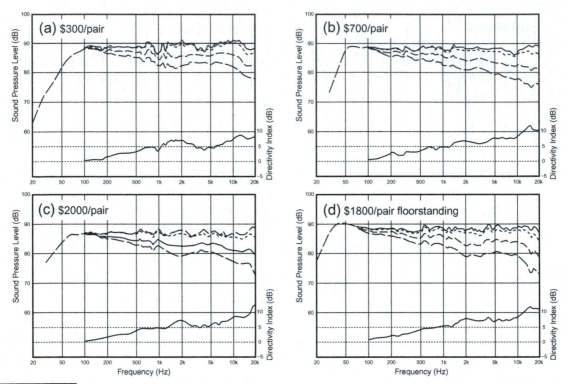

FIGURE 18.16 *(a), (b), and (c) Examples of three good bookshelf loudspeakers. The high price of (c) is explained by its elegant visual presentation, materials, and high-power drivers. (d) An example of a very well-balanced design; some may call it the "point of diminishing returns" in that it has the bandwidth and sound quality to be very rewarding.*

ers. Add a subwoofer to any of the bookshelf designs shown in Figures 18.16a, (b), and (c), and set the surround processor to "small," engaging an electronic high-pass filter, and the combination is capable of delivering a truly high-end, high-sound-level performance. Loudspeaker (a) is a remarkable performer at the price ($200/pair during sales) and sets a standard for sound quality that is hard to improve on at any price. A first-time buyer is off to a good start. At almost seven times the price, the bookshelf loudspeaker (c) has slightly smoother curves, features significantly better bass, can play much louder, and has an elegant appearance (in other words, pay more, get more). The full-range floor-standing loudspeaker in (d) needs no subwoofer, and it achieves its rather excellent performance-per-dollar status by being well engineered and visually inert: a simple-to-manufacture tall, rectangular box in simple vinyl woodgrain wrapping.

Figure 18.17 shows samples of two excellent, high-priced loudspeakers, that do almost everything well. To these should be added loudspeakers "R" and "I" in Figure 18.14. Collectively, these are examples of the present-day "kings of the hill." There are others, of course, but the measurements do not look very different. When they are put against each other in double-blind tests, the audible differences are small, somewhat program dependent, and listener ratings tend to vary slightly and randomly around a high number. In the end there may be no absolute winner that is revealed with any statistical confidence; the differences in opinion are of the same size as those that could occur by chance.

18.4.2 Horizontal Center-Channel Loudspeakers

A direct-view video display is a challenge for the center loudspeaker. Some people give up in frustration and use the stereo "phantom" center. DON'T DO IT! (See Section 9.1.3.) Most people use a horizontal loudspeaker system commonly configured in one of the two options shown in Figure 18.18. The simple one, often called the "midrange-tweeter-midrange" or MTM, arrangement is usually found in entry-level products but also, occasionally, in some expensive products. In its basic configuration of both woofers operating in parallel, crossing over to a tweeter—a two-way design—it is not optimum because of off-axis acoustical interference. In Figure 18.18a it is seen that this interference is symmetrical, so both lateral reflections suffer from the same flaw, affecting sound quality.

These designs also show up in vertical arrangements, in which case the acoustical interference is heard after reflection from the floor and ceiling. An

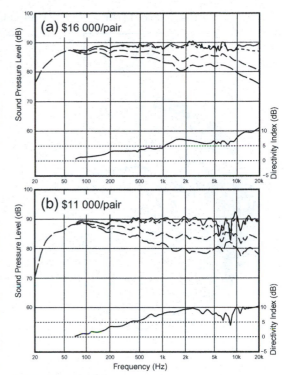

FIGURE 18.17 *(a) A high-priced, high-power floorstanding cone and dome loudspeaker. (b) A high-priced, high-power floorstanding cone woofer combined with compression-drivers and horns for the mids and highs. (a) has wider dispersion (lower directivity index) than (b), a characteristic difference between cone/dome and cone/horn designs.*

intermediate configuration, sometimes called the 2½-way, rolls off one of the woofers at a low frequency, allowing the second unit to function as a midrange. The result is a slight improvement in overall performance, but the horizontal-plane interference pattern is then asymmetrical and still not what is needed. The real solution is to add a midrange loudspeaker allowing both woofers to be crossed over at a frequency sufficiently low that the acoustical interference is avoided. The explanation is in the caption.

18.4.3 Multidirectional Surround Loudspeakers

Multichannel music is a latecomer to the sound reproduction scene. Professionals on the music side of the industry see surround loudspeakers as functionally equivalent to the front loudspeakers, placed in locations as shown in Figure 15.10a. Home theaters sometimes follow this lead, perhaps embellished with

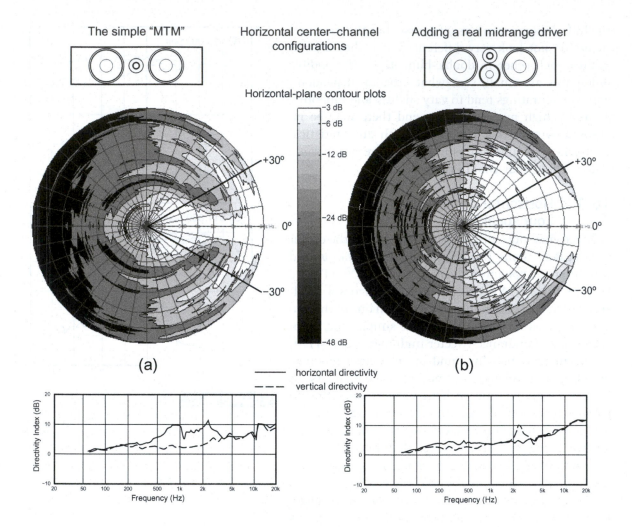

The simple "MTM"

Horizontal center–channel configurations

Adding a real midrange driver

Horizontal-plane contour plots

(a)

(b)

—— horizontal directivity

– – – vertical directivity

additional loudspeakers of the same kind in locations as suggested in the ITU layout (Figure 15.10b), the Dolby layout (Figure 15.11a) or the author's composite suggestion (Figure 15.11b). These schemes have similar conceptual foundations, and they all work well.

But the movie motivation for multichannel reproduction has a longer history. Its practices have changed with time because of evolving technologies on the professional side of the industry. Back in the very early days of home theater, we were stuck with a single, limited-bandwidth surround channel in the Dolby Stereo encoded signal. The task of the surround channel in those days was to provide a sense of envelopment and very occasionally to assist in the illusion of a flyover of some sort. There was no ability to direct sounds left or right, channel separation was limited, and sound quality in the surround channel was seriously compromised. In cinemas, several surround loudspeakers were dotted along the

FIGURE 18.18 *Two common configurations for horizontal center-channel loudspeakers. (a) The MTM design has two "midranges," which actually are woofer/midranges, that acoustically interfere with each other at increasing horizontal angles. This is because they are physically separated, and both radiate sound up to high frequencies to cross over to the tweeter. This can be seen in the polar plot displaying a horizontal view of sound level as a function of frequency and angle. Moving radially from the center, the concentric rings represent 200 Hz, 500 Hz, I kHz, 2 kHz, 5 kHz, 10 kHz, and 20 kHz. The areas in white represent sound levels within 3 dB of the axial (0°) output. The contours at the onsets of progressively darker gray areas represent sound levels at 6, 12, 24, and 48 dB below the axial output. It can be seen that moving horizontally off axis, the first interference dip in the frequency response appears at under 10° off axis. Recall that in Figures 16.6 and 16.10, it was shown that a horizontal dispersion of ±30° was required for the center loudspeaker to deliver intact direct sound to all listeners in a typical home theater. This figure shows that by 30° this loudspeaker is experiencing heavy acoustical interference, and the output has dropped seriously over a wide frequency range. This is not good. (b) depicts a much better design, in which a midrange loudspeaker has been added in a central location. This allows the widely spaced woofers to be turned off at a lower frequency, and the interference dip disappears. The superior dispersion of the small midrange driver is apparent. The result is that at 30° the sound is unimpaired; only a normal slight loss above 10 kHz due to tweeter directivity is seen (this would be seen in any loudspeaker with a conventional dome tweeter). In fact, the lateral first reflections, occurring at much larger angles, are also in reasonable condition (Figure 16.10). The bottom graphs show separate vertical and horizontal directivity indexes for these designs. The horizontal dispersion problem with the MTM layout and the virtue of the three-way design are apparent. Data provided by the creator of this informative display, William Decanio, Harman Consumer Group.*

walls and across the back of the auditorium (see Figure 15.3), which when combined with sounds reflected around the large room, provided a pleasant sense of being surrounded by sound, although those in the center of the cinema were treated to the best illusion.

When this experience was replicated in small rooms in homes, using only a single pair of surround loudspeakers, the effect was less than impressive. Inverting the polarity of one loudspeaker helped with the "in-head" localization if one sat in the exact middle, but sitting off-center led quickly to localization of the nearer surround loudspeaker. This is easily explained by the inverse square law propagation loss of conventional small loudspeakers, as illustrated in Figure 16.8b and the fact that the surround signal was mono. To lessen this effect, Shure and then THX introduced electronic decorrelation into the signal paths of the surround loudspeakers, which eased the localization problem and improved the sense of envelopment (see Section 15.6). However, the problem for off-center listeners, especially those close to a side wall, remained a problem, for the reason explained in Figure 16.8: The perception of envelopment diminishes as the left-right sounds differ in level, and localization of the nearer loudspeaker is probable.

The idea of using bidirectional out-of-phase loudspeakers was introduced by THX as a means of reducing the level of the sound aimed at the listeners and increasing the level of the sounds aimed toward the front and rear. These were called "dipoles." However, a true dipole, or doublet, consists of two sources separated by a very small distance and radiating in opposing phase (Beranek, 1986). Unbaffled diaphragms are approximations of true dipoles. The typical "dipole" surround loudspeakers employ small-box enclosures in which the front- and back-facing drivers are separated by distances that are large compared to the radiated wavelengths, and as a result the directional patterns become very disorderly. Figure 8.6 shows that sounds arriving at the listening location after reflection from front and back walls are not very effective at generating envelopment. Fortunately the loudspeakers were not true dipoles with sharp nulls, as shown in Figure 8.7; there was substantial "leakage" of direct sound in the direction of the listeners. Nevertheless, the combination was deemed to be acceptable for film sound. Because the record/playback technology of the time did not permit it, there was no intention that individual surround channels be used as stand-alone, localizable sound sources.

Then the situation rapidly improved. First there came several playback algorithms that offered a measure of separation between the left and right surround channels: Dolby ProLogic, Harman/Lexicon Logic 7, Fosgate 6 Axis, and others. Surround channels acquired a full-frequency range. Ultimately, the introduction of digital discrete delivery systems, Dolby Digital and DTS, removed all of the old restrictions. Suddenly, the original five loudspeaker layouts sounded much better, and inevitably the algorithms were elaborated to provide for six or seven channels. It was a new game. Sound designers for films could themselves control the decorrelation in the surround channels, varying the sense of envelopment, and they took advantage of the discrete surround channels by sending localizable sounds to the left, right, and rear (if available). So the surround channels acquired a new job in addition to providing enveloping ambience. It was time for the playback system to stop "editorializing" and just let the art come through. The motivation for the dipole surround loudspeaker had disappeared.

The residual problem was that it all could work well for those seated in the center of the loudspeaker array, but listeners seated close to the side walls could localize the surround loudspeakers even when they were not intended to be localized—that is, when they were delivering enveloping, ambient sounds. Elevating them helps, but the real problem, as shown in Figure 16.8, has to do with propagation loss. To deliver the correct impressions of the localized sounds, listeners must receive strong direct sounds from the surround loudspeakers so the precedence effect can work. This, as is pointed out in Figure 16.10, requires that the surround loudspeakers have uniformly wide horizontal dispersion (up to ±70° in the example).

Still, dipole surrounds continue to be used by some installers. One of the popular justifications is that they seem to lessen the localization problem. As will be shown in Figure 18.20, the explanation has to do with the greatly attenuated high frequencies in the direct sound, not the bidirectional directivity. Rolling off the high frequencies in the surround channel was used in the first-generation Dolby Stereo mixes to prevent audiences from localizing sibilant "splashes" leaking from the front channels because of playback errors in optical sound tracks.

Many manufacturers realized that the "bipole" (bidirectional in-phase) configuration delivered a satisfying set of surround illusions, and that option has become commonplace. New audio jargon began to include descriptions of different surround loudspeaker options based on how "diffusive" they are. Since diffusion is a property of the sound field, what is being referred to is how "dispersive" they are. The ±70° horizontal dispersion requirement found for the sample home theater in Figure 16.7 means that many conventional forward-firing loudspeakers may have difficulty covering a large audience if there are only two surround channels. Adding more surround channels is one solution—certainly if the loudspeakers can be aimed to optimally cover the audience and the added decorrelation contributed by the additional processed channels helps with envelopment.

Let us pause and have a look at some real loudspeakers designed for surround use. The first example is a bidirectional on-wall design that can be used in any of several configurations. First, each directional loudspeaker can be fed a separate signal. Second, only one of the loudspeakers can be used. Third, both can be used, connected in phase and fourth, both can be used connected out-of-phase. In this example product, the real engineering effort went into the performance of the individual front- and rear-firing systems and how they merged when operated together in phase. An out-of-phase setting was provided, but performance in this mode was *not* optimized.

Figure 18.19 illustrates the behavior of the three directional modes available on a particular bidirectional surround loudspeaker. In (a) the monopole mode produces a good-looking family of curves, distinguished by the tight grouping of all of the curves, from on axis, through listening window and early reflections, to sound power. Only above 10 kHz is there any misbehavior, probably because the listening axis is about 60° off of the tweeter design axis. Cabinet diffraction may be involved. The sensitivity of this configuration is low because only half of the transducers are used.

In Figure 18.19b, the bipole mode has substantially higher sensitivity and is capable of considerably more sound output because all transducers are in use and they are all operating in phase. The family of curves describes a good-sounding loudspeaker. The DI describes a loudspeaker that is almost (hemispherically) omnidirectional. This seems to fit the description of the loudspeaker required in Figures 16.7 and 16.10.

When the drivers are operated in the dipole, out-of-phase mode, sensitivity plummets because the drivers are working against each other. This acoustical interference can be seen at work in the irregular shapes of the direct sound curve and the listening window curve. This is the penalty of the physical separation of the front and back radiators. What happens within the broad null is disorderly, as the sounds move in and out of phase with each other, depending on frequency

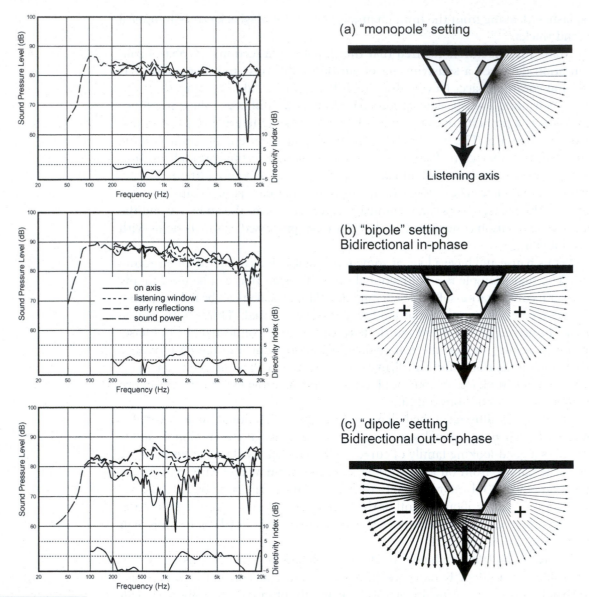

FIGURE 18.19 *Measurements on a bidirectional surround loudspeaker with switchable directivity patterns. The listening axis is perpendicular to the (side) wall. Note that the DI in these and all other diagrams in this book is calculated as the difference between the listening window and the sound power curve. Normally, the listening window curve is only slightly different from the on-axis curve. However, when the front- and rear-firing drivers of these loudspeakers are connected out of phase—as in (c)—there is a great amount of acoustical interference at and around the listening axis. This is what creates the "null." In that mode, the shape of the on-axis curve applies only to one specific microphone location. Because it is the result of acoustical cancellation, the curve shape changes considerably with small angular changes. Note the greatly different listening window curve representing a spatial average over ±30°. Every listener would experience a different directivity index if it were measured in the conventional manner. The directivity index in such a loudspeaker has little practical meaning.*

and horizontal angle. In terms of sound quality of the three directional options, it is evident that the dipole mode is not competitive. No more will be said about this performance because the next figure includes four loudspeakers that were optimized to perform in the dipole mode, that being the only mode available in them.

Figure 18.20 shows the dipole from Figure 18.19c and four other dedicated dipole surround loudspeakers, three of which have THX certification. Readers by now can probably guess how these loudspeakers sound. They are far from ideal. Apart from speculations about what their absolute sound qualities may be, it is evident that they are all very different from one another. In terms of specifying the performance of these loudspeakers, it has been common practice to use sound power, the total sound radiated by the loudspeaker, most of which must be reflected by the room surfaces before arriving at the listener. Sound power alone is not a reliable measure of sound quality. Even in an ideal, highly-reflective, frequency-neutral room, these loudspeakers cannot sound or measure the same.

The large and inconsistent differences between the on-axis curves and the listening window curves confirm the complexity of events within the acoustical interference region at and around the listening axis. It is further obvious that these loudspeakers revert to conventional wide-dispersion devices at frequencies below about 500 Hz. Because it is frequencies in the range from about 100 Hz to about 1 kHz that generate the desirable perception of envelopment, it is fortunate that much of that capability remains intact. Let us put these data into the context of what is required of surround loudspeakers:

- *Localizable sound effects directed to a single channel*. With the on-axis and listening window curves so variable and so different from each other and the other curves, sound quality suffers. These curves are also substantially attenuated in the mid-upper frequencies, so the direct sound—the one that determines direction—may not be as loud as a first reflection from a large surface. The author recalls more than one installation in which sending a signal to the side surround loudspeaker produced localization at the rear wall. Because the precedence effect is substantially nullified if the direct sound and delayed versions of that sound have different spectra, localization will be less than ideal.

- *Enveloping ambience and music*, involving both/all surround channels. Envelopment is a perception generated by sounds in the frequency range from about 100 to about 1000 Hz that arrive from the sides 80 ms or more after a similar sound from the front (see Figure 7.1). The essential concept of the dipole prevents this from happening because more sound is radiated toward the front and rear walls than toward the listener. Sounds that arrive from those directions are less productive at producing

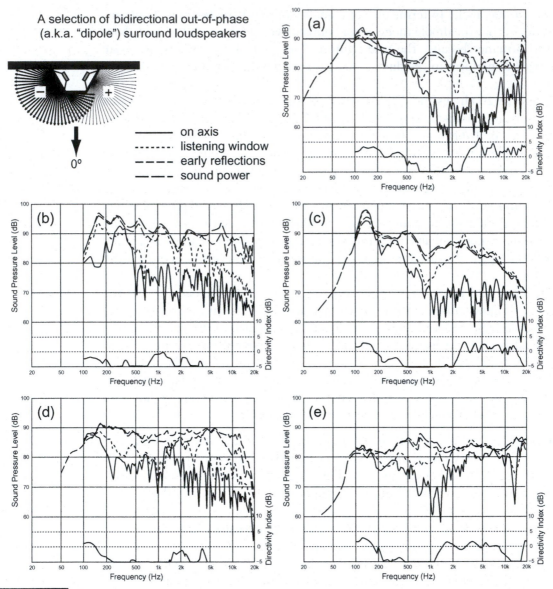

FIGURE 18.20 *Five on-wall-mounted "dipole" surround loudspeakers.*

envelopment (see Figure 8.6). However, close inspection of the curves reveals that these bidirectional out-of-phase loudspeakers combine to mono at low frequencies to preserve low-frequency output. From about 500 Hz down, they all exhibit somewhat conventional wide-dispersion behavior. As a result, some impressions of envelopment are preserved, although only at low frequencies, and sound quality has suffered in the process.

- *A vocal or instrumental component of a "middle of the band" musical recording.* If the customer decides to listen to music, the requirement is for comparably good, preferably identical performance from all loudspeakers. Dipoles don't have it.

18.4.4 The Perfect Surround Loudspeaker?

A very wide, uniform horizontal directivity pattern is needed to provide the localization cues for directed sound effects and to establish the basis for the perception of envelopment. Conventional forward firing or bidirectional in-phase on-wall loudspeakers are eminently capable of delivering those experiences, but excellence is guaranteed only for the central seating area. As listeners move toward the sides, sounds arriving from the nearer loudspeaker get rapidly louder, and those from the opposite loudspeaker get quieter. The sense of envelopment is progressively diminished, and it eventually disappears, replaced by sound emerging from the nearby loudspeaker. Figure 16.8 explains the cause—propagation loss—and proposes one solution: full-height line-source loudspeakers. However, as good as they may be, for reasons of size and cost they are not practical solutions for the mass market. A target performance for "the perfect surround" loudspeaker was also proposed: a loudspeaker with, in effect, no propagation loss.

Figure 18.3 showed sound-level contour plots for several variations of truncated, curved, and shaded-line loudspeakers. The two on the bottom (e and f), versions of the constant beamwidth transducers (CBT), were of special interest because they exhibited constant directivity (potentially good sound), and some of the contours held nearly the same sound level over long distances. Inspired by this, Figure 18.21 shows a family of contours taken from the same paper (Keele and Button, 2005) but inverted, placing the loudspeaker at the ceiling interface. The row of "heads" across the width of this imagined room intersect with only one line; they are at a nearly constant sound level from 200 Hz to 8 kHz. And as one moves even closer to the loudspeaker, at the same height, the sound level goes *down*.

There is a leap of faith required in this because the Keele and Button simulations involved only a single horizontal surface, the floor in Figure 18.3, or the ceiling in this inverted diagram. A modified solution needs to include, at the very least, a side wall. Note that sound levels drop rapidly below head level, so the floor reflection is not a factor. Incidentally, these are true line source configurations in that the transducers are small and densely packed. The narrow horizontal profile allows for a very wide and uniform horizontal dispersion.

If a variation of this design, or something else entirely, can come close to this performance target in a real room, the result would be a remarkable improvement in spatial and directional effects for the entire room. It is probable that fewer surround loudspeakers would be needed for large audiences. Whatever may happen, significant degradations from this idealized picture are possible

Data from Keele and Button (2005) showing a constant beamwidth transducer (CBT) inverted to simulate a surround loudspeaker, showing sound levels as they might be at several listening locations. This simplistic illustration ignores the fact that in reality there is a wall behind the loudspeaker, which was not part of the Keele and Button simulations. A real loudspeaker for this application would need to be modified to accommodate this constraint. The white lines are contours of equal sound level and adjacent lines differ by 3 dB.

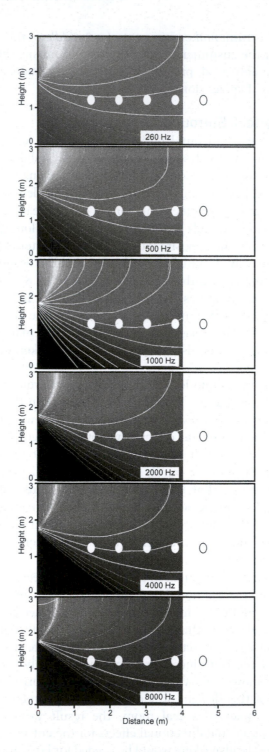

before we are worse off than we are now with our conventional, spherical-spreading, small-box loudspeakers.

18.4.5 Equalizing the Surround Channels

It began with the THX "timbre matching" feature, initially justified on the basis that the timbre of a surround channel did not match that of a front-channel loudspeaker. Section 15.6.1 discusses the topic, but the data in Figure 18.20 add to the argument, confirming three things:

- First, if the surround loudspeakers are of the "dipole" configuration, there is absolutely no doubt that the timbres don't match, nor could they possibly do so with such radical differences in acoustical performance.

- Second, the differences among these loudspeakers indicate that no single equalization curve could satisfy all "dipole" loudspeakers.

- Third, equalization can change frequency response, spectrum, and the differences among these loudspeakers and, between any of them and a front-channel loudspeaker, include directivity as well as spectrum. Equalization cannot change directivity or neutralize the timbre of its non-minimum-phase reflected sound field.

Recently, another issue has surfaced relating to the behavior of the phantom image in transition from front to surround loudspeakers (±30° to ±110°). Corey and Woszczyk (2002) examined this situation. When phantom images were placed to the side, some listeners reported split images, with frequencies greater than about 1500 Hz being localized near the front loudspeaker and lower frequencies localized to the side or rear. In spite of this, when asked, listeners responded with a single directional response. Split images are commonplace, and human perception has learned to deal with it, usually with what could be described as "compromise localizations." The analysis of multiple images in sound localization, as it happens, was part of the author's PhD thesis work (see Sayers and Toole, 1964; Toole and Sayers, 1965a, 1965b).

Looking again at Figure 7.6, which shows directional loudness in the horizontal plane, and comparing what the perceived spectral trends are at ±30° and at ±110°, it can be seen that at ±30° the high frequencies are slightly louder than the lower frequencies, and at ±110°, the reverse is true, with lower frequencies appearing to be louder. This pattern supports the listeners' descriptions of the split images when they chose to pay attention to them, which, for the most part, they did not.

It seems there isn't a serious problem, but if one wished to polish the performance of these intermediately localized phantom images, the correct remedy would be to provide for the spectral modifications in the recorded sound tracks, and then only when intermediate panning is done. Hard-panned—discrete—signals to each of the loudspeakers have no such problem, and these are extensively used for music and sound effects in movies and for ambience or individual musicians in multichannel music.

Holman (2008) discusses a similar split-image situation for sounds panned between front and side channels and recommends that the surround channels be equalized to remedy it. The suggested equalization takes the form of a dip in the frequency response in the range 2 to 6 kHz, implying that in the split-image situation, it is the high frequencies that are dominant in the surrounds, which is the reverse of the Corey and Woszczyk observation. There appears to be disagreement on the audible effects in this situation, but whatever the real truth, it seems inappropriate to permanently compromise the performance of any loudspeaker channel to cater to a possible perception during a little-used transient event. It is a psychoacoustic issue that should be handled in the creation of the recorded signals, if at all.

Front-to-side phantom images are relatively unstable phenomena, partly because the loudspeakers are separated by about 80°. They are predictable only for the single listener in the sweet spot. Because of this they are not often used in program, except briefly, to demonstrate movement in movies.

Equalization of the surround channels therefore is no different from equalization of the front channels. If the loudspeakers have been well chosen, it is most likely that the spectral issues will be at the lower frequencies having to do with adjacent boundaries (Chapter 12) and achieving a suitable acoustical crossover to the subwoofers.

18.5 EXAMPLES OF PROFESSIONAL MONITOR LOUDSPEAKERS

Monitor loudspeakers (the real thing, the ones used in recording studios, not the consumer loudspeakers that are sometimes called "monitors" or "reference monitors" for image building) are important to our industry. It is through these loudspeakers that musicians, recording and mastering engineers, and producers get to judge what they are doing. This was first discussed back in Chapter 2. There it was explained that the idea of monitoring through bad loudspeakers to get an idea of what the product might sound like to the average consumer is not a particularly useful idea because there are no standards for failing to be good. There are an infinite number of ways to sound bad. But as we have seen in this chapter, there seems to be a rather limited number of options to get high subjective ratings—that is, to sound good. A long-term look at the audio industry at all of its levels leads to the conclusion that the vast majority of loudspeaker designs aim to be flat on axis. Many fail, and they fail in many different ways. The one thing that poor-sounding, usually inexpensive, loudspeakers have in common is a lack of low bass. Thus, the suggestion is that audio professionals use the very best, most linear, most neutral loudspeaker they can find for all of their monitoring tasks. If they wish to hear how their artistic creation will sound through average inferior loudspeakers, use a high-pass filter to progressively eliminate the low bass; higher cutoff frequencies would then represent lower sound quality categories.

However, many years ago, the Auratone 5C came upon the monitoring scene. It was represented as a means of evaluating what "average" sound systems might be like (see Figure 2.6). A few samples showed that it was not a very consistent "standard" for the absence of excellence. Nevertheless, it became one of the loudspeakers that recording engineers liked to find in studios as they traveled around (familiarity is a good thing). It may also have been a factor in the popularity of near-field or close-field listening, where small loudspeakers are perched on top of the meter bridge of a recording console.

In any event, with the passage of time, other small loudspeakers found their ways to the meter bridge, and one of the products that really took hold was the Yamaha NS-10M. This was part of a series of high-quality loudspeakers, aimed at the professional but also at the high-end consumer market. The top of the line NS-1000M was distinguished by having beryllium dome midrange and tweeter diaphragms, something very adventurous in the mid-1980s. In some ways, they set new standards of performance, especially, as I recall, in terms of low distortion and the absence of resonances in the transducers.

In a discussion with one of the design engineers, the author was told that the bookshelf loudspeaker, the NS-10M, was intended to be listened to at a distance in a normally reflective room. The bass contour allowed for some bass boost from a nearby wall, and the overall frequency response was tailored for a listener in what then would be called the reverberant sound field, which, it was thought in those days, was best characterized by sound power. So the NS-10M was designed to have flat sound power. Figure 18.22a shows that they succeeded very well.

About the same time, JBL Professional made a monitor loudspeaker using drivers of much the same size, but being interested in delivering accurately balanced sound to listeners not far away, they designed their product to have a flat on-axis frequency response. Figure 18.22b shows that they succeeded very well.

Also shown in Figure 18.22 is the directivity index, and it can be seen, especially in the overlaid curves in (b), that they are almost identical. Interesting, two very similar loudspeaker systems engineered for different purposes: one of them to deliver accurate sound at short listening distances and the other at long listening distances. So which one ends up being the informal international standard near-field monitor? The one designed to be listened to at the far end of the room, the NS-10M! How could this happen? How could audio professionals be so wrong? Unfortunately, the whole truth may never be known, and

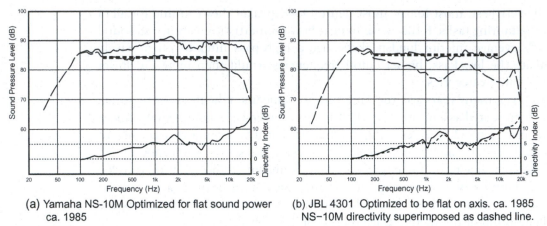

(a) Yamaha NS-10M Optimized for flat sound power
ca. 1985

(b) JBL 4301 Optimized to be flat on axis. ca. 1985
NS–10M directivity superimposed as dashed line.

FIGURE 18.22 *Two small 7- to 8-in. two-way loudspeakers with similar directivity indexes, each one optimized to a different target. (a) The Yamaha NS-10M was designed to have constant sound power. (b) The JBL 4301 was designed to have constant on-axis frequency response.*

one suspects that there are some good stories among industry insiders, but from this perspective, there is a possible reason: familiarity.

Figure 18.23a shows measurements of an NS-10M manufactured about 12 years after the one shown in Figure 18.22. It has changed and in a very interesting way. The original loudspeaker had a strong suggestion of the Auratone 5C shape in its on-axis frequency response, but this vintage is even more similar. Figure 18.23b shows that the 1997 version of the NS-10M behaves like a slightly smoother Auratone 5C with better bass extension. (Don't forget: familiarity.)

The other measurements depict monitor loudspeakers that seem to suffer from the "wandering standard" syndrome. If you like it, buy it. But isn't that what consumers do, not professional audio engineers? These are but a sample of the wide range of loudspeakers offered to audio professionals as monitors. Nowhere in the informational literature for these products is there any technical data to suggest what the loudspeaker might possibly sound like. Listen to it, and if you like it, buy it. The problem is that the performance of a monitor loudspeaker can be reflected in the recordings made while using it. The art can be corrupted in a way that may not have been intended, and, once done, it can never be restored to what it might have been. Recordings are forever.

Meanwhile, other manufacturers have pursued another goal, spectral neutrality, to give listeners in the control room an unbiased perspective on what they are doing. Figure 18.24 shows a little historical perspective on some monitor loudspeakers that were intended to hit the same performance target. Obviously, as time passed, technology and engineering methods improved, and the deviations from the target are now very small indeed.

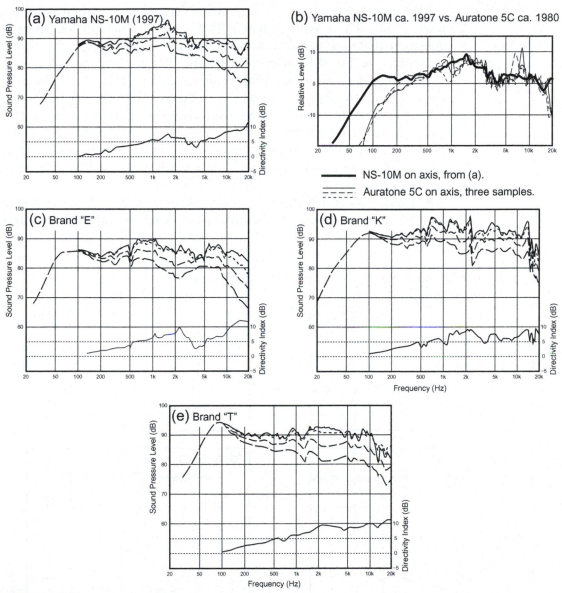

FIGURE 18.23 *(a) and (b) show that the Yamaha NS-10M performance drifted from that of the original version shown in Figure 18.22a. It is now even closer to the performance of the Auratone 5C, shown in (b), but it is slightly smoother and has more extended bass. The loudspeaker in (c) has a basic similarity, but the midfrequency emphasis is moderated. It is lower in amplitude and slightly lower in frequency. The loudspeaker in (d) also has an underlying midfrequency emphasis but adds some high-Q resonances around 600 Hz and 2.5 kHz that add an annoying "personality" to the playback. Why? (e) shows a loudspeaker that seems to be trying to be flat but fails at both frequency extremes: "punchy" bass due to the 80–100 Hz, underdamped woofer bump, and rolled-off high frequencies. The fact that all of these monitors, some more than others, exhibit high-frequency roll-off is puzzling in an age when arguments are being made that bandwidth beyond 20 kHz is a necessity for monitoring.*

This entire chapter full of data on loudspeakers is both reassuring and disappointing. The reassuring part is that there are monitor loudspeakers in circulation that can allow recording personnel to anticipate, very accurately, what might be heard by many consumers. The performance objectives for the professional monitor loudspeakers shown in Figure 18.24 are the same as can be seen in the consumer products in Figure 18.14 (R and I), 18.16, and 18.17. Those products were a sample of many readily available products from several major manufacturers, covering a wide price range. All, especially the lower-priced ones, are very popular, so one can be assured that increasing numbers of consumers are equipped to hear the art very much as it was created. This is good news.

The sad news is that there are loudspeakers at all price levels that perform badly. There is no excuse for this inadequate behavior; the "bill of materials"

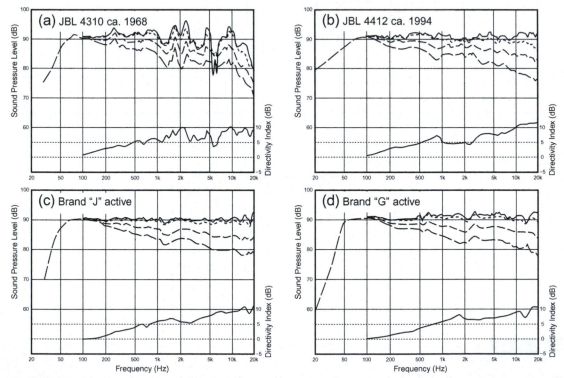

FIGURE 18.24 *Examples of loudspeakers designed to be as spectrally neutral as possible. (a) A classic of the industry, the JBL 4310, from 1968. The performance objectives for this loudspeaker are no different from those for excellent loudspeakers today: flat on-axis frequency response and constant directivity. It was limited by the technology of the period. (b) Evidence of 26 years of progress in transducer and system design. (c) and (d) Recent models from two manufacturers an ocean apart who clearly agree on what the performance target is. Both of these manufacturers reveal enough anechoic data on their loudspeakers for consumers to anticipate the excellent performances depicted here.*

would have permitted loudspeakers of much higher performance to have been built. Was it deliberate, a misguided notion of what sounds good? Was it incompetence, knowing what is wanted but not being able to get there? In a few cases, it seems to be simple carelessness.

18.5.1 Professional-Audio Loudspeaker Performance Objectives

Consumer audio is what it is, a mixture of marketing and engineering, and inconsistency is to be expected. However, the professional side also has its share of issues. There is more than a hint of complacency, and perhaps condescension, among professionals: Newell and Holland (2007) said, "Whilst professionals tend to work with standardized, known, and objectively designed equipment, domestic equipment tends to be individualistic, and marked by diversity more than commonality." In the scientifically based, engineering-driven mainstream of consumer audio loudspeakers, that is a perspective lacking evidentiary support. The loudspeaker measurements in this chapter identify excellence and mediocrity in both professional and consumer camps. The large survey reported in Figure 2.4 can only be interpreted with alarm—essentially identical high-quality loudspeakers that in the hands of professionals exhibit an enormous range of performances. The very popular monitor shown in Figure 2.6b was not only accepted in spite of its substantial imperfections, but a new fashion in control room design assisted in its rehabilitation by creating a "dead end" and absorbing most of the (terrible) off-axis sound.

In professional worlds, standards are often relied on to maintain levels of excellence. However, standards often describe the way things are, as decided by a committee of practitioners, not the way they should be. For certain kinds of standards, those dealing with the normalization of physical dimensions, labeling, and so forth, things are straightforward. However, in the development of a standard, if a controversial opinion enters the discussion, progress ceases. Only agreement moves the document forward, and so the content gravitates toward a comfortable middle-of-the-road position, often representing long-standing traditions in the industry. It is rare for a standard to espouse the state-of-the-art because, inevitably, it is not widespread within the industry and, equally inevitably, there will be detractors. In fact, it is common for such documents to be issued as recommendations not as standards because of lack of universal support. The result is that many of these standards and recommendations may prevent truly bad things from happening, but excellence is not assured, nor is consistency from location to location.

I thought about using such a standard as an example and examining it in detail using perspectives taught earlier in this book. When this was done with one of the more popular international recommendations for choosing and using "reference monitor loudspeakers," the result was greatly disappointing; there was very little left intact. All of these documents are revised periodically, so rather than criticize a particular example, I have chosen to describe what I perceive as serious shortcomings that are commonly found in such documents and hope that future revisions may incorporate improvements in these areas.

- First, it is common for all acoustical measurements to be detuned to 1/3-octave resolution, so all medium- and high-Q resonances and sharp discontinuities will be substantially attenuated, if not rendered invisible. In existing documents this is done for anechoic measurements on loudspeakers (inexcusable), as well as measurements of these loudspeakers as installed on site, measured at the listening location (understandable).

- Anechoic measurements frequently focus on a narrow frontal perspective, rarely exploring beyond 30° off axis. As has been seen in this chapter, this fails to provide *any* insight into directivity, early reflections, sound power—all of the information that can help us to understand how the loudspeaker may sound in a room. Question: where might ordinary audio professionals find such data, even in the inadequate form specified? Answer: it should be (but often is not) provided by the loudspeaker manufacturers, as discussed in Section 18.2.7, as evidence of their qualification to play in this league. Ideally, these data should be available in a standardized format, with standardized frequency resolution, and so on.

- All of the preceding measurements are commonly allowed to vary within tolerances as generous as ±3 dB, or even more, meaning that there are no assurances of sound quality whatsoever.

- Measurements made at the reference listening location are required to fall within a tolerance that is never less than ±3 dB and that increases in the downward direction as the upper- and lower-frequency extremes are approached. The generosity of the tolerances is much appreciated by studio owners, who should be able to qualify with minimal effort (except see Figure 2.4 for examples of installations that appear not to have even made an attempt).

- Mounting variations—adjacent-boundary effects—are rarely discussed. Interestingly, a loudspeaker that passes the anechoic test for axial flatness may fail this test if it is installed in a soffit or 2π wall mounting (see Figure 12.8) or even close to a wall. It will have different problems if it is placed on the meter bridge.

- Equalization is almost always required, and the topic is rarely discussed. It needs to be at the core of the document, because it changes the requirements for anechoic performance and performance at the listening position. Anechoic data should be used to eliminate loudspeakers with unacceptable resonances and nonuniform directivity. Broad inconsistencies in frequency response can be corrected with equalization. Performance below about 300 Hz can only be addressed after installation in the room.

- Measurements made at the reference listening location—room curves— are subject to all of the misgivings expressed in Section 18.2.6 in connection with the "X" curve. They include aspects of room acoustics that are normally not adequately specified.

- In control rooms, attenuation of early reflections, particularly those from the side walls, is usually a requirement. The need for this has been discussed in detail earlier, and it is an option, not a requirement. All too commonly, the requirement to attenuate these reflections applies to frequencies above 1 kHz, meaning that 1-in.-thick (25 mm) fiberglass board or slab foam will suffice, and all that is accomplished is an attenuation of first reflections from tweeters. All sound from 1 kHz down is fully reflected, and, as shown in Figure 6.18, the audible threshold of the reflection has been negligibly changed. The spectral balance of the sound has been altered, and possibly a good loudspeaker has been made to sound less good. A review of Chapters 5 through 9 offer persuasive arguments that there is more to this than is represented in typical documents. Figure 21.9 shows that traditional random-incidence acoustical measurements of absorption coefficient do not describe what happens in a "first-reflection" situation.

- Most recommendations set limits on room proportions implying that compliance with the dimensional requirements leads to audible advantages. As shown in Section 13.2.1, without specific knowledge of listener and loudspeaker locations within the room, dimensional proportions are of little value, and having five full-range loudspeakers adds complication that none of the normal predictive calculations account for, making them totally useless.

- The option of using subwoofer/satellite systems is almost never mentioned, even though it is the most commonly used and most likely the best possible configuration for multichannel playback. It is also the most likely configuration to achieve a transfer of a *high-quality* listening experience from the control room to the home. At the very least it needs to be included as an option so recording personnel can hear their product as it is likely to be heard in home listening rooms and theaters.

- When subwoofer/satellite systems are used, the *acoustical* crossover from the subwoofer(s) to each of the satellite loudspeakers must be individually measured and the low-pass and high-pass filter characteristics adjusted to achieve smooth summing in the crossover region. Doing that requires high-resolution transfer-function measurements (amplitude and phase) and a flexible electronic crossover customization routine because each installation will be different. It is not sufficient to rely on fixed-slope electronic filters.

- Stereo programs need to be evaluated through the upmix algorithms commonly used in homes.

- Reverberation time is always specified, and its importance is exaggerated by requiring a precision and frequency-dependent consistency that are excessive for this application. Often there are no requirements for how it should be measured (see Chapter 4).

This list could be extended, but if even these points could be considered for incorporation into an agreed-upon recommendation, the industry will have made a substantial step forward.

18.6 OTHER MEASUREMENTS: MEANINGFUL AND MYSTERIOUS

So far we have talked about frequency response as if it were all that matters. Actually, it is almost true, in the sense that if the set of frequency response measurements that has been displayed does not look "right," almost nothing else matters. This assumes that nonlinear distortion is not gross and that neither loudspeakers nor amplifiers are hitting their limits. These frequency response curves tell us how the products will sound, within reason, but there are other dimensions to be looked at.

18.6.1 Phase Response—Frequencies Above the Transition Zone

The combination of amplitude versus frequency (frequency response) and phase versus frequency (phase response) totally defines the linear (amplitude independent) behavior of loudspeakers. The Fourier transform allows this information to be converted into the impulse response, and, of course, the reverse can be done. So there are two equivalent representations of the linear behavior of systems: one in the frequency domain (amplitude and phase) and one in the time domain (impulse response).

To put it in slightly different terms, the accurate reproduction of waveforms is possible only if the signal is delivered to the listener's ears with perfect amplitude and phase responses. The obvious question is: do we hear waveforms? All of the evidence in this chapter indicates that listeners are attracted to linear (flat and smooth) amplitude versus frequency characteristics. Toole (1986) shows phase responses for 23 loudspeakers arranged according to subjective preference ratings. The most obvious relationship was that those with the highest ratings had the smoothest curves, but linearity did not appear to be a factor. The agreement that smoothness is desirable argues that listeners were attracted to loudspeakers with minimal evidence of resonances because resonances show themselves as bumps in frequency response curves and rapid up-down deviations in phase response curves. The most desirable frequency responses were also horizontal straight lines. The corresponding phase responses had no special shape other than the smoothness. This suggests that we like flat amplitude spectra and we don't like resonances, but we tolerate general phase shift, meaning that waveform fidelity is not a requirement.

Loudspeaker transducers, woofers, midranges, and tweeters behave as minimum-phase devices within their operating frequency ranges (i.e., the phase response is calculable from the amplitude response). This means that if the frequency response is smooth, so is the phase response, and as a result, the impulse response is unblemished by ringing. When multiple transducers are combined into a system, the correspondence between amplitude and phase is modified in the crossover frequency ranges because the transducers are at different points in space. There are propagation path-length differences to different measuring/listening points. Delays are non-minimum-phase phenomena. In the crossover regions, where multiple transducers are radiating, the outputs can combine in many different ways depending on the orientation of the microphone or listener to the loudspeaker.

The result is that if one chooses to design a loudspeaker system that has linear phase, there will be only a very limited range of positions in space over which it will apply. This constraint can be accommodated for the direct sound from a loudspeaker, but even a single reflection destroys the relationship. As has been seen throughout Part One of this book, in all circumstances, from concert halls to sound reproduction in homes, listeners at best like or at worst are not deterred by normal reflections in small rooms. Therefore, it seems that (1) because of reflections in the recording environment there is little possibility of phase integrity in the recorded signal, (2) there are challenges in designing loudspeakers that can deliver a signal with phase integrity over a large angular range, and (3) there is no hope of it reaching a listener in a normally reflective room. All is not lost, though, because two ears and a brain seem not to care.

Many investigators over many years have attempted to determine whether phase shift mattered to sound quality (e.g., Greenfield and Hawksford, 1990; Hansen and Madsen, 1974a, 1974b; Lipshitz et al., 1982; Van Keulen, 1991). In every case, it has been shown that if it is audible, it is a subtle effect, most easily heard through headphones or in an anechoic chamber, using carefully chosen or contrived signals. There is quite general agreement that with music reproduced through loudspeakers in normally reflective rooms, phase shift is substantially or completely inaudible. When it has been audible as a difference, when it is switched in and out, it is not clear that listeners had a preference.

Others looked at the audibility of group delay (Bilsen and Kievits, 1989; Deer et al., 1985; Flanagan et al., 2005; Krauss, 1990) and found that the detection threshold is in the range 1.6 to 2 ms, and more in reflective spaces.

Lipshitz et al. (1982) conclude, "All of the effects described can reasonably be classified as subtle. We are *not*, in our present state of knowledge, advocating that phase linear transducers are a requirement for high-quality sound reproduction." Greenfield and Hawksford (1990) observe that phase effects in rooms are "very subtle effects indeed," and seem mostly to be spatial rather than timbral. As to whether phase corrections are needed, without a phase correct recording process, any listener opinions are of personal preference, not the recognition of "accurate" reproduction.

In the design of loudspeaker systems, knowing the phase behavior of transducers is critical to the successful merging of acoustical outputs from multiple drivers in the crossover regions. Beyond that, it appears to be unimportant.

18.6.2 Phase Response—The Low Bass

In the recording and reproduction of bass frequencies, there is an accumulation of phase shift at low frequencies that arises whenever a high-pass filter characteristic is inserted into the signal path. It happens at the very first step, in the microphone, and then in various electronic devices that are used to attenuate unwanted rumbles in the recording environments. More is added in the mixing process, storage systems, and playback devices that simply don't respond to DC. All are in some way high-pass filtered. One of the most potent phase shifters is the analog tape recorder. Finally, at the end of all this is the loudspeaker, which cannot respond to DC and must be limited in its downward-frequency extension. I don't know if anyone has added up all of the possible contributions, but it must be enormous. Obviously, what we hear at low frequencies is unrecognizably corrupted by phase shift. The question of the moment is, How much of this is contributed by the woofer/subwoofer, is it audible, and if so, can anything practical be done about it? Oh, yes, and if so, can we hear it through a room?

Fincham (1985) reported that the contribution of the loudspeaker alone could be heard with specially recorded music and a contrived signal, but that it was "quite subtle." The author heard this demonstration and can concur. Craven and Gerzon (1992) stated that the phase distortion caused by the high-pass response is audible, even if the cutoff frequency is reduced to 5 Hz. They say it causes the bass to lack "tightness" and become "woolly." Phase equalization of the bass, they say, subjectively extends the effective bass response by the order of half an octave. Howard (2006) discusses this work and the abandoned product that was to come from it. There was disagreement about how audible the effect was. Howard describes some work of his own: measurements and a casual listening test. With a custom recording of a bass guitar, having minimal inherent phase shift, he felt that there was a useful difference when the loudspeaker phase shift was compensated for. None of these exercises reported controlled, double-blind listening tests, which would have delivered a statistical perspective on what might or might not be audible and whether a preference for one condition or the other was indicated.

The upshot of all this is that even when the program material might allow for an effect to be heard, there are differences of opinion. It all assumes that the program material is pristine, which it patently is not, nor is it likely to be in the foreseeable future. It also assumes that the listening room is a neutral factor that, as Chapter 13 explains, it certainly is not. However, if it can be arranged that these other factors can be brought under control; the technology exists to solve this residual loudspeaker issue.

18.6.3 The Loudspeaker/Amplifier Interface: Impedance, Wire, and Damping Factor

This topic is on the borderline for the topic area of this book; these are hard electrical engineering issues, but they routinely get elevated to different planes of thought. The reason for talking a little about them here is because as a result of them, frequency responses of loudspeakers get altered.

Impedance: 8 ohms. This is the kind of specification one sees for loudspeakers. It is an invented number. For a few, very, very few loudspeakers, it is a good approximation, but for the vast majority, it is a dreadful description of reality. Figure 18.25a shows an example of an impedance that varies substantially with frequency and that crosses the rated impedance at a few places only. The variations are normally of no concern.

Most power amplifiers are designed to be constant-voltage sources, so unless an unfortunate interaction between amplifier and loudspeaker provokes limiting or protection, all is well. Sadly, there have been some notable examples of high-end loudspeakers having impedances that dipped to small fractions of an ohm. This is a problem of incompetent loudspeaker design. However, sensing a market, amplifier designers responded with monster "arc-welder" devices that can drive these problem loudspeakers, and anything else, but it is overkill for most circumstances. It was amusing, at the time, to read that these incompletely designed loudspeakers "revealed" differences between power amplifiers, as if it were a virtue.

But there is a situation in which the varying impedance becomes an issue. Going straight to the problem, Figure 18.25b shows the kind of change in loudspeaker frequency response that can be caused by variable impedance; it is easily audible. The culprit? In this case, a tube

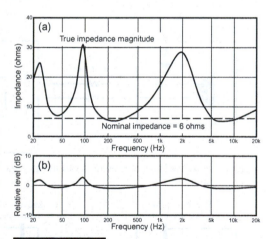

FIGURE 18.25 *(a) An impedance curve for a loudspeaker compared to the nominal impedance rating chosen by the manufacturer for it. (b) The change in frequency response of this loudspeaker caused by driving it with a tube amplifier having a large output impedance. Note that the shape of the frequency-response error is the same as the loudspeaker impedance curve.*

power amplifier with a large output impedance. The explanation is in Figure 18.26a and (b). The output impedance of the power amplifier and the resistance of the loudspeaker wire are components in a voltage divider circuit. When combined with the frequency-dependent impedance of the loudspeaker, it means that the "flat" frequency response voltage at location "A" inside the power amplifier acquires a shape following that of the impedance curve at location "B." Because this is the voltage driving the loudspeaker, the overall performance of the loudspeaker—that is, all of its frequency response curves—are modified by this amount. Different loudspeakers have different impedance curves; some are strikingly variable, others change little.

The amount of the change in frequency response depends on the total voltage drop across the combined amplifier output impedance and wire resistance, meaning that minimizing both of these is desirable. For solid-state power amplifiers, output impedances tend to be very small: typically 0.01 to 0.04 ohms. Those for tube power amplifiers are much higher: typically 0.7 to 3.3 ohms. These numbers come from a survey of *Stereophile* magazine amplifier reviews over several years. (My thanks to *Stereophile* for doing useful measurements.)

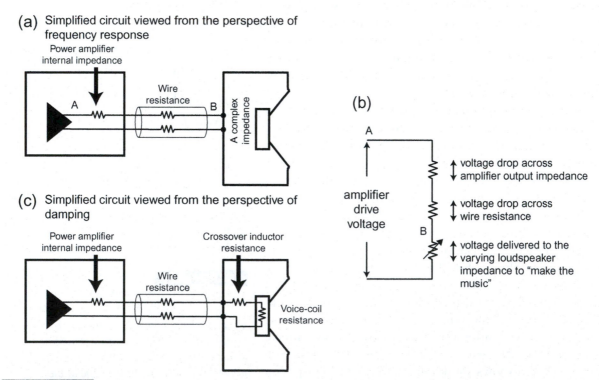

FIGURE 18.26 *Schematic diagrams showing (a) and (b) the electrical circuit explaining how amplifier and wire impedances cause variations in loudspeaker frequency response, and (c) how they affect loudspeaker damping.*

To reviewers, these are moderately discomfiting numbers because the inevitable conclusion is that tube power amplifiers, as a population, cannot allow loudspeakers to perform as they were designed. Different reviewers handle it in different ways. Some ignore it, and others have danced around the issue, concluding that it is just one more uncertainty in sound reproduction. Rarely is it acknowledged to be what it is.

The Infinity Prelude MTS, now discontinued, had an impedance of 4 ohms ±1 ohm, almost constant, as a result of deliberate design. This loudspeaker, and the few others with this property, can perform with remarkable consistency in spite of significant losses in the upstream signal path. Rarely, though, is impedance ever discussed as a virtue or a problem. One well-known high-end loudspeaker specified that it should be used with wire in which the resistance is less than 0.2 ohms. This conscientious behavior is admirable, but it was probably not interpreted as implying that if the total upstream resistance cannot exceed 0.2 ohms, the restriction is violated as soon as any tube amplifier is connected, no matter what wire is used.

Table 18.1 shows the resistances per unit length of stranded copper wire. The numbers are for *both* wires in the circuit, so just measure the length of the two-conductor wire and multiply by these numbers. If you do not see a gauge rating for a loudspeaker wire, be very suspicious. Some exotic cables use small wire for seriously mistaken reasons.

Minimizing wire resistance is easy: use large wire (low gauge numbers) or, better yet, just use less wire (see Table 18.1). If there is a risk of radio-frequency signal pickup, it is important to know that unshielded wires act as antennas. A great deal of mystique has evolved around loudspeaker wires, attempting to elevate this simple device to impossible heights of importance. Notions that they behave as transmission lines persist, but Greiner (1980) offers persuasive arguments that this is unrealistic. There are other beliefs, some of which are impossible (e.g., directional wires), and most of which remain unproven because of the cost of running double-blind tests. At prices that can exceed $20 000 for a pair of 8-ft (2.4 m) loudspeaker wires, one expects a lot. Enough said. Wire is a good product for the industry: totally reliable, inexpensive to manufacture,

Table 18.1 Resistances per Unit Length of Two-Conductor Stranded Copper Wire

AWG wire gauge	Resistance per ft (ohms) BOTH conductors	Resistance per m (ohms) BOTH conductors
10	0.0020	0.0067
12	0.0032	0.0106
14	0.0052	0.0169
16	0.0082	0.0268
18	0.0148	0.0483

highly profitable, and, if you like what you hear, an excellent investment, so long as you did not pay more than you needed to (aye, there's the rub!).

One of the universal compliments attached to audio products, including wires, is that it results in "tighter bass." In the case of loudspeaker wire, it seems as though there might be some truth to it because of its role in the loudspeaker/amplifier interface and damping. Damping unwanted motion of a loudspeaker diaphragm is undoubtedly a good thing.

In 1975, I wrote an article for *AudioScene Canada* called "Damping, Damping Factor, and Damn Nonsense." I still like the title because it is a succinct statement of reality. The point of the article is summarized in Figure 18.26c. The internal impedance of the power amplifier is used to calculate something called the damping factor (DF) of the amplifier (DF = 8/output impedance); the number 8 was chosen because it is the nominal load (resistive) used to measure the power output capability of amplifiers. The logical inclination is to think that larger is better. Solid state amplifiers have damping factors ranging from about 200 to 800, using the impedances quoted earlier in this section. Tube amplifiers in my survey ran from 2.4 to 11.4 because of their high output impedances.

Figure 18.26c also shows the complete circuit involved in the electrical damping of loudspeakers. It does not mysteriously stop at the loudspeaker terminals. Current must flow through components and devices inside the enclosure. The first component to be encountered is typically an inductor, part of the low-pass filter ahead of the woofer in a passive system. Then inside the woofer is the voice coil. The inductor resistance is commonly around 0.5 ohm, and the voice coil resistance can have different values but is commonly around 6 ohms. So let us examine all of the resistances in the circuit to arrive at the following progression of damping factor changes:

Amplifier internal impedance: 0.01 ohm	DF = 800
Add wire resistance: 10 ft of 10-gauge	
Both conductors: 0.02 ohm	DF = 266
Add crossover inductor resistance:	
0.5 ohm (typical)	DF = 15
Add voice-coil resistance: 6 ohms (typical)	DF = 1.2

Obviously, the resistances inside the loudspeaker are the dominant factors. Even eliminating the inductor and driving the woofer directly changes things only slightly. The article (Toole, 1975) shows oscilloscope photographs of tone bursts of various frequencies and durations while the damping factor of the amplifier was varied from 0.5 to 200. At damping factors above about 20 (internal impedance less than 0.4 ohms), no change was visible in any of the transient signals, and changes in frequency response were very much less than 1 dB, and then only over a narrow frequency range. On music, no change in sound quality could be discerned, including attentive listening for "tightness." Because 0.4 ohms is at least a factor of 10 higher than internal impedances found in typical solid-state amplifiers, it means that from the perspective of damping the transient behavior of loudspeakers, the wire resistance can be allowed to creep up substantially. However, as just shown, doing so can change the frequency response of the loudspeaker, and that, we know, *is* audible.

In summary, with tube amplifiers, the internal impedance is already so high that damage is done to the frequency responses of loudspeakers having normal impedance variations. Added losses in wire simply make the situation worse. To hear the loudspeakers that the manufacturer made, it is necessary to seek out those with very constant impedance as a function of frequency. Damping of the loudspeaker is also marginally impaired by the high internal impedance.

With solid-state amplifiers, internal impedances are negligibly low, so wire resistance must be controlled to minimize corrupting the frequency response of loudspeakers. How low? It depends on the *variations* in the impedance of the loudspeakers being used and how low those impedances are; wire resistance represents a higher percentage of low impedances. For example, a loudspeaker ranging from 3 ohms to 20 ohms (not unusual for consumer loudspeakers and a moderately demanding situation) would experience about 0.6 dB variations in a system with 0.2 ohm wire resistance. The next chapter will show that this is slightly higher than the detection threshold for low-Q spectral variations in quiet anechoic listening. Twelve-gauge wire would allow for a run of 0.2/0.0032 = 63 ft (19 m). Obviously, this is not very restrictive. Loudspeakers having nearly constant impedance can tolerate large wire losses, sacrificing only efficiency up to the resistance at which damping is affected. If compelled to do better than this suggestion, more copper, shorter runs, or higher-impedance loudspeakers are the solutions.

18.6.4 Observations on Sensitivity Ratings and Power Amplifiers

Years ago, loudspeaker sensitivity was rated as the sound level at a distance of 1 m for an input of 1 watt. Power input is voltage^2/resistance. Because loudspeakers do not have the same impedance at all frequencies, a sensitivity rating would apply only at a single frequency (and sometimes a few). Figure 18.27

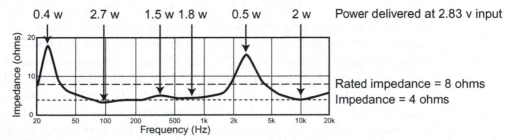

FIGURE 18.27 *A typical loudspeaker impedance curve, showing the rated impedance of the product and, at several frequencies, the input power for 2.83 volts.*

shows the impedance curve for a loudspeaker, specified by the manufacturer as an 8-ohm unit. It is 8 ohms at four frequencies, but looking at the curve, it is generally hovering at a level slightly above 4 ohms, dropping to a minimum of 3 ohms. A more realistic rating would have been 5 ohms. But that is an "odd" number in the industry, and such numbers, even if true, tend to be avoided. The 3-ohm minimum is important because many receivers and some stand-alone power amplifiers are unhappy driving these low impedances because they lack the current capacity to deliver the required power into the load.

Figure 18.27 shows the actual power delivered at a constant input voltage of 2.83 v, and it ranges from a high of 2.7 w at the impedance minimum to a low of 0.4 w at the highest impedance point. Obviously, rating sensitivity according to power input does not work well. The domination of solid-state amplifiers really provided the solution. These amplifiers are essentially constant-voltage sources, with power rated according to what they can deliver into an 8-ohm resistor. If the load impedance drops to 4 ohms, the power will double; at 2 ohms, the power quadruples, and so on, until the amplifier can deliver no more current or dissipate no more heat. Amplifiers that are optimized to meet a specification sheet deliver their rated power into 8 ohms but may fail to deliver double power into 4 ohms. This is a major differentiating factor among power amplifiers. Those big, heavy monoblocks with massive heat sinks are the ones that are able to drive huge currents into very low impedances, and they tend to double their output into halved impedances. How much of this is necessary? The correct answer is *enough*. If the loudspeaker load is well behaved, inexpensive amplifiers work just fine. Powered loudspeakers have a big advantage: the power amplifiers needed to drive individual transducers can be much less ostentatious devices because the details of the load they drive are known and well defined. It is the uncertainty of the load that forces us to buy amplifiers that can drive anything we connect to them. There is more to this tale than is revealed here. Benjamin (1994) and Howard (2007) add much more perspective.

Returning to the theme of sensitivity ratings, the present circumstances allow us simply to define an input voltage, not an input power. The selected standard voltage is 2.83 volts, the voltage that delivers one watt into 8 ohms. The loudspeaker used for Figure 18.27 is shown in Figure 18.10b, where it can be seen that the sound pressure level on axis undulates around and slightly above the 90 dB level. All of the measurements shown in this book were made with loudspeakers driven at 2.83 volts. Measurements were made at 2 m, a distance that safely represents the far field for small loudspeakers, although it is borderline for very large ones (see Figure 18.1). The SPL is adjusted to show what it would have been at 1 m, which is the standardized distance (see the end of Section 18.1.1). The manufacturer specified the sensitivity for this loudspeaker to be 91 dB, which, as can be seen, is a reasonable average sound output level. Not all manufacturers are so accurate in their sensitivity ratings. However, as shown in Figure 18.27, at only four frequencies is the input power 1 watt—those where the curve crosses the 8-ohm line. That is why SPL @ 1 w @ 1 m, is a specification relegated to history.

18.6.5 To Be Continued

There are other specifications to be discussed, notably nonlinear distortion, but understanding the issues requires a foundation of understanding that goes beyond the theme of this chapter. The following chapter picks up the story after some psychoacoustic background.

CHAPTER 19

Psychoacoustics—Explaining What We Measure and Hear

Figure 5.1 showed "the central paradox" as described by Arthur Benade (1994). In Figure 19.1, it has been elaborated on to show a parallel path related to sound reproduction because the same situation exists. We can make measurements of many dimensions of sound as it is represented in sound waves, and we find that the numbers and graphs are not always simply or logically related to what we hear. Often they suggest that there should be problems, but we listen to the sound and find only a pleasurable experience. This is one of the principal tasks of psychoacoustics: to find ways to make more relevant measurements and ways to process the measured data so they relate more directly to hearing perceptions.

Chapter 18 set an optimistic tone in that there were trends in certain kinds of measurements on loudspeakers that showed good visual correlations with subjective ratings. The next challenge is finding a way to numerically process that same data so subjective ratings can be anticipated from measurements.

19.1 LOUDNESS AND THE BASICS OF HEARING

Large portions of large books have been devoted to the topic of loudness, so this is an attempt only to introduce some of the key concepts that relate to our interests. Figure 19.2 illustrates some typical sound-pressure levels, some of their audible effects, an indication of the risks involved in prolonged and repeated exposures, and very rough approximations of what amplifier power might be required to reproduce those sound levels in a listening room. The sound sources identified with sound-pressure levels in the left-hand column have been collected over many years; not all were measured with the same weighting, and many of the sources identified are extremely variable, so these are only approximations.

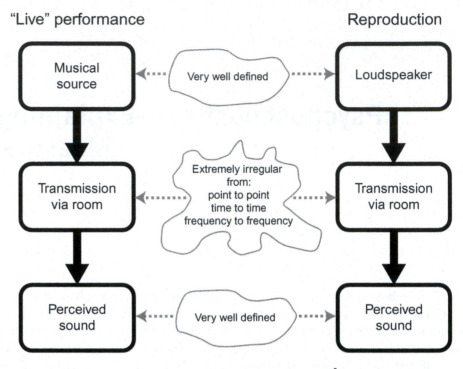

FIGURE 19.1
"The central paradox" of sound in rooms (Benade, 1984) expanded to include reproduced sound.

The column related to hearing damage is more serious. For many years, there have been guidelines for preserving hearing in the presence of workplace noises. Most people naturally think of these guidelines, which differ slightly in different countries, as representing what is "safe." It is important to note that in setting these guidelines, the intent was not to prevent hearing loss. The intent was that at the end of a normal working life, there would be enough hearing left for basic functionality—example, to be able to understand conversation across a table. Hearing the subtleties of reproduced sound would be a pleasure long gone. Sections 17.4 and 17.5 discussed consequences of hearing loss that are simply not considered in setting standards for occupational noise exposure, to which must be added nonoccupational noise exposures: rock concerts, shooting, motorcycles, lawnmowers, power tools—all of which contribute to the lifelong accumulation of hearing damage. Hearing that is considered "normal" by your audiologist may already exhibit significant losses when it comes to appreciating or evaluating sound quality.

The column at the far right shows a guess at what amplifier power may be needed to deliver various sound levels in typical rooms. Obviously, this is fundamentally dependent on loudspeaker sensitivity, how many are operating, the frequencies being radiated, and the acoustical properties of the room. In spite

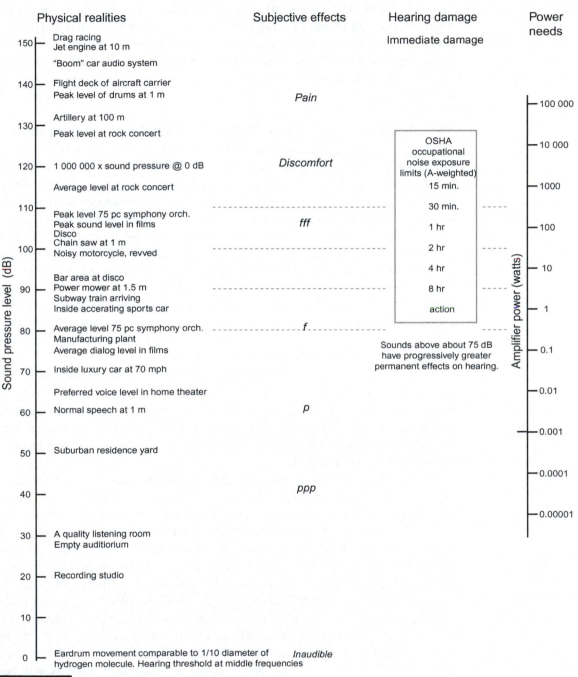

FIGURE 19.2 *Some dimensions of sound.*

of these uncertainties, the *relative* quantities are correct; the overall vertical scale may slide up or down a few dB. This is shown to give an appreciation for the reality that a 3 dB change in sound level requires a doubling or halving of power, and a 10 dB change translates into 10 times the power. When one is talking about high sound levels from domestic loudspeakers having typically moderate sensitivity (85–90 dB @ 2.83 v @ 1 m), one is also talking about very large amounts of power.

19.1.1 Equal-Loudness Contours and Loudness Compensation

Loudness is the perceptual correlate of sound level. It is dependent on frequency, sound level, incident angle, the duration of the signal, and its temporal envelope. It is another of those psychoacoustic relationships that defies a simple description. Yet, one must start somewhere, so most investigators focused on pure tones in acoustically simple circumstances, like headphone or anechoic listening.

Fletcher and Munson (1933) were among the first to evaluate the sound levels at which we judged different frequencies to be equal in loudness. Starting with a reference pure tone at 1 kHz, other tones at different frequencies were adjusted until they appeared to be equally loud. With enough of these measurements, it was possible to draw a contour of equal loudness identified by the sound-pressure level of the reference tone, but called *phons* rather than decibels, to denote its subjective basis. Fletcher and Munson used headphones, and there has been concern about the calibration thereof.

Several more recent studies, notably the work of Robinson and Dadson (1956), used pure tones presented to listeners in an anechoic chamber; their results were the ISO standard for several years. Even then, errors were identified in the contours at low frequencies. Stevens (1961) employed fractional-octave bands of steady-state sounds in reflective spaces as a basis for calculating the loudness of complex sounds.

Figure 19.3a shows a sample from the current ISO standard (ISO 226, 2003), the result of an international collaboration. Figure 19.3b shows a comparison of it and two earlier contours, revealing how very different they are in shape, especially at low frequencies. These are obviously *not* hard engineering data. These curves are averages over many listeners; there can be large intersubject differences. About the only feature they share is the crowding together of the curves at low frequencies—the very familiar and easily audible rapid growth and decline of loudness at bass frequencies when the overall volume is changed.

The curves tell us that different frequencies at the same sound level may be perceived as having different loudness. This is not a message that anything needs correcting. We live with these characteristics from birth, and they are a part of everything we hear, whether it is live or reproduced. That is why audio equipment must exhibit flat-frequency responses—uniform output at all audible frequencies—so the sounds we perceive have the correct relative loudness at all frequencies, *assuming they are reproduced at realistic sound levels*.

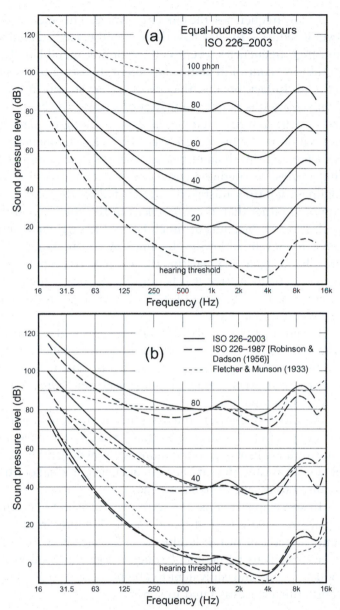

FIGURE 19.3 *(a) Curves of equal loudness for pure tones auditioned in an anechoic space according to the most recent internationally agreed standard: ISO 226 (2003). Each curve is created with reference to a pure tone at 1 kHz and is identified by the sound-pressure level at that frequency but expressed in "phons." (b) A comparison of the current standard with two other determinations of equal-loudness contours. © ISO. This material is reproduced from ISO 226 with permission of the American National Standards Institute (ANSI) on behalf of the International Organization for Standardization (ISO). No part of this material may be copied or reproduced in any form, electronic retrieval system or otherwise or made available on the Internet, a public network, by satellite or otherwise without the prior written consent of the ANSI. Copies of this standard may be purchased from the ANSI, 25 West 43rd Street, New York, NY 10036, (212) 64 4900. http://webstore.ansi.org.*

At low frequencies, the curves converge. This means that as the overall sound level is reduced, the low bass decreases in perceived loudness faster than middle and high frequencies, eventually becoming inaudible. This is the basis for the "loudness" control in audio equipment, boosting the bass as the overall level is reduced. In fact, because the curves rise so steeply at low frequencies, loudness compensation is as much a matter of keeping the bass sounds above the threshold of detection (i.e., audible) as it is in trying to maintain equal loudness. At normal domestic listening levels, the lowest frequencies in music may already be below the threshold of hearing. Because the contours are all parallel at high frequencies, no loudness correction is required in this frequency range. Many loudness controls mistakenly try to follow the *shapes* of the equal loudness contours rather than the *differences* in the shapes, and they boost the highs as well. If one is listening at much below the original sound level, it makes more sense to elevate all low-level sounds by compressing the dynamic range of the broadband signal and boosting the low bass to keep it above threshold. An argument can be made that it is better to hear something in an abnormal form than to not hear it at all. Most of the music we hear in background systems was not recorded for that purpose.

In terms of our ability to hear differences in loudness, it seems that something like 1 dB is a good estimate of the smallest audible change in overall loudness level: turning the volume up or down. As we will see later, we can discern even smaller differences in spectrum *shape*; 3 dB is an easily-perceived change in sound level.

At middle and high frequencies, where the equal-loudness curves are approximately parallel, a change of 10 dB is perceived, on average, to represent a doubling or halving of perceived loudness. Crowding of the curves at low frequencies obviously means that at these frequencies, a smaller change in sound level is required to double or halve loudness.

In some applications, the term *sone* is used to describe loudness. One sone equals the loudness at 40 phons. Each doubling in the number of sones represents a factor of two in loudness, so 2 sones is twice as loud as 1 sone, and it is equivalent to 50 phons; 4 sones is twice as loud as 2 sones; four times as loud as 1 sone and equivalent to 60 phons; and so on. The noise output of some ventilating devices is rated in sones.

19.1.2 Equal-Loudness Contours and Deteriorated Hearing

The lowest contour is the threshold of hearing. Sadly, it and all of the other contours are not constant with age. Figure 19.4a shows my own hearing thresholds deteriorating over four decades. It seems that at age 30, I performed better than average up to about 4 kHz, but even then, some high-frequency deterioration had set in. Since that age, I have been careful to take precautions—ear defenders and plugs—against unnecessary loud sounds (not music and movies!), yet this has not prevented further degradation. Nevertheless, I seem to be doing slightly better than the average for my age.

Robinson and Dadson (1956) show that along with elevated thresholds, age brings with it an accompanying elevation of all equal loudness curves at higher phon levels. So when a hearing threshold is elevated, it is not only that we lose the ability to hear the smallest sounds, but all sounds at all levels at those frequencies are also perceived to be less loud.

Figure 19.4b puts hearing loss into an easily understandable, and disturbing, context. Plotted on top of the ISO 226 equal-loudness contours are the hearing threshold measurements of listeners who exhibited high variability in their ratings of loudspeaker sound quality. All of these listeners were audio professionals, and many were part-time musicians. The message here is that these listeners could not hear large portions of the lower-level spectrum. All of the listeners lost at least 10 dB of the lowest-level sounds (just above threshold), whereas others lost increasingly more up to the few who lost the ability to hear all of the sounds in the shaded area. One obvious reason for variability in opinions expressed by these listeners is that they simply were not hearing all of the music: the good—small details and timbral subtleties—and the bad—distortions. Also apparent is the reduction in effective dynamic range experienced by these listeners. Unfortunately, hearing deterioration usually has other dimensions as well:

- Increased sensitivity to loud sounds (which reduces the dynamic range even further) or to all sounds, which makes normal life difficult.

- Degraded binaural discrimination (more difficult to focus on one sound while discriminating against sounds arriving from other directions—for

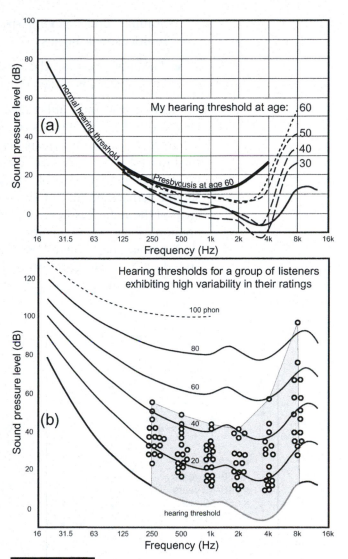

FIGURE 19.4 *(a) A history of the author's hearing threshold, compared with the statistical normal threshold and with the normally anticipated hearing loss as a function of age (presbycusis). (b) The hearing thresholds for listeners who had high variations in repeated sound-quality ratings of loudspeakers. Adapted from Toole, 1985, Figure 7. Equal-loudness contours from ISO 226 (2003). © ISO (see copyright conditions in Figure 19.3 caption).*

example, understanding speech in restaurants, parties, and in multichannel movies).

■ Tinnitus (noises in the ears, which once were temporary reminders of a loud concert, become permanent).

These are unfortunate afflictions, especially for audio professionals and product reviewers whose judgments are no longer representative of normal hearing listeners. No amount of experience can compensate for the inability to hear the lowest 20 or 40 dB of musical dynamics, timbral subtleties, distortions, and noises. Seeking opinions of younger ears is always a good idea. An audiogram should perhaps be part of the personal résumé of people in certain sensitive areas of the audio business, displaying evidence of why anyone should trust their opinions about sound quality.

Martinez and Gilman (1988) report results of an audiometric survey of 229 attendees at an AES convention. Figure 19.5 and the explanation in the caption reveal that audio professionals suffer slightly more than the normal amount of

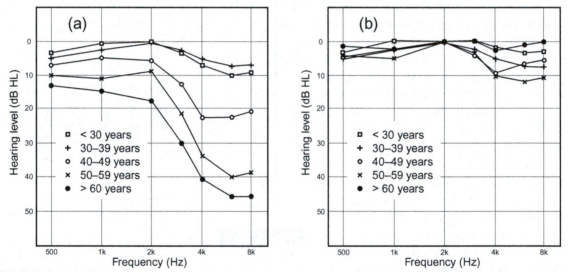

FIGURE 19.5   *(a) Results of an audiometric survey of 229 attendees at an AES Convention. These plots are conventional "hearing level" plots in which the amount by which the hearing threshold is elevated above the statistical average is shown by an increasing downward shift of the curves. These curves, therefore, show hearing loss compared to normal hearing. The small shift at very low frequencies is suspected to be attributable to low-frequency noise from the convention leaking into the measurement booth. (b) The same data corrected for the normally expected amount of hearing loss for persons of those age groups. If these audio people experienced the same hearing loss as a function of age as the normal population, the curves would be flat and at 0 dB. These results show that all age groups of audio professionals, except those over 60, suffer slightly higher hearing loss than the normal population. The nature of the loss suggests that it is the result of exposure to high sound levels. From Martinez and Gilman, 1988.*

hearing loss for persons in their age categories. Only those over 60 appear to have avoided that trend. Were occupational and recreational noise exposures less in the past?

19.1.3 Loudness as a Function of Angle

The contours we have been discussing are not the complete story of loudness. First, they apply only to isolated tones, auditioned in quiet, anechoic space, arriving from frontal incidence. In spite of those severe limitations, they routinely are interpreted as having important meaning in listening to music, in reflective spaces, with sounds arriving from many angles of incidence.

These days it is usually the head-related-transfer-functions (HRTFs) that are displayed to explain differing sensitivities to sounds arriving from different angles. They explain the physical acoustic aspect of how sounds are modified on their way to a single eardrum as they arrive from different incident angles. However, we have two ears, and there is a brain interpreting the (often very different) sounds at both of them together, so the final perception of sound is a binaural combination. Robinson and Whittle provided useful data of this kind (Figure 19.6). (See also Figure 7.6, and Sivonen and Ellermeier, 2006.) It is interesting confirmation of why sounds that arrive from different directions have different perceived spectra—for example, elevated high frequencies in sounds from the sides and above.

19.1.4 Basic Masking and the Auditory Reflex

Masking occurs when the presence of one sound inhibits the perception of another. The most common experience is that of simultaneous masking, when both sounds coexist. There are other opportunities for masking in the time domain, mainly forward-temporal masking (when a brief sound reduces the audibility of sounds immediately following it) and backward masking. Figure 19.7a gives an impressionistic view of how a 500 Hz pure tone can simultaneously mask a lower-level 2 kHz tone. It would also mask nonlinear distortions produced by the 500 Hz tone and other sounds and noises that fall within the shaded areas. Figure 19.7b shows data on masking generated by a 50 Hz tone at different sound levels. The upward masking effects are substantial.

However, there is also the *auditory reflex* to be considered: the tiny muscle in the middle ear that, when tightened, reduces the audibility of low-frequency sounds. It is sometimes described as a protective device for the ears, but it does little to attenuate high frequencies and therefore cannot protect ears from damage (high frequencies are much more harmful than low frequencies). It is also slow to activate, meaning that loud transient sounds pass through unmodified. Finally, being a muscle, it fatigues, and whatever effects it has eventually fade. This means that the loud, low bass line in a rock selection

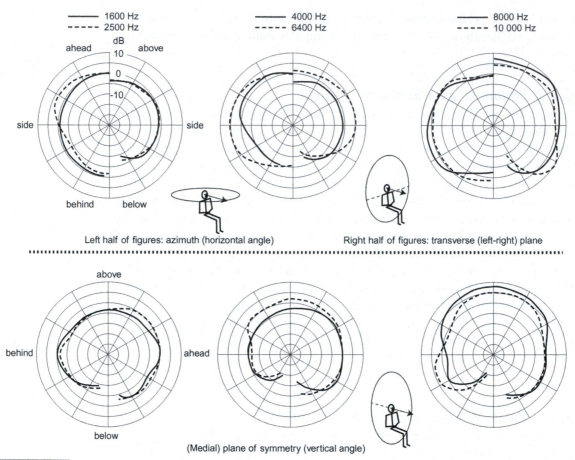

FIGURE 19.6 *Hearing thresholds as a function of angle for three rotational axes. The greater the distance from the center of the polar plot, the greater is the perceived loudness. From Robinson and Whittle, 1960.*

may actually sound better after the middle-ear reflex has had time to fatigue, and relax.

It is a reflex activity, but some individuals can also voluntarily activate it. So why does the auditory reflex exist? It is not clear that anyone knows for certain. Because it automatically activates when talking and eating, it seems possible that one role is that of a protective device for the whole human, allowing us to hear the approach of foes and large flesh-eating animals while enjoying food and conversation. According to Fielder (personal communication, 2007), it is possible that some of the apparent masking shown in Figure 19.7b could be attributed to the middle-ear reflex. Whatever the cause, the masking effect of loud low-frequency rumbles and booms in movies and the bass components of music are powerful agents of masking. So also are the low frequency drive-train, aerodynamic, and road-surface rumbles in motor vehicles.

19.1.5 Criteria for Evaluating Background Noises

A simple sound-level meter is commonly used to provide a number representing loudness. In making recordings, VU (volume-unit) or other program-level meters are employed. They are simple to use and provide rough guidance. A-weighted sound levels are used to assess the acceptability—and often the legality—of workplace, environmental, and neighborhood noises. Soulodre and Norcross (2003) provide interesting data and insights into several measures (see Figure 17.4 for various weighting curves). However, in determining the acceptability of background noises in recording and critical-listening spaces, it is common to start with octave-band analysis of the noise spectrum and then to compare this spectrum to one of several criterion curves purporting to describe the acceptability of the background noise for different specific purposes.

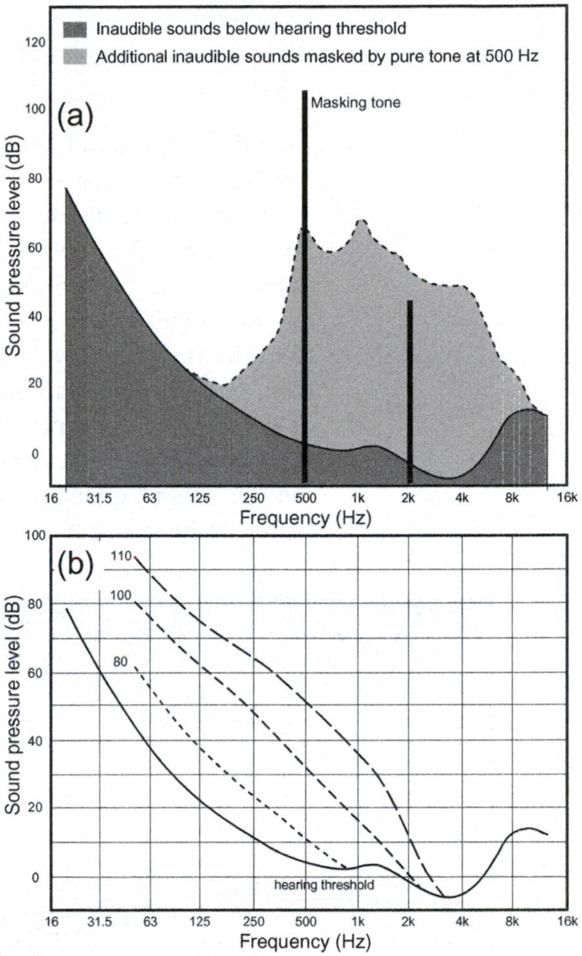

Early measures of acceptable background noise levels focused on the issue of speech interference. Measures driven by such concerns remain at the core of the popular criteria used in setting acceptable background-sound levels in listening and recording spaces, even though speech interference is not a problem. In those situations, one is likely to be more concerned with whether the noises are annoying or pleasant. In North America, the NC curves are widely applied to define acceptable background-noise levels for audio environments. According to Beraneck (2000), as reported in Tocci, (2000), "[The NC curves] are intended to be octave band noise levels that just permit satisfactory speech communication without being annoying." Warnock (1985) states the following:

FIGURE 19.7 (a) A simple view of simultaneous masking of a tone at 2 kHz by another, louder one at 500 Hz. It shows that the masking effect spreads substantially upward in frequency and only slightly downward. At lower sound levels, the masking effect exhibits less spreading. (b) The substantial masking effects of very low frequencies, especially at high sound levels. From Fielder, Figure 1.60 in Talbot-Smith, 1999.

NC contours are not ideal background spectra to be sought after in rooms to guarantee occupant satisfaction but are primarily a method of rating noise level. There is no generally accepted method of rating the subjective acceptability of the spectral and time-varying characteristics of ambient sounds. In fact, ambient noise having exactly an NC spectrum is likely to be described as both rumbly and hissy, and will probably cause some annoyance.

This does not sound like an objective for an expensive studio facility or recreational listening space.

Tocci (2000) and Broner (2004) provide lucid surveys of the numerous families of contours proposed by acousticians for various purposes. If the sound quality of the background noise is important—and it should be in listening environments—there is agreement that it is worth considering something other than the traditional NC contours where the shape of the spectrum is not evaluated, only the penetration of the highest point in the spectrum into the family of curves (the tangency criterion). Among the options, Beranek (1989) has updated the NC curves to the Balanced Noise Criterion (NCB) curves, attenuating the "hissy" quality of the NC curves and extending the low frequencies to better evaluate rumbles. Some examples are shown in Figure 19.8.

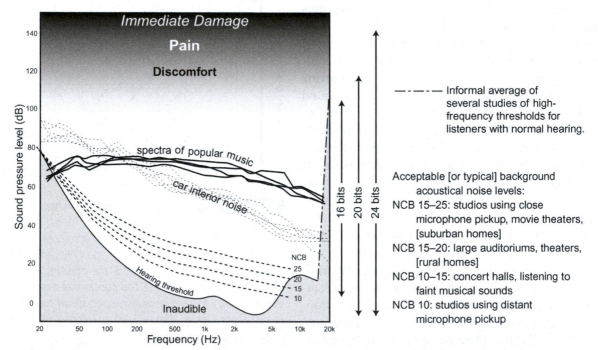

FIGURE 19.8 *A simplified two-dimensional display of what we can hear. Starting at the bottom, below threshold, we proceed upward through examples of typical background sounds in recording and listening environments, including cars at highway speeds, to high sound levels where things become uncomfortable and then permanently damaging. Running across the middle of the white area representing the useful listening region are examples of long-term spectra of popular music at foreground listening levels (from Olive, 1994). On the right are shown estimated dynamic ranges of perfect digital record/playback systems and explanations of the Balanced NC (NCB) curves (Beranek, 1989).*

The new Room Criterion Curves (RC Mark II) appear to be an attractive option in that the process evaluates the amount by which measured spectra fluctuate creating some assurance of a palatable sound quality. Full details, plus much helpful guidance for sound and vibration control can be found in the ASHRAE (American Society of Heating, Refrigerating and Air-Conditioning Engineers) Handbook (2003), Chapter 47 "Sound and Vibration Control"; available at www.ASHRAE.org. Individual chapters may be purchased).

As for what sound level is required, some very low numbers are needed for recording studios, especially those in which distant microphone pickups are used. Listening rooms are much more tolerant. In fact, it happens occasionally that the background noise in a recording is higher, on replay, than the ambient sound in the listening room. A very quiet room, though, allows one to make that judgment. It is also very impressive for a customer to demonstrate how quiet the room is, whether or not it matters once the movie or music is under way. Home theaters that are acoustically isolated from the rest of the house to prevent the escape of theatrical dynamics almost always end up being very quiet; the sound transmission loss works both ways.

19.1.6 The Boundaries of What We Can Hear

Figure 19.8 summarizes some aspects of sound and hearing that collectively bear on sound reproduction. The bottom of the display begins with a shaded area representing sounds that are below the threshold of audibility. The short wavelengths at high frequencies make it difficult to be sure of the sound pressure level at the eardrum when using conventional audiometric headphones, so audiometric measurements rarely go above 8 kHz. However, there have been some independent investigations focusing on these frequencies. The dot-dashed curve on the right side is my attempt to summarize results of efforts to measure hearing thresholds at high frequencies. There is considerable individual variation among subjects, as well as some issues related to calibrations of the measurements; anyone seriously interested in this topic should examine Stelmachowicz et al. (1989) and Ashihara (2007) and references cited therein.

Moving up the display, some examples of NCB curves are shown with examples of sound environments corresponding to various levels. These represent background-sound spectra that might find their way into recordings or that might be present during their playback.

Much higher is a collection of background noise measurements made inside cars at highway speeds that, when combined with the concept of masking described in Figure 19.7, explains why under these circumstances music loses much of its bass, timbral subtlety, and spatial envelopment. Only in the parking lot or in stop-and-go traffic can good car audio systems reveal their true excellence. Quiet cars are highly desirable. It also explains why car audio systems are often balanced to have rather more bass than would be usual in the home and why elaborate ones incorporate volume adjustment and/or bass boost that is activated according to road speed and/or interior noise. Surround channel levels would similarly benefit from that kind of automated adjustment.

All of this is made directly relevant by the set of music spectra slicing through the middle of the display. These come from measurements done by Olive (1994) and represent four programs exhibiting unusually flat and extended spectra. These are long-term average spectra, which ignore whatever dynamic range the program had. Today, in much of the music, movies, and television we are exposed to, dynamic range is diminishing. Part of it has to do with highly compressed audio-delivery formats that sacrifice both bandwidth and dynamic range, and some of it is that programs are tailored for where we listen: cars, earphones on the street, in buses, the subway, and so on. Every once in a while we need to sit down in quiet surroundings, put on an old-fashioned (relatively) uncompressed music source, and be reminded what dynamic range sounds like. It is not "all loud all the time," which so much of available programming seems to be. Sadly, it is part of the "dumbing-down" of audio. Compressed programs sound tolerable when played through inexpensive audio systems that are incapable of playing loud. In this technology-intensive era, it seems that it should be possible to supply programs and playback devices that could satisfy a number of different audiences or the same audience in different circumstances—at home, in the car, walking the dog (metadata?).

At the top of Figure 19.8, we run into regions of discomfort and pain and the absolute barrier of deafness. These are the reasons why turning up the volume is not the answer to elevated background noise levels. It becomes tiresome.

On the right are the dynamic ranges of some digital record/playback systems, showing that the conventional 16-bit CD has the potential to be a satisfactory delivery format for most program material; 20 bits would be better, allowing leeway for some imperfection in the process, and 24 bits would permit the encoding of everything from "I can't hear it" through to "I'll never hear it again." Taking full advantage of wide-system dynamics requires some serious power from amplifiers and loudspeakers, as shown in Figure 19.2. In reality, we make do with (and enjoy) much less dynamic range than is possible. The reason that occasional loud sounds emerge from modest audio systems is that we can tolerate up to 6 dB of clean amplifier clipping with remarkably little complaint (Voishvillo, 2006); 6 dB is 4× power. Those experiments used music. In movies, *very* loud events are mostly sound effects with no real "fidelity" issues, since the original could possibly be the sound of an elephant trumpeting, equalized, and played at quarter-speed, backward.

19.1.7 The Benefits of High-Resolution Audio

Figure 19.8 really portrays the amplitude and bandwidth requirements for a record/replay system. In the amplitude dimension, a 24-bit system is far more than is necessary; 20 bits would do nicely, but 16 bits, fully utilized, should get the job done. As we know from experience, not everything works as intended, so a little headroom in the system is probably a good thing. With the combined eccentricities of recording and playback, trusted sources report that "16-bit" programs have been known to perform more like 14 bits or even less. All of this assumes that the program material actually has some dynamic range to communicate, which, as time passes, seems to be diminishing.

It is bandwidth—the audible merits of high-frequency extension—that provokes the most animated arguments. All of the audiometric tests that I know of suggest that only a tiny part of the population can hear sounds above 20 kHz, and then most likely when they are young—before their first rock concerts or hunting trip. Personally, according to Figure 19.4a, I must have done something acoustically indiscreet early in life because although my midfrequency thresholds were better than normal for my age, my high-frequency thresholds were already elevated by age 30, and they have continued to get worse. Consequently, none of what follows is a personal commentary on the audible importance of very high frequencies. In fact, anyone with gray hair, especially if they are in the professional audio business, should be considered suspect as an arbiter (see Figure 19.5).

Back in 1980, Plenge et al. compared the audibility of 15 kHz and 20 kHz upper limits in music and concluded that 15 kHz was adequate, since none of their listeners could distinguish a difference at a rate better than chance. Obviously this result has been totally ignored. Miyasaka (1999) investigated the audibility of very high frequencies and noted that some studies reporting positive results had ignored distortions in the audible frequency range generated by transducers, and because of this, their findings were dubious. Nishiguchi et al. (2003) found that their 36 listeners were not able to detect the presence of sound above 21 kHz with statistical significance. In a different kind of test, Blech and Yang (2004) compared DVD-Audio (24 bit/176 kHz) with SACD in 100 surround comparisons and found no audible advantage to either. In further tests, they found that 4 of the 110 listeners in 4 of 145 tests were able to perform better than chance; however, achieving this involved listening in stereo over headphones.

Meyer and Moran (2007) took the straightforward approach of interrupting a high-resolution stereo signal path with a CD standard (16-bit/44.1 kHz) A-to-D conversion, followed by a corresponding D-to-A conversion. The test was to see if listeners could hear degradation due to the A/D/A conversion sequence at the reduced data rate. The answer after many A/B/X comparisons with many listeners over several months was "no." However, it was noticed that "virtually all the SACD and DVD-A recordings sounded better than most CDs—sometimes much better." Because of the test procedure (everything was, at some point, restricted by the same A/D/A bottleneck), this could not be attributed to the high-resolution recording processes, but the authors were able to confirm that recording engineers were using more dynamic range (less compression) than they would for the same or comparable program material released on CDs. The reason? It was thought, probably correctly, that customers who buy such recordings have better than average playback equipment, listen more attentively, and may appreciate the additional care that went into preparing the recordings. So these tests concluded that high-resolution recordings may sound better than CDs but not because they have high resolution.

The audio industry long ago matured in the sense that it was good enough to be remarkably entertaining. Equipment was affordable, simple to use, and highly reliable. As a hobby, it lost almost any requirement for "participation." While one segment of audiophiles has reverted to LPs and tube electronics for stimulation, another component looks to expanded digital data space to provide enhancements to the listening experience. Apart from movies, most program material continues to be in stereo. It remains a puzzle why the multichannel alternatives have not found a larger following.

Advocates of the new formats are probably unconvinced by scientific evidence suggesting that the expanded amplitude and bandwidth dimensions are not audible in music in the form it is delivered to consumers. If there are advantages to such systems in the recording studios, minimizing degradations in multiple layers, and generations of processing and mixing, that is a separate matter entirely, but that is not being raised as the prime argument.

Equipment manufacturers desperately want something new to sell, and journalists yearn for new topics to write about, so perhaps something good may yet happen. New HD video delivery formats promise attractive capabilities in audio channel count, dynamic range and bandwidth. It remains to be seen who uses it, and how. As discussed earlier, any tendency to stop compressing the dynamic range in music is welcomed. We don't need new delivery formats to achieve that, but if the price is right, so be it. It will simply equate to slightly longer download times over the Internet.

19.2 HEARING TILTS, PEAKS, DIPS, BUMPS, AND WIGGLES

Frequency response is the key measurable parameter of any audio device, and the general objective is a smooth, flat curve extending over the audible frequency range. However, that is a fantasy yet to be universally achieved in our imperfect physical world. Consequently, especially in loudspeakers, we must learn to evaluate the audible consequences of different kinds of deviation from the ideal.

The simplest deviation from flat is probably a spectral tilt. There is some evidence that we can detect slopes of about 0.1 dB/octave, which translates into a 1 dB tilt from 20 Hz to 20 kHz—not much. Such a spectral error, if small, is likely to be quite benign and subject to adaptation: we simply would get used to it.

19.2.1 The Audibility of Resonances

Beyond that, we get into peaks and dips. Buchlein (1962) conducted one of the first investigations. His subjects listened through headphones (which, as noted in Chapter 9, is not the most revealing circumstance), and he used peaks and dips with equivalent but inverted shapes. It is tidy, but it does not acknowledge the physical mechanisms that cause such shapes to occur in the real world. The most common mechanism for generating peaks in loudspeaker frequency response curves is resonances. If a dip were to have the same shape, only inverted, it would indicate the presence of something that functions as a powerful resonant absorber of energy. Such occasions are rare. More likely the cause of a dip is a destructive acoustical interference, in which case the dip will not have the appearance of an inverted "hump" like a resonance but rather a very sharp, possibly very deep, dip at the frequency where the interfering sounds cancel. Buchlein concluded that dips are less noticeable than peaks and that narrow interference dips would be the least noticeable of all. Wide peaks and dips were easier to hear than narrow ones. He also found that they were difficult to hear with solo instruments as test sounds, since they were audible only when the frequency of the defect and the musical tones coincided. Broadband sounds, such as noise, were more revealing of these irregularities in frequency response.

Fryer (1975, 1977) reported detection thresholds for resonances of different Q and frequency added to different kinds of program material. He found that detection thresholds fell with decreasing Q and that frequency (at least from 130 Hz to 10 kHz) was not a strong factor. As for program, the lowest thresholds were found with white noise, the next lowest for symphonic music and the highest thresholds for a female vocalist with jazz combo. It seems that spectral complexity matters; the more dense the spectrum and the more continuous the sound, the lower the thresholds.

Toole and Olive (1988) repeated some of these tests, confirming the results, and then went further. As reported in Chapter 9, sounds that are more continuous are more revealing of resonances than isolated transient sounds. The repetitions (reflections) necessary to lower thresholds (increase our sensitivity to resonances) can either be in the recorded sounds themselves or contributed by the listening room. The conclusion reported there was that we appear not to be sensitive to the ringing in the time domain (at frequencies above about 200 Hz at least) but to the spectral feature: the peak. As discussed earlier, at low bass frequencies, depending on the program, we can hear both the spectral feature and/or the ringing.

The least revealing sounds are, as Buchlein found, solo instruments and voices, especially those recorded in an acoustically "dry" or close-miked situation (lacking "repetition"). It is interesting that Olive (1994) found simple vocal and instrumental recordings not to be the most revealing of differences between loudspeakers. The most useful recordings had broad, dense spectra, such as those shown in Figure 19.8. This confirms in a different way our observation over many years that if a loudspeaker has an audible problem, there is a high likelihood that resonances are involved. It also suggests a strategy for choosing program material: for demonstrations intended to impress, use simple sounds, solo voice, guitar, small combos, and so on, especially with little reverberation—and, if possible, use a relatively dead room. To look for problems, listen to complex orchestrations with wide bandwidth and reverberation—and listen in a room with some reflections.

Taking a very different approach, one based on trying to understand the inner workings of the hearing process, studies of "profile analysis" have yielded results that appear to have important parallels. Studies by the prime investigator, D. M. Green and his colleagues, are summarized nicely in Moore (2003) pp. 105–107. It was found that a single frequency could be detected against a background of many other frequencies at threshold levels that decreased as the bandwidth of the background increased. The threshold is lower if the background spectrum is uniform and there are many components within it (although there appears to be a limit).

In the context of the present discussion, a pure tone being auditioned against a background of a complex of many tones covering a band of frequencies seems like a parallel to a high-Q resonance being detected in a "background" of a dense musical spectrum. Green and his colleagues are reported to have found that listeners could hear 1 to 2 dB changes in the relative level of the target tone against the background. As we will see, these numbers are in the range of detection thresholds for resonances. There is an important difference, though. Adjusting the level of a tone against a steady-state background is very different from detecting activity in a resonance that is energized by the varying temporal and spectral content of music. One would expect the amplitude of the "equivalent" resonance to be higher because only when it is driven by a frequency-matched signal of appropriate duration can it reach full output amplitude. In music this is a matter of probability, not certainty. It could be interesting to take this further, as there is a "natural" connection to resonances; they are the fundamental building blocks of timbre in voices and musical instrument sounds. It would not be surprising if humans were in some ways optimally able to perceive, and extract meaning from, such phenomena.

So what do just-audible resonances look like in frequency response curves? Figure 19.9 shows examples of deviations from flat for high- (50), medium- (10), and low- (1) Q resonances at three frequencies when they were adjusted to the audible threshold levels using pink noise in an anechoic chamber and for the 200 Hz resonances detected when listening to typical close-miked, low-reverberation pop and jazz. In the latter case, the absence of repetition

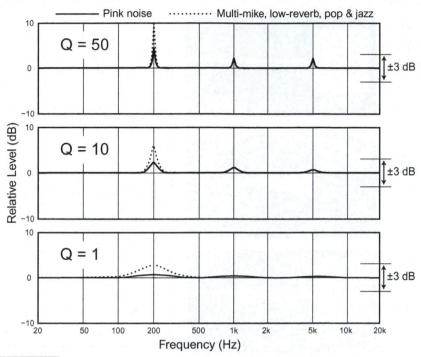

FIGURE 19.9 *Deviations from flat caused by resonances of three different Qs that were just detectable when listening to pink noise (all three frequencies) and to close-miked, low-reverberation, pop/jazz recordings (200 Hz only). Also shown is a ±3 dB tolerance, indicating that it is too tolerant of medium- and low-Q resonances and unnecessarily restrictive of some high-Q resonances. Adapted from Toole and Olive, 1988.*

(reflections/reverberation), and possibly also spectral complexity, has relaxed the requirements.

These data show that we are able to hear the presence of very narrow (high-Q), low-amplitude, spectral aberrations. Measuring these, especially at the lower frequencies, requires free-field, anechoic circumstances. The popular digital FFT/TDS measuring systems that are based on time-windowed data can have severe limitations when used in reflective spaces, as shown in Figure 18.9.

Selecting results for the most revealing signal, pink noise, at 200 Hz, Figure 19.10 shows what happens when the system, just at the threshold of presenting an audible problem, is driven by different kinds of signals. On the left are shown frequency responses taken from Figure 19.9. In the middle, we see the effect of a concentrated drive signal, a tone burst tuned exactly to the resonance frequency with sufficient duration to drive the high-Q (Q = 50) resonance to rise significantly in amplitude. It takes some time, as can be seen, which is why these resonances show up as being higher in amplitude than lower-Q resonances; for them to be energized, a musical tone must be at the correct frequency for a sufficient time, and that is a probabilistic event. Once excited, though,

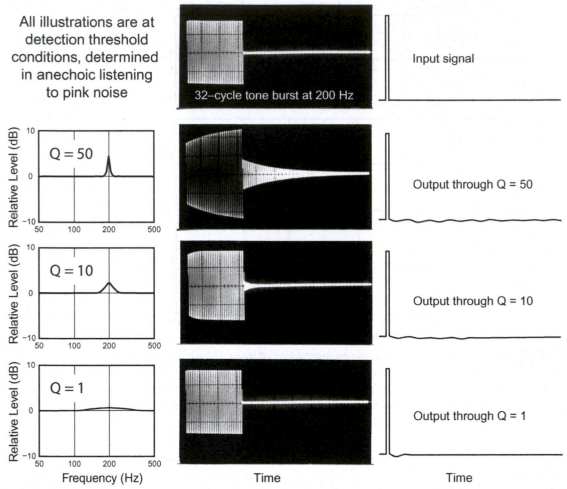

FIGURE 19.10 *On the left are shown resonances at the threshold of detection using pink noise, from Figure 19.9. In the center column are oscilloscope pictures of how this system responded to a tone burst at the frequency of the resonance. The right column shows system responses to a brief impulse. It is important to note that all of these represent threshold conditions, and therefore each of the resonances is perceptually equivalent in spite of the large visual differences in frequency and time domain displays of their behavior. Adapted from Toole and Olive, 1988.*

they take a similar amount of time to decay: the ringing. The impulse on the right is a broadband signal, so it energizes resonances at all frequencies equally but not very strongly. The ringing is of low amplitude but long in duration.

The medium-Q (Q = 10) resonance responds quickly to the tone burst, reaching full amplitude in a few cycles and, correspondingly, decaying in a few cycles. The impulse initiates the same response magnitude in the first cycle of ringing, but the ringing dies away quickly. Obviously, one reason why we can detect medium-Q resonances having lower amplitudes than high-Q resonances is that it takes only a few cycles of signal to get them to respond to full amplitude. Since they have some bandwidth, the signal does not even need to be precisely on frequency. This is even more true of low-Q resonances.

The low-Q (Q = 1) resonance responds to the tone burst within one cycle and decays within a cycle. The impulse elicits only a one-cycle of overhang. All of this is an intuitive illustration of the concept expressed earlier: a flat frequency response in a minimum-phase system (like a loudspeaker) results in correct time response. The next step beyond this low-amplitude, low-Q resonance would be no resonance—a straight line frequency response—which is what is needed to produce the images at the top of this figure: the input signal itself.

Reviewing these data, it is clear that neither the measured amplitude of the spectral aberration nor the duration of the ringing is a direct correlate of what we hear. So let us enjoy those attractive waterfall diagrams. Use them as evidence of the presence of resonances, but don't rely on them as indicators of audibility. Our ability to hear a resonance depends on the ability of the driving signal—the music or movies—to excite it. If the signal is at the correct frequency and lasts long enough, one can hear high-Q resonances; if not, we don't. Medium- and low-Q resonances have wider "footprints" in the frequency domain, so more frequencies can excite them, and they need not be very long to achieve full effect. Consequently, we end up hearing them more readily. The duration of the ringing, except at low-bass frequencies, appears to be just an interesting engineering observation (see discussion and data in Section 9.2.1).

The only measured quantity that seems to be constant in all three conditions, and thereby might provide a clue as to what is actually at the basis of these perceptions, is the initial amplitude of ringing after the impulse. It is not the steady-state amplitude of the resonance, nor is it the duration of the ringing. This prompted a hypothetical explanation, a possible mechanism, discussed in Toole and Olive (1988).

In summary, some facts need to be emphasized:

1. The amplitudes of the resonances shown in frequency responses are the *steady-state* measured changes in the playback system caused by the presence of the resonances that have been adjusted to the detection-threshold level while listening to different kinds of program. This is *not* the amplitude of the output from the resonance when listening to musical program material because music is not a steady state signal. That amplitude is likely to be much lower. The fact that the resonant peaks are higher for the chosen pop/jazz examples is a reflection of the fact that the program material exhibits a lower probability of exciting the resonance than pink noise, a spectrally dense, steady-state signal. Not shown, but interpretable from the Fryer (1975) data, shown in Toole and Olive (1988), is the fact that symphonic music, which is

both spectrally complex and reverberant, exhibits thresholds that are between the two shown in Figure 19.9 at 200 Hz.

2. High-resolution measurements are necessary to reveal high-Q features in the frequency responses. Because they exist at all frequencies, including the lower-frequency regions it means that anechoic, or long-window simulated-anechoic measurements are necessary to reveal them.

3. Any tolerance applied to a frequency-response curve needs to take into consideration the bandwidth/Q of the deviations that are being described. The conventional ±3 dB, or other tolerances, have no meaning without being able to see the curve(s) that are being verbally described.

4. Finally, all of these threshold determinations were done in anechoic listening conditions. As shown in Chapter 9, these thresholds may be even lower when listening in reflective rooms. However, if the thresholds were determined using signals incorporating significant repetitions (e.g., reverberation), the effect of the listening environment is minimal.

19.2.2 Critical Bands, ERB$_N$'s, and Timbre

One of the traditional justifications for 1/3-octave spectral analysis has been that this is a rough approximation of perceptual "critical bands" over much of the middle- and high-frequency ranges. Some people even have argued that we "hear" in critical bands, that this is the "resolution" of the hearing system. Such statements are simplistic and misleading.

Readers who would like to understand details of the process are referred to Moore (2003), where the evolution of the concept leads to a new definition: the equivalent rectangular bandwidth, the ERB$_N$ (see Figure 9.4). The quick explanation is that these bands define the bandwidth over which spectral information is summed for estimates of loudness and in the simultaneous masking of tonal signals by broadband noise. They also help define the separation required for two adjacent tones to be individually identifiable. However, within these bands, multiple tones (which can be tones and overtones of musical sounds) beat with each other, and this influences a perceptual quality called "roughness." Differences in roughness contribute to the distinctiveness of sounds—their timbre—so timbre can be changed if the reproducing system has spectral variations occurring within a single critical band or ERB$_N$. A full understanding of the performance of loudspeakers requires measurements with high resolution—of the order of 1/20-octave—in the frequency domain. That said, there are situations where spectrally smoothed measurements are useful for portraying more general trends in spectral balance, but contemporary understanding suggests that 1/3-octave bands are not optimal (Moore and Tan, 2004).

Moore (personal communication, 2008) summarizes, "The auditory filter bandwidths (ERB_N values) are about 1/6- to 1/4-octave at medium to high frequencies. And . . . within-band irregularities in response can have perceptual effects." As has been found in other aspects of perception, it is difficult to find a "one-curve" description for what we hear.

19.3 NONLINEAR DISTORTION

The memory of distortions in 78 rpm recordings and early LPs is still vivid in my memory. They ranged from inaudible to intolerable, and all recordings had them. LPs and their playback devices improved, and at their best they became highly enjoyable—except for pesky inner-grove crescendos, the exciting wind-up of a symphony or opera that one has carefully been prepared for by the composer.

I recall testing phono cartridges as part of the effort to improve playback quality (Toole, 1972). This exercise is substantially a test of test records, which really is a test of the entire LP mastering, pressing, and playback process. I participated in the creation of a test record. It was impossible to replay from an LP the signal that was delivered to the (carefully chosen) mastering lab. When using pure tones or bands of noise, distortions were easily measured and easily audible. They registered in whole (sometimes high) percentages much of the time, and this applied to both harmonic and intermodulation versions. Yet, when the signal was music, the experience was enjoyable. What is going on here?

Masking: conventional simultaneous masking of a smaller sound (the distortion) by a larger sound (the same musical signal that created the distortion) as is shown in Figure 19.7. Simple test signals leave the distortion products spectrally exposed so they can be measured, and sometimes enough of them are unmasked that we can hear them. The wide-bandwidth, dense spectrum of music is a much more effective masking signal, despite the fact that it is at the same time a generator of much more complex distortion products. It is also almost useless as a measurement test signal.

To understand what is going on, first consider what the real problem is: nonlinear behavior in the audio device. This means that the relationship between the input signal and the output from the device changes with level; it is not linear. If the input increases by a certain percentage, the output should change by the same percentage—no more, no less. However, the percentage does change; the shapes of audio waveforms are altered from a little to a lot. When we hear these distorted waveforms, we notice that new sounds (distortion products) have been created. When the nonlinear device is driven by a known, well-specified signal, we can measure how different the output is from the input, and obtain a measure of the magnitude of the distortion for that specific signal. But instead of measuring the nature of the nonlinearity itself—as an input-to-output transfer relationship—we have chosen to try to quantify what it does to audio signals. Put a known signal into the device and spectrally analyze what comes out.

Engineers have invented several ways to quantify the magnitude of the nonlinearity by measuring the distortion products and expressing them as a percentage of the original input signal or of that signal plus distortion and noise. If the input signal is a pure tone, the distortion products show up as harmonic overtones of that signal: harmonic distortion. If the input signal is two tones, each of the tones will generate a set of harmonic distortions, but they will also interact with each other in the nonlinearity and create combination tones: intermodulation distortions. These are sum and difference multiples of the two input signals and their harmonics that extend upward and downward in frequency. Because masking is much more effective in the upward-frequency direction (see Figure 19.7), it is the difference frequencies that end up being more audible and, because they are inherently unmusical, more annoying.

Thus began the legend that harmonic distortions are relatively benign and intermodulation distortions are bad. It is true in the context of these test signals, but they are both simply different ways of quantifying the *same* problem (the nonlinearity in the loudspeaker), and neither test signal (one tone or two) is even a crude approximation of human voices or music. Contributing to the mismatch between perception and measurement is the fact that such a technical measurement totally ignores masking. Included in the numbers generated by the measurements are distortion components that, to humans, are partially or completely masked. The numbers are wrong.

The end result of this is that traditional measures of harmonic or intermodulation distortion are almost meaningless. They do not quantify distortion in a way that can, with any reliability, predict a human response to it while listening to music or movies. They do not correlate because they ignore any characteristics of the human receptor, an outrageously nonlinear device in its own right. The excessive simplicity of the signals also remains a problem. Music and movies offer an infinite variety of input signals and therefore an infinite variety of distorted outputs.

Taking advantage of advances in psychoacoustic understanding and using the analytical and modeling capabilities of computers, some new investigations are attempting to identify some of the underlying perceptual mechanisms and develop better test methods. Voishvillo (2006) provides an excellent overview of the past, present, and possible future of distortion measurements. Geddes and Lee (2003), Lee and Geddes (2003, 2006), and Moore et al. (2004) provide additional perspectives.

In loudspeakers it is fortunate that distortion is something that normally does not become obvious until devices are driven close to or into some limiting condition. In large-venue professional devices, this is a situation that can occur frequently. In the general population of consumer loudspeakers, it has been very rare for distortion to be identified as a factor in the overall subjective ratings. This is not because distortion is not there or is not measurable, but it is low enough that it is not an obvious factor in judgments of sound quality at normal foreground listening levels.

19.4 POWER COMPRESSION

Turn the volume up, and the sound gets louder, more amplifier power is delivered to the loudspeakers, power generates heat in the voice coils of the transducers, increasing temperatures result in increased resistance in the voice coils, increased resistance results in less current passing through the voice coil, and less sound being produced. The more the voice coil temperature rises, the greater is the reduction in sound output; this is power compression. Different transducers exhibit different amounts of power compression depending on their sensitivities (how much power is required to produce a certain sound level) and on their abilities to dissipate heat. High-efficiency transducers, like horns, use less power and are therefore less susceptible to this problem. That is one reason why horns are so popular in systems designed to fill large rooms or to play at high sound levels in smaller rooms. The fact that different transducers exhibit different amounts of power compression means that in a multiway loudspeaker, as the volume is turned up, different portions of the spectrum are affected by different amounts; the frequency response, and timbral character, changes.

Power compression has long been a matter of fundamental interest in professional audio (Gander, 1986) because of the physical abuse those transducers are subjected to on a routine basis. Only rarely are consumer products tested at high sound levels, but it happens. In the early 1980s, my wife and I hosted our annual New Year's Eve party. Animated dancing occurred, and sound levels rose accordingly. The loudspeakers in the dance room were examples in a continuous series of "borrowed" products for "personal evaluation." At a point in the evening it became apparent that the treble had disappeared; only small amounts of distorted sound emerged from the tweeters. The system was shut down, and amidst a good deal of grumbling, people reverted to conversation for entertainment. It was assumed that the tweeters had been destroyed. Some time later, music again emerged from the room; somebody had decided to turn it back on. A moment of listening revealed that the treble was miraculously back.

This was an extreme example of power compression. What happened, as was confirmed a few days later back at the lab, was that the tweeter voice coil got so hot that it softened the mechanical interface between the voice coil and the dome. The voice coil continued to move, but the radiating area of the dome was unable to follow. Amazingly, in this case the process was reversible and, in the lab, repeatable. Leading up to the catastrophic loss of treble was a progressive degradation during which this loudspeaker took on many different personalities.

At moderate listening levels, it was comfortable, but it could not survive "party" mode.

If the loudspeaker is active with internal power amplifiers, there is an added dimension to power compression: the effects of clipping and protection (of the amplifier itself and/or the transducer(s) it drives). Figure 19.11a illustrates the effect of turning up the volume on a loudspeaker in which the power amplifiers are still functioning linearly at the higher sound level. The moderate differences in the shapes of the curves are due primarily to resistive heating effects—that is, conventional power compression. Figures 19.11b, (c), and (d) show systems that are operating outside of their linear ranges and are exhibiting combined amplifier and transducer misbehavior. These are all closely competitive professional monitor loudspeakers, three of which tell different versions of the "truth" at different sound levels. Because products of this kind nowadays find their way into home theater installations, these issues become matters of more general interest. A full evaluation of loudspeakers cannot ignore these effects.

19.4.1 Any Port in a (Turbulent) Storm?

Finally, we need to dwell a moment on another special contributor to power-dependent misbehavior: bass-reflex ports or vents. In simple models and calculations, the port in a bass-reflex system is assumed to function as a lumped element: a mass in a spring-mass resonating system intended to augment bass output and reduce distortion over a narrow band of very low frequencies. All of this is embodied in well-known equations for designing woofer and subwoofer systems. The reality is that often such systems are designed for performance at low sound levels and then required to operate at high sound levels.

The aerodynamic behavior of ports can be complex, as might be imagined. Abandon notions that there is anything resembling laminar (smooth, nonturbulent) flow when air is pulled and pushed through this tube during a high level bass guitar riff combined with a kick drum. The flow rapidly degrades into varying degrees of turbulence and eventually chaos. In this process, all of the careful measurements and calculations that led to the original design of the bass-reflex woofer/subwoofer system fall by the wayside. The tuning (i.e., the Q and frequency of the resonance) is changed, distortion rises, and we hear turbulent "chuffs" as air is alternately pushed and pulled through the orifice.

Salvatti et al. (2002) describe experiments, systematically investigating the variables of ports, inlet and outlet shapes, size, surface texture, and so on. The results support some of the common practice, but intuition is not always a reliable guide and there are a few surprises.

Coming as no surprise is the fact that a straight-sided tubular or rectangular port is not optimum. Large ports function better than small ports. The inlet and outlet need to be radiused, and the entire port may benefit from being contoured throughout its length, but the amounts of both are very much depen-

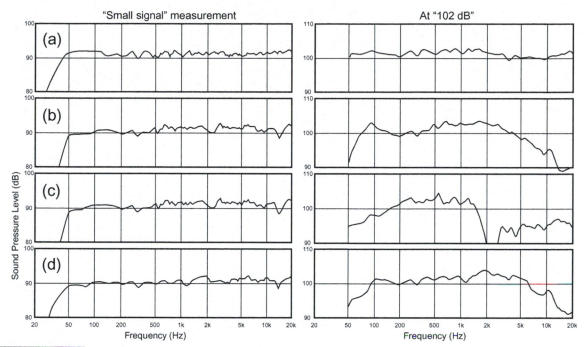

FIGURE 19.11 *The on-axis frequency responses of four active professional monitor loudspeaker systems of similar size and driver configuration. Their on-axis frequency responses were measured conventionally under what is called "small signal" conditions and again at an elevated sound output level. To achieve the latter condition, it was necessary to depart from the standard measurement procedure, resulting in detail differences between curves on the left and curves on the right. However, the differences that matter are not small details.*

dent on sound level (velocity of the air in the port). There are many versions of contours that function satisfactorily; the best of them yield substantial reductions in "port compression" and audible distortions and noises. Texturing the port walls is ineffectual at the air velocities encountered in normal ports. The paper has a lot of data that designers will find interesting and all of it is worthwhile because a well-designed bass reflex system has distinct advantages for many applications.

CHAPTER 20

Closing the Loop: Predicting Listener Preferences from Measurements

Psychoacousticians dream of being able to insert measured numbers into an equation representing a model of a perceptual function and accurately predicting a subjective response. It has not been straightforward for those perceptual dimensions that appeared to be simple, much less for more complex ones like loudness, incorporating all of the possible interactions with different sounds. Something as multidimensional and abstract as preference in the sound quality from loudspeakers presents a new level of challenge.

Nevertheless, there have been several attempts, all of which have shown some degree of correlation with perceptions. Olive (2004a) discusses several of them. The oldest investigations suffer from a lack of truly good loudspeakers. Differences were audible and measurable, but the differences were of such magnitude that almost any measurement would have shown a correlation with perceptions. Loudspeakers have improved, and our understanding of how we perceive their sound has advanced. It is now evident that more than a single curve (e.g., sound power or a room curve) is necessary to explain or describe how a loudspeaker might sound in a room. It is further evident that 1/3-octave or critical-band measurements are insufficient.

20.1 THE KLIPPEL EXPERIMENTS

Klippel (1990a, 1990b) summarized his elaborate investigations, conducted in stereo, of perceptual dimensions and their relationships to measured quantities: "All dimensions perceived in the performed listening tests correspond with features extracted from the sound pressure response of the diffuse and direct field at the listener's position. There was no hint of a relation to phase response or to nonlinear distortion." According to this, the relevant loudspeaker measurements, therefore, are the anechoic on-axis frequency response and the sound

457

power response—or at least a sufficient collection of off-axis measurements to describe the reflected sounds arriving at the listening location.

Of special interest was his finding that what he called "feeling of space" figured prominently in listener responses. In evaluating listener perceptions, Klippel assessed the listener responses against what was judged to be the ideal quantity of each perceptual parameter. Thus, the perceived quality is evaluated as a defect:

$$\text{Defect} = [\text{basic measure} - \text{ideal value}]$$

A defect can therefore indicate that there is too much or too little of a perceived dimension; listeners responded according to what seemed to them to be appropriate.

When Klippel analyzed the factors that contributed to the perception of "naturalness," one of the general measures of quality, he found the following:

- 30% was related to inappropriate discoloration (sound quality).

- 20% was related to inappropriate brightness (which is explained as a 70% excess of treble and 30% lack of low frequencies).

- 50% was associated with the "feeling of space."

For the second general measure of quality, "pleasantness," the factor weighting was the following:

- 30% was related to inappropriate discoloration (sound quality).
- 70% was associated with the "feeling of space."

Thus, sensations of sound quality and spaciousness dominated both "naturalness" and "pleasantness" in ratings of these loudspeakers and, of these, "feeling of space" held a slight lead.

As we saw in Chapters 7 and 8, it is mainly laterally reflected sounds that contribute to perceptions of space. In small rooms, it is improbable for natural reflections to initiate impressions of envelopment, but sensations of ASW (apparent source width), image broadening, and early spatial impression are very real and desirable. Klippel chose as his objective measure of "feeling of space" (R) the difference between the sound levels of the multidirectional reflected sounds and the direct sound at the listening location:

$$R = L_{diffuse} - L_{direct}$$

Thus, the difference between the anechoic on-axis frequency response (L_{direct}) and total sound power (L_{diffuse}), representing the late reflected sound field at the listening location, provides a measure of the potential a loudspeaker has for generating perceptions of "feeling of space."

Figure 20.1 shows that there is an optimum amount of reflected sound; there can be too much and too little, depending on the nature of the program.

The smallest amount is required to provide a satisfying setting for speech, and more is required for music. The optimum amount by which the combined reflected sounds should exceed the direct sound is about 3 dB for speech, 4 dB for a mixed program, and 5 dB for music. There is no frequency dependence considered in these numbers, and we know that loudspeakers do not exhibit constant directional behavior at all frequencies. There is a parallel with data shown in Figures 7.2 and 7.3 that indicated that listeners preferred listening to speech and music with individual reflections at levels above those of natural room reflections, and other data (see Figure 8.2) that indicated that reduced interaural cross-correlation (IACC) is a favored condition for recreational listening.

Let us examine the Klippel evidence in the context of data from a real loudspeaker configured to show the sounds that it delivers to a listener in a room. Figure 20.1 indicates a requirement for a 3 to 5 dB difference between the reflected and direct sound fields (as represented by the sound power and on-axis measurements of a loudspeaker) for optimum perception of "feeling of space." Figure 20.2 shows data for a loudspeaker having a good

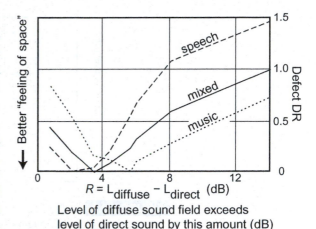

FIGURE 20.1 *The relationship between the Defect DR—the subjective evaluation of "feeling of space"— on the vertical axis, and the objective measure R (the estimated difference in sound level between the direct and reflected sounds at the listening position) on the horizontal axis. Moving down the vertical scale corresponds to increased subjective satisfaction; moving to the right corresponds to increased reflected sound in proportion to the direct sound. The minimum point of each curve describes the optimum level of diffuse sound compared to the direct sound, for each program. Curves are shown for speech, music, and a mixed program. From Klippel, 1990a, Figure 5.*

on-axis frequency response but poorer sound power because of inconsistent directivity. The light shaded area exhibits differences that fluctuate between about 1 dB and 5 dB below 5 kHz, and above about 5 kHz, there is a darker shaded area indicating no potential for the desirable spatial effect. Although there will be a "feeling of space," this is not a particularly attractive result; yet, it is not a "bad" loudspeaker.

A good loudspeaker for this purpose would therefore be one that has two qualities: wide dispersion, thereby promoting higher levels of reflected sound, and a relatively constant directivity index so that the direct-sound and reflected-sound curves have similar shape. This would be revealed here as a larger light shaded area, and the essence of good design in this respect would be to deliver the optimum proportion of reflected sound for the program being auditioned. An associated requirement of considerable importance is that at least some of the off-axis sounds be allowed to reflect, and any that are acoustically modified by absorption or diffusion are treated in a spectrally neutral manner. In other words, any reflected or scattered sounds should convey the spectral balance of the incident sounds, only at a lower sound level.

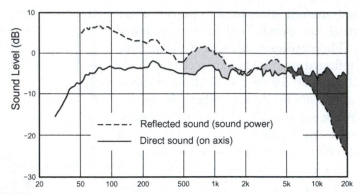

FIGURE 20.2 *Using data from Figure 18.5b, originally from Toole, Part 2, 1986, the shaded areas show the difference between the axial on-axis sound (representing the direct sound at the listening position) and the sound power (a rough approximation of the reflected sound field at the listening position) for a particular loudspeaker as it is located in a "typical" listening room having reflective side walls at the points of first lateral reflections. The light shaded area is that which would be effective at creating "feeling of space" in Klippel's terms, or ASW, image broadening and early spatial impression in terms used in this book. The condition for this is that the reflected sound field is higher in level than the direct sound. The dark shaded area is a region that is not effective at creating "feeling of space" because the direct sound is higher in level than the reflected sound field. The shaded areas extend from about 500 Hz to 20 kHz, which is the frequency range over which ASW/image broadening is believed to be most effective (see Figure 7.1). As discussed in the context of Figure 18.5, this loudspeaker has a problem with frequency-dependent directivity. The on-axis curve is quite flat, but the sound power is uneven. In respect of how this loudspeaker might perform in generating the desirable "feeling of space," it very much depends on frequency. It fails in a broad band around 2 kHz.*

This same loudspeaker was used in the stereo versus mono listening tests in Chapter 8, where it was highly rated compared to a narrow-dispersion loudspeaker. It was concluded then that wide, but uneven, dispersion appears to be preferable to narrow dispersion; a flawed "feeling of space" is preferable to one that is insufficient. More recent loudspeaker designs, having wide *and* uniform dispersion, such as those seen in Figures 18.16 and 18.17, would exhibit a "feeling of space" that is more uniformly represented as a function of frequency and over a wider frequency range. The comparison shown in Figure 18.14 seems to be a particularly good example of two loudspeakers with wide and relatively uniform dispersion (R and I) being preferred to one that has wide but nonuniform dispersion (B) and one that has narrow dispersion (M). As noted in Chapter 8, spatial impression ("feeling of space") does indeed rank with sound quality (lack of "discoloration"), and, as Klippel finds, the two factors account for almost all of the important general subjective ratings of "naturalness" and "pleasantness." It seems relevant to note again that none of this is revealed in steady-state room curves, making such measurements poor metrics by which to judge the audible excellence of loudspeakers.

As shown in Figure 8.1, simply absorbing the first reflections can have a large effect on diffusivity within the room, and Figure 8.2 shows that the first lateral reflections have large effects on IACC and perceived spaciousness. These data apply to stereo and mono sources. What about multichannel? As has been stated many times before, in movies and television the center channel functions as a monophonic source most of the time, and much, if not most, of the ambient music in movies is stereo, with some leakage into the surrounds. We may be iterating our way to a true multichannel world, but we are not there yet.

20.2 THE OLIVE EXPERIMENTS

For two decades, Sean Olive has been a participant in many of the NRC and Harman International research projects reported in this book. Both of us have been exposed to hundreds of loudspeaker measurements and double-blind listening tests, and, as you can see in samples of results shown in Chapters 17 and 18, it is not difficult to recognize from an inspection of measurements the loudspeakers that are likely to rate highly in listener preferences and those are not. But this is not a numerical prediction or a model that interprets measured data and outputs a prediction of a subjective judgment.

Over the years, information has been accumulating about the audible consequences of various measured quantities, and, inevitably, the time came when it seemed right to combine that knowledge into a predictive model of listener preference. An obvious model for comparison purposes is that used by Consumers Union (CU) in the loudspeaker evaluations published in their magazine *Consumer Reports* over the past 30 years. It is based on 1/3-octave measurements of sound power that, after manipulation, yield an accuracy score out of 100, indicating how far the tested loudspeakers deviate from their notion of an ideal performance. Apparently, no formalized subjective evaluations are involved. The product ratings have a significant influence on the North American market, but if there has been a proper validation of the method, it has not been made public.

Using the collection of anechoic measurements described in Figure 18.6, Olive (2004a, 2004b) undertook an evaluation of 13 loudspeakers that had recently been reviewed by CU. It began with a fully balanced, double-blind listening test (every loudspeaker was auditioned against all others the same number of times) by a group of selected and trained listeners. The result is shown in Figure 20.3a and compared with the corresponding accuracy scores for the same loudspeakers published by Consumers Union. The contrast between the two sets of data is striking, with loudspeaker 1 exhibiting both "best" and "worst" evaluations. Overall, there was no correlation between the two results, indicating that if the subjective data are correct, the CU method of evaluation is based

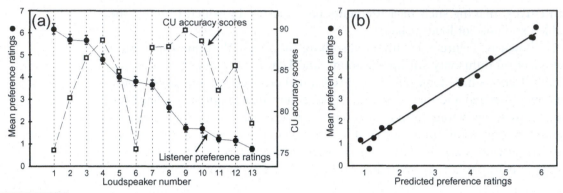

FIGURE 20.3 *(a) A comparison of preference ratings from double-blind listening tests and accuracy scores from evaluations of the same loudspeaker models published by Consumers Union (CU). The correlation between the two was small and negative (r = –0.22) and not statistically significant (p = 0.46). The 95% confidence intervals are shown for the subjective preference ratings, indicating high confidence in several of the rating differences, although there were clusters of loudspeakers that were in statistical ties: 2 & 3; 5, 6, & 7; 9 & 10; and 11 & 12. Data from Olive, 2004a, Figure 3. (b) The comparison of the same subjective data and predictions from a model created by Olive, using anechoic data of the kind shown in Figure 18.6. In this case, the correlation was 1.0 (r = 0.995), and the statistical significance a very high (p = <0.0001). Data from Olive, 2004b, Figure 4.*

on a faulty premise: a sound power measurement alone contains information necessary to describe listener perceptions of sound quality in small rooms. In fact, the flaw in this logic was indicated in much earlier results by Toole (1986). In Chapter 4 of this book, it is argued from an analysis of sound fields in small rooms that sound power could not be the dominant factor because the sound fields are not sufficiently diffuse. It is important to note that, as this is being written, Consumers Union is in the process of revising its evaluation process for loudspeakers.

Employing the much more comprehensive, higher-resolution data generated routinely in the Harman International tests, Olive identified a model that described the subjective preference ratings in a near-perfect manner, as shown in Figure 20.3b. The precision exhibited by this model implies a complex analysis, and this is indeed the case. The statistics derived from the different curves of the kind exemplified in Figure 18.6 are as follows:

- AAD: *absolute average deviation* (dB) relative to the mean sound level between 200 and 400 Hz, in 1/20-octave bands from 100 Hz to 16 kHz.

- NBD: average *narrow-band deviation* (dB) in each 1/2-octave band from 100 Hz to 12 kHz.

- SM: *smoothness* (r^2) in amplitude response based on a linear regression line through 100 Hz to 16 kHz.

- SL: *slope* of the best-fit linear regression line.

All of these statistics were applied to all of the measured curves (see Figure 18.6):

- On-axis (ON).

- Listening window (LW).

- Early reflections (ER).

- Sound power (SP).

- Both directivity index curves (ERDI and SPDI).

- Predicted-in-room (PIR), a proportioned combination of ON, ER, and SP that approximates measured room curves in several typical rooms.

Then there are two statistics that separately focus on low-frequency performance:

- LFX: low-frequency extension (Hz), the 6 dB down point relative to listening window (LW) sensitivity in the range 300 Hz to 10 kHz.

- LFQ: absolute average deviation (dB) in bass frequency response from LFX to 300 Hz.

When the dominant contributing factors to this model were analyzed they were as follows:

- Smoothness and flatness of the on-axis frequency response: 45%
  —AAD_ON (18.64%) + SM_ON (26.34%)

- Smoothness of sound power, SM_SP: 30%

- Bass performance: 25%
  —LFX (6.27%) + LFQ (18.64%)

On-axis flatness and on- and off-axis smoothness account for 75% of the preference estimate. Bass, on it own, accounts for 25% of the preference estimate, something not to be dismissed.

As impressive as this result is, there is a common factor that may prevent the model from being generalized: all 13 of the loudspeakers were bookshelf models, a natural basis for a comparative product review in a magazine. Many of them had common features: enclosure size, driver complement and configuration, and so on. The population needed to be expanded to include larger products, using different driver configurations and types. This was done in a second test that involved 70 loudspeakers from many origins, in a wide range of prices and sizes. The penalty in this test was that the listening had been done in a fragmented manner, 19 independent tests over an 18-month period, as happens in the normal course of business. All loudspeakers were evaluated in comparison with some other models, but there was no overall organization to ensure that the comparisons were balanced, each model against all other models. It would be unreasonable to expect the same high precision in the subjective-objective correlations as has just been seen.

Still, the result was impressive: predicted preference ratings correlated with those from listening tests with a correlation of 0.86, with a very high statistical significance ($p = <0.0001$). These are remarkable numbers, given the opportunities for variation in the listening tests, meaning that the listeners themselves are highly stable "measuring instruments" and that the strategy of always doing multiple (usually four products) comparisons is a good one. The model that for these products and these subjective data produced the best result was different from the earlier one. Here, the dominant factors were as follows:

- Narrow-band and overall smoothness in the on and off-axis response: 38%
  —NBD_PIR (20.5%) + SM_PIR (17.5%)
  —Since PIR incorporates on-axis, early-reflection, and sound power curves it is evident that all three must exhibit narrow and broadband smoothness.

- Narrow-band smoothness in on-axis frequency response (NBD_ON): 31.5%

- Low-bass extension (LFX): 30.5%

Again, on- and off-axis smoothness and the lack of narrow-band deviations account for most of the preference estimate (38 + 31.5 = 69.5%). Bass again is a big factor at 30.5%. This time, perhaps because there was a mixture of large and small loudspeakers, listeners focused on bass extension more than smoothness. Since the bass smoothness metric does not include the influence of the room (it is based on anechoic data), the issue may be debatable, but extension alone seems fairly straightforward.

As part of the first tests, listeners were required to "draw" (using sliders on a computer screen) a frequency response, describing what they thought they heard in terms of spectral balance. This is a task that obviously could not be asked of average listeners, but these listeners had been through a training program (Olive, 1994, 2001) and were able to estimate the frequencies at which audible excesses and deficiencies occurred. When these data were compiled, the results indicated that all of the highly-rated loudspeakers had been judged to have very flat curves. Of great interest, of course, is which one of the several measured curves for these loudspeakers does this correspond to, the logical candidates being sound power (the CU model), the PIR (the in-room response relied upon by most room-equalization schemes), or the on-axis curve?

Figure 20.4 tells an important story. Let us focus on the region above the shaded area—in other words, above the transition frequency. This is where the room is not the dominant factor and also where, back in Figure 4.10 (from Toole,

Part 2, 1986), it was shown that the loudspeaker was the principal factor in what was measured and heard.

From empirical evidence accumulated over many years (some of it shown in Chapter 18), it is evident that loudspeakers with flat and smooth on-axis frequency responses and well-behaved directivity are the ones that win listening tests. Such loudspeakers yield smooth room curves above the transition frequency, but they are not flat because with typical cone and dome designs, directivity index (DI) rises gradually with frequency, meaning that room curves and sound power both fall gradually with frequency. Thus began the guessing game of which "target" room curve is ideal for which circumstance (see Section 18.2.6 for a discussion of the "X" curve). Obviously, it must depend on the directivity of the loudspeaker and the reflective properties of the room. However, listeners have a remarkable ability to "listen through" rooms and to be able to extract key information about sound sources, whether they are voices, musical instruments . . . or loudspeakers (see the discussion of adaptation in Chapter 11).

In these experiments, trained listeners were able to draw curves of spectral trends—crude frequency responses—describing what they heard in the listening room. All of the high-scoring loudspeakers were described as having flattish spectra (above the transition frequency), a trend that matches the flattish on-axis/listening window curves for all of the corresponding anechoic measurements.

Loudspeakers with lower preference ratings were described as having either insufficient bass or excessive treble, or both. These trends tend to yield a flatter sound power curve and thereby higher accuracy scores in CU reviews. However, it is a trend that runs contrary to the interests of high sound quality as judged by these listeners and many others.

Below the transition frequency, in the gray shaded area in Figure 20.4, it is evident that listeners responded to something closer to the low-frequency response in the room than to the anechoic frequency response (see the comments in the figure). This is the ultimate limitation of all such models, in that the listening room and the arrangements within it determine the bass level and quality. With bass accounting for about 30% of the overall subjective rating, this is a matter of great importance, and it implies that if there is to be parity among listening situations—both professional and consumer—there must be substantial control over the low frequencies in the room. Chapter 13 provides guidance.

20.3 AN INTERIM SUMMARY

I say "interim" because I hope that work of this kind will continue. It is the only way to ensure that the "circle of confusion" shown in Figure 2.2 is broken. We may or may not ever eliminate the need for subjective evaluations, but it

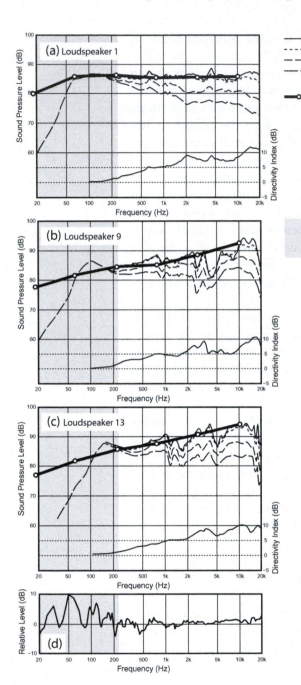

—— on axis
------ listening window
— — early reflections
— — sound power

⊸⊙⊸ subjective estimate of spectral balance in the room during the listening test. The vertical scaling of this rating has been adjusted by this author to match the anechoic curves. It is approximately half of that shown by Olive (2004a) and is the same in all three curves. The vertical locations of the curves were also determined by appearance. Open circles indicate center frequencies of bands used by listeners to describe the shapes of the spectra.

These shaded areas are below the transition frequency indicating that the frequency response is dominated by the room. The curve shown in (d) is the average difference between measured and predicted room curves (Olive 2004b) indicating that above about 250 Hz the predictions are excellent, but below about 250 Hz these loudspeakers as heard in this listening room exhibit much more bass—about 10 dB more at 50 Hz and 6 dB more around 35 Hz—than is predicted by the anechoic data. These differences appear to be somewhat reflected in the spectral balance estimates of the listeners.

FIGURE 20.4 *Figures (a), (b), and (c) show the familiar set of anechoic data for 3 of the 13 loudspeakers tested (results in Figure 20.3). The dark line superimposed on each curve is the subjective estimate of the spectral balance given by the listeners while in the listening room. This is a "dimensionless" curve on the vertical axis, reflecting subjective impressions of boost and cut relative to the personal ideals of the listeners. Consequently, this author has taken the liberty of adjusting the vertical scale and the vertical position of these curves until they appeared to fit the pattern of anechoic data. The chosen vertical scale is used in all three curves. As explained in the figure, below about 250 Hz, the anechoic data cease to reflect accurately what is heard in the listening room; it will be less smooth and significantly higher in amplitude in the room. This is shown in (d), the average error between PIR (from anechoic data) and in-room measurements. From Olive, 2004b, Figure 7. The most obvious visual correlation with the subjective spectral balance is with the on-axis or listening window curves.*

is greatly reassuring to know that there are measurements that add value to the examination of a product. In fact, one may venture a challenge that an examination of the right set of anechoic measurements may well be more reliable as an indicator of sound quality than a "take-it-home-and-listen-to-it" subjective evaluation. In fact, it is likely that methods of subjective loudspeaker evaluation can be improved through an examination of these data. Among the information gained is guidance about what factors should be listened for in the music and what to look for in the measured data. It is in the combination of both subjective and objective data that progress will come. The problem now is that comprehensive anechoic data of the kind shown in this book are available only to a privileged few inside certain companies and laboratories. That needs to change.

It is time to look again at Figure 19.1, the so-called "central paradox," in which the communication of sound, live or reproduced, through a room appears to be a chaotic mess, and yet two ears and a brain make sense of it. It is wise not to underestimate the power of the perceptual process. The data in Figure 20.4 show that listeners were able to draw curves that almost uncannily echo the shapes of the on-axis responses of loudspeakers. This tells us that the first sound to arrive at the ears is of substantial importance. It is not the full story, of course. We know from other tests that off-axis misbehavior can detract from listener preferences. However, evidence from Klippel's work suggests that this contribution to preference may be in the domain of "feeling of space," image broadening, early spatial impression, as much as or more than sound quality. It is a reminder of ideas raised in experiments described in Section 8.2, where it simply was not clear that listeners could totally separate issues of imaging from issues of sound quality. The concept of "preference" embodies them both, without prejudice.

Smoothness, rather than flatness, is the prime requirement for all curves other than the on-axis curve, and this message is strong in both studies. Wide dispersion appears to be desirable, but how much tolerance there is on differences in underlying shape or tilt of off-axis curves remains to be determined.

However, there is no doubt about the importance of minimizing coloration due to spectral undulations and narrow-band bumps associated with resonances.

The strong message is that sound power and in-room curves are, by themselves, imperfect guides to loudspeaker timbre and spatial impression. There is simply insufficient information in them. The Klippel findings suggest that wide dispersion is desirable to deliver a "feeling of space" that listeners like. This is a repeated observation throughout this book. Why is it so? It can only be speculation, but perhaps the supremely clever two ears and a brain find stereo (two channels) or conventional multichannel (five channels) to be less than perfectly capable of reconstructing a realistic three-dimensional listening experience from conventional recordings without some help from the playback environment. The nagging question is how much of this is due to non-optimum techniques used in the making of "conventional recordings"?

Figures 15.8 and 15.9 indicated that the right choice of five channels could create a very convincing illusion of a fully enveloping sound field. However, those experiments employed signals that had been optimally synthesized or recorded. The real world of recordings is rarely so good. Much more likely is an impression of clusters of instruments or voices delivered by single loudspeakers. Such stark spatial incompatibilities occur routinely in movies, where much of the on-screen sound, not just dialogue, is delivered by the center channel, and in stereo where pan-potting and/or spaced microphones end up cramming multiple musicians into a single channel. In these instances, as was discussed in Section 8.2, listeners vote for wide-dispersion loudspeakers to generate image broadening (ASW), which softens the spatial incongruity.

This message includes guidance for room acoustical treatment as well, in that the absorption of first lateral reflections—a ritual in control rooms and a (bad?) habit in home listening spaces—in effect destroys the potential for loudspeakers to generate the "feeling of space" and, if the absorber is 1-in. (25 mm) or 2-in. (50 mm) fiberglass board, it also degrades the sound quality by disproportionately attenuating the high frequencies.

More research is needed. In the meantime, it is important to recognize some limitations of research of this kind. Constructing models to predict perceptions must begin with listening tests to generate the perceptions to be modeled. In a very real sense, this is a case of "you get what you ask for." Realizing this at an early stage, the author developed elaborate questionnaires for listeners, not so much to obtain detailed results in all response categories but to force listeners to pay attention to aspects of the sound that they may otherwise have ignored (Toole, 1985, Figures 2 & 3). Some listeners admitted to paying attention only to certain instruments or voices or types of music when forming opinions, thus possibly missing useful insights to what is being heard.

Olive (2007) and Olive and Martens (2007) reported results suggesting that listeners who had extensive experience in evaluations of sound quality tended to be less responsive to the effects of room acoustics. They did what they were trained to do—very well and sometimes in spite of differences in rooms. A fully balanced test, therefore, may need to include listeners representing many perspectives and, perhaps, prompting them to pay attention to all (known) important variables.

Klippel found that the important general ratings of naturalness and pleasantness were 50% and 70%, respectively, related to his "feeling of space" perception. In the Olive data, there was no explicit evidence of spatial impressions being important except for one multidirectional loudspeaker that distinguished itself in a negative sense. Could it be that these listeners didn't hear a "feeling of space"? This seems unlikely; if it was a factor, it is more likely that they responded to the cues in a different manner—on a different "scale." But it may also have been that the population of loudspeakers to which they were exposed were so similar in that particular respect that spaciousness was not a significant variable. For example, if all of the loudspeakers in a test had smooth and flat axial frequency responses and differed only in overall directivity, perhaps spaciousness would be the only significant variable in the listening tests.

Obviously, all of the answers are not known, but much seems to have been proved beyond reasonable doubt. Most of the evidence fits together in a logical pattern, and although not simple, it is eminently comprehensible. The greatest encouragement is that the basic rules for designing good sounding loudspeakers seem to be sitting in front of us.

CHAPTER 21

Acoustical Materials and Devices

The development of glass fiber and mineral wool as sound absorbers had a significant influence on audio. As discussed in the early chapters, a combination of these materials and close-microphone recordings had a lot to do with the elimination of "space" from reproduced sound—reflections became enemies of what was then fashionable as "good sound" except, fortunately, in live, unamplified performances. A common practice among acoustical consultants has been to walk into a room, stand, clap hands, listen, furrow the brow, and announce that there are serious problems that require an expert's help and possibly some expensive repairs. This is superb marketing, but it is not analysis of a kind that matters to good reproduced sound. (Actually, my prize for the best acoustical marketing exhibition goes to the consultant who climbed a stepladder in the middle of a room to demonstrate that clapping hands close to the ceiling produced an audible flutter echo.)

One frequently hears and reads words to the effect that "acoustics is a very complicated topic," and it may appear to be if one is uninformed. Acoustics has multiple facets. If there is an aspect of the subject that requires significant special knowledge and an ability to deal with the real world, it is in reducing unwanted noises inside and outside of a listening space: sound isolation and noise reduction. The ability to achieve adequate sound transmission loss in the sometimes challenging circumstances of the real world is a skill in which field experience is highly beneficial, and a successful result may require some real engineering.

However, when it comes to deciding how to treat the interior surfaces and furnishing of the room, the space shared by listeners and loudspeakers, there is evidence of more marketing than science. There are acoustical challenges, but in practice they may fade to insignificance compared to the negotiations with the interior decorator. Those who have read through the book this far know that the requirements for good sound start with a need for good loudspeakers. If this is accomplished, the rest is greatly simplified. No amount of "room acoustics" effort or expense can compensate for poor loudspeakers. However, as we will soon see, some conventional acoustical practices have the potential to degrade sound from good loudspeakers. In a very real sense, the overriding rule for interior room acoustics should be "do no damage." Humans have a great natural ability to extract detailed information from sound sources in rooms (see Figure 19.1).

As described in detail in Chapter 13, the bass is likely to need significant effort and, quite possibly, some expense either in multiple subwoofers or in *effective* low-frequency absorbers, or both. But as far as the overall impressions of sound quality are concerned, as has been said a few times now, the contribution of the room is greatest when we listen to a single loudspeaker, less in stereo and even less in multichannel. The catch is that much of the time we are listening to a single loudspeaker, the center channel, and it delivers the dialogue in movies and television, so it, more than any other channel, deserves special attention.

Many people choose to have entertainment rooms that serve other purposes, like my own shown in Figure 3.2 and Figure 21.2. Such rooms tend to be furnished as living spaces, with the paraphernalia of life: books, cabinets, art, chairs, tables, and lamps. There is probably carpet on the floor, drapes on some of the walls, openings to other rooms, and so on. The first portion of this chapter provides some insight into how to deal with such spaces by giving examples of the acoustical performances of those common materials. With skill and a little luck, success may be achieved with a suitable arrangement of normal furnishings and materials.

Then there are custom rooms that are designed to provide escape and isolation from the rest of the world, to be used primarily or exclusively for home theatrical performances. Some of these are elaborately ornamented, and they all need acoustical treatment. It is in the choice of these acoustical materials that care is necessary, not because the challenge is so complex but because it is so very difficult to find useful and truthful performance data on some of the materials that are in the marketplace. If one is starting with an empty room, there is no excuse for less than excellent sound, so the pressure to succeed is great.

21.1 KEY ACOUSTICAL VARIABLES AND MEASUREMENTS

Figure 21.1 illustrates the very basic principles of sound absorption and transmission. Historically, acoustical measures of both have been based on speech frequencies, so it is rare to find data about the absorption or transmission properties of materials or structures that include the bass frequency range below 100 Hz, just where much of the greatly enjoyable but potentially problematic sound energy is located. The summary number that describes absorption is the

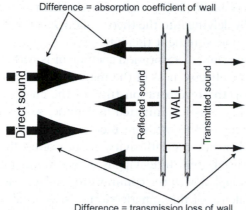

Difference = absorption coefficient of wall

Difference = transmission loss of wall

FIGURE 21.1 *The largest arrows depict direct sound from a source propagating toward a wall. When it strikes the wall, some of the energy will be transferred to the structure. This is especially noticeable at low frequencies when one can feel vibrations in the wall. The remaining portion of the sound will be reflected back into the room. The difference between the magnitude of the direct sound and that of the reflected sound is a measure of the absorption of the wall and/or materials placed on the wall: its absorption coefficient. The vibrations in the wall are communicated by the studs, and the air in the cavity, to the opposite side and that surface is set in motion. The sound generated by that vibrating surface radiates into the adjacent room as transmitted sound. The difference in the magnitudes of the direct and transmitted sounds is called the transmission loss. Absorbing material placed on the outside of the wall typically has minimal effect on transmission loss. (Fiberglass and acoustical foam have little transmission loss.) Absorbing material placed between the walls will attenuate the sound reflecting around the cavity, but that will have only a slight effect in a wall of this kind because most of the energy is coupled mechanically through the studs. If, however, there is a mechanical break between the two wall surfaces, such as is provided in "double-stud" walls where each wall surface has its own supports, then absorbing material can be a significant benefit.*

NRC (noise reduction coefficient), which is the average of absorption coefficients measured in the octave bands centered on 250, 500, 1000, and 2000 Hz. The summary number that describes sound transmission loss is STC (sound transmission class), and it is derived from measurements done in 1/3-octave bands from 125 Hz to 4 kHz. These are useful numbers for many purposes—for example, in factories, offices, and schools, where speech comprehension and privacy are issues. They are seriously limiting in the design of home entertainment spaces with thundering subwoofers. We need better specifications.

Books have been written on these topics, and it is beyond the scope of this one to delve into the measurement methodologies and meanings of these and other measures of acoustical performance. The classic books on the topic include Cremer and Müller (1982), Beranek (1986), Harris (1991), and Kuttruff (1991), to which are added contemporary coverage from Everest (2001) and Long (2006).

Cox and D'Antonio's book (2004) explains the theory of absorbers but truly distinguishes itself by delving into the theory of engineered diffusing devices and issues related to the measurement thereof.

The traditional measure of absorption coefficient is performed in a reverberation chamber, and the effectiveness of the material is calculated from the effect a sample of it has on the reverberation time of the chamber. This application of classic diffuse-field acoustics, initiated by Sabine, works well, but there is a problem. The accuracy of the measure increases with the size of the sample placed in the chamber, but the diffusivity of the sound field decreases as the sample size is increased, and diffusivity is a requirement for accuracy. Many clever minds have grappled with the conundrum, and a perfect solution has still not been found. Evidence of the imperfection is seen in absorption coefficients that exceed unity. This, of course, is impossible, but it is something the industry has learned to live with. A moment of thought will lead to the conclusion that this same problem exists in the calculation of reverberation time using these measured numbers. The predictions work quite well in large reverberant spaces, but as rooms get smaller and have more absorption in them, the sound field cannot be diffuse, and the calculations become less precise. Schultz (1963) reported that "[in small rooms] the very concept of reverberation time may be inappropriate, since the statistical conditions upon which the concept depends cannot be established." It hasn't stopped people from trying, though. There has been a succession of ever more complex equations attempting to perfect something that cannot be generalized; they work better in some rooms but not all rooms (Dalenbäck, 2000).

The idea that the sound field in listening rooms is not diffuse was explained in Chapter 4 and is emphatically illustrated in Figure 21.2. In these cases, random incidence absorption and scattering data are not the primary interest. They are data intended for use in large, reverberant performance spaces, not small acoustically damped listening rooms. These are more "Sabine," diffuse-field, concepts being misapplied. For our purposes, we need to know the performance of materials for sounds arriving from very specific angles, not an average of all possible angles. This is data not to be found in normal manufacturers' literature—yet. In the examples shown later, we will see that it is very instructive.

Transmission loss is more complex, as it is influenced by many factors other than the wall itself, especially what are called "flanking paths": sounds leaking over the wall through the ceiling or attic space, the floor, and through HVAC services. Then there are the compromises to the wall itself, through electrical or plumbing service penetrations, lightweight doors with poor door seals, and so forth. This is where acoustical expertise is needed in the design phase and where absolutely fastidious supervision is necessary during construction.

The Institute for Research in Construction, a division of my old employer the National Research Council of Canada, provides a remarkable amount of

Views from the center channel loudspeaker in
a room showing:

(a) what the eye sees

(b) what the loudspeaker sees at frequencies
above about 300–500 Hz—leather furniture

(c) what the loudspeaker sees at frequencies
above about 300–500 Hz—fabric upholstered
furniture

FIGURE 21.2 *A view of a listening space as "seen" by the center channel loudspeaker. It shows that much of the direct sound from the loudspeaker is absorbed at the first encounter with a surface, thereafter being unavailable to contribute to later reflections. (b) The leather-covered furniture is shown as being translucent because there will be some high-frequency reflection from the surface. With fabric upholstery (c), even this is gone. There is really not much hard, reflecting, or scattering surface area left, although a wider-angle lens would have shown the first sidewall reflections. In this room, there is no sidewall on the left side; it is an opening to the rest of the house, and on the right side, this lost energy is balanced by velour drapes over a window wall. Each room will be different. It is often instructive to put yourself at the loudspeaker location and imagine what happens to the radiated sound.*

useful data to the public, and most of it is available on their website (http://irc. nrc-nrc.gc.ca/pubs/bsi/2002/pubs_e.html.) In particular, look at BS185 for a good discussion of the basics and to the set of publications—CTU1, IR 693, IR 761, and IR 832—for data on many gypsum-board wall constructions at frequencies that go beyond the basic range. Other reports deal with concrete walls, floors, and other building elements.

The building materials industry also provides an abundance of data on their own products and some general guidance. The Noise Control Design Guide (available at www.owenscorning.com) is a good example of a general guide, and www.rpginc.com provides white papers explaining some of the science underlying their products, especially the diffusers. Check also www.jm.com, www.roxul.com, or any of the numerous other sources of acoustical materials that an Internet search will turn up.

21.2 THE MECHANISMS OF ABSORPTION

When sound is absorbed, it is basically converted into heat. The most common mechanisms are resistive absorbers and membrane or diaphragmatic absorbers. *Resistive absorbers* are best described as "fibrous tangles," in which air molecules are forced to move through a three-dimensional combination of fibers and tiny orifices. Glass fiber, mineral wool, and other man-made and natural fibers all are used, as well as reticulated foam—plastic foam in which the thin walls of the bubbles have been eaten away, leaving a skeleton through which air can move. It is available in various, quite-well-defined densities (hole sizes) because one of its uses is as a filter medium. Interestingly, with larger orifices it is useful as an acoustically transparent, visually translucent material: a grille cloth. Fiberglass and mineral wool exist because of their thermal insulation properties, so they are available in many forms from flexible batts to compressed rigid boards of different densities. The boards have advantages in building construction because of their mechanical strength, which also makes it convenient to wrap them in fabric for decorative acoustical purposes. The random-incidence absorption coefficients of all forms of these fibrous materials, from soft batts to rigid boards, as well as slab versions of acoustical foam, are quite similar for a given thickness. High-density boards tend to become reflective for high-frequency sounds arriving at angles approaching grazing incidence. As the material gets thicker, the low-frequency performance improves. However, to be effective at bass frequencies, these devices become impractically large (approximately 1/4-wavelength deep). We need another way to absorb energy.

Membrane (mechanically resonant) absorbers are the most popular absorption mechanism for low frequencies. This happens when a surface moves in response to sound, mainly low-frequency sound. Energy is expended to move the surfaces, and less is returned to the room. Normal floors and walls therefore are useful membrane absorbers, and, to a limited extent, it is possible to optimize their contributions in controlling room resonances. Obviously, it is also possible to design stand-alone devices tuned to vibrate at, and therefore to absorb, specific ranges of frequencies. Such devices are the perfect complement to resistive absorbers because they are useless at high frequencies. Several of the handbooks mentioned earlier give a simple equation based on the mass-per-unit area of the thin membrane and the volume of air behind (the spring-mass resonant system). The BBC has issued several reports over the years describing the design and construction of such devices (e.g., BBC Research Department, 1971; Fletcher, 1992). They also show absorption coefficient measurements. However, common experience is that often the frequency is wrong or the membrane breaks up into modes, some of which absorb nothing. Some amount of trial and error may be necessary to make them work as desired. This may be a reason to purchase ready-made units from a trusted source (one that publishes independently-measured performance data). Damping of the resonant system is normally accomplished with resistive absorbing material in the volume, although it is also possible to damp the membrane itself with elastomeric materials. Membranes can be thin plywood, MDF/hardboard, or limp vinyl sheeting (Voetmann and Klinkby, 1993).

Helmholtz (acoustically resonant) absorbers are acoustical spring/mass resonant systems that use the mass of air in holes or in slots between closely spaced boards, resonating with the spring in the air in the volume behind. Tuning these tends to be quite reliable, and, again, equations and graphs for their design can be found in standard handbooks. They tend to be useful in the upper-bass to lower-midrange frequency range. Resistive absorbent in the volume normally supplies damping. Cox and D'Antonio (2004) discuss these at length. See also Section 13.3.1 for more discussion of these principles and devices as they apply to controlling low-frequency resonances in rooms.

21.3 ACOUSTICAL PERFORMANCES OF SOME COMMON MATERIALS

This section discusses materials that can be used in rooms designed and decorated as normal living spaces. The next sections describe engineered materials that can be added, if necessary, to provide some fine tuning or can be organized to provide the total acoustical treatment in a dedicated listening space.

21.3.1 Typical Domestic Materials

Figure 16.1 showed the profound effect that wall-to-wall carpet had on the reverberation time of an empty room. Figure 21.3 shows why; a good carpet on a good underlay is an effective acoustical absorber, and it is easy to justify a lot of it in a room. As a resistive absorber, it is most effective at middle and high frequencies, which is why it is important to maximize the effective thickness by using an acoustically useful (as opposed to just comfortable under foot) underlay. Because it is all on one surface, we need sound-scattering/diffusing objects and surfaces to redirect sound into the floor. Again, Figure 16.1 shows that adding nonabsorbing but scattering "stuff" to the room substantially reduced the reverberation time. So furniture is important. The acoustical properties of carpets vary substantially, and this is not a published property of the product. Harris (1955) provides a survey of some of the traditional carpet weaves that may be helpful. The basic rule is that if absorption is required, avoid the looped-pile rubber-backed industrial "indoor/outdoor" style of product.

As shown in Figure 21.4, drapes can also be effective sound absorbers if they are of the right kind, and that is simply described as "heavy." Heavy velour, the

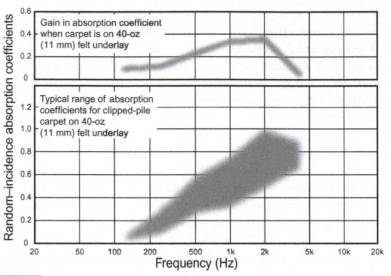

FIGURE 21.3 *At the top is shown the isolated influence of a good carpet underlay—40 oz/sq yd (1.4 kg/m²) hair felt, which is typically about 0.43 in. (11 mm) thick. The same thickness of common rubber, plastic, or foam cannot deliver this level of acoustical performance. There may be other materials that are comparably good, but check for acoustical measurements. The shaded area in the bottom curve combines this underlay with different kinds of high-quality clipped-pile woven carpets (with porous backing because the sound must be allowed to penetrate into the felt underlay).*

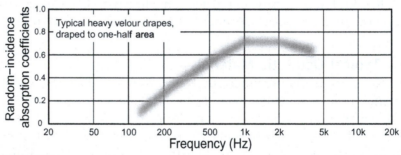

FIGURE 21.4 *A fuzzy curve showing approximate absorption properties of heavy velour drapes, draped to one-half of their flat area. The drapes should be hung on a track located 4–6 in. (10–15 cm) from the wall to ensure good low-frequency absorption.*

heavier the better, is probably the most popular choice because in its draped form it constitutes a resistive absorber of significant thickness. Other heavy fabrics with the right porosity/flow resistance also work well, but lightweight, open-weave fabrics (easy to blow air through) are useful only for decorative purposes.

Upholstered seating performs two important functions: absorption and scattering. Chairs, especially of the home-theater variety, are substantial obstacles

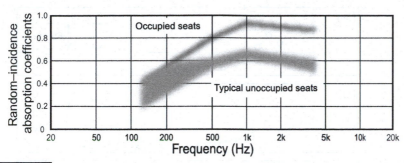

FIGURE 21.5 *The acoustical absorption coefficient of seats of the kind used in large auditoriums. In those circumstances, the rooms are so large that the seats can be treated as a "layer" on the floor; their vertical dimensions are insignificant factors. So these numbers are of use only to indicate the frequency dependence of large collections of seats with and without occupants. Not explained in the graph is the fact that leather-covered seats were responsible for the elevated absorption at lower frequencies—a probable membrane-absorber effect. Based on data from Long, 2006.*

in the middle of small rooms, and some of the sound falling on them is absorbed and some redirected by reflection and scattering elsewhere in the room. As can be seen in Figure 21.2, they interact with a significant fraction of the direct sound radiated from the loudspeaker, and as shown in Figure 21.5, they effectively absorb sound over a large frequency range, including lower frequencies.

The most natural place to look for absorption to help with low-frequency room resonances is in the room itself. Figure 21.6 shows that significant absorption is available from normal gypsum-board-on-stud constructions. The absorption coefficients are not very high (may be higher at lower frequencies), but there is a lot of wall area. A second consideration is that there will always be some of the absorption in the correct location to provide damping for all room modes (must be at high-pressure points in the standing-wave patterns). If sound isolation is not a consideration, there is little to ponder: just use one of these convenient structural designs. If sound isolation is important, another wall outside this one will be needed (room within a room design). Remember, in multiple-wall structures, the distance between the separate wall surfaces is a prime determinant of sound attenuation (along with mass), so be sure to allow several inches between the outer surface of the inner "room" and the inner surface of the outer "room," or omit the inner surface entirely, adding more layers of gypsum board to the outside. Good handbooks and guides (some mentioned earlier) will show examples with measured attenuations. In many situations, though, there is no choice but to make the inner surface of the room massive and rigid, to contribute to sound isolation, in which case low-frequency absorption must be added by special devices.

Random-incidence absorption coefficients tell part of the story, but what really matters is the effect on resonances in real rooms. A few years ago, an

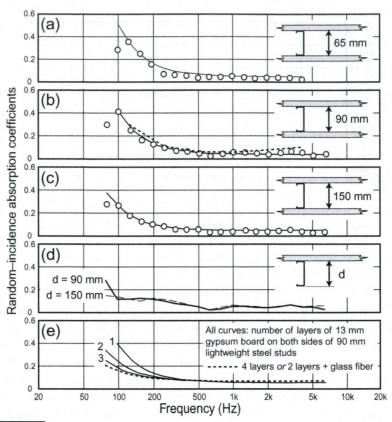

FIGURE 21.6 *Absorption coefficients for various wall constructions using 0.5-in. (13 mm) gypsum board on lightweight steel studs. (a), (b), and (c) show the effect of increasing stud depth from 2.5 in. (65 mm) to 3.5 in. (90 mm) and then to 6 in. (150 mm). The effect is that larger studs reduce absorption. The dashed curve in (b) is for the same structure using wooden studs (from Long, 2006), indicating that there is no important difference. (d) shows that a single layer of gypsum board on one side of the studs is less effective as a low-frequency absorber than a layer on each side. (e) shows the expected effect of adding more layers of gypsum board; the absorption is diminished, but there is another interesting effect: filling the cavity with fiberglass reduces the low-frequency absorption. Data from Bradley, 1997.*

opportunity presented itself to find out. A listening room had been built, and the friendly builder thought he was doing us a favor by adding an extra layer of gypsum board on the interior surface. The "favor" was discovered too late, and the room was used in that condition for a couple of years, during which it acquired a reputation for having somewhat boomy bass and upper-bass/lower-midrange coloration. The evidence was in measurements. Eventually, we had the interior of the room stripped and the originally intended single layer of gypsum board installed. Before and after measurements are shown in

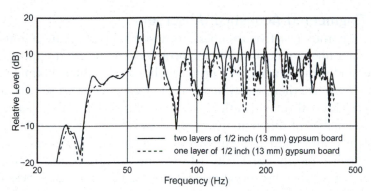

Measurements made with a subwoofer placed in a front corner and a microphone located at the prime listening location in a 3000 cu ft (85 m³) listening room. The solid curve shows results with two layers of gypsum board, and the dashed curve shows results for a single layer of gypsum board mounted normally on 3.5-in. (90 mm) wooden studs. It is clear that the single layer provides a substantial increase in acoustical damping at frequencies below about 200 Hz. Measurements courtesy of Allan Devantier, Harman International.

Figure 21.7. The benefits were measurable and audible. The surprise was how high in frequency the effects extended.

This is a common error in listening rooms. Rooms have been built with, believe it or not, sand-filled walls, having heavy plywood and multiple layers of gypsum board on the exterior surfaces for strength. In the example I encountered, the idea was to keep the bass in, to make it "tight." It succeeded in keeping it in, with a vengeance, and the result was a room with enormous undamped bass resonances. Far from being "tight," it boomed mercilessly. The solution required the addition of costly and bulky low-frequency absorbers to return the room to a state that could have been achieved with normal household construction materials and methods. As might be expected, the sound transmission loss through the massive wall was considerable, but, ironically, in this industrial building, it was not a requirement.

If steel studs are used, extra precautions are needed to avoid buzzes and rattles. At the very least, it is necessary to substantially increase the number of attachment screws and preferably to run a bead of acoustical caulk down each stud before the gypsum board is applied. This done, such walls are eminently satisfactory, and there is the additional advantage that steel studs offer slightly better sound isolation than wood because they flex.

Scattering, and the diffusion of the sound field that results, is the missing ingredient. In normally furnished rooms, this is provided by cabinets, tables, chairs, bookcases, fireplaces, even angled pictures hanging on the wall. Any objects that cause sound to change direction or that break up large, flat sections of wall are useful.

21.3.2 Engineered Acoustical Absorbers

The normal porous resistive absorbers, as described earlier, are fibrous tangles, of which glass fiber, mineral wool, and acoustical foam are the most common. Glass fiber and mineral wool exist primarily because of their thermal insulation properties. They are therefore available in many forms, from soft, flexible batts or rolls to rigid boards in which the fibers are compressed with an adhesive. The latter are available in different thicknesses and densities rated in pounds per cubic foot (pcf) and kg/m³. Measured in the traditional reverberation chamber fashion, which typically covers the 125 through 4 kHz octave bands, there is little difference in the absorption coefficients for the same thickness of any of the materials, soft or hard fibrous materials or slab foam.

Figure 21.8 shows random-incidence sound absorption coefficients for various thicknesses of a low-density rigid panel. Examining similar data for panels of different fibrous or foam materials will reveal differences, but they are small and inconsistent. In fact, the largest variable in comparing materials from different

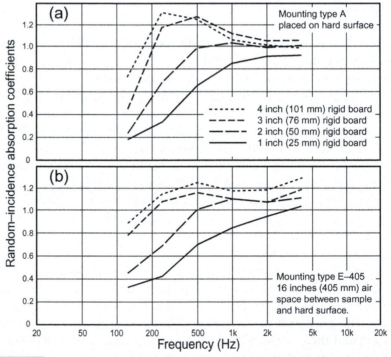

FIGURE 21.8 *Random-incidence absorption coefficients for 1.5 pcf (24 kg/m³) rigid fiberglass board when mounted (a) directly on a rigid backing and (b) 16 in. (405 mm) away from the reflecting surface (e.g., in a dropped-ceiling installation). It is clear that the low-frequency performance increases with thickness of the material and that the addition of the air space has a beneficial effect on performance at very low frequencies—extrapolating below the lowest frequency at which measurements were made. Data from www. owenscorning.com, for their Fiberglas® 700 series insulation.*

manufacturers may be the laboratory at which the measurements were made. If the lab is certified, these differences are merely evidence of the fact that such measurements are not exact. Nobody is being deceptive. Some data eliminate numbers in excess of 1.0 on the basis that this is somehow untidy, but that prevents comparisons of *relative* performances of those materials. In the end, it really does not matter because for our purposes in small rooms, these data are of limited usefulness. They are measurements optimized for the calculation of reverberation times in large rooms with highly diffuse sound fields. Small listening rooms do not have diffuse sound fields; reverberation time is a factor of much diminished importance and achieving satisfactory values is straightforward. For our purposes, we need to have other data, and in a form that allows us to anticipate what the absorbing material might do to the sound from a loudspeaker. We need *directional* absorption coefficients, measures of the attenuation experienced by sounds reflected from different angles—individually.

Figure 21.9 shows data of a kind we need to see more of. As part of a program of characterizing the performance of the engineered surfaces popularly called diffusers, the people at RPG Diffusor Systems Inc. diverted some effort to providing interesting and useful data on absorbers (a note about spelling: *diffuser* is the correct spelling for the generic device; *diffusor* is the spelling chosen for RPG products). This figure confirms what has been predicted in mathematical models that, as incident sounds move away from normal incidence, the reflections experience different amounts of attenuation at different frequencies. In this example, the effect is all in the direction of increased absorption for the 45° sound. The difference is substantial. These curves are for 2-in. (50 mm) rigid boards; those for 1-in. (25 mm) boards are, as might be anticipated from Figure 21.8, similar but moved roughly an octave higher up the frequency scale.

All of this information is lost in the random incidence data shown in Figure 21.8. Random-

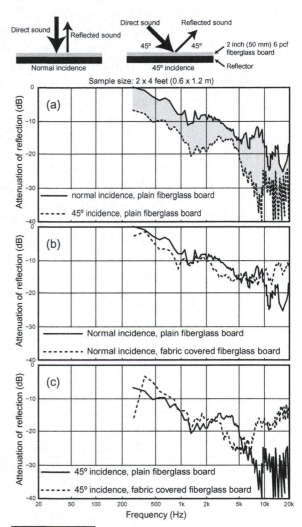

FIGURE 21.9

Comparisons of the sound reflected from a 2 × 4 ft (0.8 × 1.2 m) rigid panel and that panel covered by a 2-in. (50 mm) rigid 6 pcf (100 kg/m³) fiberglass board. The difference is shown as the attenuation experienced by the sound because of the absorbing panel. (a) The result for normal incidence and for 45°. The difference, shown by the shaded area, indicates that sounds arriving from an angle experience greater absorption because the sound is forced to travel through more of the material. Figures 21.8b and (c) show the effect of covering the fiberglass board with the popular Guilford of Maine FR701 fabric. Data courtesy of D'Antonio, 2008.

incidence absorption coefficients give the impression that reflected sounds are attenuated by similar amounts above some "cut-off" frequency. Clearly the inference is wrong. Instead, sounds reflected from different angles experience differing amounts of attenuation at different frequencies. This is especially evident in the very uneven 45° curve. Predictions of performance at even greater angles suggest less attenuation (more reflection) at mid and high frequencies and more attenuation at low frequencies, as the sound is forced to travel through more of the fibrous material (Cox and D'Antonio, 2004, Figure 5.19). For sound reproduction purposes, it is also helpful to have data measured to 20 kHz, which is beyond the range of normal absorption data.

These curves are fundamentally important to us because they describe modifications to the frequency responses of loudspeakers after they encounter such surfaces in a room. It is difficult enough to design loudspeakers to have relatively constant directivity; it does no good if the results of good design are corrupted by acoustical treatments on the room boundaries. The 45° curve in Figure 21.9a could describe the effect on a first reflection, an important factor in the early reflected sound field of a small room. (See Figure 16.6 for examples of reflection angles.) One can anticipate effects on impressions of space, as discussed in Chapter 8 and Section 20.1, and on impressions of sound quality, as is suggested by the importance of wide dispersion and relatively constant directivity in Chapter 18. All subsequent reflections of the sound within the room will perpetuate the distorted spectrum. Sadly, this is a common reality made all the more regrettable because, as has been pointed out several times in earlier chapters, there seems to be no persuasive requirement to attenuate the first lateral reflection.

Figures 21.9b and (c) show the effect of covering the fiberglass panel with a commonly used fabric designed for acoustical applications. The effect of this fabric is to slightly modify the amount of absorption at different frequencies, ending up with significant elevations in the levels of reflected sounds at high frequencies, especially for the 45° sound. None of this is apparent in random-incidence absorption coefficient data that do not include the highest frequencies.

It will be interesting to follow progress in this area because it is clear that some combination of these angular absorption/attenuations and the direction-dependent frequency responses of loudspeakers will help to explain perceptions in sound reproducing systems. The general trends exhibited in Figure 21.9, somewhat foreshadowed in the data of Figure 21.8, indicates that if it is one's wish to absorb sound, thick, resistive absorbers are needed. This does not mean 1- to 2-in. (25 to 50 mm) material but much thicker, 3 to 4 in. (75–100 mm) *or more*. Attenuating *all* the sound down to the transition frequency by a large amount is a reasonable objective. Nonuniform attenuation of the kind shown here has the potential to degrade the sound quality of good loudspeakers. There is another interesting possible consequence. As discussed in Chapter 11, the precedence effect appears to be disrupted if the spectra of the direct and reflected sounds are sufficiently different. Could it be that by removing only the high frequencies we are causing what is left of the reflected sound to be more spatially influential than it normally would be?

Sculptured acoustical foam materials have random-incidence absorption coefficients roughly equivalent to slab materials half their thickness, so there is a suggestion that it would be advisable to augment the acoustically effective thickness of these materials with additional fibrous or foam materials placed behind them. The air space, shown in Figure 21.8b, will have an improved effect on absorption at specific angles if the space is filled with absorbing material.

It is worth remembering that low-density (and low-cost) fibrous materials, such as the flexible batts used for thermal insulation, have excellent acoustical performance and are eminently useful for such volume-filling tasks. It is obviously necessary to avoid versions with paper or plastic moisture barriers, unless these materials are placed against a wall. In Figure 21.2a, the large triangular area at the top of the rear wall, through which the front-projector fires, was framed with 2 × 4-in. wooden studs and the cavities filled with 6-in. fiberglass batts, which were then covered with acoustically translucent fabric stretched in place to form a slightly "pillowed" appearance. It all made for an attractive and effective broadband absorber, preventing sound from the front loudspeakers from being reflected back into the high ceiling space and the dark color preventing light reflection back on to the screen.

In recent discussions with Dr. Peter D'Antonio, he said the following:

For optimal absorption, a porous absorber should offer a surface impedance with a low-flow resistivity, which matches that of air to remove reflections, while offering a high internal acoustic attenuation. When attempting to control reflections with a single density material, it is fair to say that thin fiberglass panels should not be used, and in my view lower density is preferred over higher density. In addition, thicker panels and a rear air cavity both contribute to extending the absorption to lower frequencies.

When it comes to sound absorption, it could be useful to ask what is used in anechoic chambers, where the goal is *complete* elimination of sound above a certain "cut-off" frequency. In traditional designs, the wedge-shaped absorbers must be at least 1/4-wavelength long at the lowest frequency of interest. In good chambers, this is typically 3 to 4 ft (0.9 to 1.2 m). The material used in these wedges is usually compressed glass fiber with a density of about 3 pcf (48 kg/m^3). According to one study, that density appears to have been a choice of convenience (structural rigidity and particle shedding considerations) because a lower-density material (1.1 pcf, 18 kg/m^3) was closer to the acoustical target performance (Koidan et al., 1972). Rasmussen (1972) describes an anechoic chamber constructed of suspended cubes, somewhat randomly arranged with small (2-in., 50 mm) low-density (1.9 pcf, 30 kg/m^3) cubes farthest from the wall and progressively larger, denser cubes placed closer to the wall, with the inner layer being

6 to 7.5 pcf (100–120 kg/m³) stacked against the wall. It apparently worked superbly.

In summary, all of this suggests that the surface of the absorbing material—the interface with the sound field—should have relatively low density and that to achieve performance at lower frequencies, one may need to seriously consider how much real estate can be devoted to the task because materials that work well have appreciable thickness. The problem is the damage inadequately thick materials do to the sound quality of reflected sounds from loudspeakers. The appropriate fabric covering appears to have little effect at low and mid-frequencies, but it becomes reflective at high frequencies, especially for sounds approaching from angle, as in the case of sidewall reflections.

The typical "acoustical" fabrics are not grille-cloth; they are acoustically translucent, not transparent. This means that loudspeakers should not be placed behind them. If they are stretched across a section of wall, even a blank wall, and spaced away from the wall, they will function as absorbers—admittedly, not very good ones. Some installations, however, have large areas treated in this manner, so the audible effect can be significant. In home theaters, it is not uncommon for walls to be sheathed in fabric chosen by an interior decorator, with little or no concern for what it may do to the performance of acoustical devices and materials underneath. I suspect that this is an explanation for the excessive "deadness" of so many custom home theaters.

Summarizing, it may be time to optimize absorbers for listening room applications rather than use only what is convenient from the thermal insulation catalog. We need also to expand the catalog of acoustically acceptable fabrics, and to have some sound-transmission specifications for them. Finally, let us add to the wish list a request for more acoustical absorption coefficient measurements of a useful kind.

21.3.3 Engineered Acoustical Diffusers

Diffusion, as has been stated earlier, is a property of the sound field. It describes a condition in which sounds arrive at a point with equal probability from all directions and where this condition exists everywhere in the space. A perfectly diffuse sound field may or may not exist, but an empty reverberation chamber gets close. Any room within which we live or amuse ourselves—in live concerts, cinemas or our home entertainment rooms—is not perfectly diffuse, but there is a measure of diffusion. Figure 4.14 shows measurements of directional diffusion in a small room, and it shows that the diffusivity can change in both amount and direction with time.

In performance spaces, diffusion is a very good thing, providing important feedback to musicians in the orchestra, blending their sounds, and delivering them to all parts of a large audience. In small rooms for sound reproduction, there is no necessity for, nor merit in, having a diffuse sound field. However, objects that scatter or diffuse sound, sending it in many different directions, are very useful alternatives to absorbers because they don't destroy sound energy.

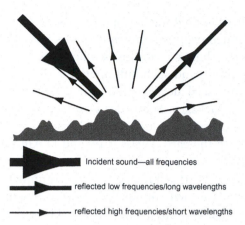

Incident sound—all frequencies

reflected low frequencies/long wavelengths

reflected high frequencies/short wavelengths

FIGURE 21.10 *The basic concept of a diffusing surface. At frequencies where the wavelength is long compared to the protruding features of the surface, there is no scattering, and the sound is simply reflected as if the surface were flat. At higher frequencies, shorter wavelengths, sounds are scattered, diffused, in many directions simultaneously. Because the sound is spread over a large angular range, the sound sent in any single direction is substantially attenuated compared to the original, incident, sound. Each diffuser, depending on its design and depth, will have a frequency below which it does not scatter the sound. If, for example, the diffuser has a lower limit of 1 kHz, it means that the sound sent off in the direction of the main (specular) reflection arrow will have a lot of energy up to the diffusion limit, and then the high frequencies will be attenuated. In other directions, listeners will hear the reverse: high frequencies with insufficient spectral energy in the lower frequencies. It all depends on the specific circumstances, to be sure, but one cannot ignore the fact that only a portion of the spectrum is being manipulated by these devices. Ideally, when substantial amounts of loudspeaker energy are at issue, as in the case of first reflections, diffusers should exhibit comparable diffusion coefficients down to the transition-region that begins around 300 Hz in small rooms. For alleviating flutter echoes, thinner devices will suffice.*

In fact, they make whatever absorbers there are work harder. But just as absorbers need to function over a wide frequency range, so do diffusers (see Figure 21.10 and the discussion in the caption).

In the early days of room acoustics, it was common to add projecting shapes to walls of recording studios. These could be convex curves, hemicylinders, rectangular boxes, pyramids, or prisms of various artistically inventive shapes. With some creative lighting, such walls could be very attractive. Gilford (1959) describes some BBC experiments in which they concluded that to be effective, projections from walls must be 1/7-wavelength or greater. This means that to address voice resonances (100+ Hz), they must be of the order of 20 in. (50 cm) deep. Rectangular prisms were found to be more effective than either triangular or hemicylindrical shapes. They also conducted a comparison of box-like projections and the equivalent area of modular absorbers and found both were about equally effective. Summarizing his findings, in small studios, there is no advantage to nonparallel walls or diffusing projections as long as many small-area absorbers are distributed over as many surfaces as possible.

This is interesting, but a recording studio is not a control room, home theater, or listening room, and such bulk methods that involve the entire room are inappropriate. We need devices that can deliver predictable amounts of diffusion in compact packages. Schroeder proposed a number-theory basis for designing diffusing surfaces, an idea expanded upon by D'Antonio and various colleagues beginning in the early 1980s (Cox and D'Antonio, 2004; D'Antonio and Cox, 1998). Others joined in, and the result, today, is a wide choice of what I call "engineered surfaces" that provide various combinations of diffusion and absorption, providing scattered sounds that have different directional orientations and operate over different frequency ranges. These devices have now joined the repertoire of options available to acoustical designers.

Let us look at the performance of some diffusing devices. For this, the "normalized diffusion coefficient" will be used. It is the subject of a new industry standard and is a measure of how uniformly the device distributes the sound over a semicircular arc compared to a flat surface of the same size. The normalization corrects for the effect of the comparison reflector, so the resulting number is a measure of the diffuser alone.

I can recall a time when some otherwise serious-minded people thought that textured paint made a difference to the sound of a room. This might be true if you are a bat, but at human audio frequencies, this was just another audio fantasy. Diffusers must manipulate sounds with significant wavelengths and must therefore themselves be of significant size. One of the early popular shapes was the hemicylinder or, in a slightly flattened form, the polycylindrical absorber/ diffuser. Large ones were covered with thin membranes so they could serve also as absorbers in the upper bass region.

Figure 21.11 shows the normalized diffusion coefficient for a single and then for multiple hemicylinders with a 1 ft (0.3 m) radius. Obviously, in isolation the device works very well, and in this thickness it is effective down to a usefully low frequency. However, when combined with others, it loses the ability to diffuse low frequencies. The geometric regularity is also visible in the cyclical pattern in the curve. The solution is obviously to space them, probably by random distances, and possibly to vary the depth. Changing angles is another option. As Gilford (1959) noted, other protruding, curved, or faceted convex shapes also work well. In home theaters such shapes can be incorporated into interior designs as vertical columns.

The arrival of the Schroeder diffuser was exciting. Peter D'Antonio, an early enthusiast, has been involved in developing both products and standardized measurement processes for this industry (Publication AES-4id-2001, to be incorporated into ISO 17497-2). Other manufacturers produce diffusers, and no prejudice is intended by the author in employing only RPG products and data in these examples. The purpose is to demonstrate some of the basic acoustical

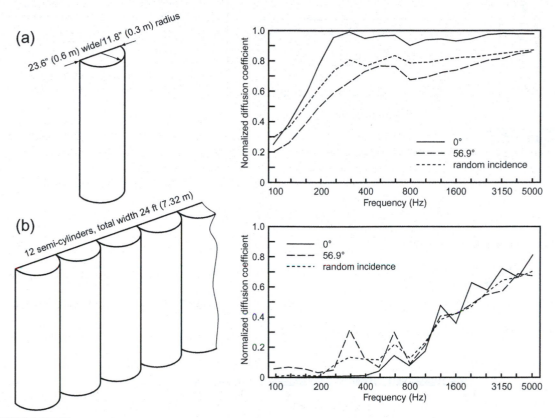

FIGURE 21.11 *(a) The normalized diffusion coefficient for a single hemicylinder and (b) for a collection of them covering an area. Data courtesy D'Antonio, 2008.*

properties of such devices and the ways in which these properties can be displayed. As with any product, it is up to manufacturers to provide performance data—in this case, preferably using methodologies developed under the auspices of international standards-writing bodies.

Figure 21.12 shows a comparison of a standard version of a Schroeder diffuser with an RPG Modffusor in two configurations. The advantages of the newer devices over the standard Schroeder unit and over the continuous array of hemi-cylinders shown in Figure 21.11b are clear. These designs get better as they get larger. At 7.9 in. (0.3 m) deep and exhibiting useful diffusion down to about 300 Hz, they appear to meet the requirements for wideband diffusers and, interestingly, are very close to Gilford's 1/7-wavelength estimate for depth requirements (Gilford, 1959).

It is one thing to produce a number indicating a level of performance. It is another to examine the details of what is happening. Figure 21.13 shows one example from many that must be measured to compute the diffusion coefficient.

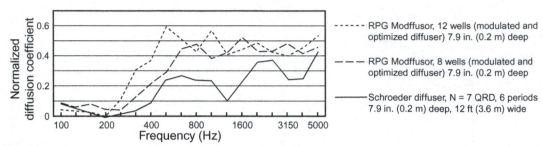

FIGURE 21.12 *A comparison of a classic Schroeder QRD diffuser with an evolved form, the RPG Modffusor that has been modulated and optimized for improved performance. In contrast with Figure 21.11, it can be seen that the Modffusor actually gets better when used in larger forms. Data courtesy D'Antonio, 2008.*

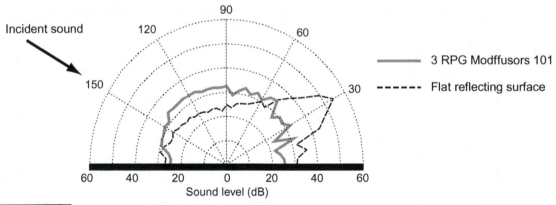

FIGURE 21.13 *The raw data in the 4000 Hz 1/3-octave band showing the measured pattern of sound reflected from a flat reflecting surface and the same area covered by three RPG Modffusors. The flat surface sends a strong specular reflection away at the mirrored angle of the incident sound, and the diffuser spreads the energy very uniformly over the semicircle, showing little evidence of angular favoritism.*

In this case, it shows, in measurements, what was shown in conceptual form in Figure 21.10.

21.3.4 Acoustically "Transparent" Projection Screens and Fabrics

Figure 21.14 shows examples of transmission loss data on a few well-known screens. All of them perform superbly at large off-axis angles, only showing differences in the direct sound angular range 0° to 30°. The ClearPix product by itself is acoustically sufficiently transparent that it requires no compensation. With a good acoustically transparent scrim behind it, it is only slightly more lossy. Perfection might demand that the latter combination and the Microperf product be equalized, but all the effects are at very high frequencies. It is interesting that the screens change the directivity of the loudspeakers. The losses,

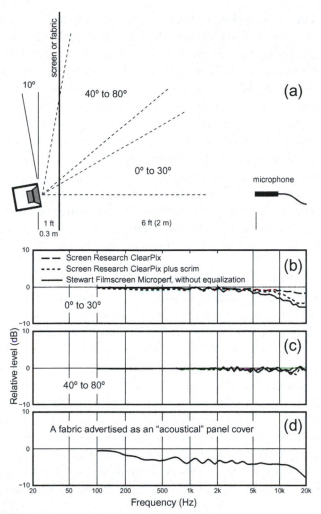

FIGURE 21.14 *(a) The measurement setup in the anechoic chamber. The loudspeaker was angled to minimize comb filtering; the angling can be in any direction. The screens are slightly reflective, and the front panel of the loudspeaker is greatly reflective, and reflections can occur between them. The more transparent the screen, the less the need for angling the loudspeaker. (b) The transmission losses over the 0° to 30° angular range of three screen configurations, a woven screen alone (Screen Research ClearPix), the same screen with a dark scrim behind it to minimize optical reflections from objects behind, and the Stewart Filmscreen Microperf. (c) The transmission losses averaged over the 40° to 80° range. (d) The transmission loss of a popular "acoustical" fabric intended to cover absorbers but that has, on occasion, been placed in front of loudspeakers for ornamental reasons, or used as a scrim. Data courtesy Allan Devantier, Harman International.*

when they occur, happen close to the principal axis. If the high-frequency losses are compensated for by equalization, the on-axis sounds are made flatter, but off-axis sounds are elevated. In practice, this could be construed to be an advantage, as it assists the normally deteriorating off-axis high-frequency performance of most loudspeakers. Whatever option is chosen, it is evident that placing loudspeakers behind well-engineered screens is not a problem.

However, during these measurements it came to our attention that some people were using fabrics created for other purposes as grille cloths or optical scrims. Loudspeakers radiated their sound through them. A look at these fabrics should be enough to raise suspicions; they are quite dense, fibrous fabrics. They don't look like grille cloth. They appear to be innocuous when covering fiberglass panels, as least as measured in the random-incidence reverberation chamber method. However, as shown in Figure 21.9, for sounds reflected from specific angles of incidence and at higher frequencies, their effects are not negligible.

Figure 21.14d shows that this particular fabric is not acoustically transparent; in fact, it reduces the radiated sound level by 3 dB and more at frequencies above about 300 Hz. In terms of amplifier and loudspeaker power requirements, this is a factor of 2 or more—not negligible. Any significant area of this fabric in a room, whether it is in the direct sound path from a loudspeaker or not, has a significant effect on the sound of the entire room. It cannot be treated as a "neutral" factor. Whether with this fabric or another one selected by an interior designer, many rooms end up with large portions of walls covered with it. The performance of the loudspeakers and acoustical devices behind are degraded, and the fabric itself behaves as an absorber with an air space behind. Such home theaters often end up being excessively dead. At the present time, sound-transmission data on fabrics is virtually impossible to find, so it is up to individuals to measure it. This is one more missing element in the acoustical puzzle.

21.4 FLUTTERS, ZINGS, AND THE LIKE

I was walking through a hotel garden one day, talking to a friend, when suddenly my attention was diverted by the unmistakable sound of a flutter echo. Looking around, it was evident that we were standing in the middle of a masonry arch over the path and that sounds were bouncing back and forth between the bases. Clapping hands produced the predictable "zing-g-g-g." It was a wonderful example of the phenomenon. To hear a flutter echo, all one needs is a pair of parallel surfaces (and they need not be very large), stand between them, and clap hands. Nowhere else in the vast garden were there any flutters.

In a room, however, there are many, many opportunities for flutter echoes, but we are not usually aware of them. They exist, and if we could strip away everything except single pairs of parallel surfaces, we could arrange demonstrations of them. What prevents this from being a disaster in normal life is that rooms have lots of reflections, and the sounds that might in isolation be flutters simply join the extensive parade of delayed sounds arriving at our ears. It has long been known that in performance spaces and recording studios, it is important to treat all the surfaces somewhat similarly. If front and back walls are "dead" and the side walls are "live," it is likely that flutter echoes will be heard between the more reflective surfaces.

However, to hear them, sounds need to be created in the right locations, and the listener must also be well situated. Strictly speaking, in a home theater the only sound sources that *really* matter are the loudspeakers, and the only listener locations that matter are in the audience. So to test if there are problem flutters, have someone go to each loudspeaker location and clap hands while someone else listens from seats in the audience area.

However, custom home theaters are "designed" spaces, and it may be that customers' expectations include being able to stand anywhere in the room, clap hands, and not hear a telltale "zing-g-g-g." One hopes that they stop short of using a stepladder. In any event, be sure that the loudspeakers cannot initiate flutters, and use your judgment beyond that. To eliminate a flutter, it is necessary to interrupt the back-and-forth ricocheting of impulsive sounds. Typical methods include angling one surface (a large conventionally tilted painting may do), absorbing some energy (a drape, wall hanging, or a patch of fiberglass or foam), and scattering some energy (a bookcase, display case, or a diffuser). This is an instance where shallow diffusers are useful because hand claps and bothersome impulsive sounds are biased to the high-frequency end of the spectrum. It is not difficult.

21.5 SUMMARY

It is clear that the traditional method of specifying absorption coefficient, the random-incidence reverberation room method, provides incomplete information so far as using these devices in small listening rooms. When we look at examples of how sounds arriving from specific angles are modified, a very different picture emerges, and it is one that strongly suggests that a naked wall may be a better option than thin absorbing panels. I may exaggerate, but it is difficult to be restrained when attempting to counter a practice that has gone on for many years and actually is encouraged by some international standards (see Section 18.5.1). In the days of mediocre loudspeakers, the effects were perhaps not obvious, but now there are increasing numbers of excellent loudspeakers that have quite uniform and wide dispersion. These products have no opportunity to exhibit their inherent excellence in rooms with areas of absorbing panels of the kind shown in Figure 20.9, or thinner, especially if those panels are placed at the points of first reflection. One can debate the pros and cons of first reflections, but if the decision is to eliminate such a reflection, then there are two options: absorb it with a *thick* resistive absorber (not less than 3–4 in. (75–100 mm)), or diffuse it with a *thick* diffuser (not less than about 8 in. (200 mm)). The criterion in both cases is not the thickness itself but the need to maintain high levels of acoustical performance down to 300 Hz or below.

Finally, we need more and more appropriate data on the acoustical materials and devices used in the audio—sound reproduction—industry. We are not designing concert halls.

CHAPTER 22

Designing Listening Experiences

The purpose of this book has been to assemble and review the science relevant to sound reproduction and use this as guidance in designing loudspeakers, rooms, and multichannel systems—in other words, listening experiences. For those readers who have struggled through the saga to this point, the following recommendations will come as logical conclusions. For those who have started here, there may be some surprises, some points of difference with acoustical rituals and practices of an industry that has evolved without a lot of scientific leadership. The science has been there, but some of it has been undiscovered, and, until this book, little of it has been organized to be useful from an audio perspective.

It is difficult to summarize the contents of hundreds of pages, dozens of diagrams, and concepts having multiple variations and interpretations and to deliver succinct recommendations, some of which are certain to differ from common practice. What follows is, I believe, free from commercial biases, but it is not completely free from bias. The overriding bias is that what we hear is what matters, and it is up to science to understand and to explain those perceptions. The task is to identify the key variables and explain the psychoacoustic relationships to physical acoustical events. Finally, here, we try to describe what is necessary to create the physical sounds that are most likely to provide pleasurable listening.

All of the perspectives outlined here have been arrived at after an examination of what I believe is the best technical and scientific information available. Aspects of it are incomplete, and it is hoped that someone will pick up where others have left off and add to the common body of knowledge. There are some areas where it is time for discussion, not directives. Yet, it is necessary to make proposals based on interpretations of what we think we know now, and that is what follows.

The philosophy driving this part of the book has been to do the following:

- Define what we want to deliver to listeners in the way of directional and spatial perceptions.

- Understanding that all of this must happen in a small room, decide on an arrangement of loudspeakers that can deliver those perceptions to a group of listeners within the room.

- Determine what the loudspeakers need in terms of directional sound radiation patterns to satisfy the previous two requirements, taking into account the influences of room boundaries.

- Consistent with the directional and spatial perceptions, define what is further required of the loudspeakers to deliver high sound quality. This includes identifying a statistical understanding of listener preferences, their abilities to hear differences, and so on.

- Attempt to reduce these collective requirements to a set of measurements that are sufficient to allow loudspeakers to be differentiated in terms of sound quality.

- Combine all of the preceding into recommendations about what should and should not be done in furnishing and/or acoustically treating listening rooms/home theaters to ensure the satisfactory delivery of all perceived directional, spatial, and sound quality dimensions.

22.1 CHOOSING THE MULTICHANNEL DELIVERY SYSTEM

At the present time, the two common delivery modes are stereo and 5.1 multichannel, although new high-definition formats allow for 7.1 discrete channels. In the following discussions, it is assumed that listening in two channels is an option, employing the LF and RF loudspeakers. Contrary to some beliefs, there is absolutely no difference between the loudspeakers required for stereo music and home theater uses, other than the need for the latter to play at high momentary sound levels for massive sound effects in movies.

Stereo programs can be upmixed for playback through five or seven channels. The basic 5.1-channel system can be expanded by upmixing to incorporate one or two extra surround channels (6.1 and 7.1 channels), and a few movies are encoded to take advantage of this. There is also a 6.1-channel digital discrete option. A sensible approach, if space and budgets allow, would be to aim for a three front/four surround, 7.1-channel system. Large audience areas may benefit from the addition of more loudspeakers to provide adequate coverage, but these nonstandard schemes may need some experimentation.

Let us begin by defining the duties of a surround-sound system:

- *Localization.* The perception of direction: where the sound is coming from. The minimum number of locations is the number of discrete or steered channels in the system. Beyond that, we rely on phantom images floating between pairs of loudspeakers. Those across the front are familiar because of stereo, which are even more stable with a center channel. In multichannel systems, other opportunities exist—for example, between the front and sides. These are rarely used except to convey a brief sense of movement because these capricious illusions move around, depending on where one is sitting relative to the active loudspeakers. Anyone seated away from the sweet spot will hear a distorted panorama of phantom sound images.

- *Distance.* The addition of delayed sounds (reflections) in the recordings can create impressions of distance, moving the apparent locations well beyond the loudspeakers themselves.

- *Spaciousness and envelopment.* The sense of being in a different space, surrounded by ambiguously localized sound. This important effect is the principal reason for multichannel audio. In movies, television, and games this is expanded to include impressions of being immersed in crowd sounds, natural sounds of forests and jungles, and mood-setting music.

- *Sound good and play loud* (enough).

All of these effects require a strong direct sound of pristine quality from every loudspeaker in the system delivered to every listener. The necessary information should be in the recordings, and very little is required of the listening room, although some targeted acoustical treatment may be able to provide assistance. From Chapter 13, it is clear that in the bass frequency range there is work to be done, decisions to be made, and, perhaps, money to be spent. In compensation, at frequencies above about 300 Hz, if the loudspeakers have been well chosen, the main task of the listening room is simply to stay out of the way, in the background. Let them do their work.

The possibilities for expanding localization and spatial effects are limited by the number of channels the industry feels customers will buy and install in their homes. Anything beyond the present number seems like a difficult "sell." Frontal sounds associated with on-screen action are greatly helped by the powerful "ventriloquism" effect, and the hours spent by moviegoers listening to a monophonic center channel as action moves around on the screen suggests that it works very well. Sounds originating off-screen are usually momentary sound effects for which no real precision is demanded (nor delivered in the cinema situation). Other off-screen sounds fall into the broad "ambiance" category, where ambiguity of location is desirable, which brings us to impressions of distance, spaciousness, and envelopment.

Here it is again necessary to emphasize that reflections that occur in small rooms cannot *alone* generate a sense of true envelopment. Envelopment requires the strong sounds delayed by 80 ms or more that are in the multichannel programs and that are reproduced through suitably located surround loudspeakers. Additional room reflections of those greatly delayed signals may enhance the impression, but the initial delay and the appropriate directions are provided by the recorded sound and delivered by playback loudspeakers. All of the sounds important to these perceptions need to arrive at the listening locations from the sides, not from the front and rear. It is the differences in the sounds at the ears that create the perceptions.

Figures 15.8 and 15.9 show results of elaborate experiments that looked at many loudspeaker arrangements, asking the questions, how many channels are necessary, and where do we place the loudspeakers for the best impression of LEV, listener envelopment (that sense of being in another, usually large space)? The following summarizes the findings:

- Two-channel stereo does not fare well; more channels are needed.

- *Symmetrical* front-back arrangements contribute nothing to envelopment, but they add two more locations for special effects sounds in movies and voices or instruments in music.

- A center-rear loudspeaker is useful only for localized sound effects.

- All combinations of a center-front, a pair of loudspeakers at ±30°, and another pair of loudspeakers at angles from ±60° to ±135° perform superbly, but if there are only two surround loudspeakers avoid ±150° or whatever angle identifies the spread of the front loudspeakers (i.e., symmetry). (It is probably safe to assume that systems with four surround loudspeakers, two sides and two rears, may use the ±150° locations for the rears.)

- Four loudspeakers behind the listener do not perform as well as four in front, at the same reflected angles. Lesson: Placing side surround loudspeakers ahead of listeners can be advantageous if there are rear surround loudspeakers to provide that directional effect.

- The five-channel arrangement described in ITU-R BS.775–2 (2006) with loudspeakers at 0°, ±30°, and ±120° performed about as well as any other configuration that was tested. This is obviously good news because this is the arrangement promoted almost universally within the industry.

Caution: All of these tests were done for a single listening location. Listeners seated away from this location will experience various forms of degradation, increasing with distance. Additional channels and loudspeakers provide some compensation, but they do not address the fundamental problem of propagation loss (see Figure 16.8). Never forget that there is a prime listening location and that every theater should place a listener—the paying customer—in that seat. It is recommended that as many listeners as possible be seated on or close to the symmetrical left-right axis of the room. Concerns about bass problems on this axis are eliminated by using two or more subwoofers (see Figures 13.11 and 22.4), with bass equalization as and if necessary. Multichannel audio is a great improvement over stereo in providing impressive entertainment for multiple listeners, but it is wrong to suggest to customers that all seats are equal.

22.2 LAYING OUT THE ROOM

There are several interactive factors involved in this process, and how one goes about dealing with them depends greatly on how much flexibility there is in choosing room dimensions. Are we designing a room to fit the requirements or adjusting the requirements to fit the room? Is the room a simple rectangle, or is it part of a larger space? In the following discussion I will attempt to address the important factors, but it is up to you to decide the sequence in which they are addressed and the importance given to each. The design will be iterative, in that some early decisions may need to be adjusted later because of other considerations. Fortunately, there is a fair amount of tolerance in the design objectives, with the possible exception of how one contrives to deliver similarly good bass to several listeners.

- Video—getting the right sized screen for the viewing distances that are possible in the room and arranging seats to provide good viewing angles horizontally and vertically, avoiding obstruction by heads, and so on.

- Audience size and seats. The reclining chairs designed for home theaters can be deceptively large, especially when access space is added. Make this decision at an early stage.

- Then comes the layout of the loudspeakers. The front channels must coordinate with the video display. The surrounds must be arranged to deliver the directional and spatial cues. Along with this come some necessary and some optional acoustical treatments. A certain baseline amount is necessary to tame excessive reflections and flutters in an empty room. Others are specifically targeted to enhance spatial and enveloping effects.

- The bass-managed low frequencies are next, and it is very convenient that certain arrangements of subwoofers concentrate the acoustical energy in a small number of room modes. Knowing where the nulls are located means that we can manipulate the room dimensions and/or the seating locations to avoid them for many, if not all, of the seats. If this

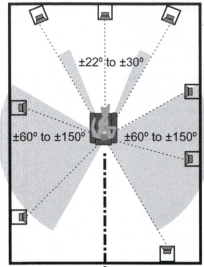

For 3/4, 7.1-channel arrangements. NOTE: side/rear channels should be decorrelated, e.g., by delays (a feature of 7.1-channel processors).

For 3/6-channel arrangements. NOTE: additional side/rear channels should be decorrelated, e.g., by added delays.

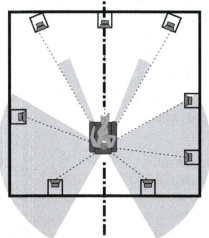

FIGURE 22.1 *Summary recommendations for 3/4- and 3/6-surround arrangements in rooms of two different proportions. The prime location is shown; place as many other seats as possible on the axis of symmetry.*

is unsuccessful or impractical, there are other techniques involving signal processing that can be brought to bear.

Missing from this list is a very common prerequisite: choosing the dimensions and proportions of an "ideal" room. It is absent because in our context, there is no such thing as a generic "ideal" room. The most important issues involving room dimensions are (1) fitting everything and everybody in and (2) delivering good bass to everybody—which is part of the last point in the preceding list. (See Figure 13.6 and the associated text.) Beyond this, efforts to find optimum distributions of room modes fall into the category of a possible, not a probable, benefit. Do it if you want to, or can, but even if you don't, it probably won't matter.

The first tasks focus on video concerns.

1. Use the data in Figure 16.3 to choose a display or to adapt whatever display may exist into a satisfactory viewing experience.

2. If there is a large audience, decide on whether staged seating is necessary, and adjust the video display accordingly (see Figure 16.4).

3. Arrange the seating for good viewing angles (see Figure 16.5a).

Then come the audio considerations for loudspeakers operating above the subwoofer crossover frequency:

4. Overlay the angular recommendations on the room floor plan, and decide where the loudspeakers should be placed (see Figures 16.5b and (c), and Figure 22.1). Although there is no requirement to put loudspeakers on walls, in most instances this is the most convenient and, if the loudspeakers were designed for on-wall or in-wall operation, one of the acoustically optimum locations (see Figure 12.8). Full-range, free-standing loudspeakers at a constant radius from the prime location have a traditional attraction, but the acoustic reality is that the bass is compromised, although the rest of the spectrum is fine (see Figure 4.11). So in such situations, use smaller loudspeakers plus a subwoofer array in a bass-managed scheme (see Figure 13.10). Extra care must be exercised in obtaining proper *acoustical* crossover transitions:

low-pass (from the subwoofer(s)) and high-pass to *each* of the satellites. Measurements must be made at the prime listening location.

5. The three front loudspeakers should be at or close to the same height (as judged by the tweeter locations) and, for large front-projection displays, probably not higher than about 60% of screen height. Because of the ventriloquism effect, errors of this kind are rarely noticed unless they are very large. The left and right loudspeakers should be at the same height, and the center loudspeaker should not be more than 5° or 6° higher or lower (1 to 1.25 inches per foot of viewing distance, or 88 to 105 mm per m, all measured from the prime location).

6. The front loudspeakers should be close to seated ear height for the prime location or as close to that as is possible to provide line-of-sight communication from the tweeters to each listener's head. In large theaters, this may require placing at least the center loudspeaker behind the front-projection screen, in which case it must be acoustically "transparent." In fact, many are translucent, exhibiting small high-frequency losses that should be compensated for by equalization (see Figure 21.14).

7. The front loudspeakers should be aimed approximately at the prime listening location with minor adjustments to best cover the audience.

8. Surround loudspeakers should be roughly 2 feet (0.6 m) above seated ear height and mounted so that sound radiates freely over a horizontal dispersion angle that embraces all the listeners. In some layouts, this will be a challenge for in-wall designs, so consider on-wall alternatives and be sure to give them enough forward offset that their acoustical "field of view" is not obscured by wall treatments. A flat surface behind and around each loudspeaker is a good idea because this is the condition they should have been designed for.

Figure 22.1 illustrates my summary recommendations for 3/4- and 3/6-surround arrangements, based on the observations from the experiments discussed above, and recommendations of Dolby and the ITU (see Figure 15.11b). Use the basic 3/2 (5.1) arrangement only if constrained by room or budgetary issues. The guidance is in the form of angular ranges within which each of the loudspeakers is placed. There is no requirement for them to be on any specific wall or on a wall at all. Angular location is the guide, but, as stated earlier, this is not precision engineering, so adaptable humans will find pleasure in many variations. If the size of the audience requires more than four surround loud-speakers, there arises the decision about which pairs to connect in parallel because at this time, surround processors do not provide separate, electrically delayed, or decorrelated signals for more than four loudspeakers. This may require some experimentation.

In general, a solo center-rear loudspeaker is not an attractive idea because movies are already center-dominated by the relentless front-center channel. Lateral expansion, spaciousness, or envelopment is a welcome relief. Using a pair of rear loudspeakers is the preferred option, because for the vast majority of the time, they will improve the sense of envelopment in upmixed programs, and they are there for 7.1 discrete program content. Even when they are called upon to do duty as a true "rear" channel, they are not significantly compromised. Because the steered signal is monophonic, listeners on the center line will hear a phantom image in the middle of the back wall. Listeners to the left or right will hear an image to the left or right of the center in the back wall. In all cases, the image, if it is a flyover, will simply proceed to pan to the front of the room as it should. All of these sounds tend to be very brief because moviemakers avoid localizable sounds that dwell at off-screen positions.

22.3 LOUDSPEAKER DIRECTIVITY AND THE ACOUSTICAL TREATMENT OF INTERIOR SURFACES

The first requirement of any loudspeaker in a home theater is to deliver strong, high quality direct sounds to all of the listeners. These sounds define the directions of sounds steered to the various channels, and of the phantom images existing between them. As can be seen in Figure 22.2, the front channels meet this requirement with a moderate ±30° horizontal dispersion. The surround channels in the example arrangement need much wider dispersions to reach all members of the audience shown in this example. First lateral reflections can be useful to the front channels, and if it is desired to take advantage of them, the dispersion requirement for these loudspeakers expands considerably. For the direct sounds, variations over the angular spread should be minimal, but the lack of reliable measurement data on loudspeakers makes a technical specification of the allowable variation futile. Looking to the future, perhaps a measure based on the difference between the on-axis response and an average over the appropriate angular window(s) could be developed for loudspeakers in the different roles (e.g., see listening window data in Figure 18.6). Surround loudspeakers require a uniform horizontal dispersion exceeding that which can be delivered by many forward-firing loudspeakers. The on-wall bidirectional in-phase design exemplified by the product in Figure 18.19b would appear to be a good choice for this application, although in many theater configurations wide dispersion conventional forward-firing designs would comfortably suffice, especially if they were aimed for optimal coverage.

As described in Figure 16.8 for the surround channels there would seem to be advantages to using loudspeakers that exhibit less than the −6 dB/double-distance propagation loss of small-box loudspeakers. Line source loudspeakers,

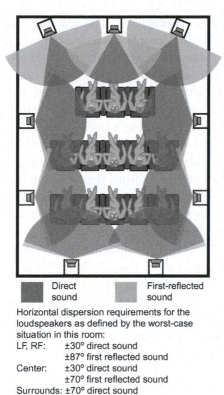

Direct sound

First-reflected sound

Horizontal dispersion requirements for the
loudspeakers as defined by the worst-case
situation in this room:

LF, RF: ±30° direct sound
           ±87° first reflected sound
Center: ±30° direct sound
           ±70° first reflected sound
Surrounds: ±70° direct sound
           ±87° first reflected sound
Loudspeaker are assumed to have symmetrical
left/right dispersion.

FIGURE 22.2 *This is a repeat of a portion of Figure 16.10, which shows a summary of the horizontal-plane angular dispersions required of the loudspeakers to deliver direct sounds of comparable quality and level to all listeners and, for the front channels, to deliver similar sounds to the wall surfaces from which the first reflections occur. There is no doubt that a very wide horizontal dispersion is a requirement if reflections are to be encouraged.*

as discussed in Section 18.1.2 are an existing option, but there is the tantalizing possibility of something different as discussed in Section 18.4.4.

22.3.1 Side-Wall Reflections from Front L, C, and R Loudspeakers

The matter of side-wall reflections of L, C, and R loudspeakers warrants some discussion because of the widespread belief that these reflections should be eliminated as a matter of ritual. The ritual had its origins in recording control rooms—listening in stereo—encouraged by alarmist cautions about comb filtering (see Chapter 9) or degraded speech intelligibility (see Chapter 10) or masking of other reflections within recordings (Olive and Toole, 1989). When examined, none of these turn out to be problems. The real factor appears to be spaciousness (ASW/image broadening), and the possibility that recording engineers, like musicians, are many times more sensitive to it and the reflections causing it than ordinary people (see Section 8.1). Even though many (but not all) recording professionals feel that their recording work is hindered by lateral reflections, most of them prefer to have them in place for recreational listening.

So what does one do with them in a home listening room? If the loudspeakers have good off-axis performance, and especially if the customer likes to listen to stereo music, my recommendation is to leave some blank wall at the locations of the first lateral reflections from the front loudspeakers. An area with a small dimension of at least 4 feet (1.2 m) centered on the reflection path is sufficient. Figure 16.6 shows reflected pathways for one room. Providing reflections for the front rows is probably sufficient.

If one chooses to eliminate the reflection, one look at Figure 21.9 should be enough evidence that much more than 2-in. (50 mm) panels are necessary. If absorption is chosen, *all* of the sound down to 200–300 Hz should be absorbed. If only part is absorbed, then the performance of the loudspeakers is compromised, and some of the reflection remains; there is just no point. Doing it correctly requires not less than 3–4 in. (76–101 mm) of depth. The problem with absorbers is that they don't "turn down the volume" uniformly at all frequencies. Section 21.3.2 explains this and more.

To reduce the level of the reflection, use diffusers. Like absorbers, to be effective at lower frequencies, they need to be thick. Even properly engineered surfaces may need to be about 8 inches (0.2 m) thick.

There appears to be no evidence in the now substantial literature that these first-order lateral reflections are problems in normally furnished or the equivalent moderately treated rooms. If the surround channels are active, it is probable that the modest spatial contributions of these front-channel reflections will be masked. If only the front channels, especially the center channel, are active, it is possible that a small spatial effect may be beneficial. In the grand scheme of things, these are factors but not the dominant factors.

It is difficult to ignore the advantages, without apparent disadvantages, of using normal forward-firing loudspeakers with wide dispersion, good off-axis behavior, and allowing the relevant areas of side walls to reflect. However, in a multichannel context, this is an issue where the customer and/or the consultant can express some free will.

22.3.2 The Surround Channels and Opposite-Wall Reflections

The surround channels working in collaboration with the front channels can generate a remarkably realistic sense of envelopment. As shown by the experiments described in Section 22.1, five loudspeakers in an anechoic chamber driven by the right signals are enough to generate a sense of envelopment that can compete with arrays of 12 or 24 channels (see Figures 15.8 and 15.9). No assistance from room reflections was required. However, that was to satisfy a single listener in the sweet spot.

To cater to other listeners in the room, we need more sources of sound and possibly different kinds of sources. It can be seen in Figure 22.2 that uniform dispersion over a huge horizontal angle—almost ±90°—is required to deliver similarly good sound in the direct and reflected pathways to all of the listeners. This is a significant challenge. It is achievable with some existing wide-dispersion designs but not with "dipole" surround configurations.

As shown in Figure 8.6 and discussed there, the surrounds do two jobs:

- *Provide momentary localizable sound effects in films and stationary localizable sound sources in music.* For these tasks, the direct sound is of paramount importance. Natural reflections may or may not be the only spatial accompaniment to the sound because one expects that the recording engineers would provide for these circumstances with delayed and/or reverberant sounds in the other channels. Specific acoustical treatment is therefore unnecessary. The precedence effect ensures a persuasive directional impression.

- *Provide impressions of immersion or envelopment by reproducing delayed versions of sounds originating in the front soundstage (such as reflections in the recordings).* In this, which is a primary function of the surround channels, reflections from the opposite wall should be able to assist the envelopment illusion by making it appear that there are more surround loudspeakers than physically exist. They also arrive from the correct (lateral) directions. Be very careful not to create an opportunity for flutter echoes (see Section 21.4) between the opposing walls. Because it is the frequencies in the range 1 kHz down to about 100 Hz that generate envelopment (see Figure 7.1), *broadband* diffusers alone or in combination arrangement with reflecting and absorbing surfaces would be suitable. This requirement denies the use of low-profile diffusers for this application. Reflecting surfaces would generate the most energetic reflections, but at the same time they address a restricted listening area and they create the highest risk of flutter echoes. A secondary function of these strong early reflections is providing directional distraction from the surround loudspeakers, perhaps making them less localizable by listeners seated near the perimeter of the room.

22.3.3 Treating the Interior of a Room

Combining the ideas expressed to this point, Figure 22.3 shows practical suggestions that are intended to achieve the desired results. Acoustical treatments are not very attractive, and so it is common to stretch fabric over large portions of walls. This is not a problem if the fabric is acoustically transparent, but many

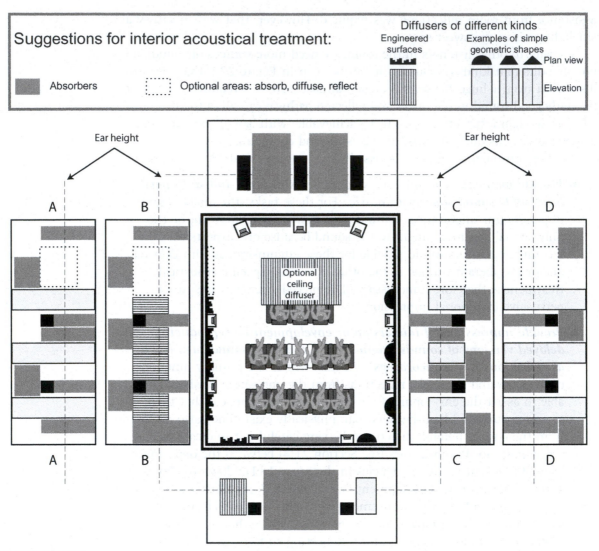

FIGURE 22.3 *This figure illustrates a few of the many possible ways to combine acoustical materials within a home theater. It shows a floor plan with walls folded down to show how materials might be arranged on them. The white seat in the middle is the prime listening location. Wall B shows a long array of engineered diffusers in a band around ear level. Wall C shows a version that uses semi-cylindrical geometric shapes intermixed with reflecting surfaces and absorbing panels. Walls A and D show mirrored treatments in which the diffusing shapes have been extended floor to ceiling for visual effect. Absorbing panels have been placed in staggered locations to avoid flutter echoes. Many artistic variations are possible and changes will be necessary to accommodate different numbers of surround channels. The dashed lines that identify the ear height of seated listeners must be adjusted to follow the floor contours of staged seating.*

are not, in which cases a lot of careful design, and the sound of the room, are degraded when the "finishing touch" is added.

The basic requirements can be summarized as follows:

- The **floor** is covered with wall-to-wall clipped-pile carpet on 40 oz/sq yd felt underlay.

- The center portions of the **front and rear walls** are mostly absorbing, with scattering devices toward the sides of the rear wall. All absorbers, wherever they are located, must be not less than 3 to 4 in. deep.

- The **side walls** are a mixture of reflection, absorption and scattering/diffusing devices. Diffusers designed to scatter the sound horizontally should be located in the region about one foot below and about three feet above ear level.

- Diffusers can be of the engineered-surface type, that are attached to the walls, or they can be simple geometrical shapes that can be attached to walls or integrated into structures, constructed on site using conventional building materials. To assist listeners in the perception of immersion or envelopment, engineered surfaces should be not less than about 8 in. deep, and of the type that disperses sound horizontally. Many curved and multifaceted convex geometric shapes will work, so considerable artistic freedom exists, as long as they are not less than about a foot deep. Again they should be designed to disperse sounds horizontally. For visual effect it is possible for the geometric shapes to extend floor to ceiling, imitating columns. Engineered diffusing surfaces can be arranged side-by-side to cover an area, as shown in B, but to work well geometric shapes must be separated as shown in A, C, and D. Both types of diffusers can be mixed.

- Absorbing material and blank reflecting areas should be arranged so that walls facing each other do not present opportunities for flutter echoes (in the drawing facing walls A and D and B and C are designed with staggered areas of absorption and reflection). The total amount of absorbing material in the room must be sufficient to meet the reverberation time criterion discussed in Section 22.3.4.

- Use geometry or a mirror to determine the best location for the optional diffuser on the **ceiling**. It should scatter sounds from the front loudspeakers that would normally be reflected to the head locations of audience members. This is most effective if it is an engineered surface designed and positioned to scatter sounds toward the sides of the room.

- The locations of the first side-wall reflections at the front of the room are specified as areas for optional treatment. Leaving these areas as flat wall surfaces provides an open and spacious soundstage for those customers who listen in stereo. In television and movies these reflections will "soften" the image of the commonly dominant center channel. Well-designed wide-dispersion front loudspeakers will generally sound better in the presence of lateral reflections. When multiple channels are operating simultaneously, these reflections are swamped by the recorded sounds and become neutral factors. So, the effects of these side-wall reflections range from neutral to slightly beneficial. In any event, they are not large effects, so the choice can be left to the designer.

■ The corners of the room are available for low-frequency absorbers. These are preferably of the membrane/diaphragmatic/panel type, because they are located in high-pressure regions of the low-frequency standing-wave patterns.

22.3.4 Other Surfaces—Reverberation Time

Conventional acoustical design would have had us calculating reverberation time (RT), measuring it, and fussing to get it right. If we were designing a concert hall, that would be justified, but in the context of home listening spaces, the correct RT is a generous target, ranging from "too dead" (below about 0.2 s) to "too live" (above about 0.5 s), both of which are easily recognizable by walking into the room and carrying on a conversation. That commentary is probably a little facetious, but it is done to compensate for the sometimes overly fastidious attention devoted to RT in the audio industry. "It can be measured, so it must be important" seems to be the rationale; see Section 4.3.4.

Let us compute where we stand using the room in the example shown in Figure 22.3, specifically using wall treatment configuration "B" (and its corresponding opposite with staggered areas of absorption) because it offers some interesting discussion. For this $20 \times 24 \times 9$ ft ($6.1 \times 7.3 \times 2.7$ m) room, the Sabine reverberation time (see Chapter 4) would be calculated (in imperial units) as RT = 0.049 V/A. V = 4320 ft^3, so the total amount of absorption (A in sabins) required in the room to achieve a midfrequency (say, 500 Hz) RT of 0.4 s is 530. The carpet on the floor has an absorption coefficient of about 0.5 (see Figure 21.3), so we multiply that area by the absorption coefficient, giving us $20 \times 24 \times 0.5 = 240$ sabins. The absorbing areas specified for the front and rear walls add up to about 106 sq ft $\times 1.0 = 106$ sabins (they are thick resistive absorbers that absorb perfectly at 500 Hz). Scattered patches of absorption on the side walls yield 110 sabins. If RPG Modffusors (exposed absorption coefficient = 0.2, fabric covered it is 0.6) were used along the side and rear walls, and they would contribute about $128 \times 0.2 = 26$ sabins, but if they were covered with fabric for appearance, that would increase to 77 sabins. Adding them up: $240 + 106 + 110 + 26$ (77) = 482 (533)—close to the target of 530 sabins. The RT would be a maximum of 0.44 (0.4) s. The additional absorbing surface area of the seating and the people in the seats (see Figure 21.5) would further reduce the RT, as would the additive effects of scattering by the furniture and diffusers. If geometric or other kinds of diffusers are used on the side walls, simply mix in enough pure absorption with the hard scattering shapes to approximate the numbers in the preceding prediction.

Reverberation times less than 0.5 s are not likely to degrade speech intelligibility. Music and multipurpose rooms might drift toward 0.4 s, and dedicated movie rooms might drift toward 0.2 s. For home entertainment of the highest quality, there is a comfortably large acceptable range. The key is not the number, but the sound. The room should *sound* comfortable for conversation. The ultimate speech test is for one person to stand where the center channel is located and carry on a conversation with someone else who is moving around the audience area (facing each other, of course). It is highly improbable that small-room acoustics will be a problem for intelligibility (see Chapter 10), but there is the matter of "ambiance." The most common problem in custom home theaters is that they are too "dead." In conversation, voices sound muffled, and more than the usual amount of vocal effort is required. It is not a relaxing situation. However, I know of designers who have done this deliberately to make the theater seem "special."

Feel free to add or take away materials until it sounds "right"; in small listening rooms, RT numbers are not highly predictive. Once it sounds "right," you have created your own signature design, and it is certain that there are many variations on these suggested schemes that will work. In this respect, the design is done, and, once done, it is easy to imagine that it will not be necessary to repeat it for every new job.

22.4 SUBWOOFERS, SEATING AND ROOM DIMENSIONS

Discussions to this point have focused on sounds above the subwoofer crossover frequency—about 80 Hz. Chapter 13 delves into the details of how sounds behave at low frequencies in rooms. Cutting through the detail to deliver the essential messages, we arrive at the following points:

1. One subwoofer used with high-resolution measurements (at least 1/10 octave) and parametric equalization can deliver good bass to one listener. All other listeners in the room will take their chances because of standing waves. Seat-to-seat variations are large.

2. Or, using any of several forms of spatial averaging (measurements at multiple locations), more people can get to hear better bass, but all will be different and none may be truly optimum. Seat-to-seat variations are large.

3. To deliver similar and similarly good bass to multiple listeners *in rectangular rooms*, two or four subwoofers are necessary. Certain

arrangements of these subwoofers result in most of the bass energy being concentrated in three room modes, which share portions of the same standing-wave pattern. This means that we know where the nulls are located, and through a combination of dimensional adjustment and/ or juggling seat locations, several listeners can be treated to similarly good bass. Equalization will be necessary, with *all* subwoofers operating simultaneously. The subwoofers must be identical, placed similarly with respect to room boundaries and adjusted to the same output levels. Seat-to-seat variations are small.

4. If the room is not perfectly rectangular, or if all of the subwoofer locations are not available, or there are openings to the rest of the house, there are more elaborate solutions. One of these, called Sound Field Management, was described in Chapter 13. It can deliver excellent results in difficult circumstances, but at time of writing, it was not yet commercially available. Seat-to-seat variations are small.

Elaborating on number 3, Figure 22.4 shows how the freedom to adjust room dimensions allows us to increase the number of listeners who get to hear similarly good bass. The amount of low-frequency damping in the room will determine the magnitude of bass frequency response variations (shown in Figures 13.13 and 13.15) and therefore the amount of equalization that will be needed to produce decently flat frequency responses at the listening locations. Damping is *always* a good idea. Multiple subwoofers bring a simple structure to the modes that need to be damped and equalized.

As shown in Figure 13.15, if the two-subwoofer arrangements are to have the same sound-level capability as the four-in-the-corners arrangement, it will be necessary to stack two at each of the locations.

Most installations must be created within existing room boundaries, so the scheme shown in Figure 22.4 must be used as a basis for adjusting seat locations so that listeners' heads are not in null locations. If perfect success for all listeners is not possible, at least the customer knows where the good and less good seats are.

As stated previously, damping of room modes is advised. In some circumstances, it is possible to let the room boundaries be the primary low-frequency absorbers. However, there are many situations where this is not possible, such as existing rooms with masonry walls or where it has been necessary to add mass to the interior surface of the wall to obtain adequate sound isolation. When this happens, it becomes particularly important to add absorption, especially at frequencies below about 100 Hz. This requires special devices, and it will be necessary to look for authentic measurements of absorption coefficients. A wedge of acoustic foam in the corner does not qualify. There are several good products on the market, some quite clever, and they all work, and they all advertise properly obtained absorption coefficients that remain usefully high at

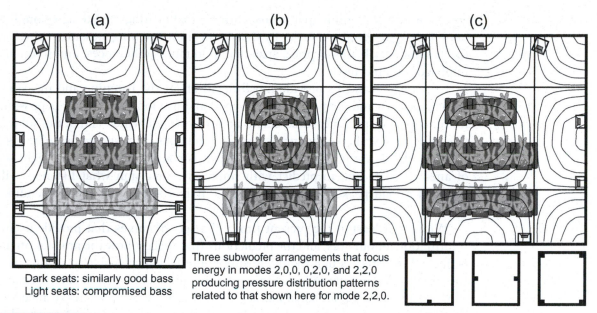

(a)

(b)

(c)

Dark seats: similarly good bass
Light seats: compromised bass

Three subwoofer arrangements that focus energy in modes 2,0,0, 0,2,0, and 2,2,0 producing pressure distribution patterns related to that shown here for mode 2,2,0.

FIGURE 22.4   *In a rectangular room, any of the subwoofer arrangements shown will concentrate most of the acoustical energy in the modes 2,0,0; 0,2,0; and 2,2,0, as explained in Figures 13.13, 13.14, and 13.15. Modes 2,0,0 and 0,2,0 have sharp nulls running side-to-side and front-to-back, respectively, at locations 25% from the walls. As shown here, mode 2,2,0 has all of these simultaneously. These diagrams show sound pressure contours, as described in Figure 13.4, and lines dividing the regions at the null locations 25% from the walls. The nulls are regions where sound is severely attenuated; bass notes will be missing. Listeners in the central areas of the blocks avoid all the nulls and therefore hear similar bass. Those whose heads are close to the null lines will hear less good bass; these seats are shown in a lighter shade. In (a), 6 of the 13 listeners receive good bass. In (b), the room has been shortened to move the null away from the rear row heads. Three more listeners now receive good bass, but four are still in the front-to-back null lines. In (c), the room is widened to move those nulls away from the heads so all the listeners are now in positions to hear similar bass. Figure 13.7 shows the measured axial mode pressure distribution across a typical room. It shows that in moving 2 ft (0.6 m) from the null location, the sound level can rise by about 10 dB, so small movements can be greatly advantageous. Note that this simple analysis does not apply to nonrectangular rooms. Even in rectangular rooms, unequal wall constructions, openings in walls, large cavities, and so forth will distort the patterns, so the only real proof of success is to make acoustical measurements at each of the seats with all subwoofers operating simultaneously. If room dimensions are fixed, seats can be relocated to avoid the nulls. At the very least, the customer will know where the good seats are.*

frequencies below 80 Hz. Locate theses devices in the high-pressure points in the standing waves. Corners are choice locations or, in the example shown here, also the midwall locations. In Figure 22.3 corners were left vacant for this purpose.

22.5 CHOOSING THE LOUDSPEAKERS

The first requirement of a loudspeaker is to *sound good*, and about that there has been a lot of disagreement over the years. However, it is now known that good sound is something that people with normal hearing tend to agree on, but only if they are given a chance. By that is meant a listening situation in which the nonauditory biases of price, size, brand, loudness differences, and so on are removed—a blind test.

Chapters 17 and 18 provide much detail about the listening and measurement processes and many examples of good and bad loudspeakers. It is possible for readers to inspect measured curves and to conclude which end of the subjective preference scale the product is likely to end up at after such a listening test. Chapter 20 takes it further and shows that it is possible to process the right set of anechoic data and to predict with impressive precision what listeners will think of a loudspeaker when they hear it. It is a story with a happy ending, right?

Wrong—at least at the present time. The sets of curves that provide useful information for our eyes, and the predictions that suggest how the product might be judged by listeners in a room, are simply not widely available. It is long past the time in this mature industry that manufacturers of loudspeakers, especially those aimed at the professional market, need to provide comprehensive anechoic data on their products. A few already do, and that is commendable. A reasonable description of how the product performs must be the "price of entry" to this marketplace. It should not be up to the customer to discover information that can be freely available. During the design of the product, this information was presumably available to the engineers who designed the product. If such data were *not* available to those engineers, then one is left to contemplate the competence of the source of the product. The descriptions of acoustical performance offered by many of the significant players in the loudspeaker business are simply insulting in their inadequacy. They do it because they can get away with it.

Ask manufacturers for *real* high-resolution (1/20-octave) anechoic measurement data on their products; many examples are shown in Chapter 18. A minimum of on-axis and several off-axis curves extending to 60° or more off-axis, vertically, and horizontally—or, as shown in the chapter, computed estimates of early reflections. Sound power and directivity index are useful additions. If they have the data, they should provide it. None of this is difficult or mysterious; I started doing such measurements in the early 1980s (Toole, 1986), and, as we saw in Chapters 17 and 18, an engineering degree is not required to interpret the curves.

22.5.1 Front Loudspeakers

These are the "bread-and-butter" products of the loudspeaker industry. Most of them are floor-standing or bookshelf units that were designed to function away from room boundaries. However, in home theaters they are often placed against a boundary (on-wall) or in a boundary (in a wall or constructed baffle) or in a cavity in a cabinet. (See Figures 12.9 through 12.11 for measurements of a loudspeaker placed in each of these mounting options.) In adapting loudspeakers to these locations, unfortunate things can happen to the sound quality. In none of them can the loudspeaker perform as it was designed. Part of the design of a freestanding loudspeaker is the acoustical diffraction that occurs at the edges of the enclosure. In some designs this is a significant factor and when such a loudspeaker is flush-mounted, that is changed; the loudspeaker has been "redesigned" in a way that may not be correctable with equalization. The bass output is increased by the boundary effect, but that change can be easily equalized.

In today's marketplace, there are other options. If a loudspeaker is to be mounted flush with a flat surface, use an in-wall loudspeaker designed for that purpose. If it is to be mounted on a large flat surface, select an on-wall design. If they have been competently designed, the excessive bass will have been compensated for and other adjacent-boundary issues accounted for in the design. A caution: if the walls are acoustically treated be certain that the in-wall loudspeakers are not buried in recesses created by the treatments. This is especially important for the surround loudspeakers because they may need very wide dispersion to deliver their sound to all parts of the audience.

No loudspeaker can perform well in an empty shelf or cavity, so at the very least, fill it with solid or absorbent material and, preferably, close the openings around the front. The best in-wall and on-wall designs compete favorably in sound quality with freestanding designs, although attention must be paid to mechanical isolation from the structure and to providing the correct volume behind the unit if it is not already enclosed (normally accomplished with a back box).

These loudspeakers must be able to generate high sound levels, of the order of 105 dB for short durations, if one plays movies at "reference"—0 dB—level. In combination with the other channels, and with the subwoofers engaged, the total sound level is even higher. Many people find that to be a bit loud—in both cinemas and their homes—but it is a target to aim for. If achieved, it means that everything is loafing at lower sound levels, leading to long life and low distortion.

In rooms about 4000 cubic feet and under, well-designed conventional cone-dome loudspeakers should have no trouble achieving the required sound levels. In larger rooms, one must be more selective because some loudspeakers may exhibit distress or transducer failure. However, there are also some highly refined designs that can deliver substantial amounts of sound (see Figure 18.17a).

In very large rooms and in rooms where the customer feels a need for high sound levels much of the time, horn-loaded loudspeakers are recommended. Some of these offer sound quality competitive with the best available (see Figure 18.17b). With large woofers and compression-driver horns, the low distortion and an absence of power compression can lead to volume settings that put one's hearing at risk and it still does not sound "loud." With subwoofers of commensurate capability, music and movies become what I call "whole-body experiences" and they are *very* impressive.

Getting back to normal matters, there is an issue with horizontal center-channel loudspeakers of the simple Midrange-Tweeter-Midrange (MTM) configuration because of their atrocious horizontal off-axis performance. However, horizontal center loudspeakers designed with a midrange unit to accompany the tweeter can be excellent, so look closely. See Figure 18.18.

22.5.2 Surround Loudspeakers

The history of multichannel developments includes the humble beginnings of the surround channel as a single band-limited signal. It was tolerable in large cinemas, with many surround loudspeakers, but in homes where there were only two, it was more a novelty than anything impressive. That situation fortunately did not last long, and soon we had expanded bandwidth, more sophisticated upmixers, and eventually, discrete left-right surround channels (see Chapter 15). It was in the seriously imperfect early stages of this story that THX contributed the idea of using bidirectional out-of-phase loudspeakers (which they called dipoles) in the surround channels to add some acoustical confusion to the sound field. This, it was thought, would lessen the obviousness of the two loudspeakers as the sources of sound. They also added electronic decorrelation, which was quite successful at doing the same things. However, with the passage of time and the evolution of recording and playback techniques, it transpired that the surround channels began to be used as individually addressed sources of localized sound, as discussed in Section 22.3.2. Recording engineers could also add whatever amount of decorrelation they felt was appropriate to the surround channels. Then came upmixers that were capable of driving four surround loudspeakers with decorrelated signals. It was a very different situation, but the "dipole" surround loudspeakers remained in the marketplace and continued to be promoted.

Meanwhile, others experimented with the basic idea of multiple loudspeakers housed in the same small box-on-wall style and a series of variations evolved: monopole, bipole, tripole, quadrapole, and perhaps others, all spreading sounds in many directions but in different patterns. Marketing departments frequently felt that it was necessary to offer a selection of directivity options, so multidirectional surround loudspeakers with switchable directivity patterns appeared (see Figure 18.19). In time, some manufacturers decided on one of the patterns and stayed with it. So now there is a choice.

When viewed from the perspective of what a surround loudspeaker is intended to do, the dipole configuration falls short. In terms of generating envelopment, the significant sounds arrive at the listener from ineffective directions: the front and rear walls (see Figure 8.6). In terms of delivering strong, high-quality, direct sounds to all listeners for purposes of localizing sound effects, the dipole attenuates the very sounds that are needed to initiate the precedence effect. In terms of sound quality, the inconsistent destructive and constructive acoustical interference in the direction of listeners is not amenable to delivering high sound quality, and the directivity is very inconsistent with frequency (see Figure 18.20). It is time to move on to other designs.

Figure 22.2 shows that surround loudspeakers need to have uniformly dispersion over a very wide horizontal angle. This is a challenge for conventional forward-firing designs, but for many installations it is an achievable goal. The very wide dispersion, constant directivity, and respectable frequency responses offered by the bidirectional-in-phase design, an example of which is shown in Figures 12.13 and 18.19b, appear to be a sensible choice for demanding installations.

22.5.3 Localizing the Surround Loudspeakers, Envelopment, and Propagation Loss

For special effects, it is important that we localize the surround loudspeakers. However, for most of the sounds they reproduce, we want them to disappear. It is an interesting problem. All conventional loudspeakers, front or surround, of whatever directional pattern, radiate a direct sound that falls off at a rate of approximately −6 dB per double distance. Steady-state sounds (the combination of direct and all reflected sounds integrated over a time interval) fall at a lower rate, about −3 dB/double distance, as explained in Figure 4.13. However, it is the direct sound that establishes localization. As shown in Figure 16.8, the result is that only the prime listening location and any other listeners on the left-right symmetrical axis receive balanced left- and right-side sounds to create the sense of envelopment. Moving toward the side of the room, the amplitude imbalance increases rapidly, the sense of envelopment is progressively reduced, and the near-side surround loudspeaker becomes progressively more obvious as a localizable sound source, even with decorrelated sounds radiated from both sides.

The traditional solution is to turn down the treble. This was done in the original Dolby Surround implementation, where surround channels were rolled off around 7 kHz to alleviate problems with optical sound tracks (see Section 15.5). It happened again in the "dipole" surround loudspeakers, which attenuate frequencies above about 500 Hz in the direct sound aimed at listeners. However, there were negative side effects of compromised sound quality and the possible incorrect localization of special effects (see Figure 18.20).

A solution that addresses the problem without any apparent compromise is a loudspeaker radiating a wavefront that attenuates less rapidly with distance. The obvious choice is a line source (a very tall, approximately floor-to-ceiling, effectively continuous source) that radiates a cylindrical wavefront exhibiting a fall rate of −3 dB/double distance (Figures 16.8c and 18.2). A possible solution involves wavefront shaping of a kind that results in a relatively constant sound level over a range of distances at ear level (Figures 16.8d and 18.21), but these are not available yet. Line sources are available now but do not confuse genuine line sources with simple "tall" arrangements of loudspeakers, which are very different devices (see Section 18.1.2). In the meantime, using conventional loudspeaker designs, the middle of the room remains a "sweet region."

22.5.4 In-Wall, In-Ceiling Options

In-wall and in-ceiling loudspeakers have come a long way. From their humble origins as sources of distributed sound and voice announcements in commercial establishments, the best of them now compete favorably in sound quality with their free standing relatives. As seen in Figures 12.8 and 12.9, if the excessive bass is corrected, a 2π, half-space mounting allows a loudspeaker to perform exceptionally well. If such loudspeakers can be located in the positions described earlier for freestanding or on-wall loudspeakers, there is no reason not to use them. If not, the magnitude of the compromise must be evaluated and discussed frankly with the customer.

The most objectionable situation involves the front L, C, and R loudspeakers in the ceiling. In the beginning, downward-firing units provided decent sound only for the family pet lounging in front of the TV. Then came steerable tweeters that delivered some high frequencies to the listening area. More recently, designs have appeared that make an effort to radiate both mid- and high frequencies toward the listeners, further improving the sound quality, but a problem (I call it the "voice of God" problem) remains. For family members, adaptation and the ventriloquism effect can do wonders to enable them to ignore the fact that the sound is coming from above—sometimes *far* above—the video display. Visitors will notice it right away. Aiming the loudspeakers at the listeners does not change the ability to localize where the sound is coming from. Fortunately, our ability to localize in the vertical plane is not very good. That said, the best that can be done is to choose loudspeakers that do direct their sound output toward the audience and place the video display as high as is tolerable. In spite of the popularity of these installations and a great deal of marketing hype, this is obviously not "real" home theater.

Placing surround-channel loudspeakers in the ceiling can be tolerable, but it is still a compromise. Place them close to the locations that one would use if they were "real" loudspeakers. Use all of your authority and powers of persuasion to convince the customer to move the loudspeakers closer to ear level.

22.6 LEVEL AND TIME ADJUSTMENTS AND EQUALIZATION

22.6.1 Level and Time

Once all of the installations are complete, it is time to calibrate the system. Receivers and surround processors provide internal band-limited signals for this purpose, and some even provide microphones. Measurements (with a sound level meter, if necessary) are made at the head location of the prime listener. Aim the microphone toward the ceiling and set the meter to C-weighting and slow response. This ritual does not guarantee perfectly equal loudness for all possible sounds from all loudspeakers, but it will be close enough for these purposes. See "relative loudness" in Section 17.3.1 and Figure 17.4. Common test signals are bandpass-filtered pink noise in the frequency range of 500 to 2000 Hz.

It is always a good idea to evaluate timbral similarity among the channels using broadband pink noise from a test disc. To do this, sit in the prime listening location and turn to face each loudspeaker as it reproduces the test signal. (If you sit facing forward as the sounds move around the room, there will be definite changes in timbre caused by the directional effects of the head and ears. This is part of normal hearing, and it is *not* a problem.) Only if the loudspeakers are identical can you expect the timbral signatures to be very close. It is common for the surround loudspeakers to have wider dispersion than the fronts. If so, they will sound slightly different simply because of the different reflective sound fields in the room. This is normal. If it is practical in an installation, use closely similar or identical loudspeakers in all locations.

If the loudspeakers are not all at the same distances from the listeners, it will be necessary to introduce delays into the signal paths to compensate. Again, the processor instructions should explain how this is to be done, and, again, all of the corrections are made for the prime listening location.

Perceptive readers may have noticed that the "prime" listening location is getting a lot of attention. Given this, and the earlier discussions about propagation loss and envelopment, there is every reason to be certain that somebody gets to sit in this location. Yet, there are many installations where this location is between seats, or a table for popcorn and drinks. Pity.

22.6.2 Equalization

The word *equalization* suggests a process of making things equal, and it is fair to ask what, for whom, and why? The answer is complicated. In common parlance, we use the word to describe a process of changing the frequency response to make it conform to some presumably ideal shape. Equalization can be applied to any audio device, of course, but in this context, it is frequently called room equalization, and there is truth to the proposition that we are able to reduce the often dramatic differences between bass sounds in different rooms (see Figure 13.9). However, the situation is very different at frequencies above about 300 Hz, the transition frequency.

Chapter 13 presented the details of what rooms can to do low frequencies and the inequities that standing waves generate: no two people hear exactly the same bass quantity or quality. Seat-to-seat variations can be enormous (see Figures 13.18a and 13.19a). Global equalization cannot change this physical fact, but measurements and equalization can allow us to improve the situation. Because low-frequency room resonances behave as minimum-phase systems, we need high-resolution measurements (1/10- to 1/20-octave) and parametric equalizers to work with them.

A parametric equalizer can attenuate the amplitude and reduce the audible ringing from gross resonances (see Figure 13.24). This is a great benefit, but it only works at the location of the measurement microphone. This should be done for the prime listening location (yet another bonus!). In general it is recommended to adjust the parametric equalizer to match the shape of and to reduce the amplitude of any upward thrusting peaks in the frequency response. Narrow dips should be left alone, but broad depressions may be boosted if the amount of boost is not more than about 6 dB.

Alternatively, one can make measurements at several locations throughout the listening area and employ some form of spatial averaging to find a compromise for those listeners. The seat-to-seat variations are still there, but it can be possible to tame the most egregious booms. These schemes work for a single subwoofer or for multiple subwoofers that have been positioned arbitrarily.

Once the curve has been smoothed the overall level of bass can be adjusted to provide the most pleasing effect. This is very much program dependent, especially if the program is music, because of inconsistencies in recordings. In general most listeners prefer a slightly elevated bass frequency response. This is not necessarily an error, because a great many recording control rooms exhibit similar elevations—one is simply hearing what the recording engineer heard. The author had a bass-level control programmed into the comprehensive home theater remote control to allow for convenient up or down adjustments of bass. It gets used. New customers tend to be impressed by slightly excessive bass levels. Perhaps we need a clock-driven bass level control so that the system can be handed over to the customer with the bass in an elevated (impressive) state, and over time it gradually reduces to a more tasteful level.

If the room is rectangular, and multiple subwoofers are arranged in certain manners, as discussed in Figure 22.4, then a special kind of equalization occurs: spatial equalization, wherein multiple listeners can get to hear similar bass. Then the normal spectral equalization can be added to make that bass good. This is the "icing on the cake." A high-technology solution exists for nonrectangular rooms or other difficult situations (see Figures 13.18 and 13.19).

The situation is very different at frequencies above the transition frequency of about 200–300 Hz. The loudspeaker itself takes on most of the responsibility for what we hear, but the room remains a factor in what we measure—and that is the problem described in Figure 19.1. Steady-state measurements—room curves—made at the listening position are not reliable indicators of very much except at low frequencies. In these measurements, a microphone collects sounds from all directions and at all times following the direct sound, adds them together, and presents them to an analyzer, traditionally a 1/3-octave analyzer. The notion that this simple process can reliably predict what is perceived by two ears and a brain is preposterous. Using this information as a basis for equalization at middle and high frequencies compounds the error.

Some elaborate equalizers make time-windowed measurements attempting to separate the direct and subsequent reflected sounds. This is a thoughtful move in the right direction, but the measurements are blind to direction: they have no idea where the sound is coming from, but the ear-brain system does. It also sacrifices frequency resolution to see into events in the time domain, meaning that the more precisely the sounds are separated in time, the less information we have about them (similar to what is shown in Figure 13.23). If we had detailed measurements on the loudspeakers to begin with, much of this would be unnecessary.

Equalization can change frequency response—that is all. As can be seen in much data shown in Chapter 18, loudspeakers can have many problems that are not revealed in room curves, and they can have directivity problems that can show up in room curves but that equalization cannot address (Figure 18.10). The only cure for a loudspeaker with directivity issues is to take it back to the engineers and tell them to redesign it.

The complex sound field in rooms can add other aberrations that the human perceptual system takes in stride, meaning that if one starts with truly excellent loudspeakers, equalization based on in-room measurements has a chance to degrade them. As stated earlier, we need to have detailed and accurate information on loudspeakers. Then and only then can we assess what the loudspeaker is doing and what the room is doing to it. Room curves bundle all of the information together. At middle and high frequencies, we learn more from an analysis of the loudspeaker than we can learn from room curves, even though both are helpful. It is time that comprehensive anechoic data on loudspeakers was widely available. Ask for it.

22.7 IN CONCLUSION

Throughout this book we have been looking at what can be done to improve the listening experience, maximizing aspects that listeners find rewarding and minimizing those that they find objectionable. In some situations, there is firm guidance about what should be done, and in others, there are suggestions and options. However, home entertainment systems need to be "future compatible," ready to deliver all forms of home entertainment: music alone, music with a picture, television, movies, and games. Entertainment patterns change. We may call our rooms "home theaters," and they may function well as personal cinemas, but they can also be home concert halls, home jazz and rock concerts, football, hockey and basketball stadiums, and in video games they can be car race tracks, imaginary worlds, and battlefields. Television programs can provide entertainment of high artistic and technical quality, and digital recorders allow us to accumulate collections of our favorites. We undersell our product if we do not present a full picture of entertainment options to the customers. A home theater should be more than a place where you sit in the dark for a couple of hours with a bowl of nibbles and a drink.

The title of the book is *Sound Reproduction*, and this means that somewhere, at some time, persons create originals to be reproduced. In any rational system, there should be some assurance that more than the mere essence of that original should be heard by customers. Contemporary recordings of music are mostly studio creations. Even if the music itself originated in a concert hall, the final tweaking is done in a control room in front of loudspeakers. Popular music is an abstraction of any physical reality. Large venue live concert performances struggle to replicate the impressions that fans have embedded in their acoustic memories, to the extent that performers occasionally resort to miming to recordings. What we hear at home may be a better approximation to the original than any other alternative, certainly if we include portable audio devices and ear-bud headphones.

Movies began in cinemas, and theatrical presentations remain an important part of the industry, but most of film industry revenue comes from DVD, free TV, pay TV, and pay-per-view. In a November 18, 2005, interview with www. hollywoodreporter.com, George Lucas commented that theatrical operations probably are not profitable, but they are still supported because of their promotional value. He sees downloaded pay-per-view as the future of the industry, with simultaneous theatrical and other media releases. Other industry insiders share the opinion.

Digital techniques have made great changes to the production of films, and digital delivery is hugely attractive, a challenge even to the term *film*. If the industry follows the money, it is obvious that homes and home theaters are the prime recipients of their product, and they need to optimize the content accordingly. What about cinemas? Well, they will manage as best they can, which is what they do now. Apart from showcase venues, the cinema experience is not what it used to be. Market research available on the Motion Picture Association of America (MPAA) website reveals that when moviegoers were asked, "What is the ultimate movie-watching experience?," 37% of them responded "Home" in 2006, up from 31% in 2005. Looking at those of age 25+, the response was 41% in favor of the home experience.

Part of the preference is the freedom that a "pause" control offers. Part of it is choosing your own loudness level, avoiding commercials and, for those with deteriorated hearing, the ability to rewind and/or to turn on subtitles. State-of-the-art home theaters also provide visual experiences that can be truly spectacular: large pictures with definition, contrast, and "snap". Audio quality can easily exceed that delivered in typical cinemas in terms of timbral accuracy and directional and spatial diversity. The new high-definition media with 7.1 channels would need to be downmixed (simplified) for reproduction in cinemas.

Experiencing a movie at home is just a starting point. Popular movies have already spawned games. Adventurous minds have created interactive movies, in which audiences get to choose plot turns. The next step may be that the staged part of the movie segues into and out of game sequences in which audience members become avatar participants in the story. Again, looking at movie industry statistics for 2006, 39% of moviegoers had a "big screen" (35+ in., 0.9+ m), 32% had what they called "home theater," and 54% had a game system. This is the current and future home theater customer. The traditional closed-plot films will remain, but who can say what variants are waiting to happen? All are likely to happen in homes not in large venues, and intensive multichannel audio will be a part of all of them.

The logical move is for the movie industry to do the final mastering of their product in a well set up home-theater-like environment. In fact, it is already happening, but inconsistently and without uniform standards. A recent interview with Margouleff, Biles, and Thiele (Reber and Richelieu, 2008) provides insight into the still dynamic situation.

The industry lacks meaningful standards for the design, equipping, and calibration of monitoring and playback circumstances, so trial-and-error experimentation and opinions are providing guidance. In principle, there is no difference between the *basic* requirements for monitoring the progress of a music or movie program and for playback in the home. An exaggeration: substitute a reclining chair and a glass of wine for an ergonomic chair and a mixing console. The electronic paraphernalia of the creative process must be accommodated of course, and that will require some clever acoustical design. But the audio playback system must, in the end, closely resemble that of a good home theater: the same number of channels and loudspeaker placement, bass management and subwoofer(s), typical upmix algorithms to audition signals with lower channel counts, and so on. Otherwise, the circle of confusion, shown in Figure 2.3 simply acquires more dimensions within which differences can occur, and customers are further distanced from the creative process.

As discussed in Section 18.2.6, the calibration procedure used within the motion picture industry leaves much to be desired, but there is at least something in place. The music industry is operating without *any* meaningful guidance—sailing without a compass. It has prospered because of good instincts and seat-of-the-pants judgment on the part of experienced engineers. Some of the existing international standards that could be used are so flawed that if one removed all elements that we know to be wrong or inappropriate, there would be almost nothing left (see Section 18.5.1). Attempts by industry organizations to establish performance metrics for consumer audio equipment and recommended practices for home theater installations are greatly encouraging. It is time somebody did something.

Our goal should be to ensure that the excellence of one listening experience has a real possibility of being replicated in other places at other times. Some changes will be necessary in the way things are done, and change is not easy. It requires strong action, strong resolve, and good organization. Any move in the right direction is beneficial. If anyone is interested, the contents of this book might help.

References

Aarts, R.M. (**1992**). "A Comparison of Some Loudness Measures for Loudspeaker Listening Tests," *J. Audio Eng. Soc.*, **40**, pp. 142–146.

AES Staff Writer (**2004**). "New Horizons in Listening Test Design," *J. Audio Eng. Soc.*, **52**, pp. 65–73.

Alexander, R.C. (**1999**). *The Inventor of Stereo—The Life and Works of Alan Dower Blumlein*, Focal Press, Oxford.

Allen, Ioan (**2006**). "The X-Curve: Its Origins and History"; *SMPTE Motion Imaging Journal*, Vol. 115, No. 7&8 (July/August).

Allen, J.F. (**2001**). "The Mythical 'X' Curve," originally in *Boxoffice Magazine*, available at www.hps4000.com/pages/general/the_mythical_X_curve.pdf.

Allison, R.F. (**1974**). "The Influence of Room Boundaries on Loudspeaker Power Output." *J. Audio Eng. Soc.*, **22**, pp. 314–320.

———. (**1975**). "The Sound Field in Home Listening Rooms, II." *J. Audio Eng. Soc.*, **24**, pp. 14–19.

Allison, R.F., and Berkovitz, R. (**1972**). "The Sound Field in Home Listening Rooms." *J. Audio Eng. Soc.*, **20**, pp. 459–469.

Ando, Y. (**1977**). "Subjective Preference in Relation to Objective Parameters of Music Sound Fields with a Single Echo," *J. Acoust. Soc. Am.*, **62**, pp. 1436–1441.

———. (**1978**). "Subjective Preference Tests of Simulated Sound Fields for Optimum Design of Concert Halls," *J. Acoust. Soc. Am.*, **63**, p.85 (abstract only).

———. (**1985**). *Concert Hall Acoustics*, Springer-Verlag, Berlin.

———. (**1998**). *Architectural Acoustics*, Springer-Verlag, N.Y.

Ando, Y., Sakai, H., and Sato, S. (**2000**). "Formulae describing subjective attributes for sound fields based on a model of the auditory-brain system," *J. Sound & Vib.*, **232**, pp. 101–127.

Angus, J.A.S. (**1997**). "Controlling Early Reflections Using Diffusion." 102nd Convention, *Audio Eng. Soc.*, Preprint 4405.

———. (**1999**). "The Effects of Specular Versus Diffuse Reflections on the Frequency Response at the Listener." 106th Convention, *Audio Eng. Soc.*, Preprint 4938.

Arau-Puchades, H. (**1988**). "An Improved Reverberation Formula," *Acustica*, **65**, pp. 163–180.

Ashihara, K. (**2007**). "Hearing Thresholds for Pure Tones Above 16 kHz," *J. Acoust. Soc. Am.*, **122**, pp. EL 52–57. JASA Express Letters.

ASHRAE Handbook (**2003**). Chapter 47, "Sound and Vibration Control," pp. 47.1–47.50.

Augspurger, G.L. (**1990**). "Loudspeakers in Control Rooms and Living Rooms," 8th International Conference, *Audio Eng. Soc.*

Backus, J. (**1969**). *The Acoustical Foundations of Music*, Norton, New York.

Bale, C. (**2006**). "1080p Does Matter—Here's When (Screen Size vs. Viewing Distance vs. Resolution)," www.carletonbale.com.

Ballagh, K.O. (**1983**). "Optimum Loudspeaker Placement Near Reflecting Planes," *J. Audio Eng. Soc.*, **31**, pp. 931–935. See also letters to the editor, **32**, p. 677 (1984).

Barron, M. (**1971**). "The Subjective Effects of First Reflections in Concert Halls—The Need for Lateral Reflections," *J. Sound Vib.*, **15**, pp. 475–494.

Barron, M. (**2000**). "Measured Early Lateral Energy Fractions in Concert Halls and Opera Houses," *J. Sound Vib.*, **232**, pp. 79–100.

Barron, M., and Marshall, A.H. (**1981**). "Spatial Impression Due to Early Lateral Reflections in Concert Halls: The Derivation of a Physical Measure," *J. Sound and Vibration*, **77**, pp. 211–232.

Baskind, A., and Polack, J.-D. (**2000**). "Sound Power Radiated by Sources in Diffuse Field." 108th Convention, *Audio Eng. Soc.* Preprint 5146.

BBC Research Department (**1971**). "Low-Frequency Sound Absorbers," Report number: 1971/15.

Bech, S. (**1992**). "Selection and Training of Subjects for Listening Tests on Sound-Reproducing Equipment," *J. Audio Eng. Soc.*, **40**, pp. 590–610.

———. (**1996**). "Timbral Aspects of Reproduced Sound in Small Rooms, II," *J. Acoust. Soc. Am.*, **99**, pp. 3539–3549.

———. (**1998**). "Spatial Aspects of Reproduced Sound in Small Rooms," *J. Acoust. Soc. Am.*, **103**, pp. 434–445.

Bech, S., and Zacharov, N. (**2006**). "Perceptual Audio Evaluation," John Wiley and Sons, Ltd., London.

Benade, A.H. (**1984**). "Wind Instruments in the Concert Hall." Text of an oral presentation at Parc de la Villette, Paris; part of a series of lectures entitled "Acoustique, Musique, Espaces," 15 May 1984 (personal communication).

———. (**1985**). "From Instrument to Ear in a Room: Direct or via Recording," *J. Audio Eng. Soc.*, **33**, pp. 218–233.

———. (**1986**). "Critique of Certain Concert-Hall Design Criteria," Notes from an oral presentation at Anaheim, CA meeting of the *Acoust. Soc. Am.*, Dec. 1986 (personal communication).

Benjamin, E. (**1994**). "Audio Power Amplifiers for Loudspeaker Loads," *J. Audio Eng. Soc.*, **42**, pp. 670–683.

Benjamin, E., and Gannon, B. (**2000**). "Effect of Room Acoustics on Subwoofer Performance and Level Setting." 109th Convention, *Audio Eng. Soc.*, Preprint 5232.

Beranek, L.L. (**1962**). *Music, Acoustics and Architecture*, John Wiley and Sons, New York.

———. (**1969**). "Audience and Chair Absorption in Large Halls, II," *J. Acoust. Soc. Am.*, **45**, pp. 13–19.

———. (**1986**). *Acoustics*, Acoustical Society of America, New York.

———. (**1989**). "Balanced Noise-Criterion (NCB) Curves," *J. Acoust. Soc. Am.*, **86**, pp. 650–664.

———. (**2004**). *Concert Halls and Opera Houses*, 2nd ed. Springer-Verlag, New York.

Bilsen, F.A (**1977**). "Pitch of Noise Signals: Evidence for a "Central Spectrum," *J. Acoust. Soc. Am.*, vol. 61, pp. 150–161.

Bilsen, F.A., and Kievits, I. (**1989**). "The Minimum Integration Time of the Auditory System," 86th Convention, *Audio Eng. Soc.*, Preprint 2746.

Bishop, J. (**2007**). "Selling the 2.35 Difference," *CEPro*, **15**, pp. 105–106.

Blauert, J. (**1996**). *Spatial Hearing: The Psychophysics of Human Sound Localization*, MIT Press, Cambridge, Mass.

Blauert, J., and Divenyi, P.L. (**1988**). "Spectral Selectivity in Binaural Contralateral Inhibition," *Acustica*, **66**, pp. 267–274.

Blech, D., and Yang, Min-Chi (**2004**). "DVD-Audio versus SACD," 116th Convention, *Audio Eng. Soc.*, Preprint 6086.

Blesser, B., and Salter, L.-R. (**2007**). *Spaces Speak, Are You Listening?*, MIT Press, Cambridge, Mass.

Blumlein, A. (**1933**). British Patent No. 394 325 "Improvements in and Relating to Sound Transmission, Sound-Recording and Sound-Reproducing Systems." Granted to Alan Blumlein and EMI, 1933. Reprinted in *J. Audio Eng. Soc.*, **6**, pp. 91–98, 130, 1958.

Bolt, R.H. (**1946**). "Note on Normal Frequency Statistics for Rectangular Rooms," *J. Acoust. Soc. Am.*, **18**, pp. 130–133.

Bolt, R.H., and Doak, P.E. (**1950**). "A Tentative Criterion for the Short-Term Transient Response of Auditoriums," *J. Acoust. Soc. Am.*, **22**, pp. 507–509.

Bonello, O.J. (**1981**). "A New Criterion for the Distribution of Normal Room Modes," *J. Audio Eng. Soc.*, **29**, pp. 597–606.

Braasch, J., Blauert, J., and Djelani, T. (**2003**). "The Precedence Effect for Noise Bursts of Different Bandwidths. I. Psychoacoustical Data," *J. Acoust. Soc. Japan*, **24**, pp. 233–241.

Bradley, J.S. (**1986**). "Acoustical Measurements in Some Canadian Homes," *Canadian Acoustics*, **14**, no. 4.

———. (**1991**). "Some Further Investigations of the Seat Dip Effect," *J. Acoust. Soc. Am.*, **90**, pp. 324–333.

———. (**1997**). "Sound Absorption of Gypsum Board Cavity Walls," *J. Audio Eng. Soc.*, **45**, pp. 253–259.

Bradley, J.S., and Soulodre, G.A. (**1995**). "The Influence of Late Arriving Energy on Spatial Impression," *J. Acoust. Soc. Am.*, **97**, pp. 2263–2271.

Bradley, J.S., Reich, R.D., and Norcross, S.G. (**2000**). "On the Combined Effects of Early- and Late-Arriving Sound on Spatial Impression in Concert Halls," *J. Acoust. Soc. Am.*, **108**, pp. 651–661.

Bradley, J.S., Sato, H., and Picard, M. (**2003**). "On the Importance of Early Reflections for Speech in Rooms," *J. Acoust. Soc. Am.*, **113**, pp. 3233–3244.

Briggs, G.A. (**1958**). *Loudspeakers*, 5th ed. Wharfedale Wireless Works, Ltd., Idle, Bradford, U.K.

Broner, N. (**2004**). "Rating and Assessment of Noise," Australian Institute of Refrigeration, Air Conditioning and Heating Conference. Available at www.AIRAH. org.au

Bronkhorst, A.W., and Houtgast, T. (**1999**). "Auditory Distance Perception in Rooms," *Nature*, **397**, pp. 517–520.

Buchlein, R. (**1962**). "The Audibility of Frequency Response Irregularities," reprinted in English in *J. Audio Eng. Soc.*, **29**, pp. 126–131 (1981).

Burgtorf, W. (**1961**). "Untersuchungen zur Wahrnehmbarkeit Verzögerter Schallsignale," *Acustica*, **11**, pp. 97–111.

Buus, S. (**1999**). "Temporal Integration and Multiple Looks, Revisited: Weights as a Function of Time," *J. Acoust. Soc. Am.*, **105**, pp. 2466–2475.

Carver, R. (**1982**). "Sonic Holography," *Audio*, **66**, March.

Case, A.U. (**2001**). "An Investigation of the Spectral Effect of Multiple Early Reflections," *J. Acoust. Soc. Am.*, **109**, p. 2303 (Abstract only).

Celestinos, A., and Nielsen, S.B. (**2005**). "Optimizing Placement and Equalization of Multiple Low Frequency Loudspeakers in Rooms," 119th Convention, *Audio Eng. Soc.*, Preprint 6545.

———. (**2006**). "Low Frequency Sound Field Enhancement System for Rectangular Rooms Using Multiple Low Frequency Loudspeakers," 120th Convention, *Audio Eng. Soc.*, Preprint 6688.

Choisel, S., and Wickelmaier, F. (**2007**). "Evaluation of Multichannel Reproduced Sound: Scaling Auditory Attributes Underlying Listener Preferences," *J. Acoust. Soc. Am.*, **121**, pp. 388–400.

Chu, W.T., and Warnock, A.C.C. (**2002**). "Detailed Directivity of Sound Fields Around Human Talkers," IRC Report 104, National Research Council Canada, http://irc.nrc-cnrc.gc.ca/pubs/rr/rr104.

Clark, D.L. (**1981**). "High Resolution Subjective Testing Using a Double-Blind Comparator." 69th Convention, *Audio Eng. Soc.*, Preprint 1771.

———. (**1983**). "Measuring Audible Effects of Time Delays in Listening Tests," 74th Convention, *Audio Eng. Soc.*, Preprint 2012.

———. (**1991**). "Ten Years of A/B/X Testing," 91st Convention, *Audio Eng. Soc.*, Preprint 3167.

Clark, M. Jr., Luce, D., Abrams, R., Schlossberg, H., and Rome, J. (**1963**). "Preliminary Experiments on the Aural Significance of Parts of Tones of Orchestral Instruments and on Choral Tones," *J. Audio Eng. Soc.*, **11**, pp. 45–54.

Coleridge, S.T. (**1817**). "Biographia Literaria," Samuel Taylor Coleridge, 1772–1834, English poet, critic, and philosopher.

Cooper, D.H., and Shiga, T. (**1972**). "Discrete-Matrix Multichannel Stereo," *J. Audio Eng. Soc.*, **20**, pp. 346–360.

Corey, J., and Woszczyk, W. (**2002**). "Localization of Lateral Phantom Images in a 5-Channel System With and Without Simulated Early Reflections." 113th Convention, *Audio Eng. Soc.*, Preprint 5673.

Cox, T., and D'Antonio, P. (**2004**). *Acoustic Absorbers and Diffusers*, Spon Press, London & N.Y.

Cox, T., D'Antonio, P., and Avis, M.R. (**2004b**). "Room Sizing and Optimization at Low Frequencies," *J. Audio Eng. Soc.*, **52**, pp. 640–651.

Craven, P.G., and Gerzon, M.A. (**1992**). "Practical Adaptive Room and Loudspeaker Equalizer for Hi-Fi Use," 92nd Convention, *Audio Eng. Soc.*, Preprint 3346. Also: UK DSP Conference, Paper DSP-12.

Cremer, L., and Müller, H.A. (Translation by T.J. Schultz) (**1982**). *Principles and Applications of Room Acoustics, Vols. 1 & 2*, Applied Science Publishers, London.

Crompton, T.W.J. (**1974**). "The Subjective Performance of Various Quadraphonic Matrix Systems," BBC Research Department Report No. BBC RD 1974/29, www.bbc.co.uk/rd/pubs.

Dalenbäck, B.-I. (**2000**). "Reverberation Time, Diffuse Reflection, Sabine and Computerized Prediction," http://www.rpginc.com/research/reverb01.htm.

D'Antonio, P. (**2008**). Personal communication, information from the second edition of Cox, T., and D'Antonio, P. (2008). *Acoustic Absorbers and Diffusers*, Taylor and Francis, London.

D'Antonio, P., and Cox, T. (**1998**). "Two Decades of Diffuser Design and Development, Part 1: Applications and Design, Part 2: Prediction, Measurement and Characterization," *J. Audio Eng. Soc.*, **46**, pp. 955–976 and 1075–1091.

D'Antonio, P., and Eger, D. (**1986**). "T_{60}—How Do I Measure Thee, Let Me Count the Ways," 81st Convention, *Audio Eng. Soc.*, Preprint 2368.

Darlington, P., and Avis, M.R. (**1996**). "Time/Frequency Response of a Room with Active Acoustic Absorption," 100th Convention, *Audio Eng. Soc.*, Preprint 4192.

Davies, W.J., Cox, T.J., and Lam, Y.W. (**1996**). "Subjective Perception of Seat Dip Attenuation," *Acustica*, **82**, pp. 784–792.

Deer, J.A., Bloom, P.J., and Preis, D. (**1985**). "New Results for Perception of Phase Distortion," 77th Convention, *Audio Eng. Soc.*, Preprint 2197.

Devantier, A. (**2002**). "Characterizing the Amplitude Response of Loudspeaker Systems," 113th Convention, *Audio Eng. Soc.*, Preprint 5638.

Djelani, T., and Blauert, J. (**2000**). "Some New Aspects of the Build-Up and Breakdown of the Precedence Effect," Proc. 12th Int. Symp. on Hearing, Shaker Publishing.

———. (**2001**). "Investigations into the Build-up and Breakdown of the Precedence Effect," *Acta Acustica*, **87**, pp. 253–261.

Dougharty, E.H. (**1973**). "Stereophony and the Musician," BBC Engineering, May, pp. 3–6.

Eargle, J. (**1960**). "Stereophonic Localization: An Analysis of Listener Reactions to Current Techniques," IRE Transactions on Audio, Sept.–Oct., pp. 174–178.

Eargle, J. (**1973**). "Equalizing the Monitoring Environment", *J. Audio Eng. Soc.*, **21**, pp. 103–107.

Eargle, J., and Foreman, C. (**2002**). "Audio Engineering for Sound Reinforcement," JBL Pro Audio Publications.

Eargle, J., Bonner, J., and Ross, D. (**1985**). "The Academy's New State-of-the-Art Loudspeaker System," *SMPTE Journal*, pp. 667–675.

Eargle, J., Mayfield, M., and Gray, D. (**1997**). "Improvements in Motion Picture Sound: The Academy's New Three-Way Loudspeaker System," *SMPTE Journal*, **106**, pp. 464–476.

EIA-J Electronic Industries Association of Japan, 4-Channel Stereophonic Study Committee (**1979**). "Listening Test on Stereophonic and Quadraphonic Reproduction Systems," document STC-004.

Engebretson, M., and Eargle, J. (**1982**). "Cinema Sound Reproduction Systems: Technology Advances and System Design Considerations," *SMPTE Journal*, **91**, pp. 1046–1057.

Everest, F.A. (**2001**). *Master Handbook of Acoustics*, 4th ed. McGraw-Hill, New York.

Fazenda, B.M., Avis, M.R., and Davies, W.J. (**2005**). "Perception of Modal Distribution Metrics in Critical Listening Spaces—Dependence on Rooms Aspect Ratios," *J. Audio Eng. Soc.*, **53**, pp. 1128–1141.

Fincham, L.R. (**1985**). "The Subjective Importance of Uniform Group Delay at Low Frequencies." *J. Audio Eng. Soc.*, **33**, pp. 436–439.

Fitzroy, D. (**1959**). "Reverberation Formula Which Seems to Be More Accurate with Nonuniform Distribution of Absorption," *J. Acoust. Soc. Am.*, **31**, pp. 893–897.

Flanagan, S., Moore, B.C.J., and Stone, M. (**2005**). "Discrimination of Group Delay in Clicklike Signals Presented via Headphones and Loudspeakers," *J. Audio Eng. Soc.*, **53**, pp. 593–611.

Fletcher, H., and Munson, W.A. (**1933**). "Loudness, Its Definition, Measurement and Calculation," *J. Acoust. Soc. Am.*, **5**, pp. 82–108.

Fletcher, J.A. (**1992**). "The Design of a Modular Sound Absorber for Very Low Frequencies," BBC Research Department. Report No. BBC RD 1992/10, www.bbc.co.uk/rd/pubs.

Flindell, I.H., McKenzie, A.R., Negishi, H., and Jewett, M. (**1991**). "Subjective Evaluations of Preferred Loudspeaker Directivity," 90th Convention, *Audio Eng. Soc.*, Preprint 3076.

Foreman, C. (**2002**). "Sound System Design." Chapter 34 in *Handbook for Sound Engineers*, 3rd ed. Ballou, G.M., editor, Focal Press.

Forsyth, M. (**1985**). *Buildings for Music: The Architect, the Musician and the Listener from the Seventeenth Century to the Present Day*, MIT Press, Cambridge, Mass.

Fryer, P.A. (**1975**). "Intermodulation Distortion Listening Tests," *J. Audio Eng. Soc.* (abstracts), **23**, p. 402.

———. (**1977**). "Loudspeaker Distortions, Can We Hear Them?," *Hi-Fi News Rec. Rev.*, vol. 22, pp. 51–56.

Gabrielsson, A., Frykholm, S.A., and Lindstrom, B. (**1979**). "Assessment of Perceived Sound Quality in High Fidelity Sound Reproducing Systems," Karolinska Institute, Stockholm. Rep. TA 93.

Gabrielsson, A., Hagerman, B., and Bech-Kristensen, T. (**1991**). "Perceived Sound Quality of Reproductions with Different Sound Levels," Karolinska Institute, Stockholm. Rep. TA 123.

Gander, M. (**1982**). "Ground-Plane Acoustic Measurement of Loudspeaker Systems," *J. Audio Eng. Soc.*, **30**, pp. 723–731.

———. (**1986**). "Dynamic Linearity and Power Compression in Moving-Coil Loudspeakers." *J. Audio Eng. Soc.*, **34**, pp. 627–646.

Gardner, M. (**1968**). "Historical Background of the Haas and/or Precedence Effect," *J. Acoust. Soc. Am.*, **43**, pp. 1243–1248.

———. (**1969**). "Image Fusion, Broadening, and Displacement in Sound Localization," *J. Acoust. Soc. Am.*, **46**, pp. 339–349.

———. (**1973**). "Some Single- and Multiple-Source Localization Effects," *J. Audio Eng. Soc.*, **21**, pp. 430–437.

Geddes, E.R. (**1982**). "An Analysis of the Low-Frequency Sound Field in Non-Rectangular Enclosures Using the Finite Element Method," PhD Thesis, Pennsylvania State University.

———. (**2002**). *Premium Home Theater: Design and Construction.* GedLee LLC, Novi, Michigan, USA. www.gedlee.com.

———. (**2005**). "Audio Acoustics in Small Rooms," a PowerPoint presentation available at www.gedlee.com.

Geddes, E.R., and Lee, L.W. (**2003**). "Auditory Perception of Nonlinear Distortion—Theory," 115th Convention, *Audio Eng. Soc.*, Preprint 5890.

Genereux, R.P. (**1992**). "Signal Processing Considerations for Acoustic Environment Correction," U.K. DSP Conference, *Audio Eng. Soc.*, Paper DSP-14.

Gerzon, M. (**1983**). "Ambisonics in Multichannel Broadcasting and Video," 74th Convention, *Audio Eng. Soc.*, Preprint 2034.

Gilford, C.L.S. (**1959**). "The Acoustic Design of Talks Studios and Listening Rooms," *Proc. Inst. of Electrical Eng.*, **106**, pp. 245–258. Reprinted in *J. Audio Eng. Soc.*, **27**, pp. 17–31, 1979.

Glasgal, R. (**2001**). "Ambiophonics," 111th Convention, *Audio Eng. Soc.*, Preprint 5426.

———. (**2003**). "Surround Ambiophonic Recording and Reproduction," 24th International Conference, *Audio Eng. Soc.*, Paper 42.

Gould, G. (**1966**). "The Prospects of Recording," *High Fidelity Magazine*, April, pp. 46–63.

Gover, B.N., Ryan, J.G., and Stinson, M.R. (**2004**). "Measurements of Directional Properties of Reverberant Sound Fields in Rooms Using a Spherical Microphone Array," *J. Acoust. Soc. Am.*, **116**, pp. 2138–2148.

Greenfield, R., and Hawksford, M. (**1990**). "The Audibility of Loudspeaker Phase Distortion." 88th Convention, *Audio Eng. Soc.*, Preprint 2927.

Greiner, R.A. (**1980**). "Amplifier-Loudspeaker Interfacing," *J. Audio Eng. Soc.*, **28**, pp. 310–315.

Griesinger, D. (**1989**). "Theory and Design of a Digital Audio Signal Processor for Home Use," *J. Audio Eng. Soc.*, **37**, pp. 40–50. See also Lexicon CP-1 owners manual, available for download at www.lexicon.com.

———. (**1997**). "The Psychoacoustics of Apparent Source Width, Spaciousness and Envelopment in Performance Spaces." *Acta Acustica*, **83**, pp. 721–731.

———. (**1998**). "General Overview of Spatial Impression, Envelopment, Localization, and Externalization," 15th International Conference, *Audio Eng. Soc.*, Paper 15-013.

———. (**1999**). "Objective Measures of Spaciousness and Envelopment." 16th International Conference, *Audio Eng. Soc.*, Paper 16-003.

———. (**2001**). "The Psychoacoustics of Listening Area, Depth and Envelopment in Surround Recordings, and Their Relationship to Microphone Technique." 19th International Conference, *Audio Eng. Soc.*, Paper 1913.

Griffin, J.R. (**2003**). "Design Guidelines for Practical Near Field Line Arrays," http://www.audiodiycentral.com/resource/pdf/nflawp.pdf.

Haas, H. (**1972**). "The Influence of a Single Echo on the Audibility of Speech," Doctoral dissertation, University of Göttingen. Reprinted in *J. Audio Eng. Soc.*, **20**, pp. 146–159, 1972. A reprint of a 1949 translation of Haas's PhD dissertation.

Hafler, D. (**1970**). "A New Quadraphonic System," *Audio*, July.

Hamasaki, E., Hiyama, K., Nishiguchi, T., and Ono, K. (**2004**). "Advanced Multichannel Audio Systems with Superior Impression of Presence and Reality," 116th Convention, *Audio Eng. Soc.*, Preprint 6053.

Hansen, V., and Madsen, E.R. (**1974a**). "On Aural Phase Detection," *J. Audio Eng. Soc.*, **22**, pp. 10–14.

———. (**1974b**). "On Aural Phase Detection: Part II," *J. Audio Eng. Soc.*, **22**, pp. 783–788.

Harris, C.M (**1955**). "Acoustical Properties of Carpet," *J. Acoust. Soc. Am.*, **27**, pp. 1077–1082.

Harris, C.M., editor (**1991**). *Handbook of Acoustical Measurements and Noise Control*, 3rd ed. McGraw-Hill, Inc.

Harvith, J., and Harvith, S. (**1987**). *Edison, Musicians and the Phonograph*, Greenwood Press, N.Y.

Hidaka, T., Beranek, L.L., and Okano, T. (**1997**). *Some Considerations of Interaural Cross Correlation and Lateral Fraction as Measures of Spaciousness in Concert Halls*. Chapter 32 in *Music and Concert Hall Acoustics*, Ando, Y., and Noson, D. editors, Academic Press, London.

Hiyama, K., Komiyama, S., and Hamasaki, K. (**2002**). "The Minimum Number of Loudspeakers and Its Arrangement for Reproducing the Spatial Impression of Diffuse Sound Field," 113th Convention, *Audio Eng. Soc.*, Preprint 5674.

Hodgson, M. (**1983**). "Measurements of the Influence of Fittings and Roof Pitch on the Sound Field in Panel Roof Factories," *Applied Acoustics*, **16**, pp. 369–391.

———. (**1998**). "Experimental Evaluation of Simplified Models for Predicting Noise Levels in Industrial Workrooms," *J. Acoust. Soc. Am.*, **103**, pp. 1933–1939.

Holman, T. (**1991**). "New Factors in Sound for Cinema and Television," *Jour. Audio Eng. Soc.*, **39**, pp. 529–539.

———. (**1993**). "Motion-Picture Theater Sound System Performance: New Studies of the B-Chain," *SMPTE Journal*, **103**, pp. 136–149.

———. (**1996**). "The Number of Audio Channels," 100th Convention, *Audio Eng. Soc.* Preprint 4292.

———. (**1998**). "The Basics of Bass Management," *Surround Professional*, **1**, October.

———. (**2001**). "The Number of Loudspeaker Channels," 19th International Conference, *Audio Eng. Soc.*, Paper No. 1906.

———. (**2007**). "Cinema Electro-Acoustic Quality Redux," *SMPTE Motion Imaging Journal*, May/June.

———. (**2008**). "Surround sound: up and running", Second edition. Focal Press, Oxford.

Howard, K. (**2005**). "Time Dilation, Part 1," *Stereophile*, **28**, Jan.; "Part 2," *Stereophile*, **28**, April. Available at www.stereophile.com/features/105kh/index.html, and www.stereophile.com/reference/405time/index.html.

———. (**2006**). "Wayward Down Deep," *Stereophile*, **29**, July. Available at www.stereophile.com/reference/706deep/index.html.

———. (**2007**). "Heavy Load: How Loudspeakers Torture Amplifiers," *Stereophile*, **30**, July. Available at www.stereophile.com/reference/707heavy/index.html.

ISO 226 (**2003**). "Acoustics—Normal Equal-Loudness Contours."

ISO 2969 (**1987**). "International Standard. Cinematography—B-Chain Electroacoustic Response of Motion-Picture Control Rooms and Indoor Theaters—Specifications and Measurements."

ITU-R Recommendation BS.775–2 (**2006**). "Multichannel Stereophonic Sound System With and Without Accompanying Picture."

JBL Professional (**2003**). "Cinema Sound System Manual," available at http://www.jblpro.com/pub/cinema/cinedsgn.pdf.

Jones, D. (**2003**). "A Review of the Pertinent Measurements and Equations for Small Room Acoustics," *J. Acoust. Soc. Am.*, **113**, p. 2273 (abstract only). Personal communication: presentation text from the author.

Julstrom, S. (**1987**). "A High-Performance Surround Sound Process for Home Video," *J. Audio Eng. Soc.*, **35**, pp. 536–549.

Kantor, K.L., and de Koster, A.P. (**1986**). "A Psychoacoustically Optimized Loudspeaker." *J. Audio Eng. Soc.*, **34**, pp. 990–996.

Kates, J.M. (**1960**). "Optimum Loudspeaker Directional Patterns," *J. Audio Eng. Soc.*, vol. 28, pp. 787–794.

Katz, B. (**2002**). *Mastering Audio: The Art and the Science*, Focal Press, Oxford.

Katz, M. (**2004**). *Capturing Sound: How Technology Has Changed Music*, University of California Press, Berkeley and Los Angeles, California.

Keele, D.B., and Button, D.J. (**2005**). "Ground-Plane Constant Beamwidth Transducer (CBT) Loudspeaker Circular-Arc Line Arrays," 119th Convention, *Audio Eng. Soc.*, Preprint 6594.

Kishinaga, S., Shimizu, Y., Ando, S., and Yamaguchi, K. (**1979**). "On the Room Acoustic Design of Listening Rooms," 64th Convention, *Audio Eng. Soc.*, Preprint 1524.

Klippel, W. (**1990a**). "Multidimensional Relationship between Subjective Listening Impression and Objective Loudspeaker Parameters," *Acustica*, **70**, pp. 45–54.

———. (**1990b**). "Assessing the Subjectively Perceived Loudspeaker Quality on the Basis of Objective Parameters," 88th Convention, *Audio Eng. Soc.*, Preprint 2929.

Koebel, A. (**2007**). "Six Myths of the High-Definition Age," *Widescreen Review*, issue 123, September, pp. 52–60.

Koidan, W., Hruska, G.A., and Pickett, M.A. (**1972**). "Wedge Design for National Bureau of Standards Anechoic Chamber," *J. Acoust. Soc. Am.*, **52**, pp. 1071–1076.

Krauss, G.J. (**1990**). "On the Audibility of Group Distortion at Low Frequencies," 88th Convention, *Audio Eng. Soc.*, Preprint 2894.

Krumbholtz, K., Maresh, K., Tomlinson, J., Patterson, R.D., Seither-Preisler, A., and Lüthenhoner, B. (**2004**). "Mechanisms Determining the Salience of Coloration in Echoed Sound: Influence of Interaural Time and Level Differences," *J. Acoust. Soc. Am.*, **115**, pp. 1696–1704.

Kuhl, W., and Plantz, R. (**1978**). "The Significance of the Diffuse Sound Radiated From Loudspeakers for the Subjective Hearing Event," *Acustica*, **40**, pp. 182–190.

Kurozumi, K., and Ohgushi, K. (**1983**). The Relationship between the Cross-Correlation Coefficient of Two-Channel Acoustic Signals and Sound Image Quality," *J. Acoust. Soc. Am.*, **74**, pp. 1726–1733.

Kuttruff, H. (**1991**). *Room Acoustics*, 3rd ed. E & FN Spon, London.

———. (**1998**). "Sound Fields in Small Rooms," 15th Conference, *Audio Eng. Soc.*, Paper 15-002.

Kwon, Y., and Siebein, G.W. (**2007**). "Chronological Analysis of Architectural and Acoustical Indices in Music Performance Halls," *J. Acoust. Soc. Am.*, Vol. 121, pp. 2691–2699.

Lee, L.W., and Geddes, E.R. (**2003**). "Auditory Perception of Nonlinear Distortion," 115th Convention, *Audio Eng. Soc.*, Preprint 5891.

———. (**2006**). "Audibility of Linear Distortion with Variations in Sound Pressure Level and Group Delay," 115th Convention, *Audio Eng. Soc.*, Preprint 6888.

Linkwitz, S. (**1998**). "Investigation of Sound Quality Differences between Monopolar and Dipolar Woofers in Small Rooms," 105th Convention, *Audio Eng. Soc.*, Preprint 4786.

Lipshitz, S. (**1990**). "The Great Debate—Some Reflections Ten Years Later." 8th International Conference, *Audio Eng. Soc.*, Paper 8-016.

Lipshitz, S., and Vanderkooy, J. (**1981**). "The Great Debate: Subjective Evaluation," *J. Audio Eng. Soc.*, **29**, pp. 482–491.

———. (**1986**). "The Acoustic Radiation of Line Sources of Finite Length," 81st Convention, *Audio Eng. Soc.*, Preprint 2417.

Lipshitz, S., Pocock, M., and Vanderkooy, J. (**1982**). "On the Audibility of Midrange Phase Distortion in Audio Systems," *J. Audio Eng. Soc.*, **30**, pp. 580–595. Comments by Shanefield, D., and authors' reply (**1983**), **31**, pp. 447–448. Comments by Moir, J. (**1983**), **31**, p. 939. More comments by van Maanen, H.R.E., Shanefield, D., and Moir, J. (**1985**), **33**, pp. 806–808.

Litovsky, R.Y., Colburn, H.S., Yost, W.A., and Guzman, S.J. (**1999**). "The Precedence Effect," *J. Acoust. Soc. Am.*, **106**, pp. 1633–1654.

Lochner, J.P.A., and Burger, J.F. (**1958**). "The Subjective Masking of Short Time-Delayed Echoes by Their Primary Sounds and Their Contribution to the Intelligibility of Speech," *Acustica*, **8**, pp. 1–10.

London, S.J. (**1963**). "The Origins of Psychoacoustics," *High Fidelity Magazine*, April, pp. 44–47, 117.

Long, M. (**2006**). *Architectural Acoustics*, Elsevier Academic Press, New York.

Lonsbury-Martin, B.L., and Martin, G.K. (**2007**). "Modern Music-Playing Devices as Hearing Health Risks," *Acoustics Today*, **3**, pp. 16–19.

Louden, M.M. (**1971**). "Dimension-Ratios of Rectangular Rooms with Good Distribution of Eigentones," *Acustica*, **24**, pp. 101–104.

Mäkivirta, A.V., and Anet, C. (**2001a**). "The Quality of Professional Surround Audio Reproduction—A Survey Study," 19th International Conference, *Audio Eng. Soc.*, Paper 1914.

———. (**2001b**). "A Survey Study of In-Situ Stereo and Multi-Channel Monitoring Conditions," 111th Convention, *Audio Eng. Soc.*, Preprint 5496.

Martens, W.L., Braasch, J., and Woszczyk, W. (**2004**). "Identification and Discrimination of Listener Percepts Associated with Multiple Low-Frequency Signals in Multichannel Sound Reproduction," 117th Convention, *Audio Eng. Soc.*, Preprint 6229.

Martinez, C., and Gilman, S. (**1988**). "Results of the 1986 AES Audiometric Survey," *J. Audio Eng. Soc.*, **36**, pp. 686–690.

Meyer, E. (**1954**). "Definition and Diffusion in Rooms," *J. Acoust. Soc. Am.*, **26**, pp. 630–636.

Meyer, J. (**1972**). "Directivity of the Bowed Stringed Instruments and Its Effect on Orchestral Sound in Concert Halls," *J. Acoust. Soc. Am.*, **51**, pp. 1994–2009.

———. (**1978**). "Acoustics and the Performance of Music," Verlag das Musikinstrument, Frankfurt am Main.

———. (**1993**). "The Sound of the Orchestra," *J. Audio Eng. Soc.*, **41**, pp. 203–213.

Meyer, E.B., and Moran, D.R. (**2007**). "Audibility of a CD-Standard A/D/A Loop Inserted into High-Resolution Audio Playback," *J. Audio Eng. Soc.*, **55**, pp. 775–779.

Meyer, E., and Schodder, G.R. (**1952**). "On the Influence of Reflected Sound on Directional Localization and Loudness of Speech," *Nachr. Akad. Wiss. Gottingen, Math. Phys. Klasse IIa*, **6**, pp. 31–42.

Miyasaka, E. (**1999**). "Consideration on Perceptual Effects of Sounds with Ultra-high Frequencies," *J. Acoust. Soc. Japan*, **8**, pp. 569–572.

Moore, B.C.J. (**2003**). *An Introduction to the Psychology of Hearing*, 5th ed. Academic Press, London.

Moore, B.C.J., and Tan, C-T. (**2004**). "Development and Validation of a Method for Predicting the Perceived Naturalness of Sounds Subjected to Spectral Distortion," *J. Audio Eng. Soc.*, **52**, pp. 900–914.

Moore, B.C.J., Tan, C-T., Zacharov, N., and Mattila, V-V. (**2004**). "Measuring and Predicting the Perceived Quality of Music and Speech Subjected to Combined Linear and Nonlinear Distortion," *J. Audio Eng. Soc.*, **52**, pp. 1228–1244.

Morimoto, M. (**1997**). "The Role of Rear Loudspeakers in Spatial Impression," 103rd Convention, *Audio Eng. Soc.*, Preprint 4554.

Morton, D. (**2000**). *Off the Record*, Rutgers University Press, New Brunswick, N.J.

———. (**2004**). *Sound Recording: The Life Story of a Technology*, Greenwood Press, Westport, Connecticut.

Moulton, D. (**1995**). "The Significance of Early High-Frequency Reflections from Loudspeakers in Listening Rooms," 99th Convention, *Audio Eng. Soc.*, Preprint 4094.

———. (**2003**). "The Loudspeaker as Musical Instrument: An Examination of the Issues Surrounding Loudspeaker Performance of Music in Typical Rooms," *J. Acoust. Soc. Am.*, **113**, p. 2215 (abstract only). Text available at www.moultonlabs.com.

Moulton, D., Ferralli, M., Hebrock, S., and Pezzo, M. (**1986**). "The Localization of Phantom Images in an Omnidirectional Stereophonic Loudspeaker System," 81st Convention, *Audio Eng. Soc.*, Preprint 2371.

Muncey, R.W., Nickson, A.F.B., and Dubout, P. (**1953**). "The Acceptability of Speech and Music with a Single Artificial Echo," *Acustica*, **3**, pp. 168–173.

Muraoka, T., and Nakazato, T. (**2007**). "Examination of Multichannel Sound Field Recomposition Utilizing Frequency Dependent Interaural Cross Correlation (FIACC)," *J. Audio Eng. Soc.*, **55**, pp. 236–256.

Nakahara, M., and Omoto, A. (**2003**). "Room Acoustic Design for Small Multichannel Studios," 24th Conference, *Audio Eng. Soc.*, Paper 40.

Nakajima, T., and Ando, Y. (**1991**). "Effects of a single reflection with varied horizontal angle and time delay on speech intelligibility," *J. Acoust. Soc. Am.*, **90**, pp. 3173–3179.

Nakayama, T., Miura, T., Kosaka, O., Okamoto, M., and Shiga, T. (**1971**). "Subjective Assessment of Multichannel Reproduction." *J. Audio Eng. Soc.*, **19**, pp. 744–751.

Neilsen, S.H. (**1993**). "Auditory Distance Perception in Different Rooms," *J. Audio Eng. Soc.*, **41**, pp. 755–770.

Newell, P.R., and Holland, K.R. (**1997**). "A Proposal for a More Perceptually Uniform Control Room for Stereophonic Music Recording Studios." 103rd Convention, *Audio Eng. Soc.*, Preprint 4580.

Newell, P.R., and Holland, K.R. (**2007**). "Loudspeakers for music recording and reproduction," Focal Press, Oxford, U.K.

Nishiguchi, T., Hamasaki, K., Iwaki, M., and Ando, A. (**2003**). "Perceptual Discrimination between Musical Sounds with and without Very High Frequency Components," 115th Convention, Audio Eng. Soc., Preprint 5876.

Niskar, A.S., Kieszak, S.M., Holmes, A., Esteban, E., Rubin, C., and Brody, D.J. (**1998**). "Prevalence of Hearing Loss Among Children 6–19 Years of Age." *JAMA*, **279**, pp. 1071–1075. Available at www.jama.com.

Nousaine, T. (**1990**). "The Great Debate: Is Anyone Winning?," 8th International Conference, *Audio Eng. Soc.*, Paper 8-015.

Noxon, A.M. (**1985**). "Listening Room—Corner Loaded Bass Traps," 79th Convention, *Audio Eng. Soc.*, Paper no. B-12, no preprint no.

———. (**1986**). "Room Acoustics and Low Frequency Damping," 81st Convention, *Audio Eng. Soc.*, no preprint no.

Olive, S. (**2001**). "A New Listener Training Software Application," 110th Convention, *Audio Eng. Soc.*, Preprint No. 5384.

———. (**2003**). "Difference in Performance and Preference of Trained versus Untrained Listeners in Loudspeaker Tests: A Case Study," *J. Audio Eng. Soc.*, **51**, pp. 806–825.

———. (**2004a**). "A Multiple Regression Model for Predicting Loudspeaker Preference Using Objective Measurements: Part 1—Listening Test Results," 116th Convention, *Audio Eng. Soc.*, Preprint 6113.

———. (**2004b**). "A Multiple Regression Model for Predicting Loudspeaker Preference Using Objective Measurements: Part 2—Development of the Model," 117th Convention, *Audio Eng. Soc.*, Preprint 6190.

Olive, S.E. (**1990**). "The Preservation of Timbre; Microphones, Loudspeakers, Sound Sources and Acoustical Spaces," 8th International Conference, *Audio Eng. Soc.*, Paper D1.

———. (**1994**). "A Method for Training Listeners and Selecting Program Material for Listening Tests," 97th Convention, *Audio Eng. Soc.*, Preprint 3893.

———. (**2007**). "Interaction Between Loudspeakers and Room Acoustics Influences Loudspeaker Preferences in Multichannel Audio Reproduction," PhD Thesis, Schulich School of Music, McGill University, Montreal, Quebec, Canada.

Olive, S.E., and Martens, W.L. (**2007**). "Interaction Between Loudspeakers and Room Acoustics Influences Loudspeaker Preferences in Multichannel Audio Reproduction," 123rd Convention, *Audio Eng. Soc.*, Preprint 7196.

Olive, S.E., and Toole, F.E. (**1989a**). "The Detection of Reflections in Typical Rooms," *J. Audio Eng. Soc.*, **37**, pp. 539–553.

———. (**1989b**). "The Evaluation of Microphones—Part 1: Measurements," 87th Convention, *Audio Eng. Soc.*, Preprint 2837.

Olive, S.E., Castro, B., and Toole, F.E. (**1998**). "A New Laboratory for Evaluating Multichannel Audio Components and Systems," 105th Convention, *Audio Eng. Soc.*, Preprint 4842.

Olive, S.E., Schuck, P.L., Sally, S.L., and Bonneville, M. (**1995**). "The Variability of Loudspeaker Sound Quality Among Four Domestic Sized Rooms," 99th Convention, *Audio Eng. Soc.*, Preprint 4092.

Olson, H.F. (**1957**). *Acoustical Engineering*, Van Nostrand, republished in 1991 by Professional Audio Journals, Inc.

Otondo, F., Rindel, J.H., Caussé, R., Misdariis, N., and Caudra, P. (**2002**). "Directivity of Musical Instruments in a Real Performance Situation," Proc. Int. Symp. on Musical Acoustics, Mexico City, pp. 312–318.

Owsinski, B. (**1999**). *The Mixing Engineer's Handbook*, MixBooks, Vallejo, Calif.

Pedersen, J.A. (**2003**). "Adjusting a Loudspeaker to Its Acoustic Environment—The ABC System," 115th Convention, *Audio Eng. Soc.*, Preprint 5880.

Pellegrini, R.S. (**2002**). "Perception-Based Design of Virtual Rooms for Sound Reproduction," 22nd International Conference, *Audio Eng. Soc.*, Paper 000245.

Perrot, D.R., Marlborough, K., Merrill, P., and Strybel, T.S. (**1989**). "Minimum Audible Angle Thresholds Obtained Under Conditions in Which the Precedence Effect Is Assumed to Operate," *J. Acoust. Soc. Am.*, **85**, pp. 282–288.

Philbeck, J.W., and Mershon, D.H. (**2002**). "Knowledge about Typical Source Output Influences Perceived Auditory Distance," *J. Acoust. Soc. Am.*, **111**, pp. 1980–1983.

Plenge, G., Jakubowski, H., and Schone, P. (**1980**). "Which Bandwidth Is Necessary for Optimal Sound Transmission?," *J. Audio Eng. Soc.*, **28**, pp. 114–119.

Pulkki, V. (**2001**). "Coloration of Amplitude-Panned Virtual Sources," 110th Convention, *Audio Eng. Soc.*, Preprint 5402.

Queen, D. (**1979**). "The Effect of Loudspeaker Radiation Patterns on Stereo Imaging and Clarity," *J. Audio Eng. Soc.*, **27**, pp. 368–379.

Rakerd, B., and Hartmann, W.M. (**1985**). "Localization of Sound in Rooms, II: The Effects of a Single Reflecting Surface," *J. Acoust. Soc. Am.*, **78**, pp. 524–533.

Rakerd, B., Hartmann, W.M., and Hsu, J. (**2000**). "Echo Suppression in the Horizontal and Median Sagittal Planes," *J. Acoust. Soc. Am.*, **107**, pp. 1061–1064.

Ranada, D. (**2006**). "Maxing Out Resolution," *Sound and Vision Magazine*, Feb./Mar., p. 20. Also at www.soundandvisionmag.com.

Rasmussen, G. (**1972**). "Anechoic Sound Chambers," a Brüel and Kjaer document sent as a personal communication.

Ratliff, P.A. (**1974**). "Properties of Hearing Related to Quadraphonic Reproduction," BBC Research Department Report No. BBC RD 1974/38. www.bbc.co.uk/rd/pubs.

Read, O., and Welsh, W.L. (**1959**). *From Tin Foil to Stereo*, Howard Sams, Indianapolis, Indiana.

Reber, G., and Richelieu, D. (**2008**). "The Mi Casa Magicians, an interview with the team behind 7.1," Widescreen Review, **17**, no. 4, April, pp. 46–58. Also available at www.micasamm.com.

Rettinger, M. (**1968**). "Small Music Rooms," *Audio*, October, pp. 25, 87.

Robinson, A. (**2005**). "The Importance of Choosing the Right Cinema Screen, Part 2." Cinema Systems, June, pp. 38–42. Available at http://www.harkness-screens. us/ds/importance_part2_pr.pdf.

Robinson, D.W., and Dadson, R.S. (**1956**). "A Re-Determination of the Equal-Loudness Relations for Pure Tones," *British J. of App. Physics*, **7**, pp. 166–181.

Robinson, D.W., and Whittle, L.S. (**1960**). "The Loudness of Directional Sound Fields," *Acustica*, **10**, pp. 74–80.

Rubak, P., and Johansen, L.G. (**2000**). "Design and Evaluation of Digital Filters Applied to Loudspeaker/Room Equalization," 108th Convention, *Audio Eng. Soc.*, Preprint 5172.

Rumsey, F. (**1999**). "Controlled Subjective Assessments of Two-to-Five-Channel Surround Sound Processing Algorithms," *J. Audio Eng. Soc.*, **47**, pp. 563–582.

Rumsey, F., Zielinski, S., and Kassier, R. (**2005a**). "On the relative importance of spatial and timbral fidelities in judgments of degraded multichannel audio quality." *J. Acoust. Soc. Am.*, **118**, pp. 968–976.

Rumsey, F., Zielinski, S., Kassier, R., and Bech, S. (**2005b**). "Relationships Between Experienced Listener Ratings of Multichannel Audio Quality and Naive Listener Preferences," *J. Acoust. Soc. Am.*, **117**, pp. 3832–3840.

Saberi, K., and Perrot, D.R. (**1990**). "Lateralization Thresholds Obtained Under Conditions in Which the Precedence Effect Is Assumed to Operate," *J. Acoust. Soc. Am.*, **87**, pp. 1732–1737.

Saldanha, E.L., and Corso, J.F. (**1964**). "Timbral Cues and the Identification of Musical Instruments," *J. Acoust. Soc. Am.*, **36**, pp. 2021–2026.

Salvatti, A., Devantier, A., and Button, D. (**2002**). "Maximizing Performance from Loud-speaker Ports," *J. Audio Eng. Soc.*, **50**, pp. 19–45.

Sato, H., Bradley, J., and Masayuki, M. (**2005**). "Using Listening Difficulty Ratings of Conditions for Speech Communication in Rooms," *J. Acoust. Soc. Am.*, **117**, pp. 1157–1167.

Sato, H., and Bradley, J. (**2008**). "Evaluation of Acoustical Conditions for Speech Communication in Working Elementary School Classrooms, "*J. Acoust. Soc. Am.*, **123**, pp. 2064–2077.

Sayers, B. McA., and Toole, F.E. (**1964**). "Acoustic-Image Lateralization Judgments with Binaural Transients," *J. Acoust. Soc. Am.*, vol. 36, pp. 1199–1205.

Schoolmaster, M., Kopco, N., and Shinn-Cunningham, B.G. (**2003**). "Effects of Reverberation and Experience on Distance Perception in Simulated Environments," *J. Acoust. Soc. Am.*, **113**, p. 2285, abstract only.

———. (**2004**). "Auditory Distance Perception in Fixed and Varying Simulated Acoustic Environments," *J. Acoust. Soc. Am.*, **115**, p. 2459, abstract only.

Schrag, R. (**1992**). "Exposing Acoustical Myths—A Collection of Time-Honored Misconceptions," *Broadcast Engineering*, March. Also at www.rbdg.com/newsroom.

Schroeder, M.R. (**1954**). "Statistical Parameters of the Frequency Response Curves of Large Rooms," *Acustica*, **4**, pp. 594–600. Translated from the German original in *J. Audio Eng. Soc.*, **35**, pp. 299–306 (1987).

———. (**1996**). "'Schroeder Frequency' Revisited," *J. Acoust. Soc. Am.*, **99**, pp. 3240–3241.

Schultz, T.J. (**1963**). "Problems in the Measurement of Reverberation Time," *J. Audio Eng. Soc.*, **11**, pp. 307–317.

———. (**1983**). "Improved Relationship between Sound Power Level and Sound Pressure Level in Domestic and Office Spaces," Report No. 5290, Am. Soc. of Heating, Refrigerating and Air-Conditioning Engineers (ASHRAE), prepared by Bolt, Beranek, and Newman, Inc.

Schultz, T.J., and Watters, B.G. (**1964**). "Propagation of Sound Across Audience Seating," *J. Acoust. Soc. Am.*, **36**, pp. 885–896.

Self, D.R.G. (**1988**). "Science v. Subjectivism in Audio Engineering," *Electronics and Wireless World*, **94**, pp. 692–696. Also at http://www.dself.dsl.pipex.com/ampins/pseudo/subjectv.htm.

Sepmeyer, L.W. (**1965**). "Computed Frequency and Angular Distribution of the Normal Modes of Vibration in Rectangular Rooms," *J. Acoust. Soc. Am.*, **37**, pp. 413–423.

Seraphim, H.-P. (**1961**). "Über Die Wahrnehmbarkeit Mehrerer Rückwürfe von Sprachschall," *Acustica*, **11**, pp. 80–91.

Shinn-Cunningham, B.G. (**2001**). "Localizing Sound in Rooms," Proc. ACM SIGRAPH and EUROGRAPHICS Campfire: acoustic rendering for virtual environments, Snowbird, Utah. http://cns.bu.edu/~shinn/pages/RecentPapers.html.

———. (**2003**). "Acoustics and Perception of Sound in Everyday Environments," Proc. 3rd Int. Workshop on Spatial Media, Aisu-Wakamatsu, Japan. http://cns.bu.edu/~shinn/pages/RecentPapers.html.

Shirley, B.G., and Kendrick, P. (**2004**). "ITC Clean Audio Project," 116th Convention, *Audio Eng. Soc.*, Preprint no. 6027.

Shirley, B.G., Kendrick, P., and Churchill, C. (**2007**). "The Effect of Stereo Crosstalk on Intelligibility: Comparison of a Phantom Stereo Image and a Central Loudspeaker Source," *J. Audio Eng. Soc.*, **55**, pp. 852–863.

Shorter, D.E.L. (**1958**). "A Survey of Performance Criteria and Design Considerations for High-Quality Monitoring Loudspeakers," Proc. Institution of Electrical Engineers (U.K.), **105 p.B**, Paper 2604R, pp. 607–621.

Sivonen, V.P., and Ellermeier, W. (**2006**)."Directional Loudness in an Anechoic Sound Field, Head Related Transfer Functions, and Binaural Summation," *J. Acoust. Soc. Am.*, vol. 119, pp. 2965–2980.

Smith, D.L. (**1997**). "Discrete-Element Line Arrays—Their Modeling and Optimization," *J. Audio Eng. Soc.*, **45**, pp. 949–964.

SMPTE 202M (**1998**). "Standard for Motion Pictures—Dubbing Theaters, Review Rooms, and Indoor Theaters—B-Chain Electroacoustic Response."

Snow, W.B. (**1953**). "Basic Principles of Stereophonic Sound," *J.S.M.P.T.E.*, vol. 61, pp. 567–587. Reprinted in IRE Transactions-Audio, vol. AU-3, pp. 42–53, 1955.

Soulodre, G.A. (**2004**). "Evaluation of Objective Loudness Meters," 116th Convention, *Audio Eng. Soc.*, Preprint 6161.

Soulodre, G.A., and Norcross, S.G. (**2003**). "Objective Measures of Loudness," 115th Convention, *Audio Eng. Soc.*, Preprint 5896.

Soulodre, G.A., Popplewell, N., and Bradley, J.S. (**1989**). "Combined Effects of Early Reflections and Background Noise on Speech Intelligibility," *J. Sound Vib.*, **135**, pp. 123–133.

Steinberg, J.C., and Snow, W.B. (**1934**). "Auditory perspective—Physical Factors," *Electrical Engineering*, vol. 33, pp. 12–17.

Stelmachowicz, P.G., Beauchaine, K.A., Kalberer, A., Kelly, W.J., and Jesteadt, W. (**1989**). "High-Frequency Audiometry: Test Reliability and Procedural Considerations," *J. Acoust. Soc. Am.*, **85**, pp. 879–887.

Sterne, J. (**2003**). *The Audible Past; Cultural Origins of Sound Reproduction*, Duke University Press, Durham & London.

Stevens, S.S. (**1961**). "Procedure for Calculating Loudness: Mark VI," *J. Acoust. Soc. Am.*, **33**, pp. 1577–1585.

Talbot-Smith, M., editor (**1999**). *Audio Engineer's Reference Book*, Focal Press, Oxford.

Theile, G. (**1986**). "On the Standardization of the Frequency Response of High-Quality Studio Headphones," *J. Audio Eng. Soc.*, **34**, pp. 956–969.

Thompson, E. (**2002**). "The Soundscape of Modernity," MIT Press, Cambridge, Mass.

Tocci, G.C. (**2000**). "Room Noise Criteria—The State of the Art in the Year 2000," available at www.cavtocci.com/portfolio/publications/tocci.pdf.

Tohyama, M., and Suzuki, A. (**1989**). "Interaural Cross-Correlation Coefficients in Stereo-Reproduced Sound Fields," *J. Acoust. Soc. Am.*, **85**, pp. 780–786.

Toole, F.E. (**1970**). "In-Head Localization of Acoustic Images," *J. Acoust. Soc. Am.*, vol. 48, pp. 943–949.

———. (**1972**). "Understanding the Phono Cartridge," *Electron Magazine (Canada)*, three parts, in Aug., Sept., and Oct.

———. (**1975**). "Damping, Damping Factor, and Damn Nonsense," *AudioScene Canada*, February, pp. 16–17.

———. (**1982**). "Listening Tests—Turning Opinion into Fact," *J. Audio Eng. Soc.*, **30**, pp. 431–445.

———. (**1985**). "Subjective Measurements of Loudspeaker Sound Quality and Listener Preferences," *J. Audio Eng. Soc.*, **33**, pp. 2–31.

———. (**1986**). "Loudspeaker Measurements and Their Relationship to Listener Preferences," *J. Audio Eng. Soc.*, **34**, pt. 1, pp. 227–235; pt. 2, pp. 323–348.

———. (**1988**). "Principles of Sound and Hearing," Chapter 1 in *Audio Engineering Handbook*, Benson, K.B., editor, McGraw-Hill, New York. Also in multiple chapters of *Standard Handbook of Audio and Radio Engineering*, 2nd ed. Whitaker, J., and Benson, B., editors, McGraw-Hill, 2002, published without consultation with the author, and therefore incorporating no updates.

———. (**1990a**). "Loudspeakers and Rooms for Stereophonic Sound Reproduction," 8th International Conference, *Audio Eng. Soc.*, Paper 18-011.

———. (**1990b**). "Subjective Evaluation: Identifying and Controlling the Variables," 8th International Conference, *Audio Eng. Soc.*, paper C-1.

———. (**2006**). "Loudspeakers and Rooms for Sound Reproduction—A Scientific Review," *J. Audio Eng. Soc.*, vol. 54, pp. 451–476.

Toole, F.E., and Olive, S.E. (**1988**). "The Modification of Timbre by Resonances: Perception and Measurement," *J. Audio Eng. Soc.*, **36**, pp. 122–142.

Toole, F.E., and Olive, S.E. (**1994**). "Hearing Is Believing vs. Believing Is Hearing: Blind vs. Sighted Listening Tests and Other Interesting Things," 97th Convention, *Audio Eng. Soc.*, Preprint 3894.

———. (**2001**). "Subjective Evaluation," Chapter 13 in *Loudspeaker and Headphone Handbook*, Borwick, J. ed., Focal Press, Oxford.

Toole, F.E., and Sayers, B. McA. (**1965a**). "Lateralization Judgments and the Nature of Binaural Acoustic Images," *J. Acoust. Soc. Am.*, vol. 37, pp. 319–324.

———. (**1965b**). "Inferences of Neural Activity Associated with Binaural Acoustic Images," *J. Acoust. Soc. Am.*, vol. 38, pp. 769–779.

Torick, Emil (**1983**). "Triphonic Sound System for Television Broadcasting." *J. SMPTE*, **92**, pp. 843–848.

———. (**1998**). "Highlights in the History of Multichannel Sound," *J. Audio Eng. Soc.*, **46**, pp. 27–31.

Vanderkooy, J. (**2006**). "The Acoustic Center: A New Concept for Loudspeakers at Low Frequencies," 121st Convention, *Audio Eng. Soc.*, Preprint 6912.

Van Keulen, W. (**1991**). "Possible Mechanism for Explaining Monaural Phase Effects of Complex Tones," 90th Convention, *Audio Eng. Soc.*, Preprint 3065.

Vermeulen, R. (**1956**). "Stereo Reverberation," *IRE Transactions on Audio*, pp. 98–105, July–August.

Viemeister, N.F., and Wakefield, G.H. (**1991**). "Temporal Integration and Multiple Looks," *J. Acoust. Soc. Am.*, **90**, pp. 858–865.

Voetmann, J., and Klinkby, J. (**1993**). "Review of the Low-Frequency Absorber and Its Application to Small Room Acoustics," 94th Convention, *Audio Eng. Soc.*, Preprint 3578.

Voishvillo, A. (**2006**). "Assessment of Nonlinearity in Transducers and Sound Systems—From THD to Perceptual Models," 121st Convention, *Audio Eng. Soc.*, Preprint 6910.

Wakefield, G.H. (**1994**). "Temporal Integration, Multiple Looks, and Signal Uncertainty," *J. Acoust. Soc. Am.*, **95**, p. 2944 (Abstract only).

Walker, R. (**1993**). "Optimum Dimensional Ratios for Studios, Control Rooms and Listening Rooms," BBC Research Department Report No. BBC RD 1993/8, www.bbc.co.uk/rd/pubs.

Warnock, A.C.C. (**1985**). "Introduction to Building Acoustics," *Canadian Building Digest—CBD-236*, published by National Research Council Canada, Institute for Research in Construction. http://irc.cnrc.gc.ca/cbd236e.html.

Warren, R.M. (**1999**). *Auditory Perception: A New Analysis and Synthesis*, Cambridge University Press, Cambridge, U.K.

Waterhouse, R.V. (**1958**). "Output of a Sound Source in a Reflecting Chamber and Other Reflecting Environments," *J. Acoust. Soc. Am.*, **30**, pp. 4–13.

Watkins, A.J. (**1991**). "Central, Auditory Mechanism of Perceptual Compensation for Spectral-Envelope Distortion," *J. Acoust. Soc. Am.*, **90**, pp. 2942–2955.

———. (**1999**). "The Influence of Early Reflections on the Identification and Lateralization of Vowels," *J. Acoust. Soc. Am.*, **106**, pp. 2933–2944.

———. (**2005**). "Perceptual Compensation for Effects of Echo and of Reverberation on Speech Identification," *Acta Acustica* united with *Acustica*, **91**, pp. 892–901.

Watkins, A.J., and Makin, S.J. (**1996**). "Some Effects of Filtered Contexts on the Perception of Vowels and Fricatives," *J. Acoust. Soc. Am.*, **99**, pp. 588–594.

Welti, T. (**2002a**). "How Many Subwoofers Are Enough," 112th Convention, *Audio Eng. Soc.* Preprint 5602.

———. (**2002b**). "Subwoofers: Optimum Number and Locations," located at www.harman.com.

———. (**2004**). "Subjective Comparison of Single Channel Versus Two Channel Subwoofer Reproduction," 117th Convention, *Audio Eng. Soc.*, Preprint 6322.

Welti, T., and Devantier, A. (**2006**). "Low-Frequency Optimization Using Multiple Subwoofers," *J. Audio Eng. Soc.*, **54**, pp. 347–364.

Willcocks, M.E.G. (**1983**). "Surround Sound in the Eighties—Advances in Decoder Technologies," 74th Convention, *Audio Eng. Soc.*, Preprint 2017.

Zahorik, P. (**2002**). "Assessing Auditory Distance Perception Using Virtual Acoustics," *J. Acoust. Soc. Am.*, **111**, pp. 1832–1846.

Zurek, P.M. (**1979**). "Measurements of Binaural Echo Suppression," *J. Acoust. Soc. Am.*, **66**, pp. 1750–1757.

Index

Page references followed by "f" denote figures; those followed by "t" denote tables

541